**The Electron Gas Gun Spacecraft Fueling System.**

**This book include novel methods for distributing spacecraft fuel. Accordingly, we consider the following.**

**Staged Electron Gas Compression Guns With Emphasis On Electron Plasma Modes For Facilitating Nuclear Fueling Of Remote Relativistic Spacecraft Propelled By A Diverse Mix Of Propulsion Mechanisms.**

**The book is divided into two main portions, each of which is self-contained in its presentation. However, each part provides notable differences in content although conceptually, there is a lot of material common to both parts.**

**The author recommends reading of both portions to obtain the benefit of the overall broader presentation of the systems thus proposed.**

To order additional copies of this book, contact:
Xlibris
844-714-8691
www.Xlibris.com
Orders@Xlibris.com

ISBN: Softcover 978-1-6698-5928-4
EBook 978-1-6698-5929-1

Print information available on the last page

Rev. date: 12/09/2022

# PART 1.

# A FIRST CONSIDERATION OF THE ELECTRON GAS GUN PELLET STREAM DISPENSING SYSTEM.

# LOOKING AHEAD AND BEYOND NUCLEAR BOMB PELLETS.

# A LEAD-IN FOR THE NEXT BOOK IN THIS SERIES BUT RELEVANT TO THE SUBJECT OF THIS BOOK.

# DRAWINGS.

# WHIMSICAL CLOSING ANECDOTES. PATHWAYS TO THE SPEED OF LIGHT.

# APPENDIX 1. PRIOR ART IN STAGED COMPRESSION GAS GUNS AND LARGE ARTILLERY.

# APPENDIX 2. ENABLEMENT BY EXTREME MATERIALS MADE FROM ORDINARY ELEMENTS

# INTRODUCTION AND OVERVIEW: EXPLORING THE LIFE GIVING POSSIBILITIES OF SAILING THE UNIVERSE.

I propose theoretical concept validation of a staged compression gun able to accelerate projectiles and spacecraft to Keplerian velocities and even to relativistic velocities.

The fundamental concept is to use a space based linear accelerator to propel small fission/fusion devices to produce a propellant runway (reference Jordan Kare: https://www.centauri-dreams.org/2017/06/16/a-fusion-runway-to-deep-space/ ) an interstellar vehicle will overtake the devices, which explode and impart energy to the vehicle roughly similar to the manner studied for the Orion nuclear pulse-propulsion (reference https://ntrs.nasa.gov/api/citations/20000096503/downloads/20000096503.pdf)

1) The proposed pellet launching gun operates via a staged explosive decompression of electronic gas. The explosive process includes thermal gas depressurization and coulombic intragas repulsion.

2) The gas gun dispenses thermonuclear explosive devices in nominal examples of 40 kilograms mass with yield of about 0.0025 of its mass converted to energy or 22.5 kilotons yield. This is about twice the yield of the explosive device dropped over Hiroshima Japan near the end of World War II. Novel research is proposed to enhance

the mass specific yield and lower the critical mass required for thermonuclear explosives.

3) In a nominal consideration, the explosive bomb energy is coupled to the spacecraft via a large pusher plate to absorb the momentum of the x-ray and soft gamma-ray radiation of the blast. The pusher plate can also be magnetically insulated or otherwise electrodynamically insulated to back reflect the hot plasma generated by the nuclear explosives. Additionally, the hot plasma from the nuclear explosions can be back-reflected by magnetic plasma bottle sails and/or by other sail type electrodynamic mirrors.

## Overview: Exploring the Life Giving Possibilities of Sailing the Universe

Sailing the universe black under nuclear power includes some simple mathematical conjectures including extreme examples of pellet fuel runway driven craft similar to that proposed under Project Medusa. Project Medusa was a British Interplanetary Society concept that involved nuclear pulse propulsion methods where nuclear bombs would be detonated behind a very large sail. A manned spacecraft would be attached by very long cables that have a very high strength to weight ratio. Medusa spacecraft systems would be capable of an $I_{sp}$ of 50,000 to 100,000 seconds or 500 to 1000 (kilo-Newton)(second)/(kg).

There would be a need to avoid excessive mechanical shock which might otherwise tear apart the sail or snap the tethering cables connecting the sail to the craft. As a result, some sort of highly elastic materials could be used in the construction of the sail and the tethers. However, the bomb sail would likely be tens to hundreds of kilometers wide and the tethers could be tens to hundreds of kilometers long, so even the use of low strain high strength materials in the construction of the sail and its tethers would seem feasible.

Some potentially excellent materials of choice for the construction of the sail and the tethers include, but are not limited to, carbon nanotubes, graphene-based fibers, boron nitride nanotubes, and perhaps some forms of still theoretical, super strong, light weight, stabilized, metallic hydrogen.

Nanotech self-repair mechanisms might be utilized to repair the sails as they are degraded by the interstellar medium. Alternatively, small scale, mesoscale, or macro-scale robotic rovers could be deployed to periodically repair, re-assemble, or otherwise maintain the sail. The microbots, mesoscale robots, or macrobots could be held in place by the Meissner Effect whereby a superconducting piece of material can be held in place by a magnetic field in a hovering position over a magnet. The craft could deploy a magnetic field for this purpose. The smaller the sail, the more intense the magnetic field can practically be since the energy density of the magnetic field can be greater. Uniform magnetic field energy density is proportional to the square of the magnetic field intensity which for high strength magnetic fields is often measured in units of Tesla. Magnetic fields as a form of bound energy would carry inertia. Therefore, a magnetic field that is too intense and widely distributed at high energy densities and thus having a high average volumetric energy density may pose problems where high craft accelerations are required.

Sails for ships traveling at such extreme gamma factors would ordinarily be immediately ionized by Doppler blue-shifted CMBR, star-light, and quasar light. Therefore, the materials of sail composition would necessarily need to be many orders of magnitude more refractory than carbonaceous materials such as carbon graphene, carbon nanotubes, and the like. Perhaps the sails can be manufactured out of neutron dense forms of matter such as stabilized, electrically chargeable forms of neutronium. A neutronium net or grid-like weave may be comprised of 3-neutron wide to 100-neutron wide fibers separated by distances perhaps as great as one meter or more. However, the spacing of the fibers might be much smaller as needed. It is conceivable that a monolithic neutronium sail that is only 100 meters wide could survive the detonation of a 100 megaton device located within 10 to 100 meters of the sail. Such a sail might have a thickness of 10 neutrons and a mass of 100,000 metric tons.

Another speculative, more refractory material, for sail construction is quarkonium such as up quark, down quark, charmed quark, strange quark, or bottom quark matter as long as the net electrical charge of the sail material is not so high as to cause the sail to fail due to coulombic self-tension. Perhaps such coulombic self-tension which might otherwise arise due to a biased construction of quarks of the same sign of electrical charge can be cancelled out by the installation of an opposite electronic charge within the sail.

Still yet another even more exotic sail material might be a composition of Higgsinos. Once again, any excessive coulombic tension otherwise produced within the sail might be cancelled out with the installation of an electronic charge of the opposite sign, or perhaps quarkonium of an opposite charge could be instilled within the Higgsinium.

We can contemplate sails composed of monopolium and of a whole assortment of other exotic materials. However, specific examples of such materials will not be described here since by now, it has likely become clear to the reader that extremely refractory nuclear matter density materials and denser materials would be applicable provided such materials can be appropriately fashioned into nuclear bomb blast sails.

Such nuclear density and super-nuclear density materials are currently a dream at best.

Note that the nuclear bombs may be fitted with an externally protruding charged member such as a charged tow line, or perhaps be fitted with nuclear electric propulsion systems for fine course adjustment. The charged tow line would be similar to the relativistic Lorentz turning force mechanism contemplated by Carl Schroeder in his Stellar Cycler concept. Here, a spacecraft that was brought up to relativistic velocities by various means would deploy an electrically charged tow line that would experience force acting at right angles to the ship's velocity vector. Such an interaction would cause the ship to trace out circulinear, or nearly so, orbit-like iterative cycles within the interstellar medium by reacting with the interstellar magnetic field.

Note that the use of a grid-like sail is especially applicable at extreme gamma factors since most of the interstellar debris such as atoms, molecules, ions, electrons, dust particles, and photons would simply pass through the openings between the sail fibers without damaging the sail.

Consider cases where both the cosmic microwave background radiation or CMBR and star light are extremely Doppler blue-shifted. Materials of sail composition could include those that are highly transmissive or transparent to the bulk of the impinging radiation. It might even be possible to construct the sail out of so-called negative electromagnetic refractive index meta-materials. Such materials in at least some forms would theoretically pull the sail forward instead of pushing against the sail. However, the active cell size of the materials, some of which are formed from microscopic inductor-like conductive elements embedded within a dielectric medium, must be roughly the same size as or smaller than the wavelength of the electromagnetic energy being negatively refracted as the energy passes through the materials. Most likely, the active cell size of the negative index material(s) would need to be at least on the same order of magnitude as the wavelength of the reactive frequencies for greatest effectiveness. Ideally, the cell size would be significantly smaller than the wave-length of the negatively refracted light.

Negative EM refraction index meta-materials are being developed at Duke University, and at other places around the globe as laboratory research items and as yet, no commercial applications for such materials are available. The main interest in such materials comes from the U.S. Department of Defense and other similar organizations because of potential military applications for such materials. Such applications might include potential uses as radar cloaks and even visible light cloaks for aircraft and land vehicles. The cloaks would not be perfect, but nonetheless, could permit much more stealthy combat operations such as the concealment of equipment in military buildups and the like.

The accelerated nuclear bombs used to drive the sail ship might exist in the form of ordinary fission, fission-fusion, or fission-fusion-fission bombs of the types commonly deployed during the Cold War and currently so by the modern nuclear military powers. Alternatively, perhaps the devices could be constructed of pure fission, pure fission-fusion, or pure fusion devices.

Pure fission devices might consist of nuclear fissile materials for which cages containing antiprotons are instilled within a bulk of highly fissionable materials. When it came time for the device(s) to be detonated, the molecular cages would be unlocked. The antiprotons would then be released to mingle with the fissile material's atomic nuclei thereby causing fission decays and a large flood of fission fragments resulting in the formation of a super-critical mass. Perhaps such pure fission devices can include batches of fissile materials that have a mass of only a few to several grams.

Pure fission-fusion devices would operate on similar principles except that a fission stage would be used to set off a fusion stage in a manner similar to modern thermonuclear weapons.

Pure fusion bombs would consist of pure fusionable materials such as dense forms of hydrogen, hydrogenic compounds, or other relatively easily fusionable materials. Accordingly, a bulk of such material would contain molecular cages in which antiprotons would be instilled. When it came time to detonate the device(s), the molecular cages would be unlocked. The released antiprotons would interact with protons thereby causing a flood of ionizing radiation and heat to be generated. Thus, a self-propagating fusion reaction may occur throughout the bomb's bulk of fusionable material(s).

The bombs itself might have a mini-magnetosphere set up around it by a permanent magnetic type of element regardless of whether such devices include traditional chemical high explosive types of triggers, or whether they take the form of pure fissile or fusionable bulk materials.

It is even possible that the bombs themselves could be driven by beam sails. Accordingly, electromagnetic energy from stations within the Solar System or generally within the Milky Way Galaxy would drive the bombs to extreme relativistic velocities through the reflection of the artificially collimated, beamed electromagnetic energy, off sails attached to the bombs. The reflected energy would push the sails and attached bombs forward.

Here, we assume that the nuclear device pellet streams could maintain an intercept heading with the craft and its bomb sail. We also assume that the craft and its sail materials could be maintained amidst the extreme Doppler blue-shifted star light and CMBR as well as the extremely energetic cosmic rays in the form of highly relativistic interstellar atoms, molecules, ions and electrons in a manner that permits net velocity gains for the ship in spite of astrodynamic drag.

Note that the shape and size of the sail(s) can be chosen for specific mission criteria. The shape and size of the sails may also be adjustable in flight as can the spacing between lines, threads, or fibers of net-like sails. Such additional morphological degrees of freedom can enable more appropriate interaction of the sail(s) with the interstellar and intergalactic medium including but not limited to natural variations in: 1) plasma distribution by density, charge, and species: 2) neutral gas distribution by density and species; 3) ambient stellar and quasar light distributions according to energy spectrum and power flux density; and 4) interstellar and intergalactic magnetic field intensity and vector field orientation. Such considerations can be important for drag reduction. For certain conditions, intentional increases in drag for re-routing and deceleration may be applicable. Such natural variations can enhance or degrade spacecraft performances including but not limited to propulsion system efficiency.

The interstellar and intergalactic matter might not significantly erode even many of highly relativistic sails of micron thickness.

The background interstellar and intergalactic matter might not erode even many of highly relativistic sail of sub-micron thickness.

The diametrical cross-sectional area of our observable universe is close to $10^{47}$ square kilometers and the total mass energy of the observable universe is equal to about $10^{50}$ metric tons of which only 4 percent is baryonic. Thus, an average column spanning the diameter of the entire visible universe would have an $H_2O$ STP density matter thickness of only about 25 micrometers for reactive matter. However, this is not a concern for the following reasons.

First, the sails could be replaceable grid sails and driven by optical, IR, microwave or radio-frequency radiation. The mass of such sails can be reduced by many orders of magnitude relative to monolithic sails that are only micrometer scales in thickness.

Second, sails having a very thick cable or thread like construction are conceivable where the cables or wires would be many times if not several orders of magnitude thicker than 25 microns. The sails could be mostly empty space to almost entirely empty space to reflect long wave radio-frequency phased array beams.

As for concerns about over burdening the conductive or super-conductive wires or cables used for such sails by extremely intense rF beams, note that such reflective members could be very conductive to superconductive to thereby yield near perfect reflection. The EM energy that was not reflected would largely pass through the openings in the sail grid.

Second, a magnetic and/or electric field based scoop or anti-scoop could divert the chargons away from the sail just as an extended electrodynamic scoop for an interstellar ramjet would. Electro-dynamic-hydro-dynamic-plasma-drive features could utilize the diverted plasma in a reactive and gainful manner.

The sail might be deployed in a manner that is orthogonal to the ship's velocity vector. The sail might be parallel to the spacecraft velocity vector and driven obliquely from behind. This way, the effective thickness of the sail could be thousands of miles and the sail could include electro-dynamic-hydro-dynamic-plasma-drive features.

Fourth, the above parallel sail could conceivably be made of negative refraction index materials that would be pulled forward by incident star light and highly blue-shifted CMBR, far infrared, and non-CMBR radio sources.

Fifth, the sail may be a deployed mag-sail or M2P2 type of sail or any other magnetic or plasma bottle sail. It is possible that a plasma affixed to the spacecraft could be driven by rf radiation. Upon attainment of extreme spacecraft gamma factors, laser light could be easily reflected by such sails. Plasma makes an excellent rF reflector even for very small plasma densities.

I have done a lot of writing on parallel sails such as negative refraction index monolithic and grid sails capable of extreme gamma factors.

Sixth, some sail materials such as any future forms of super-strong very conductive to super-conductive metallic hydrogen can be used as nuclear fusion fuel for fusion rockets upon degradation to useless levels.

Seventh, it has been proposed that very thin, metallic, very low density gas containing balloons might be used for nuclear warhead decoys which could survive 100 meter proximity detonation to a one kiloton neutron bomb in the vacuum of space. The rate of radiative cooling would be tens of billions of Kelvins per second due to the extreme thinness of the balloon membranes and most of the neutrons would pass right through the balloon without interacting or by only depositing a very small portion of the particles kinetic energy into the balloon and enclosed gas. Interstellar chargons are more reactive to electronic shell structures but not by that much.

The general idea for obliquely oriented beams involves the beamed energy incident on both sides of the sail. The sail could include a surface of hair like cilia or any other surface contour that would work so as to much more effectively grab a hold of the light.

In addition, the sail could be fabricated from photovoltaic materials in order to provide power for electro-dynamic-hydrodynamic-plasma-drives or chargon rockets, or perhaps even photon rockets.

For extreme gamma factors, the CMBR and starlight will be highly blue-shifted and will be relativistically aberrated to what would approach a point source in front of the spacecraft at $\gamma = \infty$. A sail parallel to the spacecraft velocity vector made of a suitable negative electromagnetic refraction index material will be pulled forward even by light incident on the sail at a very shallow angle from in front of the spacecraft.

To enhance the negative refraction index sails capture of EM energy, the sails may have negative index hairs or cilia distributed along its length.

Negative refraction index materials have actually been measured to be pulled on by incident light. Duke University and other academic and government labs are researching the various aspects of negative refraction index materials.

I have no problem with spacecraft being pulled forward by forward incident light. After all, the paradigm of light speed velocity limits may or may not have been shattered with any future validation or not of the CERN superluminal neutrino results. The Big Bang may have been the most recent free lunch. There is no reason why the Big Bang could not have started with miniscule quantities of mass-energy.

A good abstract for a great paper on negative super-pressure of light acting on a negative refractive index material is

Henri Lezec
(Center for Nanoscale Science and Technology, NIST)

Forty years ago, V. Veselago derived the electromagnetic properties of a hypothetical material having simultaneously-negative values of electric permittivity and magnetic permeability [1]. Such a material, denominated "left-handed", was predicted to exhibit a negative index of refraction, as well as a number of other counter-intuitive optical properties. For example, it was hypothesized that a perfect mirror illuminated with a plane wave would experience a negative radiation pressure (pull) when immersed in a left-handed medium, as opposed to the usual positive radiation pressure experienced when facing a dielectric medium such as air or glass. Since left-handed materials are not available in nature, considerable efforts are currently under way to implement them under the form of artificial "metamaterials" — composite media with tailored bulk optical characteristics resulting from constituent structures which are smaller in both size and density than the effective wavelength in the medium. Here we show how surface-plasmon modes propagating in a stacked array of metal-insulator-metal (MIM) waveguides can be harnessed to yield a volumetric left-handed metamaterial characterized by an in-plane-isotropic negative index of refraction over a broad frequency range spanning the blue and green. By sculpting this material with a focused-ion beam we realize prisms and micro-cantilevers which we use to demonstrate, for the first time, (a) in-plane isotropic negative-refraction at optical frequencies, and (b) negative radiation pressure. We predict and experimentally verify a negative "superpressure", the magnitude of which exceeds the photon pressure experienced by a perfect mirror by more than a factor of two. 1) V. Veselago, \textit{ Sov. Phys. Usp. }10, p.509 (1968).

Available at:

http://meetings.aps.org/Meeting/MAR09/Event/93172

The sail might not need to be held by guy lines. A strong magnetic field based coupling or electrical charged based connection might work.

Another option is to fabricate the sail guy lines out of graphene, carbon nanotubes, boron nitride nanotubes, graphene oxide paper, and the like. A cable constructed from such materials could stretch for about 20 to 50 kilometers yet still handle tens to hundreds of Earth G's. The tensile strength of graphene is close to 18 million PSI for perfect forms.

Materials such as solid quarkoniums and somehow stabilized neutroniums, and perhaps even Higgsiniums would be better yet, but such materials may only exist in nature in extreme mass quantity states as of the present cosmic era.

The collection area of the sail can be very, very, large. A large electro-dynamic scoop could extent very far out from the sail.

Regarding nanotech self-assembly mechanisms, just simply greatly increase the capture area of a electrodynamic scoop to collect enough interstellar materials and use most of the collected interstellar material as an EHPD, an MHPD, or a combination of the two and use the rest of the materials for sail repair.

Regarding holding M2P2 plasma affixed to the ship under high gamma factor condition, simply increase the strength of the fastening fields.

Consider the interstellar matter density near our solar system of one particle for every 10 $cm^3$. This density works out to be a layer of hydrogen or helium atoms about one atom thick for a column that is one light-year long: Not a show stopper for light sails or sails that are electro-dynamically shielded or protected!

If extreme materials are used with excellent reflectance, we could simply use a sail that has a thickness of one millimeter or more and which is monolithic, or better yet, use a sail with grid lines that are one millimeter or perhaps much greater in thickness. This way, a sail that has an area of only one square kilometer can intercept a beam having an equivalent black body temperature of several thousand Kelvins provided it is constructed of suitably refractive materials.

We could simply use electrodynamic methods of grabbing a hold of the interstellar gas and diverting around the spacecraft and sail. The power to operate the electrodynamic mechanisms can be supplied by beams. The electrodynamic methods can include lasers for ionization, or radio-frequency radiation where the gamma factors are suitably large, magnetic fields, electric fields, plasma fields affixed to the spacecraft, and the like.

Then there are always the possibilities for sails comprised of truly exotic materials such as somehow stabilized neutroniums, quarkoniums, higgsiniums, monopoliums, and perhaps even raw space-time-mass-energy forms such as the "Yelm" of mid-20th Century Big Bang theory.

Since one cubic meter of neutronium would have a mass of about $10^{15}$ tons. A 1,000 kilometer long thread of the stuff that has a cross-sectional area of 1,000,000 neutrons would have a mass of only one kilogram. A 1 kilometer long thread having a cross-sectional area of 1 billion neutrons would have a mass of only 1 kilogram. Lines made of quarkoniums could have the same length and cross-section but would be 10 to 1,000 times more massive. Higgsiniums would be all the more massive.

Provided such extreme materials could be developed, they could also serve as electric current carrying magnetic sail components.

Anyhow magnetic sails can be made of any ordinary conducting or superconducting period table materials.

It is also conceivable that a hybrid sail can be used where a current carrying magsail would deflect plasma away from a monolithic and grid-like light sail or radio-frequency sail.

Now, regarding the subject of sail erosion by exposure to interstellar or intergalactic gas, we must realize that the kinetic energy of a gas atom traveling at a velocity of 86.7 percent of the speed of light with respect to the sail would be equal to the binding energy of roughly 10 billion atoms within a sail of micron thickness. Thus, 10 billion atoms could be dislodged should all of the energy of the gas atom be deposited within the sail. Incident gas atoms having even higher associated gamma factors with respect to the starship sail could potentially knock loose even more atoms. Perhaps, there is no reason to worry about sail erosion in spite of this for the following reasons.

First, extremely relativistic particles would likely deposit only a small portion of its energy within the sail thereby greatly lessening the number of atoms that would be knocked loose. This fact would apply to chargons as well as neutral incident particles.

Second, for sails of near micron thickness, atoms that were knocked loose would likely simply be re-assimilated by the bulk sail materials. Perhaps the only chance for an atom to be knocked loose would include atoms located on the backward side of the sail. Atoms for which its bonds where broken within the bulk sail material would tend to rebond with adjacent atoms.

Third, since the incident gas or plasma particle would deposit only a small portion of its energy within the sail, the kinetic energy per particle for particles that are knocked loose may be only slightly in excess of the binding energy of the dislodged atoms. Basically, the kinetic energy of the dislodged atoms could likely be re-absorbed and/or radiated away thereby promoting rebinding of the dislodged atoms.

Fourth, for cases where the sail would completely absorb the kinetic energy of the incident gas or plasma particles such as an alpha particle, for the case of a one micron thick sail, the sail would obviously be able to completely stop the chargon without losing it. Thus, any atoms disbonded by the incident chargon would also likely be captured and prevented from leaving the sail material.

Fifth, for grid-like sails, the grid lines might be positively chargeable so that incident interstellar or intergalactic ions are pushed away from the grid lines and through the openings within the grid-like sails. The effect would be analogous to the Van der Waals force that keeps neutral atoms from being squeezed together to tightly.

Imagine that! A technology that might have destroyed human civilization and likely all life on planet Earth may end up being used to drive relativistic sail craft to extreme velocities thus opening up the cosmos, or at least the visible universe and proximate external regions for human colonization.

Consider a scenario where each star system within our universe may contain on average one habitable or at least one terra-formable planet or moon that can simultaneously support 10 billion human persons in a green and sustainable manner over cosmic time periods and even longer assuming that human life expectancies can be increased to 1,000 years. The resulting number of human persons that our Galaxy alone could support is a whopping $(10^{12})(10^{10}) = 10^{22}$ simultaneously. The number of human persons that our visible universe could support simultaneously is a whopping $(10^{24})(10^{10}) = 10^{34}$. The number of human persons that the visible portion of our universe could support over a 10 billion year period assuming non-relativistic habitats is a whopping $(10^{34})(10^{-3})(10^{10}) = 10^{41}$! The number of human persons that our universe alone could support simultaneously, a universe that may be just one of innumerable universes, is literally infinite under the condition that our universe is infinite in extent. If we assume that one solar mass can support $10^{18}$ persons simultaneously, then our visible universe could support $(10^{18})(10^{23})(10^{10})(10^{-3})$ persons having a lifetime of 1,000 years over a 10 billion year period. This is an astounding $10^{48}$ persons. This is roughly the number of atoms composing the Earth's Moon. It may be the case that the actual number of persons supportable needs to be downwardly adjusted by a few orders of magnitude. This is because a percentage of otherwise potentially habitable worlds may include pathogens, and geologic and hydrospheric instabilities that would make some planets uninhabitable.

We can likely find common ground from among all nationalities, ethnic groups, social-economic classes, faith-based systems, and political ideologies for the awesome life giving endeavor of sailing the universe black under nuclear power. Will we have the courage to set sail?

In order to archive spacecraft deceleration, electrodynamic breaking mechanisms may be employed.

For present consideration, a mechanism of choice would include one or more linear induction breaking mechanisms. Accordingly, one or more large highly conducting or super-conducting coils would be deployed. The coils would build up large currents which would then interact with the background magnetic fields to bring the spacecraft velocity down to Keplerian velocities; upon which small scale velocity correction rockets would be deployed. These rockets may include nuclear electric systems such as traditional ion thrusters or engines such as Ad Astra Rocket Company's VASIMIR Engine. See references for the VASIMIR Engine.

There is also an option for which a pre-deployed pellet stream displaced in the outer reaches of the solar system would be used to slow the craft. The breaking thermodynamics would be simply the reverse of the positive acceleration process.

To travel back to Earth, another pellet stream would be deployed to accelerate the spacecraft on a heading back to Earth. The thermodynamics of the return accelerating process would closely match that of the initial outbound acceleration.

Once again, yet another pellet stream would be used to slow the spacecraft before obtaining close proximity to Earth. The reason for the latter option is that nuclear explosions near the Earth may result in EMP effects which can damage satellites and electrical power grids on Earth.

The above proposed propulsion missions can be repeated many times and offer strong solutions to the problems of what to do in the near term with excess fissile plutonium and uranium stock piles

Once space-based mining of minerals becomes possible, the ability to find and refine uranium and thorium ores on the Moon, Mars, Mercury, the asteroids and the Kuiper Belt and Oort cloud object should enable enough fissile fuels or breeder reactor feedstocks to enable far more nuclear explosive charges to be produced than would be necessary for one or several missions.

The electron gas gun fueling of a spacecraft has merit in that relativistic spacecraft velocities can be achieved without the need for concern over otherwise storing a large quantity of nuclear explosives on the spacecraft and the lack of need of supporting architecture to store the devices on the spacecraft.

The spacecraft and the exiting fuel pellets would have fine course correction mechanisms and would inter-communicate to enable suitable temporal calibration of the spacecraft arrival in proximity to each pellet. Clock features in pellets can be based on robust quartz oscillations mechanisms since atomic clocks may not survive the launch forces.

For example, the pellets may have a charging feature that enables them to use the Lorentz turning force to adjust their flight paths as well as optional solar sails, small rockets, small explosive thrusting charges, and the like. Additionally, the pellets may have laser sails for which a beam of light from stations near Earth can induce a drive pressure on the pellets and/or spacecraft for finer course correction. The orientation of the sails can be adjusted for tacking style thrust to accelerate the pellets or spacecraft out alignment of the pellets and spacecraft velocity vectors.

These proposed methods entail a long term commitment and the development of suitable space travel industries. Once enabled, very great spacecraft Lorentz factors can be obtained with very small mass-ratios. Pellet launch rates and velocities can be up or down adjusted as needed. The above proposal is bold and costly however it is never too early to plan such missions. Once space-based mining and manufacturing develop, the feasibility will greatly increase.

In order to fabricate the number of needed nuclear charges, extremely low critical mass fissile fuels may be used for the nuclear primaries along with neutron reflectors to further reduce the critical mass of explosively compressed or combined fissile fuels during the detonation process.

The muzzle velocities of the guns can be adjusted so that the velocity, location, and timing of the ship arrival per pellet can be properly calibrated.

An interesting optimization would involve pellets fired in a direction tangential to Earth's orbit around the Sun and for pellets fired from Earth orbit, tangential to Earth orbit as well. This way, these two increased velocity components can enable reduced muzzle velocity.

Pellets may also be fired in such a flight pattern that they receive gravity assists from the gas giant planets of Jupiter and/or Saturn. Accordingly, the muzzle velocity of the pellets can be further reduced. These enhanced efficiencies can apply to many scenarios including but not limited to the ones presented in this book.

Launch rates of the pellets can be adjusted or made adjustable as can the yields of the pellets for optimal drive efficiency and abilities of the spacecraft systems to handle the nuclear blasts.

Additionally, optimized blends of fissile fuels can be used in the composition of the primaries thus offering research options to greatly reduce the required critical mass of the needed fissile fuels.

Additionally, U-238 based shots might replace conventional nuclear explosives for which neutron irradiation of the U-238 by a neutron beam originating from the ship would induce nuclear fission of the U-238 shots. The fissioning U-238 may set of a nuclear fusion fuel secondary which may have a yield 10 to 100 times that of the primary.

# Production Of Efficient Nuclear Explosives Of Small Masses.

The Ronen fissile rule states that for a heavy element with $90 \leq Z \leq 100$, its isotopes with $2 \times Z - N = 43 \pm 2$, with few exceptions, are fissile (where N = number of neutrons and Z = number of protons

The fissile rule indicates 33 isotopes as likely fissile: Th-225, 227, 229; Pa-228, 230, 232; U-231, 233, 235; Np-234, 236, 238; Pu-237, 239, 241; Am-240, 242, 244; Cm-243, 245, 247; Bk-246, 248, 250; Cf-249, 251, 253; Es-252, 254, 256; Fm-255, 257, 259. Only fourteen (including a long-lived metastable nuclear isomer) have half-lives of at least a year: Th-229, U-233, U-235, Np-236, Pu-239, Pu-241, Am-242m, Cm-243, Cm-245, Cm-247, Bk-248, Cf-249, Cf-251 and Es-252. Of these, only U-235 is naturally occurring. It is possible to breed U-233 and Pu-239 from more common naturally occurring isotopes (Th-232 and U-238 respectively) by single neutron capture. The others are typically produced in smaller quantities through further neutron absorption.

Note that in nuclear fissile primaries, it is plausible that radioactive heavy isotopes may be installed that decay from a non-fissile isotope to a fissile isotope. This process may in principle extend the useful life-time of a nuclear fissile primary.

When adding tritium as a nuclear fission primary booster, the critical mass of a primary can be further reduced.

Assuming 33 known likely fissile isotopes, the number of subsets of these isotopes is [(2 EXP 33) – 1] Assuming the isotopic blends can be adjusted in a million different fractional combinations per set of two or more isotopes per primary, the number of testable combinations for optimization becomes almost equal to [(1,000,000 EXP [(2 EXP 33) – 1]].

There are additional potential effects at play here.

First, some isotopic primary, tertiary, etc. fissile stages of nuclear explosives may designed in such manners so that these stages act like low energy accelerator collisions such as will take place at the Facility For Rare Isotope Beams or FRIB when it is completed. Accordingly, super-heavy elements might be produced that have more "exuberant" fission processes in that the resulting fission fragments will become smaller in atomic number and atomic mass to enable much higher nuclear fission energy yields.

Second, any production of transient super-heavy elements and/or isotopes under the effect of intra-radiation may stimulate the development of unstable strange, charmed, and bottom matter, and possibly a small fraction of top matter thus made from the four heavier quarks. These exotic forms of matter may have catalytic effects that can likewise transmute ordinary periodic table matter into exotic quark matter which upon decay or interaction result in a much greater percentage of explosive mass being converted to energy than even in cases of the highest mass specific energy yield available for the highest performing nuclear fusion fuels. In such cases, otherwise inert matter may be included in the explosive devices which could be exothermically processed to enable cost efficient nuclear explosives.

So, there are numerous mechanisms to consider when designing nuclear fuels for explosive pellets that can be considered to lower the cost of deployed latent nuclear energy and the energy cost of deploying the pellet streams.

Note that the composition of nuclear explosive pellets is considered more in depth later below.

Note that the shape of the chute can include elongated charged tubular configurations so as to more efficiently capture and back accelerate the plasma from the nuclear explosions. Moreover, the chute can have a pointed or tapered leading edge to reduce astrodynamics drag.

The electron gas gun fueling of a spacecraft has merit in that relativistic spacecraft velocities can be achieved without the need for concern over otherwise storing a large quantity of nuclear explosives on the spacecraft and the lack of need of supporting architecture to store the devices on the spacecraft.

The nuclear fuel pellets for missions from earth to other stars positioned along the plane of the solar system would require that the projectiles be sprayed from the gun over a small time interval once approximately every complete lunar orbit around the Earth.

For mission to destinations orthogonally displaced from the plane of Earth's orbit around the sun, the projectiles can be continuously launched.

Each projectile may be fitted with a small, low powered, fine course correction rocket.

Needed data on the location of the pellets relative to the spacecraft and relative to each other would be provided by electronic transceiver apparatus installed within the pellets. The pellet transmissions can be tracked by Earth based systems and then sent to the exiting spacecraft.

For present consideration, a mechanism of choice would include one or more linear induction breaking mechanisms. Accordingly, one or more large highly conducting or super-conducting coils would be deployed. The coils would build up large currents which would then interact with the background magnetic fields to bring the spacecraft velocity down to Keplerian velocities; upon which small scale velocity correction rockets would be deployed. These rockets may include nuclear electric systems such as traditional ion thrusters or engines such as Ad Astra Rocket Company's VASIMIR Engine.

Fine course correction of the spacecraft can also be enabled by a Stellar Cycler mechanism for which a spacecraft can attain a strong net charge and experience curvilinear acceleration by the relativistic Lorentz turning force.

For missions to destinations obliquely oriented with respect to the plane of the solar system, the Stellar Cycler method can re-route spacecraft on initial headings substantially orthogonal to the plane of the solar system. Thus, the problem of needing to fire pellets in short monthly intervals is avoided.

Once space-based mining of minerals becomes possible, the ability to find and refine uranium and thorium ores on the Moon, Mars, Mercury, the asteroids and the Kuiper Belt and Oort cloud object should enable enough fissile fuels or breeder reactor feedstocks to enable far more nuclear explosive charges to be produced than would be necessary for one or several missions. The electron gas gun fueling of a spacecraft has merit in that relativistic spacecraft velocities can be achieved without the need for concern over otherwise storing a large quantity of nuclear explosives on the spacecraft and the lack of need of supporting architecture to store the devices on the spacecraft.

These proposed methods entail a long term commitment and the development of suitable space travel industries. Once enabled, very great spacecraft Lorentz factors can be obtained with very small mass-ratios. Pellet launch rates and velocities can be up or down adjusted as needed. The above proposal is bold and costly however it is never too early to plan such missions. Once space-based mining and manufacturing develop, the feasibility will greatly increase.

In order to fabricate the number of needed nuclear charges, extremely low critical mass fissile fuels may be used for the nuclear primaries along with neutron reflectors to further reduce the critical mass of explosively compressed or combined fissile fuels during the detonation process.

# Launch Rate And Pellet Runway Characteristics.

The muzzle velocities of the guns can be adjusted so that the velocity, location, and timing of the ship arrival per pellet can be properly calibrated.

Launch rates of the pellets can be adjusted or made adjustable as can the yields of the pellets for optimal drive efficiency and abilities of the spacecraft systems to handle the nuclear blasts.

Exiting spacecraft may initially receive each shot at a rate of 1 per second to 1 every 100 seconds to enable timely acceleration. Once the velocity of the spacecraft was brought up to high enough velocity, then the frequency of the shot in the background reference frame may be reduced.

In short, the electron gas compression gun has merit in that it may enable relativistic spacecraft velocities without the need for large mass ratios and is simpler in design than traditional mass drivers.

Taken to extreme circumstances, spacecraft propelled by nuclear pellet streams may obtain velocities very close to the speed of light and extreme Lorentz factors. Tantalum-Hafnium-Carbide solutions in shielding material may in principle enable spacecraft to attain Lorentz factors at least as great as around 1,500 at which point the cosmic microwave background radiation will be relativistically blue-shifted and aberrated to an apparent black body temperature near the melting point of said carbide solutions.

Now, we consider in first order that quadrupling NASA's gas compression gun length and adding additional stages may enable projectile velocities as great as 55,000 feet per second or about 16,500 meters per second. This latter velocity is more than adequate for Mars orbit insertion and also into the main asteroid belt

Using an electron gas as final stage pressure embodiments, much greater velocities such as very mildly relativistic velocities could be obtained by guns of great length mounted on the Moon or interplanetary solar orbit. Velocities on the order of at least roughly 750,000 meter/sec or 0.239 percent of the speed of light may be obtained Conceivably, higher velocities could be achieved with the electro-gas discharge. These extreme velocities when attributed to pellets on the scale of 40 kilograms/pellet would find great use in forwardly distributed nuclear fission or nuclear fusion fuel pellet runways to enable starships to reach large Lorentz factors with nearly no onboard nuclear fuels. For a 0.0001 light-year long distributed runway and average spacecraft frame acceleration on the order of 0.5 earth g's or 4.905 $m/s^2$, a human crewed spacecraft could obtain a velocity of about 1 percent of light-speed and a Lorentz factor of 1.00005.

The above values for pellet stream velocities are in first order derived from the approximately 16.5 km/sec enabled by hydrogen compression multiplied by the square root of the ratio of the hydrogen atom mass and the electron mass which is 42.8623. This is so because the mass of the hydrogen atom is 1,837.17 times that of the electron. Since Newtonian momentum is proportional to velocity for a given mass, but assuming said square root of 1,837.17, we have said square root scaling of momentum but also said square root scaling of number of times on average a given electron will impart momentum to a gun barrel's wall per second. Here we consider pressure in the ideal gas law is a function of number of particle collisions per unit area per unit time multiplied by twice the average pre-incident momentum imparted to the barrel walls of the colliding particles while assuming elastic collisions. We thus obtain the approximately ¼ of a percent of the velocity of light for the exiting projectile.

Now, as the electron gas density increases, coulombic pressure will factor in thereby increasing the gas pressure beyond what it would be for a given average electron velocity. So, it is conceivable that an electron gas compression gun of many stages could accelerate a round to almost the speed of light. A gun barrel with a high electric flux topological reflector or insulator will make the electron-based coulombic force more concentrated thus increasing the potential velocity of the round to values even closer to the speed of light to attain extreme projectile Lorentz factors.

An anticipated method of providing electron gas is a powerful electrical discharge system such as by way of banks of super-capacitors. The charge may originate by powerful electrostatic discharge.

Other gun configurations would simply involve a long muzzle for which an electron discharge is timed to occur just as a round passes the locations of electrostatic discharge.

We assume a gun muzzle velocity of 100,000 m/s and the average time between pellet launches would be 1,000 seconds.

Assuming a pellet shot mass of 40 kilograms, the kinetic energy of each pellet upon launch would be $2 \times 10^{11}$ joules. Thus, the time averaged power pulse of the gun per shot during the acceleration process would be approximately 200 gigawatts assuming most of the electrical pulse energy is converted to pellet kinetic energy. This is roughly equivalent to the expansive gas power for large caliber artillery pieces. The time averaged power output of the gun over each 1,000 second interval is about 200 megawatts.

More specifically, for scenarios for which half of the electron gas energy is converted to heat within the barrel, the average thermal loading on the electron gas gun barrel would be about 20 megawatts per 10 meter section of barrel length over a period of one second at 1,000 second intervals. Thus, the barrel would only need thermal sinking at a rate 1,000 times less than thermal loading. Thermal sinking can be obtained by radiative elements attached to the barrel with the option of employing liquid coolants.

For pellet accelerations of 100,000 meter/$s^2$, the propulsive force would be **F**= m**a** = 4 mega-newtons which works out to approximately 1 million pounds of force and of similar order of magnitude to modern artillery. The latter force value is equivalent to the impulse imparted on the gun and is the same for all subsequently provided examples below.

Assuming an orbital location of one AU from the sun, and an adjustable reflector which reflects about 1/2 of the incident sunlight but transmits the rest, we obtain the following relation.

(4 x $10^6$ N) = <S>/c = <S>/(3 x $10^8$ m/s). So, the required sunlight power to counteract the acceleration on the gun during the firing time would be 1.2 x $10^{15}$ watts.

Assuming a barrel system mass of 5,000,000 kg, the net barrel acceleration during the firing operation would be F/m = [(4 x $10^6$ N)/(5,000,000 kg)] = 0.8 m/$s^2$ and the displacement component of the barrel during the firing process would be x = ½ **a**$t^2$ = 0.4 m.

Assuming a reverse acceleration time of 1,000 seconds, the barrel reverse acceleration need only be 8 x $10^{-4}$ m/$s^2$ and thus the required sun-light power on a sail to counteract the firing acceleration effects is a mere 1.2 x $10^{12}$ watts. The latter value can be obtained by a light-sail having collection area of a mere 1,200 $km^2$ or about 34.6 km by 34.6 km if square in shape.

It is conceivable that thin film metalized reflective graphene sheets having total thickness of one nanometer can be used to form the reflector. Assuming a areal density of $10^{-6}$ kg/$m^2$, the entire array reflector may have mass of only about 1.2 metric tons. Even for a thickness of 100 nanometers, the entire array would have mass of only 120 metric tons which required a modest correction factor to the above light-pressure calculations.

Assuming a projectile driving area of 0.01 square meters, the pressure in the barrel would be as high as 4 x $10^8$ N/$m^2$.

Required gun barrel length is ½ **a**$t^2$ or 50 kilometers.

The required power output may in principle be provided by a space-based nuclear reactor.

The required electrical power input may also be achieve is a photo-voltaic array covering about 2/3 $km^2$ assuming a radiative flux density of 1,500 w/$m^2$ and a conversion efficiency of 20 percent.

The cost to install 10 kilowatts of solar PV panels on a house is $25,300 – $31,500 according to Home Guide at https://homeguide.com/costs/solar-panel-cost. So for a 200 megawatts of installed power at the above pricing for domestic home use, the cost would be ($25,300 – $31,500)(20,000) or $506,000,000 to $630,000,000.

Ordinary domestic use PV panels may be covered in a glass or plastic enclosure for use in space. This covering would help prevent oxidation by free radicals common in interplanetary space even though the gas concentration is extremely low compared to sea level atmospheric pressure.

Assuming the above 200 megawatt instillation can include the protective coverings for a manufacturing cost of $1 billion, and that the area specific mass of the system is 10 kilograms per square meter, a fully launched and assembled array of the panels would have mass of (667,000)(10) kilograms 6,670 metric tons. This value is about one order of magnitude greater than the cost of the International Space Station (ISS). However, the cost to place materials in space is projected to reduce by one to perhaps two orders of magnitude per unit mass installed. So, the cost to install the array may be roughly that of the current accrued cost by the ISS give or take one order of magnitude. The worst case scenario would have the rigid PV array installed and operating of about $1 Trillion but as low as $10 Billion.

However, thin film PV membranes having mass of only 0.1 kilograms per square meter exist and which have conversion efficiencies on the order of 20 percent. So the mass of the PV system can be reduced accordingly to about 66.7 metric tons thus providing yet another small correction factor to the acceleration computations of the system required to maintain proper gas gun orientation and radial position with respect to the Sun.

Assuming a thin film reflector for concentrating sun-light, the mass of the rigid PV system combined with solar concentrator(s) can be only about 0.01 times that of the rigid system considered previously were sunlight is concentrated 100 fold.

Arrays operating on concentrated light can strongly reduce the capture area of the PV panels and thus the size of the array. There are some clever ways to greatly reduce the cost of concentrated sunlight systems which are discussed later below.

Turbo-electric systems powered by concentrated sunlight may also be of use. For example, a system of reflectors of capture area 1/3 $km^2$ could concentrate sunlight on a thermal mass to deliver 200 Mw usable electrical power provided a total system efficiency of 40 percent.

Note that most "solar panels are between 15% and 20% efficient, with outliers on either side of the range" according to energysage at https://news.energysage.com/what-are-the-most-efficient-solar-panels-on-the-market/

Domestic home use solar panels can easily work in space where the panels are enclosed in a low cost glass or uv-light resistant transparent plastic. The simply manufacturing process can greatly reduce the cost of the space-based PV arrays.

The thermal build up on the gas gun can be dissipated at similar rates as typical of smaller scale commercial electrical utility plants.

Assuming a barrel mass of 100 kilograms per meter of length, the total barrel mass would be 5 million kilograms or 5,000 metric tons. This is approximately one order of magnitude greater than that of the ISS. However, with anticipated launch to low Earth orbit (LEO) cost reductions by one order of magnitude or more and similarly also for placement of materials in interplanetary space for a cost of 10 million dollars per metric ton, the placement of the material composition of the gun in space would be accomplished for about $50 Billion at today's cost.

The cost of assembling the spacecraft in LEO would be roughly the same as the current instantiation of the ISS assuming a ten-fold cost reduction launch to LEO or about $100 Billion.

A nuclear reactor placed in space with the entire supporting infrastructure would have a cost similar to that accrued by the construction of the ISS at about $ 100 Billion.

So, the combined cost of the project components is estimated to be approximately $350 Billion at current dollar value. Over the life-time of the project, the cost of the project would be about $10 Billion per year.

The electron gas gun fueling of a spacecraft has merit in that relativistic spacecraft velocities can be achieved without the need for concern over otherwise storing a large quantity of nuclear explosives on the spacecraft and the lack of need of supporting architecture to store the devices on the spacecraft.

Each projectile may be fitted with a small low powered fine course correction rocket.

Needed data on the location of the pellets relative to the spacecraft and relative to each other would be provided by electronic transceiver apparatus installed within the pellets. The pellet transmissions can be tracked by Earth based systems and then sent to the exiting spacecraft.

Assume that the pellets would be detonated behind a charged space-craft chute having a collection area on the order of 100 square kilometers and attached to the spacecraft by an elastic spring like cable.

Assuming a chute mass-specific area density of 0.01 kilograms per square meter, the mass of the chute and attachment cables would be on the order of 1,000 metric tons. Moreover, the chute could be positioned 100 kilometer in front of the spacecraft for which the nuclear devices would be detonated about 10 to 20 kilometers behind the chute. Accordingly, the immediate ionizing radiation exposure to the spacecraft would be strongly reduced. Boron and other neutron capture materials can be included in a forwardly located spacecraft radiation shield. Gamma ray and x-ray absorbers can also be optimized by elemental and isotopic composition

Note that the pellets can be fired from much shorter barrels than in the systems presented herein for the three main scenarios included.

For example, gun barrels of one-hundredth the length of the ones considered herein can accelerate pellets having one-hundredth the mass of those considered herein for the primary scenarios. These smaller pellets can be fired at 100 times the frequency to thus enable spacecraft to achieve the same performance parameters per unit length of pellet runways.

Additionally, the smaller pellets can be more easily dispensed at higher invariant mass flow rates over all.

For example, a compound electron gas gun may include ten barrels affixed in a matrix, say as a pentagonal circular format, for which the pellets would be fired outward in a manner analogous to a revolver hand gun.

Accordingly, the connection of multiple barrels side by side along with cross-bracing can enable increased barrel stiffness and thus increased strength to resist bending, or torsion moments.

Such a revolver or similar activation mechanism can enable up to 1,000 pellets being fired per second in the barrel reference frame.

As another example, a matrix of 100 barrels, each 500 meters in length, may dispense pellets at rates up to 10,000 per second.

# The Blast Force.

***In this section, a presentation of formulas for blast force exerted on the pusher plates is provided.***

**SUM OF TWO VELOCITIES: GENERAL CASE.**

**Now, in the general case, the relativistic sum of two relativistic velocities V and U is**

**$V +_{(relativistic\ composition)}$ $U = \{V + U_{par} + [(alpha_v)(U_{perp})]\}/\{1 + \{[v \text{ dot } u]/[C^2]\}\}$**

**$Alpha_v = 1/[gamma\ (V)] = \{1 - \{[IVI^2]/[C^2]\}\}^{1/2}$.**

Now, for a plasma at temperature T, the kinetic energy per free chargon is defined by the following formula.

$E = (k_B)(T)$

Where $k_B$ is the Boltzmann Constant and T is the plasma temperature in kelvins.

So, the total species specific energy for a given species of plasma chargon produced by an exploded bomb is given by the following formula in the spacecraft frame.

$E_T = (k_B)(T_{ith,specie,average})(N_{ith,species})$,

where $N_{ith,species}$ is the number of chargons for the ith species of plasma particles.

The temperature, $T_{ith,specie,average}$, is the average temperature of the ith species of chargon.

Assuming the numerous plasma species are produced in a nuclear explosion in space, the total energy of the plasma produced in the spacecraft frame is equal to:

$E_{T,plasma} = \{\sum(i = 1, i = N)[\ (k_B)(T_{ith,species,average})(N_{ith,species})]\}$,

where N is the number of chargon species.

Now, the formulation for relativistic momentum for a particle or body of mass $m_0$ is as follows.

$P = (m_0)(v)/\{\{1 - [(v^2)/(c^2)]\}^{1/2}\}$.

Now, the energy of a relativistic particle as a function of momentum is provided by the following formula.

$E = \{\{[(P^2)(c^2)] + \{[(m_0)(c^2)]^2\}\}^{1/2}\}$.

So, the relativistic momentum of a particle will be:`

$P = \{\{\{(E^2) - \{[(m_0)(c^2)]^2\}\}/(c^2)\}^{1/2}\} = \{\{[(E^2)/(c^2)] - [(m_0{}^2)(c^2)]\}^{1/2}\}$.

So, the total momentum of the ith species of chargons produced in the blast as a function of blast temperature in the spacecraft frame is:

$P_{ith,chargon} = \{\{\{((k_B)(T_{ith,species,average})(N_{ith,species})^2) - \{[(m_{0,ith,species})(c^2)]^2\}\}/(c^2)\}^{1/2}\} = \{\{[((k_B)(T_{ith,species,average})(N_{ith,species})^2)/(c^2)] - [(m_{0,ith,species}{}^2)(c^2)]\}^{1/2}\}$.

The total momentum of the entire set of chargon species in the spacecraft frame is:

$P_{ichargon,total} = \{\sum(i = 1, i = N) \{\{\{((k_B)(T_{ith,species,average})(N_{ith,species})^2) - \{[(m_{0,ith,species})(c^2)]^2\}\}/(c^2)\}^{1/2}\}\} = \{\sum(i = 1, i = N)\{[((k_B)(T_{ith,species,average})(N_{ith,species})^2)/(c^2)] - [(m_{0,ith,species}{}^2)(c^2)]\}^{1/2}\}\}$.

Now, only a fraction of the chargons will impact the pusher plate and/or otherwise have their velocity vectors intercepted by the spacecraft.

So, the momentum delivered to the spacecraft in the spacecraft frame as a result of the chargons in a first order is provided by the following formula.

$P_{chargon,\ delivered} = (2)\{(A_{capture,chargon})/[(4)(\pi)(r^2)]\} \{\sum(i = 1, i = N) \{\{\{((k_B)(T_{ith,species,average})(N_{ith,species})^2) - \{[(m_{0,ith,species})(c^2)]^2\}\}/(c^2)\}^{1/2}\}\} = (2)\{(A_{capture,chargon})/[(4)(\pi)(r^2)]\} \{\sum(i = 1, i = N)\{\sum(i = 1, i = N)\{[((k_B)(T_{ith,species,average})(N_{ith,species})^2)/(c^2)] - [(m_{0,ith,species}{}^2)(c^2)]\}^{1/2}\}\}$.

Here, $A_{capture,chargon}$ is the effective chargon capture area projected on to a sphere centered at the location of the bomb explosion and r is the distance of the middle of the plate to the exploding bomb which more precisely is to be considered the effective center of the explosion. The bomb centered sphere is considered to have radius, r.

In the above formula, we assume that the chargons are reflected by an electromagnetic mirror at an angle with respect to the axis of the velocity vector of the spacecraft that is equal to the angle of incidence with respect to the axis. Here, we make no distinction between negative or positive angles of incidence nor from changes of angles as a result of momentum reversal of the chargons with respect to the spacecraft.

However, nuclear fission processes, and some nuclear fusion processes produce neutrons which impact the pusher plate.

The formula for total neutron emission kinetic energy in the spacecraft frame is

$E_{neutron,\ total} = \{\{\sum(j = 1,\ j = N_{max})\{\{(M_{0,neutron})(\gamma_j)(c^2)\} - [(M_{0,neutron})(c^2)]\}\}\ (Neutron,j)\} = \{\{\sum(j = 1,\ j = N_{max})\ \{\{(M_{0,neutron})\ \{\{1 - [(v_j^2)/(c^2)]\}^{1/2}\}\ (c^2)\} - [(M_{0,neutron})(c^2)]\}\}\ (Neutron,j)\}$.

The number of numerical integration steps is arbitrarily chosen. However, the finer the numerical integration and thus the more steps are used, the more accurate will be the results.

So, the total momentum of neutronic energy in the spacecraft frame is:

$P_{neutron,total} = \{\{\{\{\{\{\sum(j = 1,\ j = N_{neutron})\{\{(M_{0,neutron})(\gamma_j)(c^2)\} - [(M_{0,neutron})(c^2)]\}\}\ ^2\} - \{[(m_0)(c^2)]^2\}\}/(c^2)\}^{1/2}\}\ (Neutron,j)\} = \{\{\{\{\{\{\sum(j = 1,\ j = N_{neutron})\{\{(M_{0,neutron})(\gamma_j)(c^2)\} - [(M_{0,neutron})(c^2)]\}\}\ ^2\}/(c^2)\} - [(m_0\ ^2)(c^2)]\ \}^{1/2}\}(Neutron,j)\}$

$= \{\{\{\{\{\{\sum(j = 1,\ j = N_{neutron})\ \{\{(M_{0,neutron})\ \{\{1 - [(v_j^2)/(c^2)]\}^{1/2}\}\ (c^2)\} - [(M_{0,neutron})(c^2)]\}\}\ ^2\} - \{[(m_0)(c^2)]^2\}\}/(c^2)\}^{1/2}\}\ (Neutron,j)\} = \{\{\{\{\{\{\sum(j = 1,\ j = N_{neutron})\ \{\{(M_{0,neutronj})\ \{\{1 - [(v_j^2)/(c^2)]\}^{1/2}\}\ (c^2)\} - [(M_{0,neutron})(c^2)]\}\}\ ^2\}/(c^2)\} - [(m_0\ ^2)(c^2)]\ \}^{1/2}\}(Neutron,j)\}$.

Here $N_{neutron}$ = the total number of emitted neutrons thereby implying the summation is over all neutrons considered. For cases where the summation is less fine in scale, good approximations can still be obtained such as for example in cases where the neutron energy spectrum is partitioned into 100 intervals. Also, Neutron,j is the jth neutron.

So, the momentum delivered to the spacecraft in the spacecraft frame as a result of the neutrons in a first order is provided by the following formula.

$P_{neutron,total,delivered} = \{\{(A_{capture,neutron})/[(4)(\pi)(r^2)]\}\ \{\{\{\{\{\sum(j = 1,\ j = N_{max})\{\{(M_{0,neutron})(\gamma_j)(c^2)\} - [(M_{0,neutron})(c^2)]\}\}\ ^2\} - \{[(M_{0,neutron})(c^2)]^2\}\}/(c^2)\}^{1/2}\}(Neutron,j)\} = \{\{(A_{capture,neutron})/[(4)(\pi)(r^2)]\}\ \{\{\{\{\{\sum(j = 1,\ j = N_{max})\{\{(M_{0,neutron})(\gamma_j)(c^2)\} - [(M_{0,neutron})(c^2)]\}\}\ ^2\}/(c^2)\} - [(M_{0,neutron}\ ^2)(c^2)]\ \}^{1/2}\}(Neutron,j)\}$

$= \{\{(A_{capture,neutron})/[(4)(\pi)(r^2)]\}\ \{\{\{\{\{\sum(j = 1,\ j = N_{max})\ \{\{(M_{0,neutron})\ \{\{1 - [(v_j^2)/(c^2)]\}^{1/2}\}\ (c^2)\} - [(M_{0,neutron})(c^2)]\}\}\ ^2\} - \{[(\ M_{0,neutron})(c^2)]^2\}\}/(c^2)\}^{1/2}\}(Neutron,j)\} = \{\{(A_{capture,neutron})/[(4)(\pi)(r^2)]\}\ \{\{\{\{\{\sum(j = 1,\ j = N_{max})\ \{\{(M_{0,neutron})\ \{\{1 - [(v_j^2)/(c^2)]\}^{1/2}\}\ (c^2)\} - [(M_{0,neutron})(c^2)]\}\}\ ^2\}/(c^2)\} - [(M_{0,neutron}\ ^2)(c^2)]\ \}^{1/2}\}(Neutron,j)\}$.

Here, $A_{capture,chargon}$ is the effective neutron capture area projected on to a sphere centered at the location of the bomb explosion and r is the distance of the middle of the plate to the exploding bomb which more precisely is to be considered the effective center of the explosion. The bomb centered sphere is considered to have radius, r, in the spacecraft frame.

However, photon emission are also produced in the explosion.

Generally, the formula for photon momentum in the spacecraft frame is:

$P_{photon,total} = \sum(jk = 1,\ k = N_{photon})\ \{(E_k/c) = (h\nu_k/c\ ) = (h/\lambda_k)\}(Photon,k)$

where E is the photon energy, c is the speed of light, *h* is the Planck Constant, v is the photon frequency, and λ is the photon wave-length. Here, photon,k is the kth photon so summation is over the entire number of photons. However, the numerical integration can be less fine in scale of portioned, such as for instance when the photon energy spectrum is partitioned into 100 intervals.

So, the momentum delivered to the spacecraft in the spacecraft frame as a result of the photons in a first order is provided by the following formula.

$P_{photon,total,delivered}$ = {{($A_{capture,photon}$)/[(4)(π)($r^2$)]} {Σ(jk = 1, k = $N_{photon}$) {($E_k$/c) = ($hv_k/c$) = ($h/\lambda_k$)}(Photon,k)}}

Here, $A_{capture,chargon}$ is the effective photon capture area projected on to a sphere centered at the location of the bomb explosion and r is the distance of the middle of the plate to the exploding bomb which more precisely is to be considered the effective center of the explosion. The bomb centered sphere is considered to have radius, r.

Photon emissions can be categorized into direct nuclear reaction products and thermalized photons.

Thus, the following formulation can be used to characterize both forms of photonic generation for total photonic momentum delivered to the spacecraft in the spacecraft frame.

$P_{photon,total,delivered}$ = {{($A_{capture,photon}$)/[(4)(π)($r^2$)]} {Σ(jk = 1, k = $N_{photon,nuclear}$) {($E_k$/c) = ($hv_k/c$) = ($h/\lambda_k$)}(Photon,nuclear,k)}} + {{($A_{capture,photon}$)/[(4)(π)($r^2$)]} {Σ(jk = 1, k = $N_{photon,thermal}$) {($E_k$/c) = ($hv_k/c$) = ($h/\lambda_k$)}(Photon,thermal,k)}}.

The captured momentum in the spacecraft frame from the thermalized photons can be represented in first order by the following formula.

$P_{photon,thermal,total,delivered}$ = d{{($A_{capture,photon}$) [σ$T_{plasma}$ $^4$/C]}/dt= d{($A_{capture,photon}$) {{[5.670373 x $10^{-8}$] (J $m^{-2}$ $s^{-1}$ $K^{-4}$)}[$T_{plasma}{}^4$]/[3 x $10^8$ m/s]}}/dt.

Here, σ is the Stephan Boltzmann constant.

So, an alternative formulation for total photonic momentum delivered to the spacecraft in the spacecraft frame via the pusher plate is:

$P_{photon,total,delivered}$ = {{($A_{capture,photon}$)/[(4)(π)($r^2$)]} {Σ(jk = 1, k = $N_{photon,nuclear}$) {($E_k$/c) = ($hv_k/c$) = ($h/\lambda_k$)}(Photon,nuclear,k)}} + {d{($A_{capture,photon}$) {{[5.670373 x $10^{-8}$] (J $m^{-2}$ $s^{-1}$ $K^{-4}$)}[$T_{plasma}{}^4$]/[3 x $10^8$ m/s]}}/dt}.

Now, taking into account the angle of incidence and reflectance of the chargons, the momentum delivered to the spacecraft in the spacecraft frame as a result of the chargons in a first order is provided by the following formula.

$P_{chargon,\ delivered}$ = {∫[(α = 0, α = [[f(π)] $_{max,chargon}$]]{(2){($A_{capture,chargon}$)/[(4)(π)($r^2$)]} {Σ(i = 1, i = N) {{{(($k_B$)($T_{ith,species,average}$)($N_{ith,species}$)$^2$)– {[($m_{0,ith,species}$)($c^2$)]$^2$}}/($c^2$)}$^{1/2}$}}} (cos $\alpha_i$)dα} = {∫[(α = 0, α = [[f(π)] $_{max,chargon}$]]{ (2){($A_{capture,chargon}$)/[(4)(π)($r^2$)]} {Σ(i = 1, i = N) {Σ(i = 1, i = N){[(($k_B$)($T_{ith,species,average}$)($N_{ith,species}$)$^2$)/($c^2$)] – [($m_{0,ith,species}$ $^2$)($c^2$)] }$^{1/2}$}}} (cos $\alpha_i$)dα} .

Here, $\alpha_i$ is the angle of incidence of the jth chargon

In the above formula, we assume that the chargons are reflected by an electromagnetic mirror at an angle with respect to the axis of the velocity vector of the spacecraft that is equal to the angle of incidence with respect to the axis. Here, we make no distinction between negative or positive angles of incident nor from changes of angles as a result of momentum reversal of the chargons with respect to the spacecraft.

However, nuclear fission processes, and some nuclear fusion processes produce neutrons which impact the pusher plate.

Like-wise, the momentum delivered to the spacecraft in the spacecraft frame as a result of the neutrons in a second order is provided by the following formula.

$P_{neutron,total,delivered} = \{\int[(\alpha = 0, \alpha = [[f(\pi)]_{max,neutron}]]\{\{(A_{capture,neutron})/[(4)(\pi)(r^2)]\}\ \{\{\{\{\{\sum(j = 1, j = N_{max})\{\{(M_{0,neutron})(\gamma_j)(c^2)\} - [(M_{0,neutron})(c^2)]\}\}^{2}\}- \{[(M_{0,neutron})(c^2)]^2\}\}/(c^2)\}^{1/2}\}(Neutron,j)\}\ (\cos \alpha_i)d\alpha\} = \{\int[(\alpha = 0, \alpha = [[f(\pi)]_{max,neutron}]]\{\{(A_{capture,neutron})/[(4)(\pi)(r^2)]\}\ \{\{\{\{\{\sum(j = 1, j = N_{max})\{\{(M_{0,neutron})(\gamma_j)(c^2)\} - [(M_{0,neutron})(c^2)]\}\}^{2}\}/(c^2)\} - [(M_{0,neutron}{}^{2})(c^2)]\ \}^{1/2}\}(Neutron,j)\}\ (\cos \alpha_i)d\alpha\}$

$= \{\int[(\alpha = 0, \alpha = [[f(\pi)]_{max,neutron}]]\{\{(A_{capture,neutron})/[(4)(\pi)(r^2)]\}\ \{\{\{\{\{\sum(j = 1, j = N_{max})\ \{\{(M_{0,neutron})\ \{\{1 - [(v_j{}^2)/(c^2)]\}\}^{1/2}\}\ (c^2)\} - [(M_{0,neutron})(c^2)]\}\}^{2}\}- \{[(M_{0,neutron})(c^2)]^2\}\}/(c^2)\}^{1/2}\}(Neutron,j)\}\ (\cos \alpha_i)d\alpha\} = \{\int[(\alpha = 0, \alpha = [[f(\pi)]_{max,neutron}]]\{\{(A_{capture,neutron})/[(4)(\pi)(r^2)]\}\ \{\{\{\{\{\sum(j = 1, j = N_{max})\ \{\{(M_{0,neutron})\ \{\{1 - [(v_j{}^2)/(c^2)]\}\}^{1/2}\}\ (c^2)\} - [(M_{0,neutron})(c^2)]\}\}^{2}\}/(c^2)\} - [(M_{0,neutron}{}^{2})(c^2)]\ \}^{1/2}\}(Neutron,j)\}\ (\cos \alpha_i)d\alpha\}$.

Here, $A_{capture,chargon}$ is the effective neutron capture area projected on to a sphere centered at the location of the bomb explosion and r is the distance of the middle of the plate to the exploding bomb which more precisely is to be considered the effective center of the explosion. The bomb centered sphere is considered to have radius, r.

The second order momentum delivered to the spacecraft in the spacecraft frame as a result of the photons in a first order is provided by the following formula.

$P_{photon,total,delivered} = \{\int[(\alpha = 0, \alpha = [[f(\pi)]_{max,photon}]]\{\{(A_{capture,photon})/[(4)(\pi)(r^2)]\}\ \{\sum(jk = 1, k = N_{photon})\ \{(E_k/c) = (h\nu_k/c) = (h/\lambda_k)\}(Photon,k)\}\}\ (\cos \alpha_i)d\alpha\}$

Here, $A_{capture,chargon}$ is the effective photon capture area projected on to a sphere centered at the location of the bomb explosion and r is the distance of the middle of the plate to the exploding bomb which more precisely is to be considered the effective center of the explosion. The bomb centered sphere is considered to have radius, r.

Again, photon emissions can be categorized into direct nuclear reaction products and thermalized photons.

Thus, the following second order formulation can be used to characterize both forms of photonic generation for total photonic momentum delivered to the spacecraft in the spacecraft frame.

$P_{photon,total,delivered} = \{\int[(\alpha = 0, \alpha = [[f(\pi)]_{max,photon,nuclear}]]\{\{(A_{capture,photon})/[(4)(\pi)(r^2)]\} \{\sum(jk = 1, k = N_{photon,nuclear}) \{(E_k/c) = (h\nu_k/c) = (h/\lambda_k)\}(Photon,nuclear,k)\}\} (\cos \alpha_i)d\alpha\} + \{\int[(\alpha = 0, \alpha = [[f(\pi)]_{max,photon,thermal}]]\{\{(A_{capture,photon})/[(4)(\pi)(r^2)]\} \{\sum(jk = 1, k = N_{photon,thermal}) \{(E_k/c) = (h\nu_k/c) = (h/\lambda_k)\}(Photon,thermal,k)\}\} (\cos \alpha_i)d\alpha\}$.

The captured momentum in the spacecraft frame from the thermalized photons can be represented in second order by the following formula.

$P_{photon,thermal,total,delivered} = \{\int[(\alpha = 0, \alpha = [[f(\pi)]_{max,photon\ thermal}]] \{d\{\{(A_{capture,photon}) [\sigma T_{plasma}{}^4/C]\}/dt\}(\cos \alpha_i)d\alpha\}= \{\int[(\alpha = 0, \alpha = [[f(\pi)]_{max,photon,thermal}]]\{d\{(A_{capture,photon}) \{\{[5.670373 \times 10^{-8}] (J\ m^{-2}\ s^{-1}\ K^{-4})\}[T_{plasma}{}^4]/[3 \times 10^8\ m/s]\}\}/dt\}(\cos \alpha_i)d\alpha\}$.

Here, $\sigma$ is the Stephan Boltzmann constant.

So, an alternative formulation for total photonic momentum delivered to the spacecraft in the spacecraft frame via the pusher plate is:

$P_{photon,total,delivered} = \{\int[(\alpha = 0, \alpha = [[f(\pi)]_{max,photon,nuclear}]]\{\{(A_{capture,photon})/[(4)(\pi)(r^2)]\} \{\sum(jk = 1, k = N_{photon,nuclear}) \{(E_k/c) = (h\nu_k/c) = (h/\lambda_k)\}(Photon,nuclear,k)\}\} (\cos \alpha_i)d\alpha\} + \{\int[(\alpha = 0, \alpha = [[f(\pi)]_{max,photon,thermal}]]\{d\{(A_{capture,photon}) \{\{[5.670373 \times 10^{-8}] (J\ m^{-2}\ s^{-1}\ K^{-4})\}[T_{plasma}{}^4]/[3 \times 10^8\ m/s]\}\}/dt\}(\cos \alpha_i)d\alpha\}$.

The combined total momentums in the spacecraft frame thus yield results for the total bomb blast momentum portion transferred to the plates for which the factors (Chargon angular correction factor), (Neutron angular correction factor), (Nuclear photon angular correction factor), and (Thermal photon angular correction factor) are non-dimensional functions that take into account the variations in particle momentum flux incident on the pusher plate or otherwise for different angles of incidence.

Thus, the complete formulation is: $P_{combined,incident} = \{\int[(\alpha = 0, \alpha = [[f(\pi)]_{max,chargon}]]\{(2)\{(A_{capture,chargon})/[(4)(\pi)(r^2)]\} \{\sum(i = 1, i = N) \{\{\{((k_B)(T_{ith,species,average})(N_{ith,species})^2) - \{[(m_{0,ith,species})(c^2)]^2\}\}/(c^2)\}^{1/2}\}\}\}$ (Chargon angular correction factor) $(\cos \alpha_i)d\alpha\} + \{\int[(\alpha = 0, \alpha = [[f(\pi)]_{max,neutron}]]\{\{(A_{capture,neutron})/[(4)(\pi)(r^2)]\} \{\{\{\{\sum(j = 1, j = N_{max})\{\{(M_{0,neutron})(\gamma_j)(c^2)\} - [(M_{0,neutron})(c^2)]\}\}^2\} - \{[(M_{0,neutron})(c^2)]^2\}\}/(c^2)\}^{1/2}\}(Neutron,j)\}$ (Neutron angular correction factor) $(\cos \alpha_i)d\alpha\} + \{\int[(\alpha = 0, \alpha = [[f(\pi)]_{max,photon,nuclear}]]\{\{(A_{capture,photon})/[(4)(\pi)(r^2)]\} \{\sum(jk = 1, k = N_{photon,nuclear}) \{(E_k/c) = (h\nu_k/c) = (h/\lambda_k)\}(Photon,nuclear,k)\}\}$ (Nuclear photon angular correction factor) $(\cos \alpha_i)d\alpha\} + \{\int[(\alpha = 0, \alpha = [[f(\pi)]_{max,photon,thermal}]]\{d\{(A_{capture,photon}) \{\{[5.670373 \times 10^{-8}] (J\ m^{-2}\ s^{-1}\ K^{-4})\}[T_{plasma}{}^4]/[3 \times 10^8\ m/s]\}\}/dt\}$(Thermal photon angular correction factor) $(\cos \alpha_i)d\alpha\}$

or

$P_{combined,incident} = \{\int[(\alpha = 0, \alpha = [[f(\pi)]_{max,chargon}]]\{ (2)\{(A_{capture,chargon})/[(4)(\pi)(r^2)]\} \{\sum(i = 1, i = N) \{\sum(i = 1, i = N)\{[((k_B)(T_{ith,species,average})(N_{ith,species})^2)/(c^2)] - [(m_{0,ith,species}{}^2)(c^2)]\ \}^{1/2}\}\}\}$ (Chargon angular correction factor) $(\cos \alpha_i)d\alpha\} + \{\int[(\alpha = 0, \alpha = [[f(\pi)]_{max,neutron}]]\{\{(A_{capture,neutron})/[(4)(\pi)(r^2)]\} \{\{\{\{\{\sum(j = 1, j = N_{max}) \{\{(M_{0,neutron}) \{\{1 - [(v_j{}^2)/(c^2)]\}^{1/2}\} (c^2)\} - [(M_{0,neutron})(c^2)]\}\}^2\}/(c^2)\} - [(M_{0,neutron}{}^2)(c^2)]\ \}^{1/2}\}(Neutron,j)\}$ (Neutron angular correction factor) $(\cos \alpha_i)d\alpha\} + \{\int[(\alpha = 0, \alpha = [[f(\pi)]_{max,photon,nuclear}]]\{\{(A_{capture,photon})/[(4)(\pi)(r^2)]\} \{\sum(jk = 1, k = N_{photon,nuclear}) \{(E_k/c) = (h\nu_k/c) =$

($h/\lambda_k$)}(Photon,nuclear,k)}} (Nuclear photon angular correction factor)(cos $\alpha_i$)dα} + {∫[(α = 0, α = [[f(π)] $_{max,photon,thermal}$]]{d{($A_{capture,photon}$) {{[5.670373 x $10^{-8}$] (J $m^{-2}$ $s^{-1}$ $K^{-4}$)}[$T_{plasma}^4$]/[3 x $10^8$ m/s]}}/dt}(Thermal photon angular correction factor) (cos $\alpha_i$)dα}.

The total force exerted by the blast in the spacecraft frame is:

$F_{total}$ = d $P_{combined,incident}$/dt = d{∫[(α = 0, α = [[f(π)] $_{max,chargon}$]]{(2){($A_{capture,chargon}$)/[(4)(π)($r^2$)]} {Σ(i = 1, i = N) {{{(($k_B$)($T_{ith,species,average}$)($N_{ith,species}$)$^2$)– {[($m_{0,ith,species}$)($c^2$)]$^2$}}/($c^2$)}$^{1/2}$}}}(Chargon angular correction factor) (cos $\alpha_i$)dα} + {∫[(α = 0, α = [[f(π)] $_{max,neutron}$]]{{($A_{capture,neutron}$)/[(4)(π)($r^2$)]} {{{{{Σ(j = 1, j = $N_{max}$){{($M_{0,neutron}$)($\gamma_j$)($c^2$)} – [($M_{0,neutron}$)($c^2$)]}} $^2$}– {[($M_{0,neutron}$)($c^2$)]$^2$}}/($c^2$)}$^{1/2}$}(Neutron,j)} (Neutron angular correction factor) (cos $\alpha_i$)dα} + {∫[(α = 0, α = [[f(π)] $_{max,photon,nuclear}$]]{{($A_{capture,photon}$)/[(4)(π)($r^2$)]} {Σ(jk = 1, k = $N_{photon,nuclear}$) {($E_k$/c) = ($h\nu_k/c$) = ($h/\lambda_k$)}(Photon,nuclear,k)}} (Nuclear photon angular correction factor)(cos $\alpha_i$)dα} + {∫[(α = 0, α = [[f(π)] $_{max,photon,thermal}$]]{d{($A_{capture,photon}$) {{[5.670373 x $10^{-8}$] (J $m^{-2}$ $s^{-1}$ $K^{-4}$)}[$T_{plasma}^4$]/[3 x $10^8$ m/s]}}/dt}(Thermal photon angular correction factor) (cos $\alpha_i$)dα}/dt

or

$F_{total}$ = d $P_{combined,incident}$/dt = d{∫[(α = 0, α = [[f(π)] $_{max,chargon}$]]{ (2){($A_{capture,chargon}$)/[(4)(π)($r^2$)]} {Σ(i = 1, i = N) {Σ(i = 1, i = N){[(($k_B$)($T_{ith,species,average}$)($N_{ith,species}$)$^2$)/($c^2$)] – [($m_{0,ith,species}$ $^2$)($c^2$)] }$^{1/2}$}}} (Chargon angular correction factor) (cos $\alpha_i$)dα} + {∫[(α = 0, α = [[f(π)] $_{max,neutron}$]]{{($A_{capture,neutron}$)/[(4)(π)($r^2$)]} {{{{{Σ(j = 1, j = $N_{max}$) {{($M_{0,neutron}$) {{1 – [($v_j^2$)/($c^2$)]}}$^{1/2}$} ($c^2$)} – [($M_{0,neutron}$)($c^2$)]}} $^2$}/($c^2$)} – [($M_{0,neutron}$ $^2$)($c^2$)] }$^{1/2}$}(Neutron,j)} (Neutron angular correction factor) (cos $\alpha_i$)dα} +{∫[(α = 0, α = [[f(π)] $_{max,photon,nuclear}$]]{{($A_{capture,photon}$)/[(4)(π)($r^2$)]} {Σ(jk = 1, k = $N_{photon,nuclear}$) {($E_k$/c) = ($h\nu_k/c$) = ($h/\lambda_k$)}(Photon,nuclear,k)}} (Nuclear photon angular correction factor) (cos $\alpha_i$)dα} + {∫[(α = 0, α = [[f(π)] $_{max,photon,thermal}$]]{d{($A_{capture,photon}$) {{[5.670373 x $10^{-8}$] (J $m^{-2}$ $s^{-1}$ $K^{-4}$)}[$T_{plasma}^4$]/[3 x $10^8$ m/s]}}/dt}(Thermal photon angular correction factor) (cos $\alpha_i$)dα}/dt.

Provided f"(x) < 0, f has a local maximum at x.

So, we can simply set $d^2${d $P_{combined,incident}$/dt}/$dF^2$ < 0 and solve the inequality for F = force.

So then, $d^2${d $P_{combined,incident}$/dt}/$dF^2$ = {$d^2${d{∫[(α = 0, α = [[f(π)] $_{max,chargon}$]]{(2){($A_{capture,chargon}$)/[(4)(π)($r^2$)]} {Σ(i = 1, i = N) {{{(($k_B$)($T_{ith,species,average}$)($N_{ith,species}$)$^2$)– {[($m_{0,ith,species}$)($c^2$)]$^2$}}/($c^2$)}$^{1/2}$}}}(Chargon angular correction factor) (cos $\alpha_i$)dα} + {∫[(α = 0, α = [[f(π)] $_{max,neutron}$]]{{($A_{capture,neutron}$)/[(4)(π)($r^2$)]} {{{{{Σ(j = 1, j = $N_{max}$){{($M_{0,neutron}$)($\gamma_j$)($c^2$)} – [($M_{0,neutron}$)($c^2$)]}} $^2$}– {[($M_{0,neutron}$)($c^2$)]$^2$}}/($c^2$)}$^{1/2}$}(Neutron,j)} (Neutron angular correction factor) (cos $\alpha_i$)dα} + {∫[(α = 0, α = [[f(π)] $_{max,photon,nuclear}$]]{{($A_{capture,photon}$)/[(4)(π)($r^2$)]} {Σ(jk = 1, k = $N_{photon,nuclear}$) {($E_k$/c) = ($h\nu_k/c$) = ($h/\lambda_k$)}(Photon,nuclear,k)}} (Nuclear photon angular correction factor)(cos $\alpha_i$)dα} + {∫[(α = 0, α = [[f(π)] $_{max,photon,thermal}$]]{d{($A_{capture,photon}$) {{[5.670373 x $10^{-8}$] (J $m^{-2}$ $s^{-1}$ $K^{-4}$)}[$T_{plasma}^4$]/[3 x $10^8$ m/s]}}/dt}(Thermal photon angular correction factor) (cos $\alpha_i$)dα}/dt}/$dF^2$}

= {d $^2$ {d{∫[(α = 0, α = [[f(π)] $_{max,chargon}$]]{ (2){($A_{capture,chargon}$)/[(4)(π)($r^2$)]} {Σ(i = 1, i = N) {Σ(i = 1, i = N){[(($k_B$)($T_{ith,species,average}$)($N_{ith,species}$)$^2$)/($c^2$)] – [($m_{0,ith,species}$ $^2$)($c^2$)] }$^{1/2}$}}} (Chargon angular correction factor) (cos $\alpha_i$)dα} + {∫[(α = 0, α = [[f(π)] $_{max,neutron}$]]{{($A_{capture,neutron}$)/[(4)(π)($r^2$)]} {{{{{Σ(j = 1, j =

$N_{max}$) {{($M_{0,neutron}$) {{1 – [($v_j^2$)/($c^2$)]}$^{1/2}$} ($c^2$)} – [($M_{0,neutron}$)($c^2$)]}} $^2$}/($c^2$)} – [($M_{0,neutron}$ $^2$)($c^2$)] }$^{1/2}$}(Neutron,j)} (Neutron angular correction factor) (cos $\alpha_i$)dα} +{∫[(α = 0, α = [[f(π)] $_{max,photon,nuclear}$]]{{($A_{capture,photon}$)/[(4)(π)($r^2$)]} {Σ(jk = 1, k = $N_{photon,nuclear}$) {($E_k$/c) = ($h\nu_k/c$) = ($h/\lambda_k$)}(Photon,nuclear,k)}} (Nuclear photon angular correction factor) (cos $\alpha_i$)dα} + {∫[(α = 0, α = [[f(π)] $_{max,photon,thermal}$]]{d{($A_{capture,photon}$) {{[5.670373 x $10^{-8}$] (J $m^{-2}$ $s^{-1}$ $K^{-4}$)}[$T_{plasma}^4$]/[3 x $10^8$ m/s]}}/dt}(Thermal photon angular correction factor) (cos $\alpha_i$)dα}/dt}/d $F^2$}.

The total time integrated driving force provides a measure of energy used that is of propulsive form.

Thus, we obtain the following formulas for spatially integrated energy.

∫<$F_{total}$> • d r(x,y,z) = ∫<{d $P_{combined,incident}$/dt}> • d r(x,y,z) = ∫<{d{∫[(α = 0, α = [[f(π)] $_{max,chargon}$]]{(2){($A_{capture,chargon}$)/[(4)(π)($r^2$)]} {Σ(i = 1, i = N) {{{(($k_B$)($T_{ith,species,average}$)($N_{ith,species}$)$^2$)– {[($m_{0,ith,species}$)($c^2$)]$^2$}}/($c^2$)}$^{1/2}$}}}(Chargon angular correction factor) (cos $\alpha_i$)dα} + {∫[(α = 0, α = [[f(π)] $_{max,neutron}$]]{{($A_{capture,neutron}$)/[(4)(π)($r^2$)]} {{{{{Σ(j = 1, j = $N_{max}$){{($M_{0,neutron}$)($\gamma_j$)($c^2$)} – [($M_{0,neutron}$)($c^2$)]}} $^2$}– {[($M_{0,neutron}$)($c^2$)]$^2$}}/($c^2$)}$^{1/2}$}(Neutron,j)} (Neutron angular correction factor) (cos $\alpha_i$)dα} + {∫[(α = 0, α = [[f(π)] $_{max,photon,nuclear}$]]{{($A_{capture,photon}$)/[(4)(π)($r^2$)]} {Σ(jk = 1, k = $N_{photon,nuclear}$) {($E_k$/c) = ($h\nu_k/c$) = ($h/\lambda_k$)}(Photon,nuclear,k)}} (Nuclear photon angular correction factor)(cos $\alpha_i$)dα} + {∫[(α = 0, α = [[f(π)] $_{max,photon,thermal}$]]{d{($A_{capture,photon}$) {{[5.670373 x $10^{-8}$] (J $m^{-2}$ $s^{-1}$ $K^{-4}$)}[$T_{plasma}^4$]/[3 x $10^8$ m/s]}}/dt}(Thermal photon angular correction factor) (cos $\alpha_i$)dα}/dt}>• d r(x,y,z)

or

∫<$F_{total}$> • d r(x,y,z) = ∫<{d $P_{combined,incident}$/dt}>• d r(x,y,z) = ∫<{d{∫[(α = 0, α = [[f(π)] $_{max,chargon}$]]{ (2){($A_{capture,chargon}$)/[(4)(π)($r^2$)]} {Σ(i = 1, i = N) {Σ(i = 1, i = N){[(($k_B$)($T_{ith,species,average}$)($N_{ith,species}$)$^2$)/($c^2$)] – [($m_{0,ith,species}$ $^2$)($c^2$)] }$^{1/2}$}}} (Chargon angular correction factor) (cos $\alpha_i$)dα} + {∫[(α = 0, α = [[f(π)] $_{max,neutron}$]]{{($A_{capture,neutron}$)/[(4)(π)($r^2$)]} {{{{{Σ(j = 1, j = $N_{max}$) {{($M_{0,neutron}$) {{1 – [($v_j^2$)/($c^2$)]}$^{1/2}$} ($c^2$)} – [($M_{0,neutron}$)($c^2$)]}} $^2$}/($c^2$)} – [($M_{0,neutron}$ $^2$)($c^2$)] }$^{1/2}$}(Neutron,j)} (Neutron angular correction factor) (cos $\alpha_i$)dα} +{∫[(α = 0, α = [[f(π)] $_{max,photon,nuclear}$]]{{($A_{capture,photon}$)/[(4)(π)($r^2$)]} {Σ(jk = 1, k = $N_{photon,nuclear}$) {($E_k$/c) = ($h\nu_k/c$) = ($h/\lambda_k$)}(Photon,nuclear,k)}} (Nuclear photon angular correction factor) (cos $\alpha_i$)dα} + {∫[(α = 0, α = [[f(π)] $_{max,photon,thermal}$]]{d{($A_{capture,photon}$) {{[5.670373 x $10^{-8}$] (J $m^{-2}$ $s^{-1}$ $K^{-4}$)}[$T_{plasma}^4$]/[3 x $10^8$ m/s]}}/dt}(Thermal photon angular correction factor) (cos $\alpha_i$)dα}/dt}>• d r(x,y,z).

Now, all of this assumes that only one mass specific bomb yield is employed and that all of the bombs are also identical otherwise in mass and nuclear reaction chains.

Where, more than one invariant mass specific yield bomb styles are used and/or where the reaction pathways vary from bomb model with respect to another, the following formulas apply for propulsive energy or spatially integrated force. Note that the nomenclature < > indicates that the forces considered are to be interpreted as time averaged forces over the travel path of the spacecraft. This applies below in this chapter and also in the next chapter.

$\int<F_{total}> \bullet d\, r(x,y,z) = \{\sum(z_1 = 1,\ z_1 = N_{z1})\{\{\int<\{d\, P_{combined,incident}/dt\}> \bullet d\, r(x,y,z)\}_{bomb,z1}\}\} + \{\sum(z_2 = 1,\ z_2 = N_{z2})\{\{\int<\{d\, P_{combined,incident}/dt\}> \bullet d\, r(x,y,z)\}_{bomb,z2}\}\} + \{\sum(z_3 = 1,\ z_3 = N_{z3})\{\{\int<\{d\, P_{combined,incident}/dt\}> \bullet d\, r(x,y,z)\}_{bomb,z3}\}\} + \ldots + \{\sum(z_n = 1,\ z_n = N_{zn})\{\{\int<\{d\, P_{combined,incident}/dt\}> \bullet d\, r(x,y,z)\}_{bomb,zn}\}\}$

$= \{\sum(z_1 = 1,\ z_1 = N_{z1})\{\int<\{d\{\int[(\alpha = 0,\ \alpha = [[f(\pi)]_{max,chargon}]]\{(2)\{(A_{capture,chargon})/[(4)(\pi)(r^2)]\}\ \{\sum(i = 1,\ i = N)\ \{\{\{((k_B)(T_{ith,species,average})(N_{ith,species})^2) - \{[(m_{0,ith,species})(c^2)]^2\}\}/(c^2)\}^{1/2}\}\}\}(\text{Chargon angular correction factor})\ (\cos \alpha_i)d\alpha\} + \{\int[(\alpha = 0,\ \alpha = [[f(\pi)]_{max,neutron}]]\{\{(A_{capture,neutron})/[(4)(\pi)(r^2)]\}\ \{\{\{\{\{\sum(j = 1,\ j = N_{max})\{\{(M_{0,neutron})(\gamma_j)(c^2)\} - [(M_{0,neutron})(c^2)]\}\}^2\} - \{[(M_{0,neutron})(c^2)]^2\}\}/(c^2)\}^{1/2}\}(\text{Neutron},j)\}\ (\text{Neutron angular correction factor})\ (\cos \alpha_i)d\alpha\} + \{\int[(\alpha = 0,\ \alpha = [[f(\pi)]_{max,photon,nuclear}]]\{\{(A_{capture,photon})/[(4)(\pi)(r^2)]\}\ \{\sum(jk = 1,\ k = N_{photon,nuclear})\ \{(E_k/c) = (h\nu_k/c) = (h/\lambda_k)\}(\text{Photon,nuclear},k)\}\}\ (\text{Nuclear photon angular correction factor})(\cos \alpha_i)d\alpha\} + \{\int[(\alpha = 0,\ \alpha = [[f(\pi)]_{max,photon,thermal}]]\{d\{(A_{capture,photon})\ \{\{[5.670373 \times 10^{-8}]\ (J\ m^{-2}\ s^{-1}\ K^{-4})\}[T_{plasma}{}^4]/[3 \times 10^8\ m/s]\}\}/dt\}(\text{Thermal photon angular correction factor})\ (\cos \alpha_i)d\alpha\}/dt\}> \bullet d\, r(x,y,z)\}_{bomb,z1}\}\}$

$+ \{\sum(z_2 = 1,\ z_2 = N_{z2})\{\int<\{d\{\int[(\alpha = 0,\ \alpha = [[f(\pi)]_{max,chargon}]]\{(2)\{(A_{capture,chargon})/[(4)(\pi)(r^2)]\}\ \{\sum(i = 1,\ i = N)\ \{\{\{((k_B)(T_{ith,species,average})(N_{ith,species})^2) - \{[(m_{0,ith,species})(c^2)]^2\}\}/(c^2)\}^{1/2}\}\}\}(\text{Chargon angular correction factor})\ (\cos \alpha_i)d\alpha\} + \{\int[(\alpha = 0,\ \alpha = [[f(\pi)]_{max,neutron}]]\{\{(A_{capture,neutron})/[(4)(\pi)(r^2)]\}\ \{\{\{\{\{\sum(j = 1,\ j = N_{max})\{\{(M_{0,neutron})(\gamma_j)(c^2)\} - [(M_{0,neutron})(c^2)]\}\}^2\} - \{[(M_{0,neutron})(c^2)]^2\}\}/(c^2)\}^{1/2}\}(\text{Neutron},j)\}\ (\text{Neutron angular correction factor})\ (\cos \alpha_i)d\alpha\} + \{\int[(\alpha = 0,\ \alpha = [[f(\pi)]_{max,photon,nuclear}]]\{\{(A_{capture,photon})/[(4)(\pi)(r^2)]\}\ \{\sum(jk = 1,\ k = N_{photon,nuclear})\ \{(E_k/c) = (h\nu_k/c) = (h/\lambda_k)\}(\text{Photon,nuclear},k)\}\}\ (\text{Nuclear photon angular correction factor})(\cos \alpha_i)d\alpha\} + \{\int[(\alpha = 0,\ \alpha = [[f(\pi)]_{max,photon,thermal}]]\{d\{(A_{capture,photon})\ \{\{[5.670373 \times 10^{-8}]\ (J\ m^{-2}\ s^{-1}\ K^{-4})\}[T_{plasma}{}^4]/[3 \times 10^8\ m/s]\}\}/dt\}(\text{Thermal photon angular correction factor})\ (\cos \alpha_i)d\alpha\}/dt\}> \bullet d\, r(x,y,z)\}_{bomb,z2}\}\}$

$+ \{\sum(z_3 = 1,\ z_3 = N_{z3})\{\int<\{d\{\int[(\alpha = 0,\ \alpha = [[f(\pi)]_{max,chargon}]]\{(2)\{(A_{capture,chargon})/[(4)(\pi)(r^2)]\}\ \{\sum(i = 1,\ i = N)\ \{\{\{((k_B)(T_{ith,species,average})(N_{ith,species})^2) - \{[(m_{0,ith,species})(c^2)]^2\}\}/(c^2)\}^{1/2}\}\}\}(\text{Chargon angular correction factor})\ (\cos \alpha_i)d\alpha\} + \{\int[(\alpha = 0,\ \alpha = [[f(\pi)]_{max,neutron}]]\{\{(A_{capture,neutron})/[(4)(\pi)(r^2)]\}\ \{\{\{\{\{\sum(j = 1,\ j = N_{max})\{\{(M_{0,neutron})(\gamma_j)(c^2)\} - [(M_{0,neutron})(c^2)]\}\}^2\} - \{[(M_{0,neutron})(c^2)]^2\}\}/(c^2)\}^{1/2}\}(\text{Neutron},j)\}\ (\text{Neutron angular correction factor})\ (\cos \alpha_i)d\alpha\} + \{\int[(\alpha = 0,\ \alpha = [[f(\pi)]_{max,photon,nuclear}]]\{\{(A_{capture,photon})/[(4)(\pi)(r^2)]\}\ \{\sum(jk = 1,\ k = N_{photon,nuclear})\ \{(E_k/c) = (h\nu_k/c) = (h/\lambda_k)\}(\text{Photon,nuclear},k)\}\}\ (\text{Nuclear photon angular correction factor})(\cos \alpha_i)d\alpha\} + \{\int[(\alpha = 0,\ \alpha = [[f(\pi)]_{max,photon,thermal}]]\{d\{(A_{capture,photon})\ \{\{[5.670373 \times 10^{-8}]\ (J\ m^{-2}\ s^{-1}\ K^{-4})\}[T_{plasma}{}^4]/[3 \times 10^8\ m/s]\}\}/dt\}(\text{Thermal photon angular correction factor})\ (\cos \alpha_i)d\alpha\}/dt\}> \bullet d\, r(x,y,z)\}_{bomb,z3}\}\}$

$+ \ldots + \{\sum(z_n = 1,\ z_n = N_{zn})\{\int<\{d\{\int[(\alpha = 0,\ \alpha = [[f(\pi)]_{max,chargon}]]\{(2)\{(A_{capture,chargon})/[(4)(\pi)(r^2)]\}\ \{\sum(i = 1,\ i = N)\ \{\{\{((k_B)(T_{ith,species,average})(N_{ith,species})^2) - \{[(m_{0,ith,species})(c^2)]^2\}\}/(c^2)\}^{1/2}\}\}\}(\text{Chargon angular correction factor})\ (\cos \alpha_i)d\alpha\} + \{\int[(\alpha = 0,\ \alpha = [[f(\pi)]_{max,neutron}]]\{\{(A_{capture,neutron})/[(4)(\pi)(r^2)]\}\ \{\{\{\{\{\sum(j = 1,\ j = N_{max})\{\{(M_{0,neutron})(\gamma_j)(c^2)\} - [(M_{0,neutron})(c^2)]\}\}^2\} - \{[(M_{0,neutron})(c^2)]^2\}\}/(c^2)\}^{1/2}\}(\text{Neutron},j)\}\ (\text{Neutron angular correction factor})\ (\cos \alpha_i)d\alpha\} + \{\int[(\alpha = 0,\ \alpha = [[f(\pi)]_{max,photon,nuclear}]]\{\{(A_{capture,photon})/[(4)(\pi)(r^2)]\}\ \{\sum(jk = 1,\ k = N_{photon,nuclear})\ \{(E_k/c) = (h\nu_k/c) = (h/\lambda_k)\}(\text{Photon,nuclear},k)\}\}\ (\text{Nuclear photon angular correction factor})(\cos \alpha_i)d\alpha\} + \{\int[(\alpha = 0,\ \alpha = [[f(\pi)]_{max,photon,thermal}]]\{d\{(A_{capture,photon})\ \{\{[5.670373 \times 10^{-8}]\ (J\ m^{-2}\ s^{-1}\ K^{-4})\}[T_{plasma}{}^4]/[3 \times 10^8\ m/s]\}\}/dt\}(\text{Thermal photon angular correction factor})\ (\cos \alpha_i)d\alpha\}/dt\}> \bullet d\, r(x,y,z)\}_{bomb,zn}\}\}$

= {∑($z_1$ = 1, $z_1$ = $N_{z1}$){∫<{d{∫[(α = 0, α = [[f(π)] $_{max,chargon}$]]{ (2){($A_{capture,chargon}$)/[(4)(π)($r^2$)]} {∑(i = 1, i = N) {∑(i = 1, i = N){[(($k_B$)($T_{ith,species,average}$)($N_{ith,species}$)$^2$)/($c^2$)] – [($m_{0,ith,species}$ $^2$)($c^2$)] }$^{1/2}$}}} (Chargon angular correction factor) (cos $α_i$)dα} + {∫[(α = 0, α = [[f(π)] $_{max,neutron}$]]{{($A_{capture,neutron}$)/[(4)(π)($r^2$)]} {{{{{∑(j = 1, j = $N_{max}$) {{($M_{0,neutron}$) {{1 – [($v_j^2$)/($c^2$)]}$^{1/2}$} ($c^2$)} – [($M_{0,neutron}$)($c^2$)]}} $^2$}/($c^2$)} – [($M_{0,neutron}$ $^2$)($c^2$)] }$^{1/2}$}(Neutron,j)} (Neutron angular correction factor) (cos $α_i$)dα} +{∫[(α = 0, α = [[f(π)] $_{max,photon,nuclear}$]]{{($A_{capture,photon}$)/[(4)(π)($r^2$)]} {∑(jk = 1, k = $N_{photon,nuclear}$) {($E_k$/c) = (*$hv_k$/c*) = (*$h/λ_k$*)}(Photon,nuclear,k)}} (Nuclear photon angular correction factor) (cos $α_i$)dα} + {∫[(α = 0, α = [[f(π)] $_{max,photon,thermal}$]]{d{($A_{capture,photon}$) {{[5.670373 x $10^{-8}$] (J $m^{-2}$ $s^{-1}$ $K^{-4}$)}[$T_{plasma}^4$]/[3 x $10^8$ m/s]}}/dt}(Thermal photon angular correction factor) (cos $α_i$)dα}/dt}>• d r(x,y,z)}$_{bomb,z1}$}}

+ {∑($z_2$ = 1, $z_2$ = $N_{z2}$){∫<{d{∫[(α = 0, α = [[f(π)] $_{max,chargon}$]]{ (2){($A_{capture,chargon}$)/[(4)(π)($r^2$)]} {∑(i = 1, i = N) {∑(i = 1, i = N){[(($k_B$)($T_{ith,species,average}$)($N_{ith,species}$)$^2$)/($c^2$)] – [($m_{0,ith,species}$ $^2$)($c^2$)] }$^{1/2}$}}} (Chargon angular correction factor) (cos $α_i$)dα} + {∫[(α = 0, α = [[f(π)] $_{max,neutron}$]]{{($A_{capture,neutron}$)/[(4)(π)($r^2$)]} {{{{{∑(j = 1, j = $N_{max}$) {{($M_{0,neutron}$) {{1 – [($v_j^2$)/($c^2$)]}$^{1/2}$} ($c^2$)} – [($M_{0,neutron}$)($c^2$)]}} $^2$}/($c^2$)} – [($M_{0,neutron}$ $^2$)($c^2$)] }$^{1/2}$}(Neutron,j)} (Neutron angular correction factor) (cos $α_i$)dα} +{∫[(α = 0, α = [[f(π)] $_{max,photon,nuclear}$]]{{($A_{capture,photon}$)/[(4)(π)($r^2$)]} {∑(jk = 1, k = $N_{photon,nuclear}$) {($E_k$/c) = (*$hv_k$/c*) = (*$h/λ_k$*)}(Photon,nuclear,k)}} (Nuclear photon angular correction factor) (cos $α_i$)dα} + {∫[(α = 0, α = [[f(π)] $_{max,photon,thermal}$]]{d{($A_{capture,photon}$) {{[5.670373 x $10^{-8}$] (J $m^{-2}$ $s^{-1}$ $K^{-4}$)}[$T_{plasma}^4$]/[3 x $10^8$ m/s]}}/dt}(Thermal photon angular correction factor) (cos $α_i$)dα}/dt}>• d r(x,y,z)}$_{bomb,z2}$}}

+ {∑($z_3$ = 1, $z_3$ = $N_{z3}$){∫<{d{∫[(α = 0, α = [[f(π)] $_{max,chargon}$]]{ (2){($A_{capture,chargon}$)/[(4)(π)($r^2$)]} {∑(i = 1, i = N) {∑(i = 1, i = N){[(($k_B$)($T_{ith,species,average}$)($N_{ith,species}$)$^2$)/($c^2$)] – [($m_{0,ith,species}$ $^2$)($c^2$)] }$^{1/2}$}}} (Chargon angular correction factor) (cos $α_i$)dα} + {∫[(α = 0, α = [[f(π)] $_{max,neutron}$]]{{($A_{capture,neutron}$)/[(4)(π)($r^2$)]} {{{{{∑(j = 1, j = $N_{max}$) {{($M_{0,neutron}$) {{1 – [($v_j^2$)/($c^2$)]}$^{1/2}$} ($c^2$)} – [($M_{0,neutron}$)($c^2$)]}} $^2$}/($c^2$)} – [($M_{0,neutron}$ $^2$)($c^2$)] }$^{1/2}$}(Neutron,j)} (Neutron angular correction factor) (cos $α_i$)dα} +{∫[(α = 0, α = [[f(π)] $_{max,photon,nuclear}$]]{{($A_{capture,photon}$)/[(4)(π)($r^2$)]} {∑(jk = 1, k = $N_{photon,nuclear}$) {($E_k$/c) = (*$hv_k$/c*) = (*$h/λ_k$*)}(Photon,nuclear,k)}} (Nuclear photon angular correction factor) (cos $α_i$)dα} + {∫[(α = 0, α = [[f(π)] $_{max,photon,thermal}$]]{d{($A_{capture,photon}$) {{[5.670373 x $10^{-8}$] (J $m^{-2}$ $s^{-1}$ $K^{-4}$)}[$T_{plasma}^4$]/[3 x $10^8$ m/s]}}/dt}(Thermal photon angular correction factor) (cos $α_i$)dα}/dt}>• d r(x,y,z)}$_{bomb,z3}$}}

+ ... + {∑($z_n$= 1, $z_n$ = $N_{zn}$){∫<{d{∫[(α = 0, α = [[f(π)] $_{max,chargon}$]]{ (2){($A_{capture,chargon}$)/[(4)(π)($r^2$)]} {∑(i = 1, i = N) {∑(i = 1, i = N){[(($k_B$)($T_{ith,species,average}$)($N_{ith,species}$)$^2$)/($c^2$)] – [($m_{0,ith,species}$ $^2$)($c^2$)] }$^{1/2}$}}} (Chargon angular correction factor) (cos $α_i$)dα} + {∫[(α = 0, α = [[f(π)] $_{max,neutron}$]]{{($A_{capture,neutron}$)/[(4)(π)($r^2$)]} {{{{{∑(j = 1, j = $N_{max}$) {{($M_{0,neutron}$) {{1 – [($v_j^2$)/($c^2$)]}$^{1/2}$} ($c^2$)} – [($M_{0,neutron}$)($c^2$)]}} $^2$}/($c^2$)} – [($M_{0,neutron}$ $^2$)($c^2$)] }$^{1/2}$}(Neutron,j)} (Neutron angular correction factor) (cos $α_i$)dα} +{∫[(α = 0, α = [[f(π)] $_{max,photon,nuclear}$]]{{($A_{capture,photon}$)/[(4)(π)($r^2$)]} {∑(jk = 1, k = $N_{photon,nuclear}$) {($E_k$/c) = (*$hv_k$/c*) = (*$h/λ_k$*)}(Photon,nuclear,k)}} (Nuclear photon angular correction factor) (cos $α_i$)dα} + {∫[(α = 0, α = [[f(π)] $_{max,photon,thermal}$]]{d{($A_{capture,photon}$) {{[5.670373 x $10^{-8}$] (J $m^{-2}$ $s^{-1}$ $K^{-4}$)}[$T_{plasma}^4$]/[3 x $10^8$ m/s]}}/dt}(Thermal photon angular correction factor) (cos $α_i$)dα}/dt}>• d r(x,y,z)}$_{bomb,zn}$}}.

Now, a phenomenon which herein is referred to as slap-back can occur for which impinging plasma rebounding of the spacecraft pusher-plate is slapped back into the pusher plate by subsequently impinging plasma. Slap-back is likely not significantly affected by impinging neutron or impinging photons because neutrons are not easily reflected being chargeless and photons are generally within the x-ray and gamma ray range and are thus absorbed by the pusher plate because of their penetrating power instead of being easily reflected. For chargon force exertion on the plate, we will include the factor {Chargon slap-back) which modifies in abstract notation the force exerted on the plate. It is conceivable that this factor could be less than, equal to, or greater than one depending on the chargon flux patterns at or near the pusher plate.

$$\int\langle F_{total,with,slap\text{-}back}\rangle \bullet d\, r(x,y,z) = \{\sum(z_1= 1,\ z_1 = N_{z1})\{\{\int\langle\{d\, P_{combined,incident}/dt\}_{with,slap\text{-}back}\rangle \bullet d\, r(x,y,z)\}_{bomb,z1}\}\} + \{\sum(z_2= 1,\ z_2 = N_{z2})\{\{\int\langle\{d\, P_{combined,incident}/dt\}_{with,slap\text{-}back}\rangle \bullet d\, r(x,y,z)\}_{bomb,z2}\}\} + \{\sum(z_3= 1,\ z_3 = N_{z3})\{\{\int\langle\{d\, P_{combined,incident}/dt\}_{with,slap\text{-}back}\rangle \bullet d\, r(x,y,z)\}_{bomb,z3}\}\} + \ldots + \{\sum(z_n= 1,\ z_n = N_{zn})\{\{\int\langle\{d\, P_{combined,incident}/dt\}_{with,slap\text{-}back}\rangle \bullet d\, r(x,y,z)\}_{bomb,zn}\}\}$$

$$= \{\sum(z_1 = 1,\ z_1 = N_{z1})\{\int\langle\{d\{\int[(\alpha = 0,\ \alpha = [[f(\pi)]_{max,chargon}]]\{(2)\{(A_{capture,chargon})/[(4)(\pi)(r^2)]\}\ \{\sum(i = 1,\ i = N)\ \{\{\{((k_B)(T_{ith,species,average})(N_{ith,species})^2)-\ \{[(m_{0,ith,species})(c^2)]^2\}\}/(c^2)\}^{1/2}\}\}\}(\text{Chargon angular correction factor})\ \{\text{Chargon slap-back})\ (\cos \alpha_i)d\alpha\} + \{\int[(\alpha = 0,\ \alpha = [[f(\pi)]_{max,neutron}]]\{\{(A_{capture,neutron})/[(4)(\pi)(r^2)]\}\ \{\{\{\{\{\sum(j = 1,\ j = N_{max})\{\{(M_{0,neutron})(\gamma_j)(c^2)\} - [(M_{0,neutron})(c^2)]\}\}^2\}-\ \{[(M_{0,neutron})(c^2)]^2\}\}/(c^2)\}^{1/2}\}(\text{Neutron},j)\}\ (\text{Neutron angular correction factor})\ (\cos \alpha_i)d\alpha\} + \{\int[(\alpha = 0,\ \alpha = [[f(\pi)]_{max,photon,nuclear}]]\{\{(A_{capture,photon})/[(4)(\pi)(r^2)]\}\ \{\sum(jk = 1,\ k = N_{photon,nuclear})\ \{(E_k/c) = (h\nu_k/c) = (h/\lambda_k)\}(\text{Photon,nuclear},k)\}\}\ (\text{Nuclear photon angular correction factor})(\cos \alpha_i)d\alpha\} + \{\int[(\alpha = 0,\ \alpha = [[f(\pi)]_{max,photon,thermal}]]\{d\{(A_{capture,photon})\ \{\{[5.670373 \times 10^{-8}]\ (J\ m^{-2}\ s^{-1}\ K^{-4})\}[T_{plasma}^4]/[3 \times 10^8\ m/s]\}\}/dt\}(\text{Thermal photon angular correction factor})\ (\cos \alpha_i)d\alpha\}/dt\}\rangle\bullet d\, r(x,y,z)\}_{bomb,z1}\}$$

$$+\ \{\sum(z_2 = 1,\ z_2 = N_{z2})\{\int\langle\{d\{\int[(\alpha = 0,\ \alpha = [[f(\pi)]_{max,chargon}]]\{(2)\{(A_{capture,chargon})/[(4)(\pi)(r^2)]\}\ \{\sum(i = 1,\ i = N)\ \{\{\{((k_B)(T_{ith,species,average})(N_{ith,species})^2)-\ \{[(m_{0,ith,species})(c^2)]^2\}\}/(c^2)\}^{1/2}\}\}\}(\text{Chargon angular correction factor})\ \{\text{Chargon slap-back})\ (\cos \alpha_i)d\alpha\} + \{\int[(\alpha = 0,\ \alpha = [[f(\pi)]_{max,neutron}]]\{\{(A_{capture,neutron})/[(4)(\pi)(r^2)]\}\ \{\{\{\{\{\sum(j = 1,\ j = N_{max})\{\{(M_{0,neutron})(\gamma_j)(c^2)\} - [(M_{0,neutron})(c^2)]\}\}^2\}-\ \{[(M_{0,neutron})(c^2)]^2\}\}/(c^2)\}^{1/2}\}(\text{Neutron},j)\}\ (\text{Neutron angular correction factor})\ (\cos \alpha_i)d\alpha\} + \{\int[(\alpha = 0,\ \alpha = [[f(\pi)]_{max,photon,nuclear}]]\{\{(A_{capture,photon})/[(4)(\pi)(r^2)]\}\ \{\sum(jk = 1,\ k = N_{photon,nuclear})\ \{(E_k/c) = (h\nu_k/c) = (h/\lambda_k)\}(\text{Photon,nuclear},k)\}\}\ (\text{Nuclear photon angular correction factor})(\cos \alpha_i)d\alpha\} + \{\int[(\alpha = 0,\ \alpha = [[f(\pi)]_{max,photon,thermal}]]\{d\{(A_{capture,photon})\ \{\{[5.670373 \times 10^{-8}]\ (J\ m^{-2}\ s^{-1}\ K^{-4})\}[T_{plasma}^4]/[3 \times 10^8\ m/s]\}\}/dt\}(\text{Thermal photon angular correction factor})\ (\cos \alpha_i)d\alpha\}/dt\}\rangle\bullet d\, r(x,y,z)\}_{bomb,z2}\}$$

$$+\ \{\sum(z_3 = 1,\ z_3 = N_{z3})\{\int\langle\{d\{\int[(\alpha = 0,\ \alpha = [[f(\pi)]_{max,chargon}]]\{(2)\{(A_{capture,chargon})/[(4)(\pi)(r^2)]\}\ \{\sum(i = 1,\ i = N)\ \{\{\{((k_B)(T_{ith,species,average})(N_{ith,species})^2)-\ \{[(m_{0,ith,species})(c^2)]^2\}\}/(c^2)\}^{1/2}\}\}\}(\text{Chargon angular correction factor})\ \{\text{Chargon slap-back})\ (\cos \alpha_i)d\alpha\} + \{\int[(\alpha = 0,\ \alpha = [[f(\pi)]_{max,neutron}]]\{\{(A_{capture,neutron})/[(4)(\pi)(r^2)]\}\ \{\{\{\{\{\sum(j = 1,\ j = N_{max})\{\{(M_{0,neutron})(\gamma_j)(c^2)\} - [(M_{0,neutron})(c^2)]\}\}^2\}-\ \{[(M_{0,neutron})(c^2)]^2\}\}/(c^2)\}^{1/2}\}(\text{Neutron},j)\}\ (\text{Neutron angular correction factor})\ (\cos \alpha_i)d\alpha\} + \{\int[(\alpha = 0,\ \alpha = [[f(\pi)]_{max,photon,nuclear}]]\{\{(A_{capture,photon})/[(4)(\pi)(r^2)]\}\ \{\sum(jk = 1,\ k = N_{photon,nuclear})\ \{(E_k/c) = (h\nu_k/c) = (h/\lambda_k)\}(\text{Photon,nuclear},k)\}\}\ (\text{Nuclear photon angular correction factor})(\cos \alpha_i)d\alpha\} + \{\int[(\alpha = 0,\ \alpha = [[f(\pi)]_{max,photon,thermal}]]\{d\{(A_{capture,photon})\ \{\{[5.670373 \times 10^{-8}]\ (J\ m^{-2}\ s^{-1}\ K^{-4})\}[T_{plasma}^4]/[3 \times 10^8\ m/s]\}\}/dt\}(\text{Thermal photon angular correction factor})\ (\cos \alpha_i)d\alpha\}/dt\}\rangle\bullet d\, r(x,y,z)\}_{bomb,z3}\}$$

+ ... + {∑($z_n$ = 1, $z_n$ = $N_{zn}$){∫<{d{∫[(α = 0, α = [[f(π)] $_{max,chargon}$]]{(2){($A_{capture,chargon}$)/[(4)(π)($r^2$)]} {∑(i = 1, i = N) {{{(($k_B$)($T_{ith,species,average}$)($N_{ith,species}$)$^2$)– {[($m_{0,ith,species}$)($c^2$)]$^2$}}/($c^2$)}$^{1/2}$}}}(Chargon angular correction factor) {Chargon slap-back) (cos $α_i$)dα} + {∫[(α = 0, α = [[f(π)] $_{max,neutron}$]]{{($A_{capture,neutron}$)/[(4)(π)($r^2$)]} {{{{{∑(j = 1, j = $N_{max}$){{($M_{0,neutron}$)($γ_j$)($c^2$)} – [($M_{0,neutron}$)($c^2$)]}} $^2$}– {[($M_{0,neutron}$)($c^2$)]$^2$}}/($c^2$)}$^{1/2}$}(Neutron,j)} (Neutron angular correction factor) (cos $α_i$)dα} + {∫[(α = 0, α = [[f(π)] $_{max,photon,nuclear}$]]{{($A_{capture,photon}$)/[(4)(π)($r^2$)]} {∑(jk = 1, k = $N_{photon,nuclear}$) {($E_k$/c) = (*h*$ν_k$/c) = (*h*/$λ_k$)}(Photon,nuclear,k)}} (Nuclear photon angular correction factor)(cos $α_i$)dα} + {∫[(α = 0, α = [[f(π)] $_{max,photon,thermal}$]]{d{($A_{capture,photon}$) {{[5.670373 x $10^{-8}$] (J $m^{-2}$ $s^{-1}$ $K^{-4}$)}[$T_{plasma}^4$]/[3 x $10^8$ m/s]}}/dt}(Thermal photon angular correction factor) (cos $α_i$)dα}/dt}>• d r(x,y,z)}$_{bomb,zn}$}}

= {∑($z_1$ = 1, $z_1$ = $N_{z1}$){∫<{d{∫[(α = 0, α = [[f(π)] $_{max,chargon}$]]{ (2){($A_{capture,chargon}$)/[(4)(π)($r^2$)]} {∑(i = 1, i = N) {∑(i = 1, i = N){[(($k_B$)($T_{ith,species,average}$)($N_{ith,species}$)$^2$)/($c^2$)] – [($m_{0,ith,species}$ $^2$)($c^2$)] }$^{1/2}$}}} (Chargon angular correction factor) {Chargon slap-back) (cos $α_i$)dα} + {∫[(α = 0, α = [[f(π)] $_{max,neutron}$]]{{($A_{capture,neutron}$)/[(4)(π)($r^2$)]} {{{{{∑(j = 1, j = $N_{max}$) {{($M_{0,neutron}$) {{1 – [($v_j^2$)/($c^2$)]}}$^{1/2}$} ($c^2$)} – [($M_{0,neutron}$)($c^2$)]}} $^2$}/($c^2$)} – [($M_{0,neutron}$ $^2$)($c^2$)] }$^{1/2}$}(Neutron,j)} (Neutron angular correction factor) (cos $α_i$)dα} +{∫[(α = 0, α = [[f(π)] $_{max,photon,nuclear}$]]{{($A_{capture,photon}$)/[(4)(π)($r^2$)]} {∑(jk = 1, k = $N_{photon,nuclear}$) {($E_k$/c) = (*h*$ν_k$/c) = (*h*/$λ_k$)}(Photon,nuclear,k)}} (Nuclear photon angular correction factor) (cos $α_i$)dα} + {∫[(α = 0, α = [[f(π)] $_{max,photon,thermal}$]]{d{($A_{capture,photon}$) {{[5.670373 x $10^{-8}$] (J $m^{-2}$ $s^{-1}$ $K^{-4}$)}[$T_{plasma}^4$]/[3 x $10^8$ m/s]}}/dt}(Thermal photon angular correction factor) (cos $α_i$)dα}/dt}>• d r(x,y,z)}$_{bomb,z1}$}}

+ {∑($z_2$ = 1, $z_2$ = $N_{z2}$){∫<{d{∫[(α = 0, α = [[f(π)] $_{max,chargon}$]]{ (2){($A_{capture,chargon}$)/[(4)(π)($r^2$)]} {∑(i = 1, i = N) {∑(i = 1, i = N){[(($k_B$)($T_{ith,species,average}$)($N_{ith,species}$)$^2$)/($c^2$)] – [($m_{0,ith,species}$ $^2$)($c^2$)] }$^{1/2}$}}} (Chargon angular correction factor) {Chargon slap-back) (cos $α_i$)dα} + {∫[(α = 0, α = [[f(π)] $_{max,neutron}$]]{{($A_{capture,neutron}$)/[(4)(π)($r^2$)]} {{{{{∑(j = 1, j = $N_{max}$) {{($M_{0,neutron}$) {{1 – [($v_j^2$)/($c^2$)]}}$^{1/2}$} ($c^2$)} – [($M_{0,neutron}$)($c^2$)]}} $^2$}/($c^2$)} – [($M_{0,neutron}$ $^2$)($c^2$)] }$^{1/2}$}(Neutron,j)} (Neutron angular correction factor) (cos $α_i$)dα} +{∫[(α = 0, α = [[f(π)] $_{max,photon,nuclear}$]]{{($A_{capture,photon}$)/[(4)(π)($r^2$)]} {∑(jk = 1, k = $N_{photon,nuclear}$) {($E_k$/c) = (*h*$ν_k$/c) = (*h*/$λ_k$)}(Photon,nuclear,k)}} (Nuclear photon angular correction factor) (cos $α_i$)dα} + {∫[(α = 0, α = [[f(π)] $_{max,photon,thermal}$]]{d{($A_{capture,photon}$) {{[5.670373 x $10^{-8}$] (J $m^{-2}$ $s^{-1}$ $K^{-4}$)}[$T_{plasma}^4$]/[3 x $10^8$ m/s]}}/dt}(Thermal photon angular correction factor) (cos $α_i$)dα}/dt}>• d r(x,y,z)}$_{bomb,z2}$}}

+ {∑($z_3$ = 1, $z_3$ = $N_{z3}$){∫<{d{∫[(α = 0, α = [[f(π)] $_{max,chargon}$]]{ (2){($A_{capture,chargon}$)/[(4)(π)($r^2$)]} {∑(i = 1, i = N) {∑(i = 1, i = N){[(($k_B$)($T_{ith,species,average}$)($N_{ith,species}$)$^2$)/($c^2$)] – [($m_{0,ith,species}$ $^2$)($c^2$)] }$^{1/2}$}}} (Chargon angular correction factor) {Chargon slap-back) (cos $α_i$)dα} + {∫[(α = 0, α = [[f(π)] $_{max,neutron}$]]{{($A_{capture,neutron}$)/[(4)(π)($r^2$)]} {{{{{∑(j = 1, j = $N_{max}$) {{($M_{0,neutron}$) {{1 – [($v_j^2$)/($c^2$)]}}$^{1/2}$} ($c^2$)} – [($M_{0,neutron}$)($c^2$)]}} $^2$}/($c^2$)} – [($M_{0,neutron}$ $^2$)($c^2$)] }$^{1/2}$}(Neutron,j)} (Neutron angular correction factor) (cos $α_i$)dα} +{∫[(α = 0, α = [[f(π)] $_{max,photon,nuclear}$]]{{($A_{capture,photon}$)/[(4)(π)($r^2$)]} {∑(jk = 1, k = $N_{photon,nuclear}$) {($E_k$/c) = (*h*$ν_k$/c) = (*h*/$λ_k$)}(Photon,nuclear,k)}} (Nuclear photon angular correction factor) (cos $α_i$)dα} + {∫[(α = 0, α = [[f(π)] $_{max,photon,thermal}$]]{d{($A_{capture,photon}$) {{[5.670373 x $10^{-8}$] (J $m^{-2}$ $s^{-1}$ $K^{-4}$)}[$T_{plasma}^4$]/[3 x $10^8$ m/s]}}/dt}(Thermal photon angular correction factor) (cos $α_i$)dα}/dt}>• d r(x,y,z)}$_{bomb,z3}$}}

+ ... + {∑($z_n$= 1, $z_n$ = $N_{zn}$){∫<{d{∫[(α = 0, α = [[f(π)] $_{max,chargon}$]]{ (2){($A_{capture,chargon}$)/[(4)(π)($r^2$)]} {∑(i = 1, i = N) {∑(i = 1, i = N){[(($k_B$)($T_{ith,species,average}$)($N_{ith,species}$)$^2$)/($c^2$)] – [($m_{0,ith,species}$ $^2$)($c^2$)] }$^{1/2}$}}} (Chargon angular correction factor) {Chargon slap-back) (cos $\alpha_i$)dα} + {∫[(α = 0, α = [[f(π)] $_{max,neutron}$]]{{($A_{capture,neutron}$)/[(4)(π)($r^2$)]} {{{{{∑(j = 1, j = $N_{max}$) {{($M_{0,neutron}$) {{1 – [($v_j^2$)/($c^2$)]}$^{1/2}$} ($c^2$)} – [($M_{0,neutron}$)($c^2$)]}} $^2$}/($c^2$)} – [($M_{0,neutron}{}^2$)($c^2$)] }$^{1/2}$}(Neutron,j)} (Neutron angular correction factor) (cos $\alpha_i$)dα} +{∫[(α = 0, α = [[f(π)] $_{max,photon,nuclear}$]]{{($A_{capture,photon}$)/[(4)(π)($r^2$)]} {∑(jk = 1, k = $N_{photon,nuclear}$) {($E_k$/c) = ($h\nu_k/c$) = ($h/\lambda_k$)}(Photon,nuclear,k)}} (Nuclear photon angular correction factor) (cos $\alpha_i$)dα} + {∫[(α = 0, α = [[f(π)] $_{max,photon,thermal}$]]{d{($A_{capture,photon}$) {{[5.670373 x $10^{-8}$] (J $m^{-2}$ $s^{-1}$ $K^{-4}$)}[$T_{plasma}^4$]/[3 x $10^8$ m/s]}}/dt}(Thermal photon angular correction factor) (cos $\alpha_i$)dα}/dt}>• d r(x,y,z)} $_{bomb,zn}$}}.

Now chargon slap-back can occur in a manner that produces reverberations and thus imply a slapping sloshing effect. We will use the term {Chargon slap-back reverberating sloshing effect) to denote this feedback mechanism. Additionally, some of the incident neutrons may be back-reflected out of the pusher plate as may be some of the impinging x-rays and gamma rays produced by the bomb blast. Both these effects can cause energy bleeding for the respective impinging neutrons and photons. We will denote these two effects by the factors, (Neutron back-reflect with possible neutron energy bleeding), (Nuclear photon back-reflect with possible neutron energy bleeding), and (Thermal photon back-reflect with possible neutron energy bleeding). The photon term is assumed to be context specific with nuclear reaction generated photons and the thermal photons.

The resulting formula for spatially integrated force is thus the final treatment we will elaborate on in this chapter and is as follows.

∫<$F_{total,finalized}$>• d r(x,y,z)

= {∑($z_1$ = 1, $z_1$ = $N_{z1}$){∫<{d{∫[(α = 0, α = [[f(π)] $_{max,chargon}$]]{(2){($A_{capture,chargon}$)/[(4)(π)($r^2$)]} {∑(i = 1, i = N) {{{(($k_B$)($T_{ith,species,average}$)($N_{ith,species}$)$^2$)– {[($m_{0,ith,species}$)($c^2$)]$^2$}}/($c^2$)}$^{1/2}$}}}(Chargon angular correction factor) {Chargon slap-back reverberating sloshing effect) (cos $\alpha_i$)dα} + {∫[(α = 0, α = [[f(π)] $_{max,neutron}$]]{{($A_{capture,neutron}$)/[(4)(π)($r^2$)]} {{{{{∑(j = 1, j = $N_{max}$){{($M_{0,neutron}$)($\gamma_j$)($c^2$)} – [($M_{0,neutron}$)($c^2$)]}} $^2$}– {[($M_{0,neutron}$)($c^2$)]$^2$}}/($c^2$)}$^{1/2}$}(Neutron,j)} (Neutron angular correction factor) (Neutron back-reflect with possible neutron energy bleeding)(cos $\alpha_i$)dα} + {∫[(α = 0, α = [[f(π)] $_{max,photon,nuclear}$]]{{($A_{capture,photon}$)/[(4)(π)($r^2$)]} {∑(jk = 1, k = $N_{photon,nuclear}$) {($E_k$/c) = ($h\nu_k/c$) = ($h/\lambda_k$)}(Photon,nuclear,k)}} (Nuclear photon angular correction factor) (Nuclear photon back-reflect with possible neutron energy bleeding) (cos $\alpha_i$)dα} + {∫[(α = 0, α = [[f(π)] $_{max,photon,thermal}$]]{d{($A_{capture,photon}$) {{[5.670373 x $10^{-8}$] (J $m^{-2}$ $s^{-1}$ $K^{-4}$)}[$T_{plasma}^4$]/[3 x $10^8$ m/s]}}/dt}(Thermal photon angular correction factor) (Thermal photon back-reflect with possible neutron energy bleeding) (cos $\alpha_i$)dα}/dt}>• d r(x,y,z)}$_{bomb,z1}$}}

+ {∑($z_2$ = 1, $z_2$ = $N_{z2}$){∫<{d{∫[(α = 0, α = [[f(π)] $_{max,chargon}$]]{(2){($A_{capture,chargon}$)/[(4)(π)($r^2$)]} {∑(i = 1, i = N) {{{(($k_B$)($T_{ith,species,average}$)($N_{ith,species}$)$^2$)– {[($m_{0,ith,species}$)($c^2$)]$^2$}}/($c^2$)}$^{1/2}$}}}(Chargon angular correction factor) {Chargon slap-back reverberating sloshing effect) (cos $\alpha_i$)dα} + {∫[(α = 0, α = [[f(π)] $_{max,neutron}$]]{{($A_{capture,neutron}$)/[(4)(π)($r^2$)]} {{{{{∑(j = 1, j = $N_{max}$){{($M_{0,neutron}$)($\gamma_j$)($c^2$)} – [($M_{0,neutron}$)($c^2$)]}} $^2$}– {[($M_{0,neutron}$)($c^2$)]$^2$}}/($c^2$)}$^{1/2}$}(Neutron,j)} (Neutron angular correction factor) (Neutron back-reflect with possible neutron energy bleeding) (cos $\alpha_i$)dα} + {∫[(α = 0, α = [[f(π)] $_{max,photon,nuclear}$]]{{($A_{capture,photon}$)/[(4)(π)($r^2$)]} {∑(jk = 1, k = $N_{photon,nuclear}$) {($E_k$/c) = ($h\nu_k/c$) =

($h/\lambda_k$)}(Photon,nuclear,k)}} (Nuclear photon angular correction factor) (Nuclear photon back-reflect with possible neutron energy bleeding) (cos $\alpha_i$)dα} + {∫[(α = 0, α = [[f(π)] $_{max,photon,thermal}$]]{d{($A_{capture,photon}$) {{[5.670373 x $10^{-8}$] (J $m^{-2}$ $s^{-1}$ $K^{-4}$)}[$T_{plasma}^4$]/[3 x $10^8$ m/s]}}/dt}(Thermal photon angular correction factor) (Thermal photon back-reflect with possible neutron energy bleeding) (cos $\alpha_i$)dα}/dt}>• d r(x,y,z)}$_{bomb,z2}$}}

+ {∑($z_3$ = 1, $z_3$ = $N_{z3}$){∫<{d{∫[(α = 0, α = [[f(π)] $_{max,chargon}$]]{(2){($A_{capture,chargon}$)/[(4)(π)($r^2$)]} {∑(i = 1, i = N) {{{(($k_B$)($T_{ith,species,average}$)($N_{ith,species}$)$^2$)– {[($m_{0,ith,species}$)($c^2$)]$^2$}}/($c^2$)}$^{1/2}$}}}(Chargon angular correction factor) {Chargon slap-back reverberating sloshing effect) (cos $\alpha_i$)dα} + {∫[(α = 0, α = [[f(π)] $_{max,neutron}$]]{{($A_{capture,neutron}$)/[(4)(π)($r^2$)]} {{{{{∑(j = 1, j = $N_{max}$){{($M_{0,neutron}$)($\gamma_j$)($c^2$)} – [($M_{0,neutron}$)($c^2$)]}} $^2$}– {[($M_{0,neutron}$)($c^2$)]$^2$}}/($c^2$)}$^{1/2}$}(Neutron,j)} (Neutron angular correction factor) (Neutron back-reflect with possible neutron energy bleeding) (cos $\alpha_i$)dα} + {∫[(α = 0, α = [[f(π)] $_{max,photon,nuclear}$]]{{($A_{capture,photon}$)/[(4)(π)($r^2$)]} {∑(jk = 1, k = $N_{photon,nuclear}$) {($E_k$/c) = ($h\nu_k/c$) = ($h/\lambda_k$)}(Photon,nuclear,k)}} (Nuclear photon angular correction factor) (Nuclear photon back-reflect with possible neutron energy bleeding) (cos $\alpha_i$)dα} + {∫[(α = 0, α = [[f(π)] $_{max,photon,thermal}$]]{d{($A_{capture,photon}$) {{[5.670373 x $10^{-8}$] (J $m^{-2}$ $s^{-1}$ $K^{-4}$)}[$T_{plasma}^4$]/[3 x $10^8$ m/s]}}/dt}(Thermal photon angular correction factor) (Thermal photon back-reflect with possible neutron energy bleeding) (cos $\alpha_i$)dα}/dt}>• d r(x,y,z)}$_{bomb,z3}$}}

+ ... + {∑($z_n$ = 1, $z_n$ = $N_{zn}$){∫<{d{∫[(α = 0, α = [[f(π)] $_{max,chargon}$]]{(2){($A_{capture,chargon}$)/[(4)(π)($r^2$)]} {∑(i = 1, i = N) {{{(($k_B$)($T_{ith,species,average}$)($N_{ith,species}$)$^2$)– {[($m_{0,ith,species}$)($c^2$)]$^2$}}/($c^2$)}$^{1/2}$}}}(Chargon angular correction factor) {Chargon slap-back reverberating sloshing effect) (cos $\alpha_i$)dα} + {∫[(α = 0, α = [[f(π)] $_{max,neutron}$]]{{($A_{capture,neutron}$)/[(4)(π)($r^2$)]} {{{{{∑(j = 1, j = $N_{max}$){{($M_{0,neutron}$)($\gamma_j$)($c^2$)} – [($M_{0,neutron}$)($c^2$)]}} $^2$}– {[($M_{0,neutron}$)($c^2$)]$^2$}}/($c^2$)}$^{1/2}$}(Neutron,j)} (Neutron angular correction factor) (Neutron back-reflect with possible neutron energy bleeding) (cos $\alpha_i$)dα} + {∫[(α = 0, α = [[f(π)] $_{max,photon,nuclear}$]]{{($A_{capture,photon}$)/[(4)(π)($r^2$)]} {∑(jk = 1, k = $N_{photon,nuclear}$) {($E_k$/c) = ($h\nu_k/c$) = ($h/\lambda_k$)}(Photon,nuclear,k)}} (Nuclear photon angular correction factor) (Nuclear photon back-reflect with possible neutron energy bleeding) (cos $\alpha_i$)dα} + {∫[(α = 0, α = [[f(π)] $_{max,photon,thermal}$]]{d{($A_{capture,photon}$) {{[5.670373 x $10^{-8}$] (J $m^{-2}$ $s^{-1}$ $K^{-4}$)}[$T_{plasma}^4$]/[3 x $10^8$ m/s]}}/dt}(Thermal photon angular correction factor) (Thermal photon back-reflect with possible neutron energy bleeding) (cos $\alpha_i$)dα}/dt}>• d r(x,y,z)}$_{bomb,zn}$}}

= {∑($z_1$ = 1, $z_1$ = $N_{z1}$){∫<{d{∫[(α = 0, α = [[f(π)] $_{max,chargon}$]]{ (2){($A_{capture,chargon}$)/[(4)(π)($r^2$)]} {∑(i = 1, i = N) {∑(i = 1, i = N){[(($k_B$)($T_{ith,species,average}$)($N_{ith,species}$)$^2$)/($c^2$)] – [($m_{0,ith,species}^2$)($c^2$)] }$^{1/2}$}}} (Chargon angular correction factor) {Chargon slap-back reverberating sloshing effect) (cos $\alpha_i$)dα} + {∫[(α = 0, α = [[f(π)] $_{max,neutron}$]]{{($A_{capture,neutron}$)/[(4)(π)($r^2$)]} {{{{{∑(j = 1, j = $N_{max}$) {{($M_{0,neutron}$) {{1 – [($v_j^2$)/($c^2$)]}}$^{1/2}$} ($c^2$)} – [($M_{0,neutron}$)($c^2$)]}} $^2$}/($c^2$)} – [($M_{0,neutron}^2$)($c^2$)] }$^{1/2}$}(Neutron,j)} (Neutron angular correction factor) (Neutron back-reflect with possible neutron energy bleeding) (cos $\alpha_i$)dα} +{∫[(α = 0, α = [[f(π)] $_{max,photon,nuclear}$]]{{($A_{capture,photon}$)/[(4)(π)($r^2$)]} {∑(jk = 1, k = $N_{photon,nuclear}$) {($E_k$/c) = ($h\nu_k/c$) = ($h/\lambda_k$)}(Photon,nuclear,k)}} (Nuclear photon angular correction factor) (Nuclear photon back-reflect with possible neutron energy bleeding) (cos $\alpha_i$)dα} + {∫[(α = 0, α = [[f(π)] $_{max,photon,thermal}$]]{d{($A_{capture,photon}$) {{[5.670373 x $10^{-8}$] (J $m^{-2}$ $s^{-1}$ $K^{-4}$)}[$T_{plasma}^4$]/[3 x $10^8$ m/s]}}/dt}(Thermal photon angular correction factor) (Thermal photon back-reflect with possible neutron energy bleeding) (cos $\alpha_i$)dα}/dt}>• d r(x,y,z)} $_{bomb,z1}$}}

+ {∑($z_2$ = 1, $z_2$ = $N_{z2}$){∫<{d{∫[(α = 0, α = [[f(π)] $_{max,chargon}$]]{ (2){($A_{capture,chargon}$)/[(4)(π)($r^2$)]} {∑(i = 1, i = N) {∑(i = 1, i = N){[(($k_B$)($T_{ith,species,average}$)($N_{ith,species}$)$^2$)/($c^2$)] – [($m_{0,ith,species}^2$)($c^2$)] }$^{1/2}$}}} (Chargon angular correction factor) {Chargon slap-back reverberating sloshing effect) (cos $\alpha_i$)dα} + {∫[(α = 0, α = [[f(π)] $_{max,neutron}$]]{{($A_{capture,neutron}$)/[(4)(π)($r^2$)]} {{{{{∑(j = 1, j = $N_{max}$) {{($M_{0,neutron}$) {{1 – [($v_j^2$)/($c^2$)]}$^{1/2}$} ($c^2$)} – [($M_{0,neutron}$)($c^2$)]}}} $^2$}/($c^2$)} – [($M_{0,neutron}^2$)($c^2$)] }$^{1/2}$}(Neutron,j)} (Neutron angular correction factor) (Neutron back-reflect with possible neutron energy bleeding) (cos $\alpha_i$)dα} +{∫[(α = 0, α = [[f(π)] $_{max,photon,nuclear}$]]{{($A_{capture,photon}$)/[(4)(π)($r^2$)]} {∑(jk = 1, k = $N_{photon,nuclear}$) {($E_k$/c) = (*h*$\nu_k$/*c*) = (*h*/$\lambda_k$)}(Photon,nuclear,k)}} (Nuclear photon angular correction factor) (Nuclear photon back-reflect with possible neutron energy bleeding) (cos $\alpha_i$)dα} + {∫[(α = 0, α = [[f(π)] $_{max,photon,thermal}$]]{d{($A_{capture,photon}$) {{[5.670373 x $10^{-8}$] (J $m^{-2}$ $s^{-1}$ $K^{-4}$)}[$T_{plasma}^4$]/[3 x $10^8$ m/s]}}/dt}(Thermal photon angular correction factor) (Thermal photon back-reflect with possible neutron energy bleeding) (cos $\alpha_i$)dα}/dt}>• d r(x,y,z)} $_{bomb,z2}$}}

+ {∑($z_3$ = 1, $z_3$ = $N_{z3}$){∫<{d{∫[(α = 0, α = [[f(π)] $_{max,chargon}$]]{ (2){($A_{capture,chargon}$)/[(4)(π)($r^2$)]} {∑(i = 1, i = N) {∑(i = 1, i = N){[(($k_B$)($T_{ith,species,average}$)($N_{ith,species}$)$^2$)/($c^2$)] – [($m_{0,ith,species}^2$)($c^2$)] }$^{1/2}$}}} (Chargon angular correction factor) {Chargon slap-back reverberating sloshing effect) (cos $\alpha_i$)dα} + {∫[(α = 0, α = [[f(π)] $_{max,neutron}$]]{{($A_{capture,neutron}$)/[(4)(π)($r^2$)]} {{{{{∑(j = 1, j = $N_{max}$) {{($M_{0,neutron}$) {{1 – [($v_j^2$)/($c^2$)]}$^{1/2}$} ($c^2$)} – [($M_{0,neutron}$)($c^2$)]}}} $^2$}/($c^2$)} – [($M_{0,neutron}^2$)($c^2$)] }$^{1/2}$}(Neutron,j)} (Neutron angular correction factor) (Neutron back-reflect with possible neutron energy bleeding) (cos $\alpha_i$)dα} +{∫[(α = 0, α = [[f(π)] $_{max,photon,nuclear}$]]{{($A_{capture,photon}$)/[(4)(π)($r^2$)]} {∑(jk = 1, k = $N_{photon,nuclear}$) {($E_k$/c) = (*h*$\nu_k$/*c*) = (*h*/$\lambda_k$)}(Photon,nuclear,k)}} (Nuclear photon angular correction factor) (Nuclear photon back-reflect with possible neutron energy bleeding) (cos $\alpha_i$)dα} + {∫[(α = 0, α = [[f(π)] $_{max,photon,thermal}$]]{d{($A_{capture,photon}$) {{[5.670373 x $10^{-8}$] (J $m^{-2}$ $s^{-1}$ $K^{-4}$)}[$T_{plasma}^4$]/[3 x $10^8$ m/s]}}/dt}(Thermal photon angular correction factor) (Thermal photon back-reflect with possible neutron energy bleeding) (cos $\alpha_i$)dα}/dt}>• d r(x,y,z)} $_{bomb,z3}$}}

+ ... + {∑($z_n$= 1, $z_n$ = $N_{zn}$){∫<{d{∫[(α = 0, α = [[f(π)] $_{max,chargon}$]]{ (2){($A_{capture,chargon}$)/[(4)(π)($r^2$)]} {∑(i = 1, i = N) {∑(i = 1, i = N){[(($k_B$)($T_{ith,species,average}$)($N_{ith,species}$)$^2$)/($c^2$)] – [($m_{0,ith,species}^2$)($c^2$)] }$^{1/2}$}}} (Chargon angular correction factor) {Chargon slap-back reverberating sloshing effect) (cos $\alpha_i$)dα} + {∫[(α = 0, α = [[f(π)] $_{max,neutron}$]]{{($A_{capture,neutron}$)/[(4)(π)($r^2$)]} {{{{{∑(j = 1, j = $N_{max}$) {{($M_{0,neutron}$) {{1 – [($v_j^2$)/($c^2$)]}$^{1/2}$} ($c^2$)} – [($M_{0,neutron}$)($c^2$)]}}} $^2$}/($c^2$)} – [($M_{0,neutron}^2$)($c^2$)] }$^{1/2}$}(Neutron,j)} (Neutron angular correction factor) (Neutron back-reflect with possible neutron energy bleeding) (cos $\alpha_i$)dα} +{∫[(α = 0, α = [[f(π)] $_{max,photon,nuclear}$]]{{($A_{capture,photon}$)/[(4)(π)($r^2$)]} {∑(jk = 1, k = $N_{photon,nuclear}$) {($E_k$/c) = (*h*$\nu_k$/*c*) = (*h*/$\lambda_k$)}(Photon,nuclear,k)}} (Nuclear photon angular correction factor) (Nuclear photon back-reflect with possible neutron energy bleeding) (cos $\alpha_i$)dα} + {∫[(α = 0, α = [[f(π)] $_{max,photon,thermal}$]]{d{($A_{capture,photon}$) {{[5.670373 x $10^{-8}$] (J $m^{-2}$ $s^{-1}$ $K^{-4}$)}[$T_{plasma}^4$]/[3 x $10^8$ m/s]}}/dt}(Thermal photon angular correction factor) (Thermal photon back-reflect with possible neutron energy bleeding) (cos $\alpha_i$)dα}/dt}>• d r(x,y,z)} $_{bomb,zn}$}}.

Now, from Atomic Rockets under **Propulsion Shaped Charge** ©1995-2019 Winchell Chung (鄭) (ᚾᚢᚱᚠᚦ) ρ=Σ+Ψt:

at http://www.projectrho.com/public_html/rocket/spacegunconvent.php#shapedcharge

“Remember that in the vacuum of space, most of the energy of a nuclear warhead is in the form of x-rays. The nuclear device is encased in a radiation case of x-ray opaque material (uranium) with a hole in the top. This forces the x-rays to exit only from the hole. Whereupon they run full tilt into a large mass of beryllium oxide *(channel filler)*.

The beryllium transforms the nuclear fury of x-rays into a nuclear fury of heat. Perched on top of the beryllium is the propellant: a thick plate of tungsten. The nuclear fury of heat turns the tungsten plate into a star-core-hot spindle-shaped-plume of ionized tungsten plasma. The x-ray opaque material and the beryllium oxide also vaporize a few microseconds later, but that's OK, their job is done.

The tungsten plasma jet hits square on the Orion drive pusher plate, said plate is designed to be large enough to catch all of the plasma. With the reference design of nuclear pulse unit, the plume is confined to a cone of about 22.5 degrees. About 85% of the nuclear device's energy is directed into the desired direction, which I think you'd agree is a vast improvement over 1%.”

Thus, with suitably shaped charge nuclear explosives, most of the energy can be delivered to the target. However, this assumes that the bombs are substantially stationary with respect to the spacecraft.

# Shaped charge nuclear fuel pellets.

The possibility of producing a shaped-charge type of nuclear or thermonuclear device that can concentrate the explosive energy flux upon its detonation by 6 orders of magnitude above that which is possible for a standard spherically symmetric nuclear device of the same yield has been proposed in the literature. According to some sources, a concentric pattern of simultaneous detonation or a radially stepped concentric pattern of detonation of the material of either a thermonuclear and/or a nuclear fission device, where the nuclear fuel charge has a relatively simple disk like configuration, might lead to the concentration of the explosive energy in the disk’s center or axis of rotation to a level as much as 6 orders of magnitude greater than that possible with a standard spherically detonating device of the same overall yield (source not locatable).

In each of the following examples, opposing jets produced by two or more of the devices could provide the center-of-mass frame to enable the production of slower moving and more controllable strangelets, charmedlets, bottomlets, and/or toplets.

A) The first device might simply have one primary wherein a nuclear explosive disk undergoes explosive reaction simultaneously throughout. Alternatively, the reaction can be produced within differential radial portions of the nuclear explosive in a precisely timed manner so as to compound the pressure of any compressed plasma at the very center of the disk, thereby increasing the central temperature and forcing the mass-energy to exit the central axis of the detonating disk in the form of an extreme temperature and pressure jet.

In order to temporarily force the jet in one axial direction instead of producing a symmetrical bipolar jet that would otherwise result, some sort of very dense tamper or cladding might be incorporated into one face of the disk. In this manner, the super-hot jet that would otherwise be projected outward and perpendicularly to the uncladded side would be reflected back to amplify the heat and pressure of the jet formed in the opposite direction. The cladding might include a plate or ring radially extending out a distance from the center; or the cladding may consist of multiple dense materials, where the radial distribution of the cladding materials may optionally differ in density, material type, or quantity to optimize the formation of a more unipolar jet. The cladding might take the form of a very dense, perhaps neutron reflective, material. Alternatively, the jet reflection might be facilitated by the introduction of a secondary explosive jet produced by similar means as the first jet. Here, the collision and rebounding of the primary jet would act to force the back in the direction from which it originated. Note that the cladding would be reduced to plasma at essentially the speed of light and therefore the cladding would need to act on the primary jet very quickly.

Another configuration of a similar device might include cladding on both sides with an optional opening at the very center on one cladded face so that the blast pressure and temperature can be amplified, thus increasing the pressure, temperature, and density of the unipolar jet. Yet another configuration of this same basic design would entail a ring of dense material around the perimeter of a double-sided cladded device.

B) A second major configuration of large shaped charged atomic devices might include a stacked series of primary disks where each disk optionally has any of the characteristics mentioned above. Optionally, disks with dissimilar characteristics chosen from the examples given above may be incorporated within a stacked arrangement and may each be detonated simultaneously or in series or other temporal patterns to maximize the pressure, temperature, density, power density, and total jet energy.

An optional but non-limiting form of similar devices could include a long cylinder in which the progression of the nuclear reactions throughout the cylinder in a radial and/or length-wise detonation pattern would be optimized to yield the highest temperatures, pressures, densities, power densities, and/or overall energy contained within the jet. Dense cladding of appropriate materials may be included at optimal positions along the outer surface of the cylinder, or cladding in the form of plates or disks which may be distributed in a lengthwise order along the axis of the cylinder or otherwise arranged to achieve the desired thermodynamic properties of the jet.

C) In a third general configuration, multiple atomic disks would produce inwardly directed coplanar jets. The same arrangement might be used with multiple matter-antimatter primaries each having a cylindrical form such as those mentioned under section B) above. The multiplicity of devices could be detonated simultaneously so that its jets converge at the same time, producing a bipolar or unipolar jet of increased temperature, pressure, density, power density, and overall total energy relative to the single primary based configurations described under sections A) and B) above. A tamper plate or mass may be incorporated on one face within the ring like distribution of primaries so that the jet formed is preferentially concentrated in one direction.

In another form of the third basic configuration, each cylinder or each stack of plates, of which there would be several, would be wedge-like or truncated-wedge-like where the space between the cylinders or wedges would be minimized or eliminated to prevent or mitigate the divergence of the pressure pulse before it could be amplified into a unipolar jet.

D) In a fourth general configuration, a series of cylindrical arrangements where each arrangement optionally has any of the configurations described above in section C) would be stacked, one on top of the other with optionally very little or no space between each of the cylindrical arrangements, thus permitting amplified; pressures, temperatures, densities, power densities, and total energy content of the unipolar jet.

E) In a fifth general configuration, each of the substantially stacked cylindrical arrangements described in section D) could be included in multiplicity where these cylindrical components would be arranged in a manner similar to that described in section C) above, with opposing jets functioning similarly.

Note that additional hierarchies similar to those described under sections C) thru E) can be included, but configurations are not described to avoid unnecessary redundancy. However, practically any number of hierarchies can be incorporated into a single device.

F) A sixth basic configuration would include a conical wedge shaped region filled with any form of appropriate nuclear explosive which may optionally have optimized energy absorbing properties. This configuration may include any of the configurations in sections B) through E) above. The resulting devices would resemble in form, a shaped charged conventional warhead such as those commonly used in heavy armor defeating projectiles. Note that each of the substantially cylindrical arrangements described in sections C) through E) can include a similar plasma or neutral matter squeezing mechanism to enhance the shaped charge effect.

Any of the configurations described under sections B) through F) and/or some or all of the subcomponents thereof may be configured to produce a conical arrangement of jets. Therefore, the devices or sub-components can be configured to form jets that are not coplanar but converge in a conical pattern. In some designs, such an arrangement may be necessary in order to permit the jets formed from canceling out.

Shaped charge devices can more effectively propel starships that are traveling at extremely relativistic velocities compared to spherically symmetrically exploding nuclear devices. To the extent that the columniation of the jets produced by the exploding devices can be improved, the width of the jets decreased, and the temperature of the jets increased, virtually unlimited space-craft Lorentz factors are possible for spacecraft propelled by shaped charge nuclear fission and/or nuclear fusion pellets.

Future generations can establish linear, circular or spiraling fusion and/or fission pellet runways as humankind establishes itself in the cosmos.

Linear fusion pellet runways may extend the radius of the currently observable universe, and perhaps further, in consideration of the forward progression of a highly relativistic spacecraft's light cone as the spacecraft travels away from Earth. Of course, a gradual reduction in the rate of universal space-time expansion would aid in this by helping maintain the linear mass density of the pellet stream at a more uniform level, thus enabling higher spacecraft accelerations and higher terminal Lorentz factors. However, to the extent that the spacecraft could forever utilize such nuclear pellets, the Lorentz factors obtainable by the spacecraft in the depths of eternity are maximally bounded by values at least equal to Omega which is the smallest infinite ordinal.

For extreme Lorentz factor spacecraft, relativistic aberrational effects would need to be considered. However, such effects can at least initially be dealt with by the deployment of sails with a very large surface area that would capture most or all of the bomb blast energy by blast reflection and/or absorption. Robust sails made of stabilized neutronic fibers, quarkonium fibers.

Note that about 90 percent of a nuclear explosive's energy when detonated in space is soft x-ray at 80 percent and gamma rays at 10 percent. Most of the remainder is neutron energy with a small portion of the remainder being charged particle and neutrino radiation embodied energy.

## The Backward Mass Beam Motivated Interstellar Ramjet Mode.

Consider an observer in S' who measures the velocity of an object moving along the x axis at velocity u, and an observer in the S" system moving at velocity v in the x direction with respect to S'. The S" will measure the object moving with velocity u' where:

$$u' = dx'/dt' = [\gamma(dx - vdt)]/\{\gamma\{dt - [vdx/[C^2]]\}\}$$
$$= [(dx/dt) - v]/\{1 - \{[v/[C^2]](dx/dt)]\}\}$$
$$= (u - v)/\{1 - [uv/[C^2]]\}$$

A directly backwardly originating kinetic energy beam of source velocity u in the form of charged massons impinging normally on an energy extraction mechanism of a spacecraft traveling at a velocity, v, will be experienced by the space craft as having a gamma factor of:

$$\gamma rel = \{1/\{1 - [(u'/C)^2]\}^{1/2}\} = \{1/\{1 - \{\{\{(u - v)/\{1 - [uv/[C^2]]\}\}/C\}^2\}\}^{1/2}\}.$$

For cases where the beam momentum is completely extracted commensurate with full extraction of the relativistic kinetic energy of a beam with respect to the spacecraft, the spatially integrated incident beam thrust will be:

$$Egain = \{\Sigma(j= 1, j = m) \{\int\{\{\{d[(M_0 beam\ u'\ \gamma_{relj}) (\gamma_{relj})(u'/C)(A_c)]/dt_j\}(2)\} \bullet dx_j\}\}\};$$

where t is the background reference frame time, x is the background reference frame spatial coordinate of spacecraft propagation; $M_0$ and $M_{0beam}$ are the respective invariant incident masses for a unit cross-sectional area of incident mass for the ambient natural matter and beam matter over a background differential spatial interval dx, and $A_c$ is the effective capture area in number of unit cross-sectional beam area elements. Said latter differential spatial interval is not necessarily equal to $dx_j$, the spatial variable in the dot product function being integrated.

According to the preceding interpretation, the relative gamma factor of the beam with respect to the spacecraft times the relative velocity modifies the instantaneous momentum delivered to the spacecraft. However, since the incident particles are Lorentz contracted, so is the distance between the particles or unit length of beam with respect to the spacecraft. So another factoral repetition of the relative gamma factor of the beam with respect to the spacecraft is needed along with yet another factor in the form of the fractional velocity of the beam with respect to the spacecraft.

As another interpretation, in the above and below formulations including the linear energy gain terms having the factors $M_0$ and $M_0$beam, the associated background differential spatial intervals of the beam(s) is Lorentz contracted with respect to the spacecraft . So, if the beam is modeled as a continuous non-quantized massive column, the number of differential interval beam slabs incident on the spacecraft per unit of spacecraft time scales directly with ($\gamma_{relj}$)(u'/C). However, the momentum per unit slab in the spacecraft frame is equal to ($M_0$beam u' $\gamma_{relj}$). So, using this model, the momentum delivered to the spacecraft per unit spacecraft time metric is [($M_0$beam u' $\gamma_{relj}$) ($\gamma_{relj}$)(u'/C)]. Here, of course, we assume a differential slab in vanishing small if not infinitesimal and that the slab contract further to an extent implied by the beam relative gamma factor.

Taking the corresponding time derivative in the background reference frame of [($M_0$beam u' $\gamma_{relj}$) ($\gamma_{relj}$)(u'/C)] will yield the force acting on the spacecraft according to the background reference frame. Taking the dot product of [($M_0$beam u' $\gamma_{relj}$) ($\gamma_{relj}$)(u'/C)] and $dx_j$ and integrating the product with respect to dx yields the energy gain of the spacecraft over the background interval, $dx_j$.

This rather interesting analytical process does away with needing to use the spacecraft velocity explicitly in the background in computing spacecraft energy gain because of the use of the references of u' and $\gamma_{relj}$ in computing spacecraft energy gain over the background interval $dx_j$. It is possible that spacecraft computers and personnel could thus directly compute the spacecraft energy gain by knowing the relative velocity of the beam with respect to the spacecraft and the relative beam gamma factor while being aware of the beam velocity in the background or the programmed beam velocity in the background which could obviously be pre-determined by beam operation protocol.

The fictitious continuous beam slab elements can be modeled as sub-sets of atom or chargon widths to as small extent as desired for computational purposes and convenience.

Similar computations are implied for the other forms of relative beam velocities and gamma factors with respect to the spacecraft reference frame in the formulations for beamed mass below.

The arguments made in the previous few paragraphs also apply to the case of the forwardly impinging mass beams presented in the next sub-section.

Combining ISR fusion reaction thrust energy gain with the energy gain due to the absorbed backwardly incident beam energy, the spacecraft energy gain will be:

$Egaintotal = \{\Sigma(j= 1, j = m) \{\Sigma(k = 1, k = o) \{\int\{\{\{d[(M_{0beam,k} u'_{k,j} \gamma_{k,j}) (\gamma_{k,j})(u'_{k,j}/C)]/dt_j\}[A_{c,k,j}](2)\} \bullet dx_j\}\}\}\}$
$+ \{\Sigma(j= 1, j = m) \{\int\{P_{drive\text{-}ISR\text{-}fusion\text{-}energyj}\} dt_j\}\}.$

Here, we assume that the backwardly absorbed beam energy is redirected as optimized thrust.

Alternatively, expressed in terms of To;

$E_{gaintotal} = \{\Sigma(j= 1, j = m) \{\Sigma(k = 1, k = o) \{\int\{\{\{d[(M_{0beam,k} u'_{k,j} \gamma_{k,j}) (\gamma_{k,j})(u'_{k,j}/C)]/dt_j\}[A_{c,k,j}](2)\} \bullet dx_j\}\}\}\}$
$+ \{\Sigma(j= 1, j = m) \{\{P_{drive\text{-}ISR\text{-}fusion\text{-}energyj}\} \{\Delta \{(c/g) \ln \{\{[[(C^2) + (V_{0j}^2)]^{1/2}] - [V_{0j}/[[1 - [(V_{0j}/C)^2]]^{1/2} ]]\}$
$\{[(C^2) + [[(g_j)(t_j) + [V0_j /[1 - [(V0_j/C)^2]]^{1/2} ]^2]]^{1/2}] + [(g_j)(t_j)] + [V0_j/[[1 - [(V0_j/C)^2]]^{1/2}]]\} / (C^2)\}\}\}$
$\{1/\{[1 - [(v_j/C)^2]]^{1/2}\}\}\}\}\}.$

Considering massive and background photonic drags, we obtain:

$E_{gaintotal} = \{\Sigma(j= 1, j = m) \{\Sigma(k = 1, k = o) \{\int\{\{\{d[(M_{0beam,k} u'_{k,j} \gamma_{k,j}) (\gamma_{k,j})(u'_{k,j}/C)]/dt_j\}[A_{c,k,j}](2)\} \bullet dx_j\}\}\}\}$
$+ \{\Sigma(j= 1, j = m) \{\int\{P_{drive\text{-}ISR\text{-}fusion\text{-}energyj}\} dt_j\}\}$

$- \{\Sigma(j= 1, j = n) \{\int\{\{\Sigma(i = 1, i = n) \{[[f[cmbr, opt, far infrared, non-cmbr radiofrequency]]modification]_j\} \{<\{\int(0, 2\pi) \int(0, \pi/2) \int [(2)( reflected/incident) \{cmbr, opt, FIR, ncmbrrf\}_{i}] [EM_{frsfj}] \{\{[2hv^3]/C^2\}\{1/\{[e^{\{(hv)/\{kT\{cmbr, opt, FIR, ncmbrrf\}\}]} -1\}\}dv\}\sin \theta\, d\theta\, d\phi\}>/C\} \{\{1/\{\gamma_j[1 + [\beta_j \text{ cosine } [(\theta_{01i,j} + \theta_{02i,j})/2]]]\}\}4\} \{[[(\theta_{01i,j2}) - (\theta_{02i,j2})] / ((90 \text{ degrees})^2) ] \{\{[d\theta_0/d\theta_s] | (\theta_{s1i,j} , \theta_{s2i,j})\}-1\}\{\| \{\{ [dr_{bi,j}/dr_{ai,j}]\} | (r_{b1i,j} , r_{b2i,j})\} \| \}[\cos [(\theta_{01i,j} + \theta_{02i,j})/2]]\}\}\} \bullet dx_j\}\}$

$- \{\Sigma(j= 1, j = m) \{\Sigma(k = 1, k = o) \{\int\{\{\{d[(M_{0k,j} v_j \gamma_j) (\gamma_j)(v_j/C)]/dt_j\}[A_{c,k,j}](2)[Mass_{frsf,k,j}]\} \bullet dx_j\}\}\}\}.$

or alternatively, expressed in terms of $T_o$ ;

$E_{gaintotal} = \{\Sigma(j= 1, j = m) \{\Sigma(k = 1, k = o) \{\int\{\{\{d[(M_{0beam,k} u'_{k,j} \gamma_{k,j}) (\gamma_{k,j})(u'_{k,j}/C)]/dt_j\}[A_{c,k,j}](2)\} \bullet dx_j\}\}\}\}$
$+ \{\Sigma(j= 1, j = m) \{\{P_{drive\text{-}ISR\text{-}fusion\text{-}energyj}\} \{\Delta \{(c/g) \ln \{\{[[(C^2) + (V_{0j}^2)]^{1/2}] - [V_{0j}/[[1 - [(V_{0j}/C)^2]]^{1/2} ]]\}$
$\{[(C^2) + [[(g_j)(t_j) + [V_{0j} /[1 - [(V_{0j}/C)^2]]^{1/2} ]^2]]^{1/2}] + [(g_j)(t_j)] + [V_{0j}/[[1 - [(V_{0j}/C)^2]]^{1/2}]]\} / (C^2)\}\}\}$
$\{1/\{[1 - [(v_j/C)^2]]^{1/2}\}\}\}\}\}$

$- \{\int\{\{\{\Sigma(j = 1, j = m) \{[[f[cmbr, opt, far infrared, non cmbr radiofrequency]]modification]_j\} \{\{\Sigma(i = 1, i = n) \{\{<\{\int(0, 2\pi) \int(0, \pi/2) \int [(2)( reflected/incident)_{\{cmbr, opt, FIR, ncmbrrf\}\ i,j}] [EM_{frsfj}] \{I_{\{cmbr, opt, FIR, ncmbrrf\}} (v,T_{\{cmbr, opt, FIR, ncmbrrf\}} )dv\}\sin \theta\, d\theta\, d\phi\}>/C\}\}\{1/\{\gamma_j [1 + [\beta_j \text{ cosine } \{\{\{\cos^{-1} \{\{[\cos \theta_{sli,j}] - (v_j/C)\}/\{1 - [(v_j/C) \cos \theta_{sli,j}]\}\}\}+ \{\cos^{-1} \{\{[\cos \theta_{s2i,j}] - (v_j/C)\}/\{1 - [(v_j/C) \cos \theta_{s2i,j}]\}\}\}\}/2\}\}\}\}^4\} \{\{\{\{\{\{\cos^{-1} \{\{[\cos \theta_{sli,j}] - (v_j/C)\}/\{1 - [(v_j/C) \cos \theta_{sli,j}]\}\}\}^2\} - \{\{\cos^{-1} \{\{[\cos \theta_{s2i,j}] - (v_j/C)\}/\{1 - [(v_j/C) \cos \theta_{s2i,j}]\}\}\}^2\}\} / ((90 \text{ degrees})^2) \} \{\{\{d\{\cos^{-1} \{\{[\cos \theta_s] - (v_j/C)\}/\{1 - [(v_j/C) \cos \theta_s]\}\}\}/d\theta_s\} | (\theta_{s1i,j} , \theta_{s2i,j})\}^{-1}\}\{\{[\Delta (\cos$

$\theta_s$ )]/{Δ {{[cos $\theta_s$] – ($v_j$ /C)}/{1 - [($v_j$/C) cos $\theta_s$]}}}}|[cos $\theta_{s1i,j}$ , cos $\theta_{s2i,j}$]}{cos {{{cos$^{-1}$ {{[cos $\theta_{sli,j}$] – ($v_j$/C)}/{1 - [($v_j$/C) cos $\theta_{sli,j}$]}}}+ {cos$^{-1}$ {{[cos $\theta_{s2i,j}$] – ($v_j$/C)}/{1 - [($v_j$/C) cos $\theta_{s2i,j}$]}}}}/2}}}}·d$x_j$}}

– {Σ(j= 1, j = m) {Σ(k = 1, k = o) {∫{{{d[($M_{0k,j}$ $v_j$ $\gamma_j$) ($\gamma_j$)($v_j$/C)]/d$t_j$}[$A_{c,k,j}$](2)[$Mass_{frsf,k,j}$]} • d$x_j$}}}}.

Consider the scenario for which intake of backwardly incident mass beam energy is limited to a fraction of the total available energy. Further consider that a the intake of a limited fraction of backwardly incident mass beam energy is re-cycled as propulsion stream energy, while the remainder of the intake of backwardly incident mass beam energy is radiated away as more or less lost thermal and electromagnetic energy.

Velocity compositions for obliquely backwardly oriented beam incidences provide a more general and broader analysis.

In the general case, the relativistic sum of two relativistic velocities V and U is

V $+_{\text{(composition sum)}}$ U = {{V + $U_{par}$ + [($\alpha_v$)($U_{perp}$)]}/{1 + {[V • U]/[$C^2$]}}}

$\alpha_v$ = 1/[γ (V)] = {{1 – {[|V|$^2$]/[$C^2$]}}$^{1/2}$}

where $U_{par}$ and $U_{perp}$ are the components of **U** parallel and perpendicular, respectively, to **V.**

**Here V $+_{\text{(composition sum)}}$ U** is the velocity of the beam with respect to the ship where V is the velocity of the beam with respect to the source or background if stationary, and U is the is additive inverse of the velocity of the ship with respect to the source. Thus, the sign of the vector U is negative.

For cases where the beam momentum is completely extracted commensurate with full extraction of the relativistic kinetic energy of the beam with respect to the spacecraft, and where the extracted beam energy is optimally converted to directly backward thrust stream energy, the spatially integrated incident beam(s) thrust will be:

$E_{gain}$ = {Σ(j= 1, j = m) {Σ(k = 1, k = o) {∫{{{d[{$M_{0beam,j,k}$ {{{$V_{k,j}$ + $U_{par,k,j}$ + [($\alpha_{v,k,j}$)($U_{perp,k,j}$)]}/{1 + {[$V_{k,j}$ • $U_{k,j}$]/[$C^2$]}}}} $\gamma_{rel,k,j}$ } ($\gamma_{rel\ k,j}$){{{{$V_{k,j}$ + $U_{par,k,j}$ + [($\alpha_{v,k,j,}$)($U_{perp,k,j}$)]}/{1 + {[$V_{k,j}$ • $U_{k,j}$]/[$C^2$]}}}$_j$} /C}}/d$t_j$}[$A_{c\ k,j}$](2)} • d$x_j$}}}};

ti is the backward reference frame time, and xi is the backward reference frame spatial coordinate of spacecraft propagation.

Combining ISR fusion reaction thrust energy gain with the energy gain due to the absorbed, backwardly incident beam energy, the spacecraft energy gain will be:

$E_{gaintotal}$ = {Σ(j= 1, j = m) {Σ(k = 1, k = o) {∫{{{d[{$M_{0beam,j,k}$ {{{$V_{k,j}$ + $U_{par,k,j}$ + [($\alpha_{v,k,j}$)($U_{perp,k,j}$)]}/{1 + {[$V_{k,j}$ • $U_{k,j}$]/[$C^2$]}}}} $\gamma_{rel,k,j}$ } ($\gamma_{rel\ k,j}$){{{{$V_{k,j}$ + $U_{par,k,j}$ + [($\alpha_{v,k,j,}$)($U_{perp,k,j}$)]}/{1 + {[$V_{k,j}$ • $U_{k,j}$]/[$C^2$]}}}$_j$} /C}}/d$t_j$}[$A_{c\ k,j}$](2)} • d$x_j$}}}}
+ {Σ(j= 1, j = m) {∫{$P_{drive\text{-}ISR\text{-}fusion\text{-}energyj}$} d$t_j$}}.

Alternatively, expressed in terms of $T_o$:

$$E_{gaintotal} = \{\Sigma(j= 1, j = m) \{\Sigma(k = 1, k = o) \{\int\{\{\{d[\{M_{0beam,j,k}\{\{\{V_{k,j} + U_{par,k,j} + [(\alpha_{v,k,j})(U_{perp,k,j})]\}/\{1 + \{[V_{k,j} \bullet U_{k,j}]/[C^2]\}\}\}\} \gamma_{rel,k,j}\} (\gamma_{rel\ k,j})\{\{\{\{V_{k,j} + U_{par,k,j} + [(\alpha_{v,k,j,})(U_{perp,k,j})]\}/\{1 + \{[V_{k,j} \bullet U_{k,j}]/[C^2]\}\}\}_j\} /C\}\}/dt_j\}[A_{c\ k,j}](2)\} \bullet dx_j\}\}\}$$

$$+ \{\Sigma(j= 1, j = m) \{\{P_{drive\text{-}ISR\text{-}fusion\text{-}energyj}\} \{\Delta \{(c/g) \ln \{\{[[(C^2) + (V_{0j}^2)]^{1/2}] - [V_{0j}/[[1 - [(V_{0j}/C)^2]]^{1/2}]]\} \{[(C^2) + [[(g_j)(t_j) + [V_{0j}/[1 - [(V_{0j}/C)^2]]^{1/2}]^2]]^{1/2}] + [(g_j)(t_j)] + [V_{0j}/[[1 - [(V_{0j}/C)^2]]^{1/2}]]\} / (C^2)\}\}\} \{1/\{[1 - [(v_j/C)^2]]^{1/2}\}\}\}\}\}.$$

Considering massive drag and background photonic drags, we obtain:

$$E_{gaintotal} = \{\Sigma(j= 1, j = m) \{\Sigma(k = 1, k = o) \{\int\{\{\{d[\{M_{0beam,j,k}\{\{\{V_{k,j} + U_{par,k,j} + [(\alpha_{v,k,j})(U_{perp,k,j})]\}/\{1 + \{[V_{k,j} \bullet U_{k,j}]/[C^2]\}\}\}\} \gamma_{rel,k,j}\} (\gamma_{rel\ k,j})\{\{\{\{V_{k,j} + U_{par,k,j} + [(\alpha_{v,k,j,})(U_{perp,k,j})]\}/\{1 + \{[V_{k,j} \bullet U_{k,j}]/[C^2]\}\}\}_j\} /C\}\}/dt_j\}[A_{c\ k,j}](2)\} \bullet dx_j\}\}\}$$

$$+ \{\Sigma(j= 1, j = m) \{\int\{P_{drive\text{-}ISR\text{-}fusion\text{-}energyj}\} dt_j\}\} - \{\Sigma(j= 1, j = n) \{\int\{\{\Sigma(i = 1, i = n) \{[[f[\text{cmbr, opt, far infrared, non-cmbr radiofrequency}]]\text{modification}]_j\} \{<\{\int(0, 2\pi) \int(0, \pi/2) \int [(2)(\text{ reflected/incident})_{\{cmbr,\ opt,\ FIR,\ ncmbrrf\}\ i}] [EM_{frsfj}] \{\{[2hv^3]/C^2\}\{1/\{[e^{\{(hv)/\{kT\{cmbr,\ opt,\ FIR,\ ncmbrrf\}\}\}}] -1\}\}dv\}\sin \theta\, d\theta\, d\phi\}>/C\} \{\{1/\{\gamma_j[1 + [\beta_j \text{ cosine } [(\theta_{01i,j} + \theta_{02i,j})/2]]]\}\}^4\} \{[[(\theta_{01i,j}^2) - (\theta_{02i,j}^2)] / ((90 \text{ degrees})^2)] \{\{[d\theta_0/d\theta_s] | (\theta_{s1i,j}, \theta_{s2i,j})\}^{-1}\}\{ \| \{\{[dr_{bi,j}/dr_{ai,j}]\} | (r_{b1i,j}, r_{b2i,j})\} \| \}[\cos [(\theta_{01i,j} + \theta_{02i,j})/2]]\}\}\} \bullet dx_j\}\} - \{\Sigma(j= 1, j = m) \{\Sigma(k = 1, k = o) \{\int\{\{\{d[(M_{0k,j}\, v_j\, \gamma_j) (\gamma_j)(v_j/C)]/dt_j\}[A_{c,k,j}](2)[Mass_{frsf,k,j}]\} \bullet dx_j\}\}\}.$$

or alternatively, expressed in terms of $T_o$::

$$E_{gaintotal} = \{\Sigma(j= 1, j = m) \{\Sigma(k = 1, k = o) \{\int\{\{\{d[\{M_{0beam,j,k}\{\{\{V_{k,j} + U_{par,k,j} + [(\alpha_{v,k,j})(U_{perp,k,j})]\}/\{1 + \{[V_{k,j} \bullet U_{k,j}]/[C^2]\}\}\}\} \gamma_{rel,k,j}\} (\gamma_{rel\ k,j})\{\{\{\{V_{k,j} + U_{par,k,j} + [(\alpha_{v,k,j,})(U_{perp,k,j})]\}/\{1 + \{[V_{k,j} \bullet U_{k,j}]/[C^2]\}\}\}_j\} /C\}\}/dt_j\}[A_{c\ k,j}](2)\} \bullet dx_j\}\}\}$$

$$+ \{\Sigma(j= 1, j = m) \{\{P_{drive\text{-}ISR\text{-}fusion\text{-}energyj}\} \{\Delta \{(c/g) \ln \{\{[[(C^2) + (V_{0j}^2)]^{1/2}] - [V_{0j}/[[1 - [(V_{0j}/C)^2]]^{1/2}]]\} \{[(C^2) + [[(g_j)(t_j) + [V_{0j}/[1 - [(V_{0j}/C)^2]]^{1/2}]^2]]^{1/2}] + [(g_j)(t_j)] + [V_{0j}/[[1 - [(V_{0j}/C)^2]]^{1/2}]]\} / (C^2)\}\}\} \{1/\{[1 - [(v_j/C)^2]]^{1/2}\}\}\}\}\}$$

$$- \{\int\{\{\{\Sigma(j = 1, j = m) \{[[f[\text{cmbr, opt, far infrared, non-cmbr radiofrequency}]]\text{modification}]_j\} \{\{\Sigma(i = 1, i = n) \{\{<\{\int(0, 2\pi) \int(0, \pi/2) \int [(2)(\text{ reflected/incident})_{\{cmbr,\ opt,\ FIR,\ ncmbrrf\}\ i,j}] [EM_{frsfj}] \{I_{\{cmbr,\ opt,\ FIR,\ ncmbrrf\}} (v, T_{\{cmbr,\ opt,\ FIR,\ ncmbrrf\}})dv\}\sin \theta\, d\theta\, d\phi\}>/C\}\}\{\{1/\{\gamma_j [1 + [\beta_j \text{ cosine } \{\{\{\cos^{-1} \{\{[\cos \theta_{sli,j}] - (v_j/C)\}/\{1 - [(v_j/C) \cos \theta_{sli,j}]\}\}\} + \{\cos^{-1} \{\{[\cos \theta_{s2i,j}] - (v_j/C)\}/\{1 - [(v_j/C) \cos \theta_{s2i,j}]\}\}\}\}/2\}\}\}\}^4\} \{\{\{\{\{\{\cos^{-1} \{\{[\cos \theta_{sli,j}] - (v_j/C)\}/\{1 - [(v_j/C) \cos \theta_{sli,j}]\}\}\}^2\} - \{\{\cos^{-1} \{\{[\cos \theta_{s2i,j}] - (v_j/C)\}/\{1 - [(v_j/C) \cos \theta_{s2i,j}]\}\}\}^2\}\} / ((90 \text{ degrees})^2)\} \{\{\{d\{\cos^{-1} \{\{[\cos \theta_s] - (v_j/C)\}/\{1 - [(v_j/C) \cos \theta_s]\}\}\}/d\theta_s\} | (\theta_{s1i,j}, \theta_{s2i,j})\}^{-1}\}\{\{[\Delta (\cos \theta_s)]/\{\Delta \{\{[\cos \theta_s] - (v_j /C)\}/\{1 - [(v_j/C) \cos \theta_s]\}\}\}\} | [\cos \theta_{s1i,j}, \cos \theta_{s2i,j}]\}\{\cos \{\{\{\cos^{-1} \{\{[\cos \theta_{sli,j}] - (v_j/C)\}/\{1 - [(v_j/C) \cos \theta_{sli,j}]\}\}\} + \{\cos^{-1} \{\{[\cos \theta_{s2i,j}] - (v_j/C)\}/\{1 - [(v_j/C) \cos \theta_{s2i,j}]\}\}\}\}/2\}\}\}\cdot dx_j\}\} - \{\Sigma(j= 1, j = m) \{\Sigma(k = 1, k = o) \{\int\{\{\{d[(M_{0k,j}\, v_j\, \gamma_j) (\gamma_j)(v_j/C)]/dt_j\}[A_{c,k,j}](2)[Mass_{frsf,k,j}]\} \bullet dx_j\}\}\}.$$

Consider scenarios for which intake is only a limited fraction of backwardly incident mass beam energy. Further consider that intake of a limited fraction of the backwardly incident mass beam energy is re-cycled as propulsion stream energy, while the remainder of the intake backwardly incident mass beam energy is radiated away as more or less lost thermal and electromagnetic energy.

$E_{gaintotal}$ = {Σ(j= 1, j = m) {Σ(k = 1, k = o) {[(-1)(radiative loss)/(intake)]$_{k,j}$ }[(emissivity shape factor)$_{k,j}$] {∫{{{d[{$M_{0beam,k,j}$ {{{$V_{k,j}$ + $U_{par,k,j}$ + [($\alpha_{v,\ k,j}$)($U_{perp,\ k,j}$)]]}/{1 + {[$V_{k,j}$ • $U_{k,j}$]/[$C^2$]}}}$_j$} $\gamma_{rel\ k,j,}$} ($\gamma_{rel,\ k,j}$){{{{$V_{k,j}$ + $U_{par,k,j}$ + [($\alpha_{v,\ k,j}$)($U_{perp,\ k,j}$)]]}/{1 + {[$V_{k,j}$ • $U_{k,j}$]/[$C^2$]}}}} /C}}/$dt_j$}[$A_{c,\ k,j}$ ]} • $dx_j$}}}} + {Σ(j= 1, j = m) {Σ(k = 1, k = o) (2)[(intake)$_j$]($\varepsilon_{intakej}$) {∫{{{d[{$M_{0beam,k,j}$ {{{$V_{k,j}$ + $U_{par,k,j}$ + [($\alpha_{v,\ k,j}$)($U_{perp,\ k,j}$)]]}/{1 + {[$V_{k,j}$ • $U_{k,j}$]/[$C^2$]}}}$_j$} $\gamma_{rel\ k,j,}$} ($\gamma_{rel,\ k,j}$){{{{$V_{k,j}$ + $U_{par,k,j}$ + [($\alpha_{v,\ k,j}$)($U_{perp,\ k,j}$)]]}/{1 + {[$V_{k,j}$ • $U_{k,j}$]/[$C^2$]}}}} /C}}/$dt_j$}[$A_{c,\ k,j}$ ]} • $dx_j$}[$A_{c,k,j}$]} • $dx_j$}}}} + {Σ(j= 1, j = m) {Σ(k = 1, k = o) (2)[$Mass_{fbrsfj}$]{[(reflected)/(intaked)]$_j$} {∫{{{d[{$M_{0beam,k,j}$ {{{$V_{k,j}$ + $U_{par,k,j}$ + [($\alpha_{v,\ k,j}$)($U_{perp,\ k,j}$)]]}/{1 + {[$V_{k,j}$ • $U_{k,j}$]/[$C^2$]}}}$_j$} $\gamma_{rel\ k,j,}$} ($\gamma_{rel,\ k,j}$){{{{$V_{k,j}$ + $U_{par,k,j}$ + [($\alpha_{v,\ k,j}$)($U_{perp,\ k,j}$)]]}/{1 + {[$V_{k,j}$ • $U_{k,j}$]/[$C^2$]}}}} /C}}/$dt_j$}[$A_{c,\ k,j}$ ]} • $dx_j$}}}}

+ {Σ(j= 1, j = m) {∫{$P_{drive\text{-}ISR\text{-}fusion\text{-}energyj}$} $dt_j$}} - {Σ(j= 1, j = n) {∫{{{Σ(i = 1, i = n) {[[[f[cmbr, opt, far infrared, non-cmbr radiofrequency]]modification]$_j$} {<{∫(0, 2π) ∫(0, π/2) ∫ [(2)( reflected/incident) $_{\{cmbr,\ opt,\ FIR,\ ncmbrrf\}\ i}$] [$EM_{frsfj}$] {{[2h$v^3$]/$C^2$}{1/{[$e^{\{(hv)/\{kT\{cmbr,\ opt,\ FIR,\ ncmbrrf\}\})]}$] -1}}dv}sin θ dθ dϕ}>/C} {{1/{$\gamma_j$[1 + [$\beta_j$ cosine [($\theta_{01i,j}$ + $\theta_{02i,j}$)/2]]]}}$^4$} {[[($\theta_{01i,j}{}^2$) - ($\theta_{02i,j}{}^2$)] / ((90 degrees)$^2$) ] {{[d$\theta_0$/d$\theta_s$] | ($\theta_{s1i,j}$ , $\theta_{s2i,j}$)}$^{-1}$}{ ‖ {{ [$dr_{bi,j}$/$dr_{ai,j}$]} | ($r_{b1i,j}$ , $r_{b2i,j}$)} ‖ }[cos [($\theta_{01i,j}$ + $\theta_{02i,j}$)/2]]}}} • $dx_j$}} – {Σ(j= 1, j = m) {Σ(k = 1, k = o) {∫{{{d[($M_{0k,j}$ $v_j$ $\gamma_j$) ($\gamma_j$)($v_j$/C)]/$dt_j$}[$A_{c,k,j}$](2)[$Mass_{frsf,k,j}$]} • $dx_j$}}}}.

or alternatively, expressed in terms of $T_o$:

$E_{gaintotal}$ = {Σ(j= 1, j = m) {Σ(k = 1, k = o) {[(-1)(radiative loss)/(intake)]$_{k,j}$ }[(emissivity shape factor)$_{k,j}$] {∫{{{d[{$M_{0beam,k,j}$ {{{$V_{k,j}$ + $U_{par,k,j}$ + [($\alpha_{v,\ k,j}$)($U_{perp,\ k,j}$)]]}/{1 + {[$V_{k,j}$ • $U_{k,j}$]/[$C^2$]}}}$_j$} $\gamma_{rel\ k,j,}$} ($\gamma_{rel,\ k,j}$){{{{$V_{k,j}$ + $U_{par,k,j}$ + [($\alpha_{v,\ k,j}$)($U_{perp,\ k,j}$)]]}/{1 + {[$V_{k,j}$ • $U_{k,j}$]/[$C^2$]}}}} /C}}/$dt_j$}[$A_{c,\ k,j}$ ]} • $dx_j$}}}} + {Σ(j= 1, j = m) {Σ(k = 1, k = o) (2)[(intake)$_j$]($\varepsilon_{intakej}$) {∫{{{d[{$M_{0beam,k,j}$ {{{$V_{k,j}$ + $U_{par,k,j}$ + [($\alpha_{v,\ k,j}$)($U_{perp,\ k,j}$)]]}/{1 + {[$V_{k,j}$ • $U_{k,j}$]/[$C^2$]}}}$_j$} $\gamma_{rel\ k,j,}$} ($\gamma_{rel,\ k,j}$){{{{$V_{k,j}$ + $U_{par,k,j}$ + [($\alpha_{v,\ k,j}$)($U_{perp,\ k,j}$)]]}/{1 + {[$V_{k,j}$ • $U_{k,j}$]/[$C^2$]}}}} /C}}/$dt_j$}[$A_{c,\ k,j}$ ]} • $dx_j$}[$A_{c,k,j}$]} • $dx_j$}}}} + {Σ(j= 1, j = m) {Σ(k = 1, k = o) (2)[$Mass_{fbrsfj}$]{[(reflected)/(intaked)]$_j$} {∫{{{d[{$M_{0beam,k,j}$ {{{$V_{k,j}$ + $U_{par,k,j}$ + [($\alpha_{v,\ k,j}$)($U_{perp,\ k,j}$)]]}/{1 + {[$V_{k,j}$ • $U_{k,j}$]/[$C^2$]}}}$_j$} $\gamma_{rel\ k,j,}$} ($\gamma_{rel,\ k,j}$){{{{$V_{k,j}$ + $U_{par,k,j}$ + [($\alpha_{v,\ k,j}$)($U_{perp,\ k,j}$)]]}/{1 + {[$V_{k,j}$ • $U_{k,j}$]/[$C^2$]}}}} /C}}/$dt_j$}[$A_{c,\ k,j}$ ]} • $dx_j$}}}}
+ {Σ(j= 1, j = m) {{$P_{drive\text{-}ISR\text{-}fusion\text{-}energyj}$} {Δ {(c/g) ln {{[[($C^2$) + ($V_{0j}{}^2$)]$^{1/2}$] - [$V_{0j}$/[[1 - [($V_{0j}$/C)$^2$]]$^{1/2}$ ]]} {[($C^2$) + [[($g_j$)($t_j$) + [$V_{0j}$ /[1 - [($V_{0j}$/C) $^2$]]$^{1/2}$ ] $^2$]]$^{1/2}$] + [($g_j$)($t_j$)] + [$V_{0j}$/[[1 - [($V_{0j}$/C)$^2$]]$^{1/2}$]]} / ($C^2$)}}} {1/{[1 - [($v_j$/C)$^2$]]$^{1/2}$}}}}}

- {∫{{{Σ(j = 1, j = m) {[[f[cmbr, opt, far infrared, non-cmbr radiofrequency]]modification]$_j$} {{Σ(i = 1, i = n) {{<{∫(0, 2π) ∫(0, π/2) ∫ [(2)( reflected/incident) $_{\{cmbr,\ opt,\ FIR,\ ncmbrrf\}\ i,j}$] [$EM_{frsfj}$] {I $_{\{cmbr,\ opt,\ FIR,\ ncmbrrf\}}$ (v,$T_{\{cmbr,\ opt,\ FIR,\ ncmbrrf\}}$ )dv}sin θ dθ dϕ}>/C}}{1/{$\gamma_j$ [1 + [$\beta_j$ cosine {{{$\cos^{-1}$ {{[cos $\theta_{sli,j}$] – ($v_j$/C)}/{1 - [($v_j$/C) cos $\theta_{sli,j}$]}}}+ {$\cos^{-1}$ {{[cos $\theta_{s2i,j}$] – ($v_j$/C)}/{1 - [($v_j$/C) cos $\theta_{s2i,j}$]}}}}/2}}}} $^4$} {{{{{{$\cos^{-1}$ {{[cos $\theta_{sli,j}$] – ($v_j$/C)}/{1 - [($v_j$/C) cos $\theta_{sli,j}$]}}}$^2$} – {{$\cos^{-1}$ {{[cos $\theta_{s2i,j}$] – ($v_j$/C)}/{1 - [($v_j$/C) cos $\theta_{s2i,j}$]}}}$^2$}} / ((90 degrees)$^2$) } {{{d{$\cos^{-1}$ {{[cos $\theta_s$] – ($v_j$/C)}/{1 - [($v_j$/C) cos $\theta_s$]}}}/d$\theta_s$} | ($\theta_{s1i,j}$ , $\theta_{s2i,j}$)}$^{-1}$}{{[Δ (cos $\theta_s$ )]/{Δ {{[cos $\theta_s$] – ($v_j$ /C)}/{1 - [($v_j$/C) cos $\theta_s$]}}}} | [cos $\theta_{s1i,j}$ , cos $\theta_{s2i,j}$]}{cos {{{$\cos^{-1}$ {{[cos $\theta_{sli,j}$] –

$(v_j/C)\}/\{1 - [(v_j/C) \cos \theta_{sli,j}]\}\}\}+ \{\cos^{-1} \{\{[\cos \theta_{s2i,j}] - (v_j/C)\}/\{1 - [(v_j/C) \cos \theta_{s2i,j}]\}\}\}\}/2\}\}\}\}\cdot dx_j\}\} - \{\Sigma(j= 1, j = m) \{\Sigma(k = 1, k = o) \{\int\{\{\{d[(M_{0k,j} v_j \gamma_j) (\gamma_j)(v_j/C)]/dt_j\}[A_{c,k,j}](2)[Mass_{frsf,k,j}]\} \bullet dx_j\}\}\}\}$.

In the above formulations, the relativistic gamma factor of the beam with respect to the starship is;

$$\gamma = \{1/\{\{1 - \{\{\{\{V + U_{par} + [(\alpha_v)(U_{perp})]\}/\{1 + \{[V \bullet U]/[C^2]\}\}\}/C\}^2\}\}^{1/2}\}\}.$$

# Three specific examples of the primary gas gun system scenarios.

## Class 1 Electron Gas Guns Enabling Mildly Relativistic Spacecraft Velocities of 0.01 c.

***Here, we present a near term realizable scenario for which a staged compression electron gas gun enables spacecraft velocities of 0.01 c. We reiterate some of the material in the previous section regarding the 0.01 c scenario just for clarification. This is done so that further elaboration on the 0.01 c case is easily interpretable in context.***

In order to accelerate a 10,000 metric ton craft to a Lorentz factor of 1.00005 and assuming that 0.5 of the pellet explosive energy is converted to spacecraft kinetic energy, the explosive energy embodied in the pellets would need to be approximately equivalent to 0.0001 the mass of the spacecraft or one metric ton. Assuming the yield of a nuclear explosive pellet is about 0.0025 of its invariant mass equivalent, the mass of the pellet runway would need to be about 400 metric tons.

A method of converting pellet explosive energy to spacecraft kinetic energy is the deployment of a large charge sail chute that would be pushed ahead of the spacecraft but which would be attached to the spacecraft by an elastic spring system. The recoiling chute would be pulled back to the spacecraft whereupon the next pellet would detonate.

For explosive charges each having a mass of 40 kilograms, 10,000 devices would be needed. Since each device would be standardized, assuming a unit can be mass produced at a cost of $100,000, the total cost of the fuel pellets would be $1 billion.

For the anticipated 0.0001 light-year long pellet stream, each pellet would need to be separated on average by $10^{-8}$ light-years or by 100 million meters.

Assuming gun muzzle velocity of 100,000 m/s, the average time between pellet launches would be 1,000 seconds.

The time averaged thermal loading on the gun would be equal to that of the previous example.

The total time averaged power required to accelerate the pellets would be equal to that of the previous example.

The required time averaged electrical power required to operate the gun would be 200 megawatts over the 1,000 second cycle period.

Assuming an orbital location of one AU from the sun, and an adjustable reflector which reflects about 1/2 of the incident sunlight but transmits the rest, we obtain the following relation.

(4 x $10^6$ N) = <S>/c = <S>/(3 x $10^8$ m/s). So, the required sunlight power to counteract the acceleration on the gun during the firing time would be 1.2 x $10^{15}$ watts.

Assuming a barrel system mass of 5,000,000 kg, the net barrel acceleration during the firing operation would be F/m = [(4 x $10^6$ N)/(5,000,000 kg)] = 0.8 $m/s^2$ and the displacement component of the barrel during the firing process would be x = ½ **a**$t^2$ = 0.4 m.

All the rest of the details of gun thermodynamics are assumed applicable from the previous example

Here, we are considering gun placement in solar orbit but with adjustable orientation via retro-rockets, solar sails, and/or plasma sails. The orientation of the gun and its orbital radius can be re-adjusted as needed by sail tacking processes.

For a spacecraft having reached a velocity of 0.01 c such as at the end of the pellet runway, the spacing between the pellets would be about 10 times that of the case where the spacecraft still accelerating was only at a velocity of 0.001 c or about 300 km/second. For the case where the spacecraft started at a velocity of 150 km/second at the beginning of the pellet stream, the spacing of the pellets at the beginning trailing edge of the deployed runway would be about 1/60 times that provided at the leading end of the pellet stream.

Now, consider the series summation, (1 + 2 + 3 + ... + 60). The average of this series is about equal to 30. If we were to add the first Aleph 0 number of integers and take the average, the average would be ½ of the highest integer. Aleph 0 is the number of positive integer. So in the limit that the number of summed integers which are consecutive approaches infinity, the average is equal to ½ the largest integer in the series.

So, the two pellets furthest away from the Sun will be 200,000 kilometers apart. The two pellets closest to the Sun will be 3,333 km apart.

For a 0.0001 light-year long distributed runway and average spacecraft frame acceleration on the order of 0.5 earth g's or 4.905 $m/s^2$, a human crewed spacecraft could obtain a velocity of about 1 percent of light-speed and a Lorentz factor of 1.00005.

Several mechanisms might be used to reduce the effective G-forces felt by the crew. Among these is the enclosure of crew members' bodies in hydrostatically sealed breathable oxygenated liquid-containing vessels. Alternatively, perhaps nano-technology types of pressure suits could completely encase the crew members' bodies. The pressure suits might optionally pump high pressure air into the lungs of the crew members wearing them, and gradually relax the lung pressure as the rate of acceleration was reduced to more manageable levels. Intense electromagnet fields could be used to induce dipole moments in the atoms and molecules composing the crew members' bodies whereupon the magnetic field(s) would then partially "levitate" the crew members thereby effectively cancelling undue G-forces. Another option is to use nanotechnology or other precise mechanisms for the installation of an electrical charge within the crew members' bodies where an external electric field would pull on or repel the crew members' bodies thereby cancelling out excessive G-forces.

Note that we assume half of the pellet energy is converted to spacecraft kinetic energy on average.

At a dispensing velocity of 100 km/second, one pellet dispensed per 1,000 seconds on average, and 10,000 pellets dispensed, the pellet runway would take a 10 million seconds to distribute or about 1/3 years. During this time, the pellet furthest from the gun will be 1 billion kilometers distant or about 0.0001 light-years distant.

The United States in reality can support the cost of the above projects since the country was able to fund the war efforts in the Middle East and central Asia during the first two decades of the 21st century with a total cost of several trillion dollars.

A spacecraft may initially undergo an Oberth maneuver near the Sun to obtain an existing velocity of 150 km second, The required change in spacecraft velocity under the condition that the O'berth maneuver is not utilized for enhanced velocity is 50 km/second. This velocity is consistent with what can be obtained with solar thermal rockets and nuclear thermal rockets with large mass ratios. The invariant mass of the craft after the fuel has been spent will be the 5,000 metric tons originally conceived.

| Δv km/s | {1 + [(2Vesc)/(Δv)]} EXP (1/2) | Exiting Velocity km/s |
|---|---|---|
| 50 | 3 | 150 |
| 100 | 2.236067977 | 223.6067977 |
| 150 | 1.914854216 | 287.2281323 |
| 200 | 1.732050808 | 346.4101615 |
| 250 | 1.61245155 | 403.1128874 |
| 200 | 1.732050808 | 346.4101615 |
| 350 | 1.463850109 | 512.3475383 |
| 200 | 1.732050808 | 346.4101615 |
| 450 | 1.374368542 | 618.4658438 |
| 200 | 1.732050808 | 346.4101615 |
| 550 | 1.314257481 | 722.8416147 |
| 600 | 1.290994449 | 774.5966692 |
| 650 | 1.270977819 | 826.1355821 |
| 700 | 1.253566341 | 877.4964387 |
| 750 | 1.238278375 | 928.7087811 |
| 800 | 1.224744871 | 979.7958971 |
| 850 | 1.212678125 | 1030.776406 |
| 900 | 1.201850425 | 1081.665383 |
| 950 | 1.192079121 | 1132.475165 |
| 1000 | 1.183215957 | 1183.215957 |

Table 1 shows incremental velocities for various values of Δv at escape velocity of 200 km/sec.

The spacing between the pellets is best proportional to the projected velocity of the spacecraft at a given pellet-specific location along the acceleration path of the spacecraft in pellet drive mode.

## Class 2 Electron Gas Guns Enabling Mildly Relativistic Spacecraft Velocities of 0.1 c.

***Herein we consider gas guns that may enable spacecraft velocities of about 0.1 c and Lorentz factors of about 1.005. Such velocities can enable timely generation ship travel from star to star for alive and awake crew members and passengers.***

In order to accelerate a 10,000 metric ton craft to a Lorentz factor of 1.005 and assuming that 0.05 of the pellet explosive energy is converted to spacecraft kinetic energy, the explosive energy embodied in the pellets would need to be approximately equivalent to 0.1 the mass of the spacecraft or 1,000 metric tons. Assuming the yield of a nuclear explosive pellet is about 0.0025 of its invariant mass equivalent, the mass of the pellet runway would need to be about 400,000 metric tons.

For explosive charges each having a mass of 40 kilograms, 10 million devices would be needed. Since each device would be standardized, assuming a unit can be mass produced at a cost of $100,000, the total cost of the fuel pellets would be $1 Trillion.

For the anticipated 0.01 light-year long pellet stream, each pellet would need to be separated on average by $10^{-9}$ light-years or by 10 million meters.

Assuming gun muzzle velocity of 100,000 m/s, the average time between pellet launches would be 100 seconds. So, the pellet runway lay down time would be 1 billion seconds.

Assuming an orbital location of one AU from the sun, and an adjustable reflector which reflects about 1/2 of the incident sunlight but transmits the rest, we obtain the following relation.

$(4 \times 10^6 \text{ N}) = \langle S \rangle / c = \langle S \rangle / (3 \times 10^8 \text{ m/s})$. So, the required sunlight power to counteract the acceleration on the gun during the firing time would be $1.2 \times 10^{15}$ watts. By firing time, what is meant is the time interval that the round travels down the gun barrel.

Assuming a barrel system mass of 5,000,000 kg, the net barrel acceleration during the firing operation would be $F/m = [(4 \times 10^6 \text{ N})/(5{,}000{,}000 \text{ kg})] = 0.8 \text{ m/s}^2$ and the displacement component of the barrel during the firing process would be $x = ½ \mathbf{a}t^2 = 0.4$ m.

Assuming a reverse acceleration time of 100 seconds, the barrel reverse acceleration need only be $8 \times 10^{-3} \text{ m/s}^2$ and thus the required sun-light power on a sail to counteract the firing acceleration effects is $1.2 \times 10^{13}$ watts. The latter value can be obtained by a light-sail having collection area of a mere 12,000 $\text{km}^2$. This is a mere equivalent of 109.54 km by 109.54 km. For a sail having capture area of 12,000 $\text{km}^2$ and a mass specific area of one milligram per square meter, the total mass of the sail would be a mere 12,000 kilograms. A sail having area mass density of 0.01 gram per square meter would have mass of 120 metric tons.

The time averaged thermal loading on the gun would be 10 times that of the previous example.

The total energy required to accelerate the pellets would be 1,000 times as great as that of the previous example.

Additionally, the total cost to assemble and launch the pellets would be about 1,000 times greater than in the previous example.

For a spacecraft having reached a velocity of 0.1 c such as at the end of the pellet runway, the spacing between the pellets would be about 10 times that of the case where the spacecraft still accelerating was only at a velocity of 0.01 c or about 3,000 km/second. For the case where the spacecraft started at a velocity of 150 km second at the beginning of the pellet stream, the spacing of the pellets at the beginning trailing edge of the deployed runway would be about 1/600 times that provided at the leading end of the pellet stream.

Now, consider the series summation, (1 + 2 + 3 + ... + 600). The average of this series is about equal to 300.

So, the two pellets furthest away from the Sun will be 2,000,000 kilometers apart. The two pellets closest to the Sun will be 3,333 km apart.

For a 0.01 light-year long distributed runway and average spacecraft frame acceleration on the order of 0.5 Earth g's or 4.905 $m/s^2$, a human crewed spacecraft could obtain a velocity of about 10 percent of light-speed and a Lorentz factor of 1.005.

The assumption of 0.05 of the pellet explosive energy is converted to spacecraft kinetic energy may actually be quite low. Since the Project Orion craft anticipated shaped charging of the nuclear bombs propelling the spacecraft which would direct 85 percent of the explosive flux onto the pusher plate, and since 90 percent of the bomb energy would range from soft x-ray to gamma ray radiation, we can assume that about 76.5 percent of the bombs' energies would be transferred to the pusher plate and converted to spacecraft kinetic energy in first order assuming all photon energy absorbed by the plate would be converted to kinetic energy. Assuming loss of about 1/3 of the energy due to inefficiencies, we obtain a good estimate of the bomb energies converted to spacecraft kinetic energy at about 50 percent.

In order to accelerate a 10,000 metric ton craft to a Lorentz factor of 1.005 and assuming that 0.5 of the pellet explosive energy is converted to spacecraft kinetic energy, the explosive energy embodied in the pellets would need to be approximately equivalent to 0.01 the mass of the spacecraft or 100 metric tons. Assuming the yield of a nuclear explosive pellet is about 0.0025 of its invariant mass equivalent, the mass of the pellet runway would need to be about 40,000 metric tons.

For explosive charges each having a mass of 40 kilograms, 1 million devices would be needed. Since each device would be standardized, assuming a unit can be mass produced at a cost of $100,000, the total cost of the fuel pellets would be $100 Billion.

For the anticipated 0.001 light-year long pellet stream, each pellet would need to be separated on average by $10^{-9}$ light-years or by 10 million meters.

Assuming gun muzzle velocity of 100,000 m/s, the average time between pellet launches would be 100 seconds.

Assuming an orbital location of one AU from the sun, and an adjustable reflector which reflects about 1/2 of the incident sunlight but transmits the rest, we obtain the following relation.

$(4 \times 10^6 \text{ N}) = \langle S \rangle / c = \langle S \rangle / (3 \times 10^8 \text{ m/s})$. So, the required sunlight power to counteract the acceleration on the gun during the firing time would be $1.2 \times 10^{15}$ watts. By firing time, what is meant is the time interval that the round travels down the gun barrel.

Assuming a barrel system mass of 5,000,000 kg, the net barrel acceleration during the firing operation would be $F/m = [(4 \times 10^6 \text{ N})/(5{,}000{,}000 \text{ kg})] = 0.8 \text{ m/s}^2$ and the displacement component of the barrel during the firing process would be $x = \frac{1}{2} \mathbf{a}t^2 = 0.4$ m.

Assuming a reverse acceleration time of 100 seconds, the barrel reverse acceleration need only be $8 \times 10^{-3} \text{ m/s}^2$ and thus the required sun-light power on a sail to counteract the firing acceleration effects is $1.2 \times 10^{13}$ watts. The latter value can be obtained by a light-sail having collection area of a mere 12,000 $\text{km}^2$. This is a mere equivalent of 109.54 km by 109.54 km. For a sail having capture area of 12,000 $\text{km}^2$ and a mass specific area of one milligram per square meter, the total mass of the sail would be a mere 12,000 kilograms. A sail having area mass density of 0.01 gram per square meter would have mass of 120 metric tons.

For a spacecraft having reached a velocity of 0.1 c such as at the end of the pellet runway, the spacing between the pellets would be about 10 times that of the case where the spacecraft still accelerating was only at a velocity of 0.01 c or about 3,000 km/second. For the case where the spacecraft started at a velocity of 150 km second at the beginning of the pellet stream, the spacing of the pellets at the beginning trailing edge of the deployed runway would be about 1/600 times that provided at the leading end of the pellet stream.

Now, consider the series summation, (1 + 2 + 3 + ... + 600). The average of this series is about equal to 300.

So, the two pellets furthest away from the Sun will be 2,000,000 kilometers apart. The two pellets closest to the Sun will be 3,333 km apart.

For a 0.01 light-year long distributed runway and average spacecraft frame acceleration on the order of 0.5 Earth g's or 4.905 $\text{m/s}^2$, a human crewed spacecraft could obtain a velocity of about 10 percent of light-speed and a Lorentz factor of 1.005.

Again, in order to produce an average acceleration of 4.905 $\text{m/s}^2$, the pusher plate can be mounted on a spring system which slides along rails for about 66.7 seconds. The pusher plate would be accelerated to 327.16 meters per second and slide along a magnetic repulsive rail to deliver the momentum. The rail would be 10,910 meters in length. Perhaps carbonaceous super-materials can support a rail via sufficient resistance to bending moments yet still be fairly low in mass. Additionally, the pusher plate may optionally but not necessarily be electrically and/or magnetically coupled to the rest of the spacecraft but also have elastic metallic springs attached so that it is not dislodged from the craft.

Alternatively, the peak g-loading can be 49.05 $m/s^2$ and the pusher plate could slide along a rail for 6.67 seconds. The length of the rail this time will be only 1,091 meters.

Alternatively, the peak g-loading can be 490.5 $m/s^2$ and the pusher plate could slide along a rail for 0.667 seconds. The length of the rail this time will be only 109.1 meters.

Using shaped charging of the nuclear explosives assuming 0.0025 of the mass of the devices is converted to usable energy and half the blast energy is converted to spacecraft kinetic energy, I obtain terminal velocities which are a bit better than would be obtained using nuclear devices of the same invariant mass specific yield but which are carried along in equal mass quantities in the spacecraft instead of in runway pellet streams.

Specifically, for a spacecraft with a nuclear bomb based mass-ratio of 5 where the fuel mass is four times that of the final payload, the terminal velocity is 0.080273055 c and gamma is 1.003237537.

Additionally, the stress on the spacecraft is substantially reduced when the devices do not need to be carried along with the craft.

Another merit is that there is no need to shield the devices from otherwise intra-fuel supply irradiation which can become problematic when large numbers of devices are stored in a compact manner. The need to shield the devices will require rather large additional mass.

However, if it is desirable to carry a spacecraft mass-ratio to slow the spacecraft down by reverse bomb blast drive, then accelerating the spacecraft by pellet runway streams makes a lot of sense. Otherwise, the mass ratio will need to be squared in quantity.

## Class 3 Electron Gas Guns Enabling Fast Relativistic Spacecraft Velocities of 0.9 c.

***Herein we consider gas guns that may enable spacecraft velocities of about 0.9 c and Lorentz factors of about 2.2941. Such velocities can enable timely generation ship travel from star to star for alive and awake crew members and passengers.***

Next, we explore the really strong benefits are going to come for the 0.9 c case. Since about 90 percent of yield energy of the shaped charged nuclear explosives is in the soft x-ray to gamma ray range energy and about 85 percent of the yield energy can be directed toward pusher plates in a fairly narrow cone, what I have determined is that the mass of the pellets required to achieve 0.9 c is several orders of magnitude less than the invariant mass of the fuel bombs that would otherwise need to be carried along by the craft. Of course, this only applies to nuclear explosions in space. In the Earth's lower atmosphere, the initial x-ray and gamma ray flash is absorbed by the surrounding atmosphere and other objects in a manner such that about 40 to 45 percent of the yield is light and roughly the same is mechanical blast energy. The other 10 to 20 percent is prompt gamma ray radiation, neutron, fission fragment, and fusion products and residual radioactive fallout energy.

Assuming all focused beam energy can be intercepted by the craft, the only main efficiency losses that will incur will be due to non-ideal alignment of the photon flux with the velocity vector of the spacecraft due to relativistic aberration of the bomb light. I have lots of formulas I have already developed for these scenarios.

Relativistic red-shifting of the light source will be no more lossy than if a coherent single frequency source was used to push the craft forward.

The reason for this is that all of the light energy in the conical beam will be assumed to be intercepted by the spacecraft eventually. This is not the case when considering light-sails that are pushed from behind by ambient cosmic microwave background radiation which is black body in nature. In the case of the CMBR, the wave-lengths of the photons are increased in proportion to gamma so that not only do we have two dimensional reduction in the amount of photons impinging on the sail which would reduce power in ways that scale inversely with the square of gamma, but we also have an increase of time of arrival for a given number of photons of a given initial frequency that scales with gamma. Since the individual photons are also reduced in energy relative to the space craft in a manner that scales inversely with gamma, we have a reduction in drive power that scales with the inverse of the fourth power of gamma. Another way of looking at the CMBR issue is that the temperature of the CMBR from in back of the spacecraft reduces inversely with gamma. The power radiated from a black body source of constant surface area, emissivity of one, and temperature T, scales with the fourth power of temperature..

Now, we want an average acceleration of 4.505 $m/s^2$ in the spacecraft reference frame.

We also want the final Lorentz factor of the spacecraft to be 2.2941.

Now,

F=ma.

So, the force acting on a spacecraft having mass of 10,000,000 kilograms for which the acceleration is 4.505 $m/s^2$ in the spacecraft reference frame is 45,050,000 N.

Now, work = ∫F• dx. So we want the total work to be (1.2941)(10,000,000)[9 x $10^{16}$] J = 1.1646 x $10^{24}$ J.

Now, since F is constant in the spacecraft reference frame, we simply divide the work F to get the path length of required travel.

So, distance of travel = d = [1.1646 x $10^{24}$ J]/(45,050,000 N) = 2.58512 x $10^{16}$ m.

So, assuming that 0.0025 of the fusion fuel pellets is converted into energy and half of this energy is converted to spacecraft kinetic energy, the invariant mass of the pellet stream will be $\{[1.1646 \times 10^{24}](400)(2)/[9 \times 10^{16}]\}$ kg = $1.0352 \times 10^{10}$ kg.

Since each pellet has invariant mass of 40 kg, we will need $2.588 \times 10^{8}$ pellets.

The, average distance between the pellets will be $9.98887 \times 10^{7}$ m = 99,888.7 km.

Assuming gun muzzle velocity of 100,000 m/s, the average time between pellet launches would be 998.887 seconds.

Assuming an orbital location of 1 AU from the sun, and an adjustable reflector which reflects about 1/2 of the incident sunlight but transmits the rest, we obtain the following relation.

$(4 \times 10^{6}$ N$) = \langle S \rangle/c = \langle S \rangle/(3 \times 10^{8}$ m/s$)$. So, the required sunlight power to counteract the acceleration on the gun during the firing time would be $1.2 \times 10^{15}$ watts. By firing time, what is meant is the time interval that the round travels down the gun barrel.

Assuming a barrel system mass of 5,000,000 kg, the net barrel acceleration during the firing operation would be F/m = [(4 x $10^{6}$ N)/(5,000,000 kg)] = 0.8 m/s$^{2}$ and the displacement component of the barrel during the firing process would be $x = \frac{1}{2} \mathbf{a}t^2 = 0.4$ m.

Assuming a reverse acceleration time of 998.887 seconds, the barrel reverse acceleration need only be 0.00080099 m/s$^{2}$ and thus the required sun-light power on a sail to counteract the firing acceleration effects is $1.201337 \times 10^{12}$ watts. The latter value can be obtained by a light-sail having collection area of 882.686 km$^{2}$. This is a mere equivalent of 29.71 km by 29.71 km. For a sail having capture area of 882.686 km$^{2}$ and a mass specific area of one milligram per square meter, the total mass of the sail would be a mere 882.686 kilograms. For a sail having capture area of 882.686 km$^{2}$ and a mass specific area of 100 milligrams per square meter, the total mass of the sail would be a mere 88,268.6 kilograms or 88.2686 metric tons.

Now, assume that the spacecraft is pre-accelerated to 150 km/s as it passes the first pellet which we assume is traveling at 100 km second and that half of the pellet energy is converted to spacecraft kinetic energy. The energy of the pellet $(0.1)[9 \times 10^{16}]$ joules so $4.5 \times 10^{15}$ joules is transferred to the spacecraft.

So, KE = $\frac{1}{2}$ m v$^{2}$. Thus, the velocity increase of the spacecraft will be $\Delta v = [(2)(KE)/m]^{1/2}$ = 30 km/s.

So the spacecraft after passing by the first pellet will require either extreme acceleration handling ability or a long pusher plate mounted on a mechanical and/or field effect spring mechanism. A good way to provide a spring mechanism would be to include a magnetically and/or electrically suspended rod mechanism and a pusher plate that would be pushed along the rod against a magnetic bearing located on the stern of the main portion of the spacecraft.

Now, we can solve the following equation for launch time between last two pellets to be launched or the first to be intercepted by the ship. [4.505 m/s$^{2}$](t) = 30,000 m/s. Thus, t =

6,659.2 second. For a = 45.05 $m/s^2$, 450.5 $m/s^2$, 4, 505 $m/s^2$, the launch times are reduced by 0.1, 0.01, and 0.001 respectively. So, the distance between the last two pellets to be launched is ½ $at^2$ to yield 99,886.9 km, 9,988.69 km, 998.869 km, 99.8869 km, for accelerations of 4.505 $m/s^2$, 45.05 $m/s^2$, 450.5 $m/s^2$, and 4,505 $m/s^2$ respectively.

For the first two pellets to be launched, assuming acceleration of 4.505 $m/s^2$, the distance between the pellets in the background will be about (99,886.9 km) [(0.9 c)(2.2941)]/[(0.0001 c)(1)] = 2.06235 x $10^9$ km and the firing time between the pellets is 2.06235 x $10^7$ s. The spacecraft will experience the time of travel between the pellets in the spacecraft frame as 3,329.55.

For all specific cases considered, i.e., the 0.01 c, the two 0.1 c, and the 0.9 c cases considered, a much reduced mass of each pellet can enable smoother acceleration profiles

By reducing the pellet mass to 4 kg while keeping the invariant mass specific yields and fraction of pellet energy converted to spacecraft energy, we will need 10 times as many pellets. However, the acceleration of the spacecraft can be much more smooth without the need of such lengthy drive plates.

Reducing the pellet mass to 0.4 kg all else remaining the same, the pellet numbers increase by another factor of ten thus additionally smoothing out the acceleration profile and enabling still shorter drive plate motions and time intervals in the spacecraft reference frames.

By increasing the number of 40 kg projectiles anywhere from 10 to 1,000 fold, while reducing the efficiency of the kinetic energy acquisition 10 to 1,000 fold by the ship, respectively, the spacecraft acceleration can be greatly smoothed out and then so without the need for really lengthy acceleration towers.

In order to produce an average acceleration of 4.905 $m/s^2$, the pusher plate can be mounted on a spring system which slides along rails for about 66.7 seconds. The pusher plate would be accelerated to 327.16 meters per second and slide along a magnetic repulsive rail to deliver the momentum. The rail would be 10,910 meters in length. Perhaps carbonaceous super-materials can support a rail via sufficient resistance to bending moments yet still be fairly low in mass. Additionally, the pusher plate may optionally but not necessarily be electrically and/or magnetically coupled to the rest of the spacecraft but also have elastic metallic springs attached so that it is not dislodged from the craft.

Alternatively, the peak g-loading can be 49.05 $m/s^2$ and the pusher plate could slide along a rail for 6.67 seconds. The length of the rail this time will be only 1,091 meters.

Alternatively, the peak g-loading can be 490.5 $m/s^2$ and the pusher plate could slide along a rail for 0.667 seconds. The length of the rail this time will be only 109.1 meters.

For cases where the pellets are mass produced, it is conceivable that the cost per pellet might even be reduced to $1,000 at current value if the pellet manufacturing mechanisms can be completely automated by numerical artificial intelligence systems.

In general, the relativistic rocket equations are

$t = (c/a)\ \sinh(aT/c) = [[(d/c)\ ^2] + (2d/a)]\ ^{1/2}$
$d = [(c\ ^2)/a]\ [[\cosh(aT/c)] - 1] = [(c\ ^2)/a]\ \{\{[1 + [(at/c)\ ^2]]\ ^{1/2}\} - 1\}$
$v = c\ \tanh(aT/c) = (at)\ /\ \{[1 + [(at/c)\ ^2]]\ ^{1/2}\}$
$T = (c/a)$ inversesinh $(at/c) = (c/a)$ inversecosh $[[ad/(c\ ^2)] + 1]$
$\gamma = \cosh(aT/c) = [1 + [(at/c)\ ^2]]\ ^{1/2} = [ad/(c\ ^2)] + 1$

$\Delta v = C \tanh [(Isp/C) \ln (M0/M1)]$.

Now, assuming the original 0.9 c example where the mass of the pellets is included instead as a relativistic rocket fuel configuration, we obtain a mass ratio very close to 1,000. The following table provides values for mass ratio, fractional velocity of light, and gamma for an implied specific impulse of $\{[[(2)(0.0025)(0.5)] - [[(0.0025)(0.5)]^2]]^{1/2}\}$ c = 0.049984 c from the formula Isp = $\{[(2n) - (n^2)]^{1/2}\}$. Here, n is the effective fraction of fuel mass converted to ideal thrust energy.

| Mass Ratio | B = v/C | Gamma = $\{1-[(v/C)^2]\}^{-1/2}$ |
|---|---|---|
| 10 | 0.11458691 | 1.006630446 |
| 100 | 0.226203729 | 1.02660971 |
| 1000 | 0.332180524 | 1.060202734 |
| 10000 | 0.430385441 | 1.107854992 |
| 100000 | 0.519359346 | 1.170198395 |
| 1000000 | 0.59833809 | 1.248059674 |
| 10000000 | 0.667181868 | 1.342471336 |
| 100000000 | 0.726246973 | 1.454685366 |
| 1000000000 | 0.776236709 | 1.586189822 |

TABLE 2: Mass ratio, B, and gamma for Isp = 0.049984 c

As you can see, with a mass ratio of 1,000, we only obtain about ½ c and a trivial Lorentz factor. Even using a mass-ratio 1,000 times greater yet, we are still only about at 2/3 c and a not very useful Lorentz factor. So, even using 40 kilogram devices for which only 0.0005 of the yield energy is converted to spacecraft energy is a much better deal, especially when considering flights to stars within 100 years of Earth for which the crew remains awake during the journey.

In baseline cases considered, the acceleration may be increased 100 to even 1,000 fold thus reducing the length of the runway to a more manageable value as well as pellet lay-down time. The loading on the guns would be more extreme. None-the-less, solar sail mechanisms can be used to maintain position of the guns.

Additionally, more than one gun may be used to reduce loading on given gun.

In order to account for pellet need to overcome the solar gravity well at 1 AU, the actual pellet muzzle velocities may be increased by about 30 km/second to about 130 km/second. In enabling such is simply done by building modestly longer guns.

For pellets of between one and two orders of magnitude lower in mass than in the four baseline scenarios, the muzzle energy and power requirements of each gun can be reduced proportionally to pellet mass.

Pellet muzzle velocities can also be increased with lengthening of the gun barrels.

Note that pellets of 40 kilograms were chosen in the base-line scenarios because these are likely practical to manufacture and have mass specific yields that are within range of engineering abilities of contemporary nuclear weapons manufacturing processes. However, pellets of the same mass or of lower masses and mass fraction conversion to energy may be chosen.

For example, pellets having a mass of 40 kg but having yields of 225 kilotons, 22.5 kilotons, and 2.25 kilotons included in ten-fold, one-hundred-fold, and one-thousand-fold numbers relative to that of the base-line scenarios can enable the same terminal spacecraft velocities. Pellets having a mass of 4 kg but having yields of, 22.5 kilotons and 2.25 kilotons included in one-hundred-fold, and one-thousand-fold numbers relative to that of the base-line scenarios can enable the same terminal spacecraft velocities

Note that for highly relativistic velocities of the spacecraft relative to the pellets, the force of the pellet exerted on the pusher plate as the bulk of the energy blast-wave arrives can be greatly reduced as a shock. However, it would take a spacecraft Lorentz factor of about one billion to cause the blast wave to spend one second of incidence on the spacecraft assuming all blast wave energy arriving is black body electromagnetic emissions and that all focused photonic radiation impinged on the plate. Here, we assume that most of the blast energy is electromagnetic and that the bomb detonation process last about one nanosecond in the bomb reference frame. In reality, the process of detonation takes a little longer than one nanosecond but the peak detonation power lasts about one nanosecond.

The fuel pellets would be distributed along runways once each ship reached 0.9 c. The ship without the pellets would use the pellet stream for significant Lorentz factor gains relative to that associated with 0.9 c. Again, the invariant mass of the pellets would be 1.0352 x 1010 kg while the ship without pellets would have invariant mass of 10,000,000 kilograms. The pellet carrying ships would have an invariant mass of about 10,000,000 kilograms with much of the mass being pellets. The pellet dispensing gun and structural mass of the ship combined may be a small fraction of that of the four base-line scenarios considered previously.

The Lorentz factor of the ship without pellets can thus be increased to 3.5882 if just one of the pellet carrying ships dispensed its pellets. Here we assume shaped charging of the pellets.

If, say, ten ships would discharge their pellet loads, then the Lorentz factor of the pellet carrying ship could increase to 15.2351.

If, say, 100 ships would discharge their pellet loads, then the Lorentz factor of the pellet carrying ship could increase to 131.7041.

If, say, 1,000 ships would discharge their pellet loads, then the Lorentz factor of the pellet carrying ship could increase to 1,296.39.

In reality, the spacecraft Lorentz factor is only limited by the number of pellets that can be distributed. Obviously, the refractory capabilities of spacecraft shielding, collision avoidance mechanisms, and astrodynamics drag issues need to be addressed.

Here, we assume that almost all of the bomb blast energies are fully intercepted photon streams.

So, we can actually obtain matter-antimatter rocket like Lorentz factors arbitrarily large with applications of pellet runways.

Regarding methods of cooling the pusher plates, a number of mechanisms have been considered from spraying a carbonaceous residue on the plate between blasts, to magnetic and/or electric-field based shielding from bomb plasma, to cooling effluents that power turbo-electric generator to energize electrical propulsion systems, etc.

The carbonaceous residue will immediately ionize and carry phase change energy away from the surface of the pusher plate thus enabling the operation of the system without undue plate surface ablation.

Regarding shaped charging of pellets, the original Project Orion suggested systems that focus 85 percent of the blast energy within a 22.5 degree cone.

The extent of explosive flux concentration would be equal to the [[(Area of sphere)/(Area of 22.5 degree circular sector of a sphere)](0.85)] = $[(2)(90^2)/(11.25^2)](0.85)$ = 108.800

For two stage shaped charge explosive compression, the extent of explosive flux concentration would be equal to the {[[(Area of sphere)/(Area of 22.5 degree circular sector of a sphere)](0.85)]$^2$} = 11,837.44. The fraction of bomb energy compressed is $0.85^2 = 0.7225$.

For three stage shaped charge explosive compression, the extent of explosive flux concentration would be equal to the {[[(Area of sphere)/(Area of 22.5 degree circular sector of a sphere)](0.85)]$^3$} = 1,287,913.472. The fraction of bomb energy compressed is $0.85^3$ = 0.614125.

For six stage shaped charge explosive compression, the extent of explosive flux concentration would be equal to the {[[(Area of sphere)/(Area of 22.5 degree circular sector of a sphere)](0.85)]$^6$} = $1.658721111 \times 10^{12}$. The fraction of bomb energy compressed is $0.85^6$ = 0.3771495.

The above formulas for explosive compression are merely crude first order estimates.

Here we consider extremely hot or hard Ultra-violet Dark stars composed of high atomic number fissile isotopes, high atomic number radioactive non-fissile isotopes, pre-transuranic radioactive isotopes, medium mass radioactive isotopes, low mass radioactive isotopes, fusion fuels and then like. Such stars can in principle be assembled and used as light-sources for light-sail and beam sail ships.

The high atomic number isotopes would be chosen to yield radioactive isotopes which can be induced to fission on rare occasions.

Meanwhile the overall fuel blends would be chosen to produce radioactive nuclei near Iron-56. What is possible with stars may be possible with reactors and nuclear explosives.

The near Iron-56 nuclei would be planned to undergo radioactive decay to lower atomic number nuclei all the way down in mass to the lightest elements such as in radio-active isotopes.

The pre-ferrous elemental isotopes would then undergo nuclear fusion to yield more energy up to a stable state of iron-56.

So, the total energy released where all nuclear fuels spent would be

$[(2)(0.01054273) + (0.00092378)]\ mc^2 = 0.02200924\ mc^2$

The specific impulse is

$I_{sp} = C\ \{[(2n) - (n^2)]^{1/2}\}$

$= C\ \{\{[2\ (0.02200924)] - [0.02200924^2]\}^{1/2}\} = 0.208648\ C.$

The following table provides spacecraft performance data.

| Mass Ratio | B = v/C | Gamma = $\{1-[(v/C)^2]\}^{-1/2}$ |
|---|---|---|
| 10 | 0.744660026 | 1.498253116 |
| 100 | 0.958058717 | 3.489524798 |
| 1000 | 0.993749789 | 8.958129688 |
| 10000 | 0.999082803 | 23.35356664 |
| 100000 | 0.999865711 | 61.02097827 |
| 1000000 | 0.999980345 | 159.496175 |
| 10000000 | 0.999997123 | 416.9103041 |
| 100000000 | 0.999999579 | 1089.778149 |
| 1000000000 | 0.999999938 | 2848.616714 |

TABLE 3 Mass-ratio, B = v/c, and gamma for specific impulse of 0.208648 c.

Now, it is plausible that super-neutronated isotopes for atomic number elements below the transuranics might have sufficient energy whereupon decay down to low atomic number elements such as species below the heavy metals sets might yield extra energy upon decay.

In order to imply a net energy release in going to hydrogen and helium isotopes back to Iron-56, the hyper-neutronated isotopes would imply high end energy storage levels twice that of the pre-heavy element fusion fuel.

So, the actual yield of the fuel would be

$[[(2)(0.01054273)] + [(2)(0.01054273)] + (0.00092378)]]\ mc^2 = 0.0430947\ mc^2$
The specific impulse is

$I_{sp} = C\{[(2n) - (n^2)]^{1/2}\}$

$= C\{\{[2\ (0.0430947)] - [(0.0430947)^2]\}^{1/2}\} = 0.2904\ C.$

The following table provides spacecraft performance data.

| Mass Ratio | B = v/C | Gamma = $\{1-[(v/C)^2]\}^{-1/2}$ |
|---|---|---|
| 10 | 0.871032208282 | 2.035723248 |
| 100 | 0.990542606100 | 7.288338288 |
| 1000 | 0.999345231480 | 27.63835614 |
| 10000 | 0.999954853905 | 105.23975 |
| 100000 | 0.999996888074 | 400.8396554 |
| 1000000 | 0.999999785499 | 1526.75746 |
| 10000000 | 0.999999985215 | 5815.271695 |
| 100000000 | 0.999999998981 | 22149.81072 |
| 1000000000 | 0.999999999930 | 84366.42463 |

TABLE 4 Mass-ratio, B = v/c, and gamma for specific impulse of 0.2904 c.

The above results would ordinarily seem contradictory unless conservation of baryon number can be violated and a sizable but modest fraction of protons and/or neutrons can decay into non-hadronic or leptonic species.

The mass faction of the population of baryons leptonically decaying would be at least about (0.0430947/0.01054273) = .040876 o.r 4.0876 percent.

The Facility For Rare Isotope Beams may enable discovery of such reactions per collision of atomic nuclei of rare isotopes.

Discovery of large scale violation of conservation of baryon number reaction chains may open up opportunities for matter-antimatter reaction levels of mass specific energy production.

Accordingly, the associated nuclear fuels would have specific impulse values approaching 1 c.

It is plausible that fissile initiators in combination with exotic isotopes just at the edge or beyond the level of current producibility might enable the mass of bomblets to be as low as on the order of 1 gram.

Such bomblets may be designed to be carried aboard a spacecraft as a reaction material while pre-deployed bomblets would be directed in collisions with the gradually dispensed pellets carried aboard the spacecraft.

The effective specific impulse of the reaction energy where all pellets have equal invariant mass would be as much as about 2 c.

In cases where the background pellets are of k times the invariant mass of the ship board pellets but where almost all of the interaction energy is converted to spacecraft kinetic energy, the specific impulse of the fuel effectively becomes [(k)(c)] + c where k is greater than one.

For a specific impulse of 2 c fuel and large mass ratios, gamma = $(0.5)[(M_o/M_1)^2]$

For a specific impulse of 3 c fuel and large mass ratios, gamma = $(0.5)[(M_o/M_1)^3]$

For a specific impulse of 4 c fuel and large mass ratios, gamma = $(0.5)[(M_o/M_1)^4]$

In general, for a specific impulse of gc fuel and large mass ratios, gamma = $(0.5)[(M_o/M_1)$ EXP g]

The velocity would be:

$\Delta v = \tanh [(Isp/c) \ln (M_o/M_1)] = \tanh [(Igc/c) \ln (M_o/M_1)]$

or

$\Delta v = c - \{(2) [c \text{ EXP} - [(2)[\log [(M_o/M_1) \text{ EXP } g]]]]\}$

Note that the nuclear fuel pellets may optionally be standard hydrogen and helium isotopes, or a combination of fissile fuels and isotopes or simply fissile fuel that is compressed when rammed by pellets dispensed by the ship upon attaining at least mildly relativistic velocities. As such, the spacecraft can still attain effective specific impulse values many times that of c. From a practical standpoint, the carried aboard pellets may themselves be nuclear fission and/or fusion fuel based or simply inert dense pellets. In such cases, the specific impulse of the onboard craft fuel can grow to many orders of magnitude greater than c.

## The Accelerator & Pellet Fusion Reactor Rocket.

***This section includes a presentation of pellet collision induced nuclear fusion where the pellets are accelerated by mass-drive mechanisms. The kinetic energy of collision is the heat and pressure generating mechanism commensurate with nuclear fusion activation energies.***

An interesting nuclear fusion rocket concept would involve the firing of fusion fuel pellets at another stationary fusion fuel pellet. Provided the kinetic energy of the pellets was sufficiently high, the pellets may undergo temperature and pressure based nuclear fusion, even in cases for which the kinetic energy of the pellets just before collision was much lower than the reaction yield, perhaps even by a couple to a few orders of magnitude.

The pellets would be accelerated along an electrical potential or magnetic potential. For electric acceleration, the pellet need merely be electrically charged. For magnetic acceleration, the pellet would need to be effectively magnetized.

Pellet magnetization can include enclosing the pellet in a magnetic material or perhaps in iron or extremely high-strength steel cladding.

One or more layers of a dense tamper may be installed within the pellets to provide enhanced collision pressures. The tampers can be arranged in a plate like orientation along the direction of propagation and/or parallel to the direction of propagation.

Another temperature enhancing feature would include a shaped collision or kinetic energy charge mechanism that would work in a manner similar to a high-explosive shaped charge munition commonly used in antiarmor projectiles.

Another configuration would have two or more pellets simultaneously in collision. Two pellet collisions would be head on, three pellet collisions in a triangular incident, four pellet collisions in quadrangular incidence, etc.

Alternatively, the pellet collisions may be serially stacked in space and time so as to back-reflect shockwaves produced by the pellet collisions. The stacking may be substantially direct and non-staggered, or overlapping and staggered. Series-parallel collisions may also be used.

Pellets may optionally be attached to more extensive ferromagnetic webbing.

For example, the webbing may consist of a super-high strength steel foil sheet. The monolithic sheet would be pulled along by the magnetic and/or electric potential and would sweep out a much greater volume of field that would the pellet by itself. Since the sheet would be very thin, it is plausible that the invariant mass of the sheet could be substantially less than that of the fuel pellet.

The sheet may optionally consist of pure steel, or perhaps preferably, a sheet of graphene laminate, carbon nanotube fabric, boron nitride nanotube fabric, diamond fiber based fabric, carbon nitride based fabric, or any other extreme material(s) with the option of including an iron, steel, or other ferromagnetic material coating.

Alternatively, the sheet may be grid-like but have the same materials of construction as the above conjectured monolithic sheet. Again, metalized surfaces and/or interiors may be applied to any carbonaceous structural super-materials used in the grid lines.

Another option would include a coulombic magnetic string fan. Here, self-charging of a fan-like feature would result in the strings extended outward away from the fuel pellet by interstring and intrastring repulsion so that a much larger volume of electric and/or magnetic flux could be swept out.`

Yet another option would include the spaghetti magnetic flux catcher. Here, uncharged ferromagnetic fibers would be pulled forward. The fibers would be attached to the fusion fuel pellet and would thus pull the pellet forward.

Both the string flux capture mechanisms can include the same structural materials used for the monolithic and gridded sheet mechanisms as well as the same metallization materials used therein.

Assuming a fusion fuel pellet of one gram, a flux capture interface with the pellet of one square millimeter, a total flux catcher cross-sectional of one square centimeter, and a catcher length of one meter, very high mass-specific kinetic energies are possible for the system.

For example, a one micron thick catchment membrane having an area of one square meter and a length of one meter may withstand an acceleration of about 10,000,000 gs or $9.81 \times 10^7$ $m/s^2$. This result is based on the verified assumption that the strongest laboratory grade carbonaceous materials have a tensile strength of $10^7$ newtons per square centimeter.

Now F = m x a, so the force acting on the pellet would be as high as [0.001 kg] [$9.81 \times 10^7$ $m/s^2$] = 9,810 newtons. For an acceleration pathway of 100 meters, the pellet kinetic energy would be equal to 981,000 joules. Since one gram is equivalent to $9 \times 10^{13}$ Joules, the associated pellet velocity can be simply calculated from the Newtonian formula, K.E. = (1/2) $mv^2$.

Thus;

$v = [(2\ K.E.)/m]^{1/2}$ = 44.294 km/second.

This value is not quite high enough to generate fusion unless perhaps suitably dense tampers can be included within the pellets.

A 1,000 meter long acceleration pathway will produce 9,810,000 joules and a velocity with is greater by a factor of $10^{1/2}$ to yield a velocity of 140.069 km second. This might actually work if the actual analogous inertial implosion research being conducted can yield micro-pulse fusion.

Still, we would like to get the velocity a little higher. To do so we will restrict the extension of the flux catcher to 0.01 meters and reduce the pellet mass to 0.1 gram.

So the force acting on the pellet would be as high as [0.0001 kg] [$9.81 \times 10^8$ $m/s^2$] = 9,810 newtons. For an acceleration pathway of 100 meters, the pellet kinetic energy would be equal to 981,000 joules. Since 0.1 gram is equivalent to $9 \times 10^{12}$ Joules, the associated pellet velocity can be simply calculated from the Newtonian formula, K.E. = (1/2) $mv^2$.

Thus;

$v = [(2\ K.E.)/m]^{1/2}$ = 140.069 km/second.

A 1,000 meter long pathway will yield 442.94 km second. This should do the fusion trick.

Firing two pellets into each other from opposite directions will yield a relative velocity of 885.88 km/s = 0.002953 C ≈ 0.003 C.

Two pellets as such will provide reaction energy equal to about 30 tons of TNT. Ten pellet implosions per second will provide a yield of 0.3 kiloton per second which is on the rough order of that anticipated by the former Project Orion.

The pellets may be fired down a long mass-driver into a stationary target, or in a head on collision originating from two inline mass drivers. Alternatively the mass drivers may be obliquely oriented with respect to one another.

The location of the collision can be within a heat absorbing reactor for purposes of energizing turbo-electric generators, or thermo-electric generators, each of which may optionally have multi-cycle features to process w heat left over from the previous cycle. The generated electrical power may then be used to power ion rockets, electron rockets, proton rockets, photon rockets, electro-hydrodynamic-plasma-drives, magneto-hydrodynamic plasma drives, electromagneto-hydrodynamic-plasma-drives, magnetic sails, magnetic field effect propulsion systems and the like.

Alternatively, the fusion energy may be exhausted directly to space. Even highly neutronic reaction products can provide direct propulsion provided that a momentum absorbing mass is utilized to capture the neutrons. Thermal energy generated and be applied as a supplemental power source via turbo-electric and/or thermo-electric generators.

Fusion reactions which produce charged particles are ideal because the charged particles can be readily back-reflected by electric and/or magnetic mirrors.

Various fusion fuels may be used including hydrogen-1, deuterium (hydrogen-2), tritium (hydrogen-3), helium-3, helium-4, lithium-deuteride, boron, and the like. Any combination of these fuels may be used.

Alternatively, fission pellets may be used. Here, a near instantaneous realization of a super-critical mass would result in a propagation of fissions to occur throughout the pellets.

Alternatively, two stage or fusion-fission or fission-fusion pellets may be employed.

Pure fusion pellets having a invariant mass of 1 gram would yield 0.157.5 kilotons or 157.5 tons of TNT equivalent. Two such colliding pellets would yield 0.315 kilotons or 315 tons of TNT.

Fissile fuels may include Uranium-233, Uranium-235, Plutonium-239, Plutonium-241, Neptunium-237, Americium-241, Americium-242, Americium-243, Berkelium-247, Berkelium-249, and others.

The dense cladding or tampers of a fusion fuel pellet can include any of the above fissile materials. Uranium-238 is an ideal cladding because it can be made to undergo fission when stimulated by the neutrons produced in some fusion reactions. The Ulam Teller nuclear bomb design relied on this mechanism to achieve a staggering yield of about 25 megatons. The device was jacketing in a layer of U-238.

Note that all of the pellet configurations depicted below or described in this section can include analogs for which pellets carried aboard the spacecraft may collide with background pellets in either the lab or center-of-mass reference frame with respect to the spacecraft.

Coil guns are an option for mass-drivers. Coil guns involve electromagnet type coils which are switch on ahead of a projectile. The coils are then turned off once the projectile reaches the mid-point of a coil. Most coil guns have several stages.

There is a limit to the practical levels of acceleration enabled by coil guns. The limit arises when the ferromagnetic object being accelerated down the gun becomes magnetically saturated. Normally, the force applied to a ferromagnetic object in the core of a coil gun is proportional to the square of the current conducted through the coil. Upon reaching saturation, the force drops to linear dependence on current.

Using designs of greater expense and sophistication enables much higher efficiency and kinetic energy. For example, a 2 gram ring was accelerated to 5,000 m/s over an interval of 1 centimeter in the USSR by Bondaletov in 1978.

Coil guns are a subject of active research by the U.S. military.

Given the efficiency drop-off of coil guns upon reaching ferromagnetic saturation, it is tempting to question whether they would be useful for accelerating iron or steel containing pellet. The answer is likely highly affirmative. The reason this is so is such that for some fusion fuels, the fusion activation energy is a few orders of magnitude less than the reaction yield. Add extreme pressures to the scenario and the activation energy may be further reduced. As long as at least a small portion of a given pellet could fuse, perhaps the activation energy may be reduced to several orders of magnitude below the yield energy, the caveat being the propagation of the fusion wave-front in a chain reaction via heat and pressure propagation from the fused portion of a pellet to adjacent portions.

## Assembling Nuclear Explosions From Supplies Inroute.

Another mechanism would include a spacecraft carrying numerous very low yields but efficient nuclear devices which would be used as primaries to ignite inert nuclear fusion fuel pellets pellets distributed in the background.

This way, the yield energy of the carried aboard pellets may be used to trigger nuclear fusion energy yields several to many time greater than that of a pellet on a pellet by pellet bases.

Some nuclear fuels can include Lithium-6 Deuterium mixtures, deuterium, protium, as well as helium-3 and helium-4 fuels.

Once again, the reactions can occur behind pusher plates optimally designed to absorb hard ionizing radiation. The thermal masses thus heated can source turbo-electric systems to generate electrical power for energizing secondary propulsion systems.

## Ship Chambering Of Nuclear Explosive Pellets.

One important aspect to many configurations of the proposed systems considered herein is ship chambering of the nuclear explosives. After all, it is important to consider methods for enabling the energy extraction of the explosives and conversion to spacecraft kinetic energy in a safe and sustainable manner.

Nuclear pellets pre-distributed whether of self-assembly types or initially completed devices can also be of use as heating mechanism by which a very large spacecraft having a thermal-mass-lined interior cylinder would intake a pellet without capturing it nor colliding with the pellet.

The pellets would be serially detonated at appropriate intervals to heat the thermal mass which would then sink heat to turbo-electric generators which in turn would provide power to propulsion systems.

Perhaps the best types of nuclear devices for this purpose are ones that would have a large fraction of energy release as fast neutrons. Essentially, we are considering neutron bombs for this purpose.

Neutron bombs are preferable in some respects because the neutrons would gradually release energy over a thick layer of thermal mass. Thus, as long as the yield of the devices is low, the risk of melting or vaporizing portions of the thermal mass are greatly mitigated.

Pure gamma ray emitted nuclear explosives are also similarly of use with the caveat than gamma ray nuclear bombs can be designed for which almost all emissions in space are nuclear reaction sourced gamma rays.

Regardless of the specific nuclear energy release mechanisms, all of these exotic concepts are testable with current nuclear explosives technologies.

# Looking Ahead And Beyond Nuclear Bomb Pellets.

The ΛCDM model suggests that about 69.2%±1.2% of the mass and energy in the universe is a cosmological constant. Extensions to the ΛCDM model include other forms of dark energy, such as a scalar field. Dark matter comprises about 25.8%±1.1% and baryonic matter comprises about 4.84%±0.1% of the physical universe. Stars, planets, and visible gas clouds only comprise about 6% of the baryonic matter.

So, given that baryonic matter comprises only about 4.84%±0.1% of mass-energy content of our universe, while dark matter comprises 25.8%±1.1%, it is plausible that there are by symmetry, 25.8/4.84, types of dark matter in our universe or roughly 5 distinct types of dark matter.

Since stars, planets, and visible gas clouds only comprise about 6% of the baryonic matter, we may estimate by symmetry that about 6 percent of each conjectured type of dark matter similarly makes up dark matter analogs of stars, planets, and visible gas clouds.

Since dark energy makes up 69.2%±1.2% of the mass and energy content of the universe, it is plausible that there could be about two types of dark energy. This estimate is computed as 100/69 ~ 2.

Dark matter analogs of stars, planets, and visible gas of baryonic forms may in some sense be said to lie in different dimensions commensurate with the weakly interacting nature of these dark matters. As such, these dark matters would only interact with normal matter by gravitation or be gravitation and the weak force.

So, in considering prospects for extraterrestrial and so-called ultra-terrestrial intelligent technologically advanced personal life-forms, it behooves us to conjecture that these life-forms may have bodies comprised of dark energy and live on and around dark matter planets and stars.

Another prospect for the composition of dark matter includes a 4 or more spatial dimensioned manifold that encompasses galaxies, galaxy clusters, and galaxy super-clusters. As such the manifold units would envelope galaxies, galaxy clusters, and galaxy super-clusters in such a way that the manifolds would act to cause ordinary baryonic matter to condense into portions the size of galaxies, galaxy clusters, and galaxy super-clusters.

The size of the manifold enclosures would be about equal to the size of the galaxies, galaxy clusters, and galaxy super-clusters in ordinary 4-D space-times. Thus, the manifolds would make the motions of galaxies, galaxy clusters, and galaxy super-clusters appear the same or similar to the ordinary 4-D space-time cold dark matter theories.

The hyper-spatial manifolds would act to gravitationally segregate matter by higher dimensional gravitation or perhaps ironically, non-gravitational dark energy. Such long range force fields would extend into ordinary 4-D space-time so as to manifest behavior instead attributed to cold dark matter located in ordinary 4-D space-time.

Then there is a hybrid model for which the encompassing hyperspatial manifolds initially acted as gravitational seeds to start the local collections of dark matter and baryonic matter. Once the mass concentrations reached sufficient densities, not only would the matter pockets draw in more matter, but may also act to pull on the encompassing manifolds enable the manifolds to grow more in both hyperspatial density and in the potential energy they would provide in 4-D space-time for enhancing the concentration of cold dark matter and baryonic matter.

Another mechanism which may or may not be operable with the gravitational cold dark matter scenario may involve other long range forces acting on the cold dark matter and sources from hyperspatial manifolds by way of field-effect hyper-spacetime emissions or mass and energy in hyperspace leaking into ordinary 4-D space-time to produce compression effects on the dark matter. Once the dark matter became sufficiently concentrated, the dark matter self-gravitation would take over most of the compression to arrive at the currently observed mass distributions in galaxies and galactic clusters.

We need to also consider the potential for developing mechanisms to collect and exothermically process cold dark matter (CDM). Cold dark matter makes up about 85 percent of the massive component of the universe. The remainder of the massive component exists in the form of ordinary baryonic matter in the form of protons, neutrons, and electrons, most of which exist in the form of interstellar and intergalactic gas and dust. The gas and dust is mostly composed of hydrogen and helium. The rest of the mass-energy component of the universe appears to take the form of dark energy. Dark energy is a theoretical cosmological constant-like property of space that appears to be causing an acceleration of the expansion of the universe. Cold dark matter is about 6 times as plentiful as normal baryonic matter. As a result, there remains the possibility that 6 times more power can be extracted from the interstellar and intergalactic medium for a given ISR velocity relative to the case where some sort of proton fission reaction as conjectured about above would be utilized.

The rate of acceleration of an ISR for perfectly efficiently utilized interstellar and intergalactic fuel is proportional to the square root of the mass specific potential energy content of the fuel. Consequently, perfectly utilized or almost perfectly utilized cold dark matter fuel would enable accelerations as high as [(140)(6)]1/2 times that for fusion fuel alone or accelerations as great as 29 Gs or more. Accelerations as high as [(140)(6 + 1)]1/2 = [(140)(7)]1/2 = 31.3 Gs or greater should be obtainable when also including any hydrogen fission-based energy production.

The reason for my assuming the apparent "super-relativistic" energy content of any CDM-based ISR fuel is that CDM will not impose a drag on the starship, except for perhaps a CDM processing apparatus which will necessarily thermodynamically couple to the natural CDM. So the energy derived from CDM may effectively almost appear as if it came from nowhere.

Finding other reactions that liberate more energy than pure conversion of mass into energy would permit even higher accelerations, perhaps limited only to the square root of the mass specific yield of the fuel.

Since each of the above exotic ISR notions can be mathematically rationalized and can be cast in a lexicographically plausible matter, we do indeed need to remain open to the "old hat" interstellar drive concepts in spirit.

The above conjectured interstellar ramjet systems may overcome the drag issues that were initially proposed as "no go" conditions for interstellar ramjets. The issues were related to the notion that the ISRs could not be accelerated to velocities sufficient to enable collection of enough fuel for gainful acceleration. So, one solution to the drag conundrum is perhaps the use of nuclear bomb pellets to enable an ISR craft to reach critical velocities and associated Lorentz factors where the nuclear fusion ISR modes could be engaged.

Additionally, including dark matter reactions as presented above provides for potential Lorentz factors that may prove non-finitely boundable given a cosmic era of technological development.

As a researcher in the field of relativistic astronautics, I believe the dark skies of interstellar space will provide perpetual opportunities for achieving ever greater Lorentz factors.

All this being said, it is plausible that there could be dark matter planets and/or stars nearby. These planets and/or stars may serve as effective portals to other spatial dimensions even if said other spatial dimensions are merely effectively thermodynamically so and not actually topological in nature. However, if cold dark matter or what appears to be cold dark matter is manifestation of higher dimensional manifolds, then one or more of these conjectured nearby locations may serve as gateways to other topological dimensions. If such is the case, then perhaps chemical rockets can serve as transport mechanisms for probes and human crewed missions.

NASA's Artemis 1,2 and 3 missions to send humans back to the Moon are good precursor missions for rocket flight further out into the Solar System and beyond.

Ideally, nuclear powered spacecraft will provide extreme Lorentz factor travel of humans between the stars and galaxies and beyond. However, while we work to obtain politically viable solutions for high powered nuclear energy enabled space-flight, we can do a lot with chemical rockets in the mean-time. All this being said, it is plausible that there could be dark matter planets and/or stars nearby. These planets and/or stars may serve as effective portals to other spatial dimensions even if said other spatial dimensions are merely effectively thermodynamically so and not actually topological in nature. However, if cold dark matter or what appears to be cold dark matter is manifestation of higher dimensional manifolds, then one or more of these conjectured nearby locations may serve as gateways to other topological dimensions. If such is the case, then perhaps chemical rockets can serve as transport mechanisms for probes and human crewed missions.

NASA's Artemis 1,2 and 3 missions to send humans back to the Moon are good precursor missions for rocket flight further out into the Solar System and beyond.

Regarding the manifold conjecture, we have the following choices of space-times.

Ordinary 4-D space-time, in N-D-Space-M-D-Time scales where N is any integer greater than two and M is any counting number, where N is any integer greater that two and M is any positive rational number, where N is any integer greater that two and M is any positive irrational number, where N is any integer greater that two and M is any positive real number, where N is any positive rational number and M is any counting number, where N is any positive rational number and M is any positive rational number, where N is any positive rational number and M is any positive irrational number, where N any positive rational number and M is any positive real number, where N is any positive irrational number and M is any counting number, where N is any positive irrational number and M is any positive rational number, where N is any positive irrational number and M is any positive irrational number, where N is any positive irrational number and M is any positive real number, where N is any positive real number and M is any counting number, where N is any positive real number and M where N is any positive real number, where N is any positive real number and M is any positive irrational number, where N is any positive real number and M is any positive real number: for which the space-time is flat, positively curved, negatively curved, positively curved and torsioned at one or more scales in arbitrary patterns including but not limited to fractals, negatively curved and torsioned at one of more scales in arbitrary patterns including but not limited to fractals, positively super-curved, negatively super-curved, positively super-curved and torsioned at one or more scales in arbitrary patterns including but not limited to fractals, negatively super-curved and torsioned at one of more scales in arbitrary patterns including but not limited to fractals, positively super-super-curved, negatively super-super-curved, positively super-super-curved and torsioned at one or more scales in arbitrary patterns including but not limited to fractals, negatively super-super-curved and torsioned at one of more scales in arbitrary patterns including but not limited to fractals, positively super-super-super-curved, negatively super-super-super-curved, positively super-super-super-curved and torsioned at one or more scales in arbitrary patterns including but not limited to fractals, negatively super-super-super-curved and torsioned at one of more scales in arbitrary patterns including but not limited to fractals, positively super-super- ... -super-curved, negatively super-super- ... -super-curved, positively super-super- ... -super-curved and torsioned at one or more scales in arbitrary patterns including but not limited to fractals, negatively super-super- ... -super-curved and torsioned at one of more scales in arbitrary patterns including but not limited to fractals.

The above list is pretty exhaustive as a first order consideration. However, more likely the number of time dimensions in such hyperspatial manifolds would be 1 or at most 2, and the number of spatial dimensions would most likely be 4 or 5, perhaps as many as 6.

Also note that the iterated indices of "super" merely are a stand-in for very levels of so extremely curved space-times such that these differing indices indicate qualitatively and topologically distinct space-time species.

More than likely, such extreme space-times would operate on 4-D spacetime in the very early instants of the big bang and then to such an extent that large scale structures would manifest within a few million to few hundred million years after the big bang.

Ideally, nuclear powered spacecraft will provide extreme Lorentz factor travel of humans between the stars and galaxies and beyond. However, while we work to obtain politically viable solutions for high powered nuclear energy enabled space-flight, we can do a lot with chemical rockets in the mean-time.

The above conjectured interstellar ramjet systems may overcome the drag issues that were initially proposed as "no go" conditions for interstellar ramjets. The issues were related to the notion that the ISRs could not be accelerated to velocities sufficient to enable collection of enough fuel for gainful acceleration. So, one solution to the drag conundrum is perhaps the use of nuclear bomb pellets to enable an ISR craft to reach critical velocities and associated Lorentz factors where the nuclear fusion ISR modes could be engaged.

Additionally, including dark matter reactions as presented above provides for potential Lorentz factors that may prove non-finitely boundable given a cosmic era of technological development.

As a researcher in the field of relativistic astronautics, I believe the dark skies of interstellar space will provide perpetual opportunities for achieving ever greater Lorentz factors. To get started with large Lorentz factors, I believe the nuclear explosive pellet runway system disclosed herein is one of the best options we have for doing so.

A pioneering physicist and on who did much research on nuclear weapons allegedly believed that if gravity would be completely suspended, then it would take about 10 EXP 39 times the mass energy content of the universe to squeeze Earth down to a level at which it would be as dense as a black hole or about one centimeter in diameter. (Source not located but recalled from author's memory).

So, this brings us to a fascinating conjecture!

What if an Earth massed and sized body could be gradually compressed in a portion of space-time for which gravitation was greatly reduced? Alternatively, what if an Earth size body can be compressed at the same time gravitational screening materials or antigravatic materials would be instilled in the compressing body but only to the extent that the outward degeneracy pressure and super-degeneracy would be confined and safely balanced? A small amount of net gravitation would be ok outside of the otherwise black hole density Earth massed body.

Such super-compressed bodies may be extremely useful as pellets in pellet runway driven starships.

Additionally, in cases where a spacecraft had an antigravatic propulsion system and proximity to such a compressed Earth, the gravitational screening mechanism might be completely released to enable the spacecraft to ride a light-speed expanding gravitational explosion.

Since the escape velocity inside the huge black hole produced would be greater than the speed of light, it is plausible that the spacecraft being internally propelled outward and within the black hole may reach superluminal velocities.

We can also consider compression of an Earth massed body to many orders of magnitude less than one centimeter in diameter. Perhaps the compression energy can be increased to infinite levels.

As long as such concepts are intelligible to conjecture, there remains great hope for open ended possibilities for attaining extreme Lorentz factors, even ones infinite in value in light-speed impulse travel, and perhaps even superluminal impulse travel.

Obviously, the above conjectured explosives would need to occur in ways that would not harm human life, nor any other sentient life-forms in our universe. So, perhaps the best applications for the conjectured technologies would be applicable in uninhabited hyperspaces provided there again would be no risk of harming sentient life forms.

So, the run way mechanisms seem to have open ended possibilities for powering grand space travel missions for the rest of future eternity.

A fascinating prospect would involve taking 4-D space portions of more or less Standard Model matter of non-composite forms and compressing the matter down to 3-D volumes with the same linear dimensional extent as the 4-D versions.

For a Planck unit scale quantized fourth spatial dimension, a 4-D meter wide volume or one of 1 $m^4$ would compress by a factor of $|\{1/\{[h/(2\pi)]\ G/c^3\}^{1/2}\}|$ or about $10^{35}$.which would also be the density increase factor and for a quantized fourth spatial dimension at (1/a) Planck length units, the 1 $m^4$ volume would when compressed to one cubic meter would experience a density increase by $|(a)\{1/\{[h/(2\pi)]\ G/c^3\}^{1/2}\}|$.

For a Planck unit scale quantized fourth and fifth spatial dimensions, a 5-D meter wide volume or one of 1 $m^5$ would compress by a factor of $\{|\{1/\{[h/(2\pi)]\ G/c^3\}^{1/2}\}|\}^2$ or about $10^{70}$.which would also be the density increase factor and for quantized fourth and fifth spatial dimensions at (1/a) Planck length units, the 1 $m^5$ volume would when compressed to one cubic meter would experience a density increase by $\{\{|(a)\{1/\{[h/(2\pi)]\ G/c^3\}^{1/2}\}|\}^2$.

For a Planck unit scale quantized fourth, fifth, and sixth spatial dimensions, a 6-D meter wide volume or one of 1 $m^6$ would compress by a factor of $\{|\{1/\{[h/(2\pi)]\ G/c^3\}^{1/2}\}|\}^3$ or about $10^{105}$.which would also be the density increase factor and for quantized fourth, fifth, and sixth spatial dimensions of (1/a) Planck length units, the 1 $m^6$ volume would when compressed to one cubic meter would experience a density increase by $\{\{|(a)\{1/\{[h/(2\pi)]\ G/c^3\}^{1/2}\}|\}^3$.

In general, for a Planck unit scale quantized N, (N-1), (N-2), ... [N-(N-3)] spatial dimensions, a N-D meter wide volume or one of 1 $m^N$ would compress by a factor of $\{|\{1/\{[h/(2\pi)]\ G/c^3\}^{1/2}\}|\}^{[N-(N-3)]}$ or about (10 EXP 35) EXP [N-(N-3)] .which would also be the density increase factor and for quantized N, (N-1), (N-2), ... [N-(N-3)] spatial dimensions of (1/a) Planck length units, the 1

$m^N$ volume would when compressed to one cubic meter would experience a density increase by $\{\{|(a)\{1/\{[h/(2\pi)]\ G/c^3\}^{1/2}\}|\}^{[N-(N-3)]}$

Here, a, may plausibly range from about one to various infinities. In more exotic scenarios, N be a countable super-infinities.

So, the pellet stream runway concepts have extreme ramifications for those who are willing to let their imagination soar to new heights.

# A lead-in for the next book in this series but relevant to the subject of this book.

Consider the following clauses which are self-explanatory in terms of describing the underlying physics of special reactor types.

The hydrogen-helium-U-238 reactor as an emitter of neutrons, fission fragments, neutrinos, gamma rays, U-V light, visible light, IR light, electrons, positrons and the like.

The hydrogen-helium-U-238 reactor as an emitter of neutrons, fission fragments, neutrinos, gamma rays, U-V light, visible light, IR light, electrons, positrons and the like and with hydrogen and helium intake interstellar ramjet mode with supplemental intake of U-238.

The hydrogen-helium-U-238 reactor as an emitter of neutrons, fission fragments, neutrinos, gamma rays, U-V light, visible light, IR light, electrons, positrons and the like for powering craft in hyperspace.

The hydrogen-helium-U-238 reactor as an emitter of neutrons, fission fragments, neutrinos, gamma rays, U-V light, visible light, IR light, electrons, positrons and the like for powering craft in hyperspace where light-speed is the same as in ordinary 4-D space-time.

The hydrogen-helium-U-238 reactor as an emitter of neutrons, fission fragments, neutrinos, gamma rays, U-V light, visible light, IR light, electrons, positrons and the like for powering craft in hyperspace where light-speed is different from that in ordinary 4-D space-time.

The hydrogen-helium-U-238 reactor as an emitter of neutrons, fission fragments, neutrinos, gamma rays, U-V light, visible light, IR light, electrons, positrons and the like for powering craft in hyperspace where light-speed is greater than that in ordinary 4-D space-time.

The hydrogen-helium-U-238 reactor as an emitter of neutrons, fission fragments, neutrinos, gamma rays, U-V light, visible light, IR light, electrons, positrons and the like and where neutrinos are superluminal.

The hydrogen-helium-U-238 reactor as an emitter of neutrons, fission fragments, neutrinos, gamma rays, U-V light, visible light, IR light, electrons, positrons and the like and with hydrogen and helium intake interstellar ramjet mode with supplemental intake of U-238 and where neutrinos are superluminal.

The hydrogen-helium-U-238 reactor as an emitter of neutrons, fission fragments, neutrinos, gamma rays, U-V light, visible light, IR light, electrons, positrons and the like for powering craft in hyperspace and where neutrinos are superluminal.

The hydrogen-helium-U-238 reactor as an emitter of neutrons, fission fragments, neutrinos, gamma rays, U-V light, visible light, IR light, electrons, positrons and the like for powering craft in hyperspace where light-speed is the same as in ordinary 4-D space-time and where neutrinos are superluminal.

The hydrogen-helium-U-238 reactor as an emitter of neutrons, fission fragments, neutrinos, gamma rays, U-V light, visible light, IR light, electrons, positrons and the like for powering craft in hyperspace where light-speed is different from that in ordinary 4-D space-time and where neutrinos are superluminal.

The hydrogen-helium-U-238 reactor as an emitter of neutrons, fission fragments, neutrinos, gamma rays, U-V light, visible light, IR light, electrons, positrons and the like for powering craft in hyperspace where light-speed is greater than that in ordinary 4-D space-time and where neutrinos are superluminal.

These reactors may be placed behind electrodynamic mirrors to efficiently back-reflect chargons produced or otherwise emitted by the reactor.

Hydrogen, helium, and U-238 may be collected from the background and processed into ingot-like reactors.

Accordingly, for highly relativistic craft, the reaction products have backward momentum enhanced thereby enabling efficient energy collection even from neutrinos which are weakly interacting and largely unaffected by ordinary magnetic and electric fields.

The intake fuels may be slowed by electrodynamic methods for which electrical power thus generated can be inserted back into the propulsive exhaust.

Additionally, the use of nuclear fuel pellets comprised of hydrogen, helium, and U-238 can be pre-distributed in the path of a relativistic spacecraft of interstellar ramjet types.

Each fuel pellet may have a dispersive primary made of special fissionable fuels and having a yield of a few tons of TNT. These primaries would have mass on the order of one to three grams and where the subject of consideration for propelling hundreds of kilograms of small pellets during the height of the Cold War. These applications would have provided ABM hardware for destroying nuclear warheads in space and thus would have been fielded in Earth orbit.

The vaporized pellet debris would then be efficiently scooped up and fed into an interstellar ramjet reactor. In cases where the resulting vapors where ionic, the plasma would be much easier to handle than neutral gas thus facilitating the efficiency of processing the fuel for energy.

The use of nuclear explosives as a primary propulsive energy source still has merit because of the simplicity in capturing the explosive energy. However, as interstellar ramjet technology evolves, the explosive dispersal of fusion and fission fuel also had great merit especially when considering the cost of powerful nuclear explosives and the politics involved with deploying very large numbers of nuclear explosive pellets.

## Drawings.

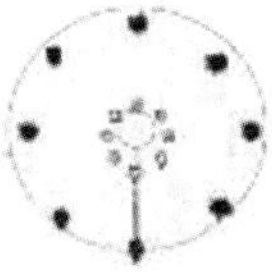

**FIGURE 2.** is a top plan view of a simple shaped charged nuclear or subnuclear explosive disk. Explosion denoted by the same color occur simultaneously. The two sets of explosions can occur simultaneously or sequentially in either of the two orders.

**FIGURE 3** is a top plan view of a simple shaped charged nuclear or subnuclear explosive disk. Explosion denoted by the same

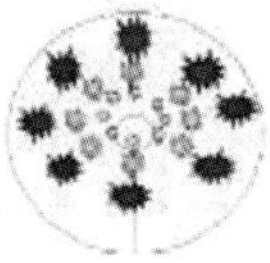

color occur simultaneously. The three sets of explosions can occur simultaneously or sequentially in any of the possible serial orders.

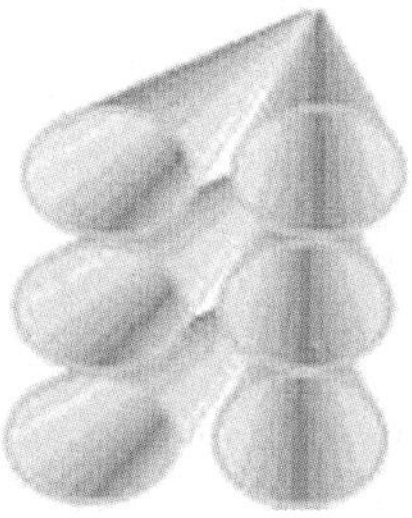

4

5

**FIGURE 4** depicts another configuration of a shaped charge nuclear or subnuclear explosive device. Note the trapezoidal partitions represent side-views of explosive disks, or partitioned wedges. The order of detonation of the partitions can be simultaneous or optionally in sequence in a manner suitable for production of maximum central temperatures and pressures.

**FIGURE 5** depicts a side elevational perspective view of an azimuthal portion of yet another configuration of shaped charged explosive device. The cylinders would be packed with radially serial arrangements of explosive disks, sub-compositions of shorter disks, or optionally be filled with a non-serially partitioned explosive mixture. The detonation pattern for each cylinder can be radially or lengthwise simultaneous throughout the cylinder; or optionally, the composition(s) can be detonated in an arbitrarily radial and/or length-wise pattern.

6

7

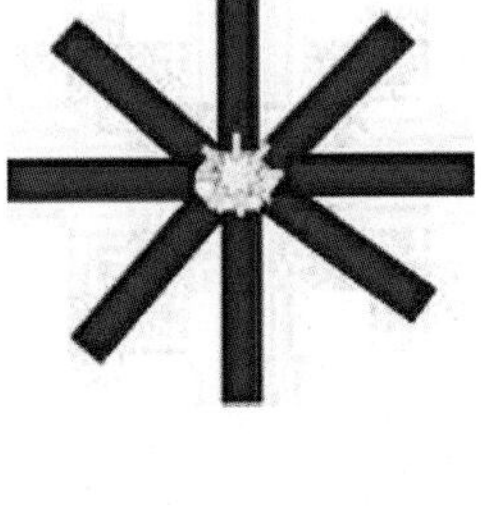

**FIGURE 6** depicts converging jets from shaped charged nuclear or subnuclear devices. The arrangement shown is planar but the relative orientation of the devices is somewhat arbitrary. A conically converging series of jets may produce higher temperatures and pressures in some cases.

**FIGURE** 7 depicts a shaped charge nuclear or sub-nuclear explosive device. The red shading indicates an arbitrary species of nuclear or sub-nuclear explosive. The blue shading or wedge shows a dense material that is squeezed by the exploding red material to produce a concentrated jet pointing to the right. The thick black lines indicate dense cladding to temporarily provide inertial and/or mechanical resistance to the exploding material. The right-triangle shape of the blue material is arbitrary since the aspect ratio of the expellant material can be chosen over a wide range of shapes and sizes.

The next five figures cover multi-level shaped charge explosive as concentric series of charges.

**FGURE** 8

**FIGURE 9.**

**FIGURE 10**

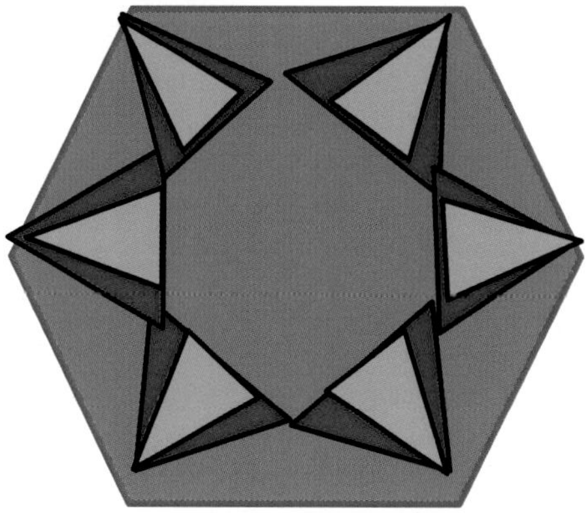

**FIGURE** 11

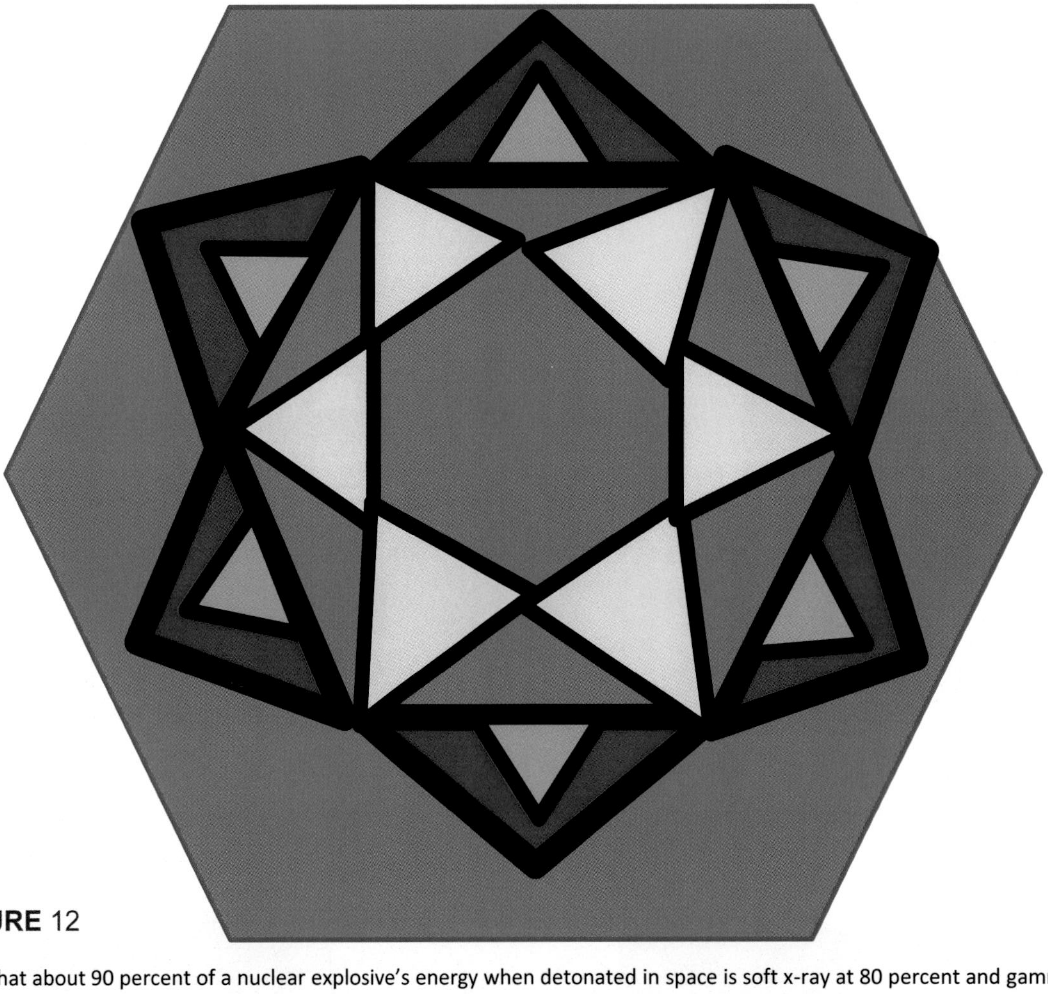

**FIGURE** 12

Note that about 90 percent of a nuclear explosive's energy when detonated in space is soft x-ray at 80 percent and gamma rays at 10 percent. Most of the remainder is neutron energy with a small portion of the remainder being charged particle and neutrino radiation embodied energy.

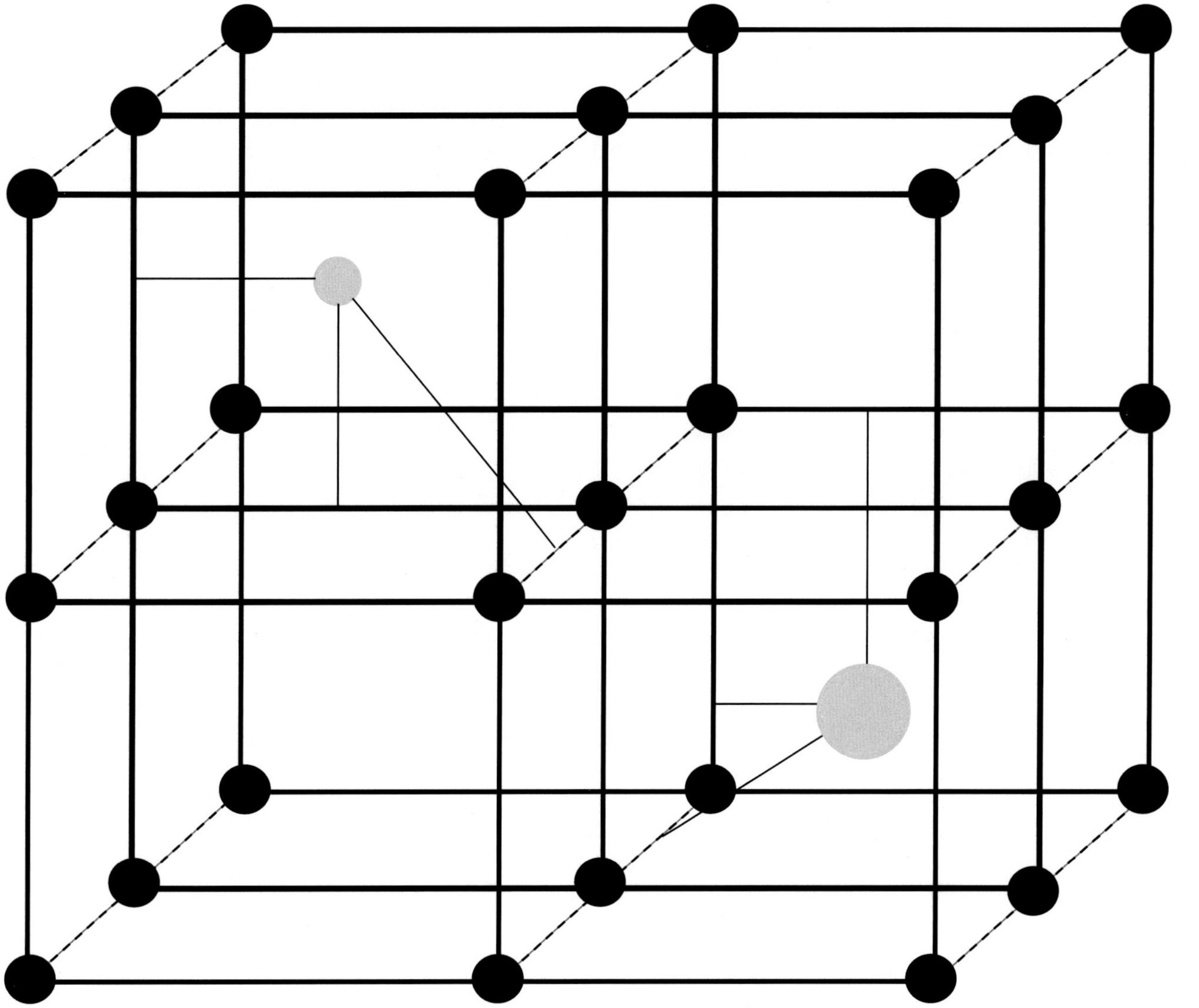

**FIGURE 13** depicts molecularly caged antimatter. The two gray dots depict a generic antimatter nucleus and an antiproton, respectively.

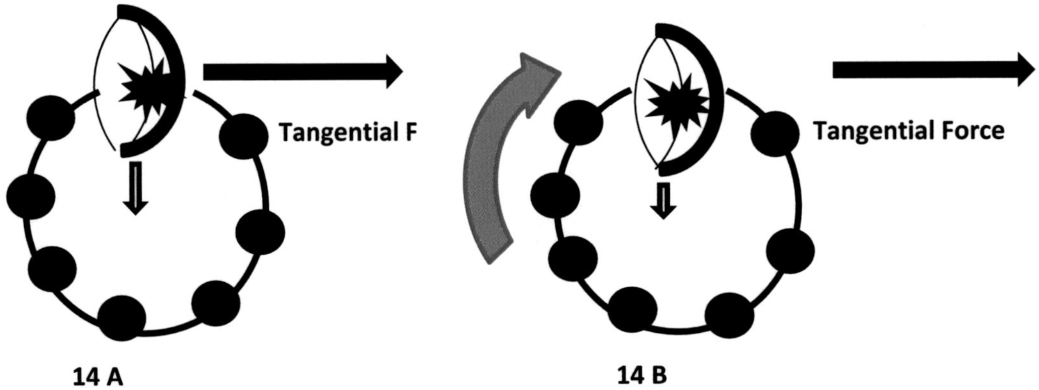


**FIGURE 14 A** depicts a stationary pellet stream fuel stream. The downward pointing arrow indicates centripetal acceleration. **FIGURES 14 B** depicts a revolving pellet stream fuel stream. The downward pointing gray arrow indicates centripetal acceleration.

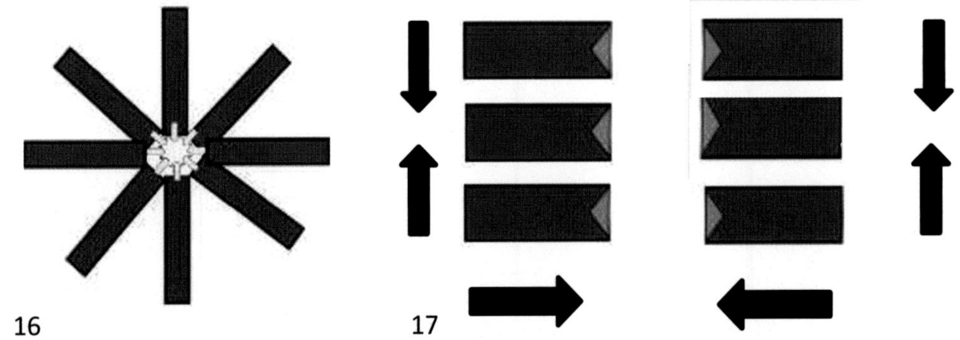


**FIGURE 15** depicts an octangular collision of eight converging fusion and/or fission pellets. **FIGURE 17** depicts collision configuration of six converging fusion and/or fission pellets.

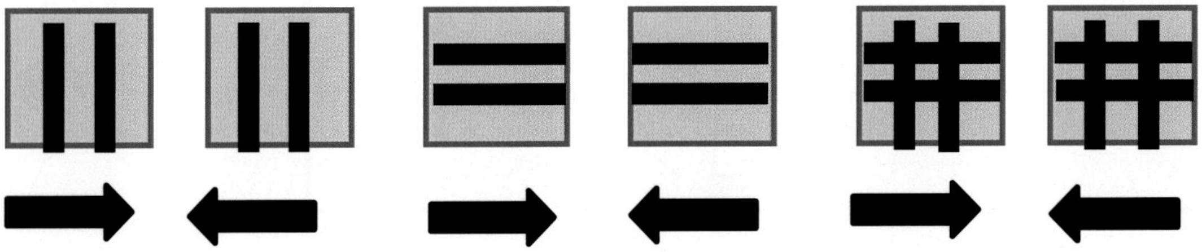

**FIGURE 16** depicts head on collisions of pellets. Three tamper options are indicated as non-limiting examples. The fusion and/or fission fuel (gray) is combined with tampers as indicated by the black bars. Arrows depict direction of incidence.

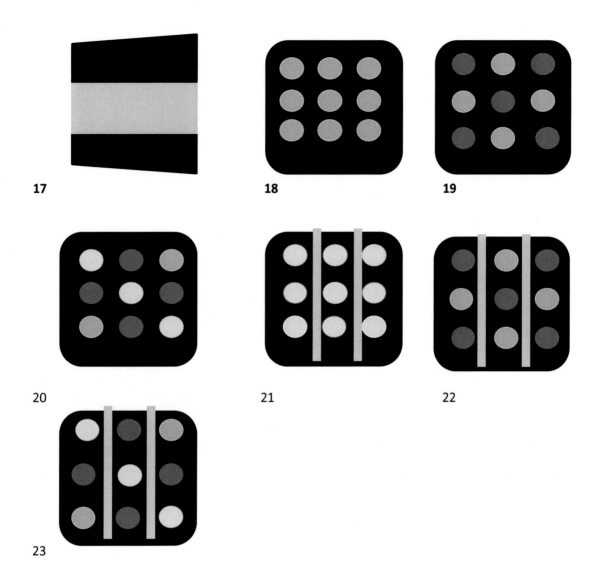

**17**

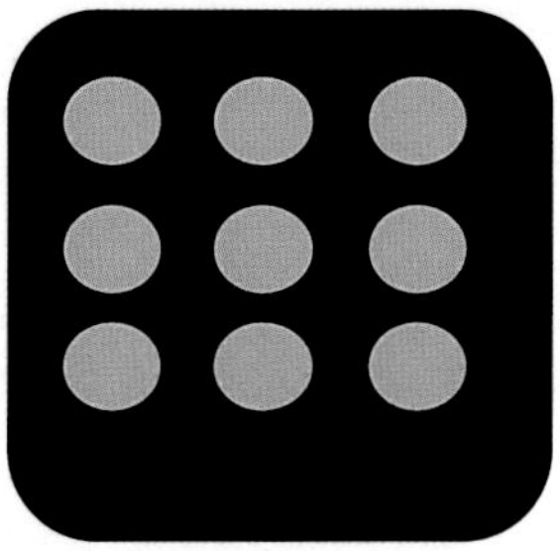

**18**

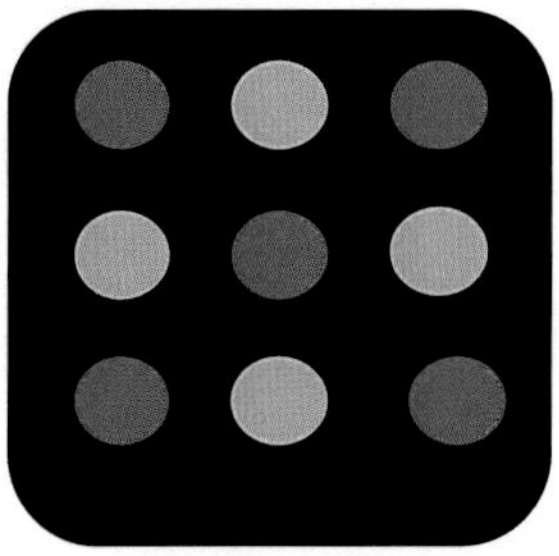

**19**

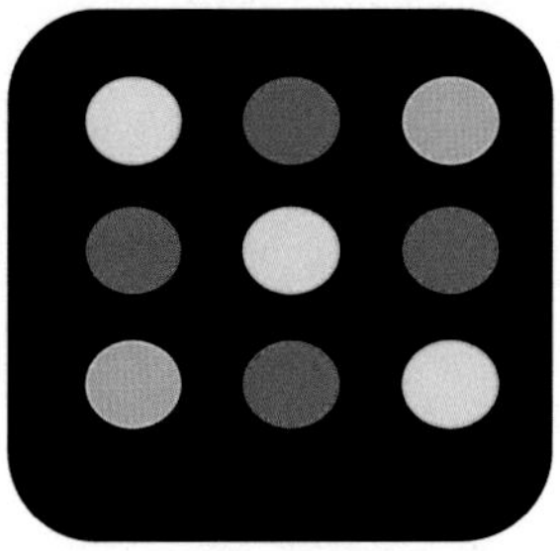

20

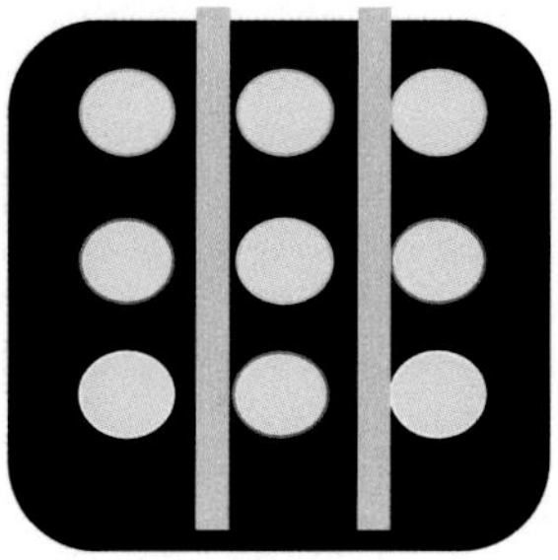

21

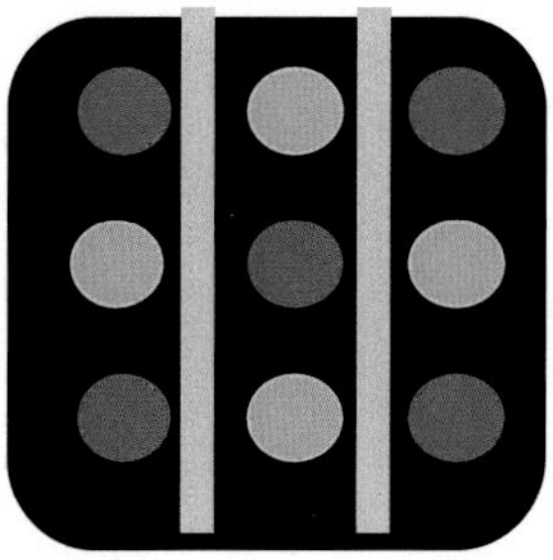

22

23

**FIGURE 17** illustrates a cross-sectional view of fusion pellet having a tapered jacket of dense material. The varying thickness of the wall may enable enhanced squeezing pressure of the fuel near the end of the pellet. **FIGURE 18** depicts a fusion fission pellet. The low density fusion fuel (black) generates neutrons when fused which then induce with then induced nuclear fission of the fissile fuel (gray). **FIGURE 19** depicts a fusion fission pellet. The low density fusion fuel (black) generates neutrons when fused which then induce with then induced nuclear fission of the fissile fuels (two gray tones). Two distinct species of fissile fuels are shown. **FIGURE 20** depicts a fusion fission pellet. The low density fusion fuel (black) generates neutrons when fused which then induce with then induced nuclear fission of the fissile fuels (three gray tones) Three distinct species of fissile fuels are shown. **FIGURE 21** depicts a fusion fission pellet with tampers (gray bars). The low density fusion fuel (black) generates neutrons when fused which then induce with then induced nuclear fission of the fissile fuel (circles). **FIGURE 22** depicts a fusion fission pellet with tampers (dark gray bars). The low density fusion fuel (black) generates neutrons when fused which then induce nuclear fission of the fissile fuels (circles). Two distinct species of fissile fuels are shown. **FIGURE 23** depicts a fusion fission pellet with tampers (dark gray bars). The low density fusion fuel (black) generates neutrons when fused which then induce nuclear fission of the fissile fuels (circles). Three distinct species of fissile fuels are shown.

24 25 26

**FIGURE 24** depicts another fusion fission pellet with tampers (gray bars). The low density fusion fuel (light gray) generates neutrons when fused which then induce nuclear fission of the fissile fuel (circles). **FIGURE 25** depicts still another fusion fission pellet with tampers (gray bars). The low density fusion fuel (lightest gray) generates neutrons when fused which then induce nuclear fission of the fissile fuels (circles). Two distinct species of fissile fuels are shown. **FIGURE 26** depicts another fusion fission pellet with tampers (black bars). The low density fusion fuel (lightest gray) generates neutrons when fused which then induce nuclear fission of the fissile fuels (circles). Three distinct species of fissile fuels are shown.

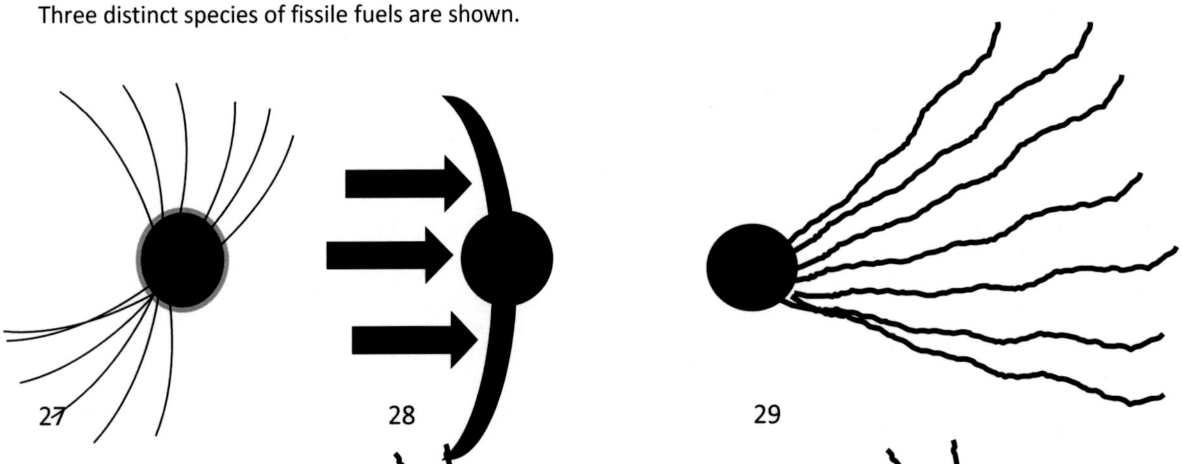

27 28 29

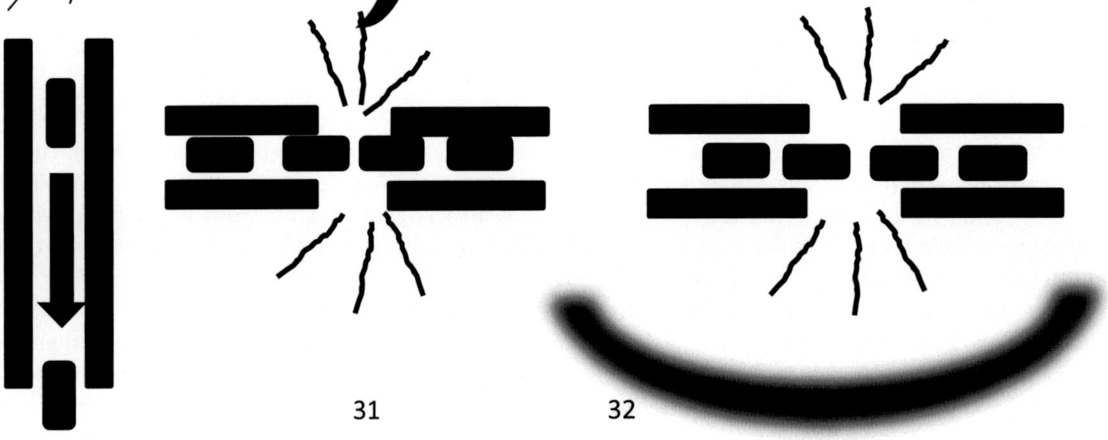

30 31 32

**IGURE 27** depicts a charged spaghetti flux capture mechanism. **FIGURE 28** depicts charged monolithic membrane flux capture mechanism. **FIGURE 29** depicts an uncharged spaghetti flux capture mechanism. **FIGURE 30** illustrates lab frame collision of fusion pellets. **FIGURE 31** illustrates CM frame collision of fusion pellets. Irregular lines depict nuclear reaction products. **FIGURE 32** illustrates CM frame collision of fusion pellets. Irregular lines depict nuclear reaction products. Soft bow shape depicts electrodynamic mirror for reflecting plasma.

## Whimsical Closing Anecdotes. Pathways To The Speed Of Light.

The pellet runway seems an ideal method for controlled spacecraft acceleration to extreme Lorentz factors. This is evident as the spacecraft mass ratio can be set very closely to unity. So, achieving extreme Lorentz factors is possible in theory without recourse to bloated mass ratios of matter-antimatter or antimatter fuel.

Consider a photon traveling in space at a velocity precisely equal to c or $[1/(\mu_0\epsilon_0)]^{1/2}$ a distance equal to one cosmic light-cone radius or 13.77 billion light-years. This works out to be $[1.377 \times 10^{10}](10^{16})\{1/[1.6 \times (10^{-35})]\}$ Planck length units= $8.521 \times 10^{60}$ Planck length units.

So, a spacecraft traveling $\{[8.521 \times (10^{60})]-e\}$ Planck length units in $1.377 \times 10^{10}$ years where e is less than one will have velocity that is cosmically locally indistinguishable from the speed of light.

The latter boundary value condition may most naturally define the transition of the spacecraft into travel in the first level of space-time foam.

$\text{Gamma} = \{1/\{1- [(v/c)^2]\}\}^{1/2}$.

$\text{Gamma}^2 = \{1/\{1-[(v/c)^2]\}\}$

$1/(\text{Gamma}^2) = \{1-[(v/c)^2]\}$

$[1/(\text{Gamma}^2)] -1 =-[(v/c)^2]$

$(1 - [1/(\text{Gamma}^2)]\}^{1/2} = v/c$

So, a Lorentz factor of 5 corresponds to about 0.98 c.

A Lorentz factor of 50 corresponds to about 0.9998 c.

A Lorentz factor of 500 corresponds to about 0.999998 c.

Now, we are considering travel at $\{c - \{\{1/[[8.521 \times (10^{60})] + e]\}\ c\}\}$.

However, consider the above patterns Lorentz factors verses velocities.

For example, $5^{-2} = 0.025$.

While $50^{-2} = 0.00025$

While $500^{-2} = 0.0000025$.

So, in first order, the number of decimal places for each digitally pentilated Lorentz factor of the examples provided above to the right of the decimal point is twice that of the associated velocity up to and including the digit, "8".

This tells us that the lower cut-off value of the Lorentz factor of a spacecraft that would travel less than c mathematically, but which would have actual velocity cosmically indistinguishable from light speed would be about $[8.521 \times (10^{60})]^{1/2} = [2.91907 \times (10^{30})]$.

So, a principally chemically fueled spacecraft or a spacecraft with chemical energy initiated background ionic, magnetic, electric, and/or electromagnetic energy uptake may in principle attain a finite Lorentz factor that is large enough to enable velocities indistinguishable from the speed of light over cosmically local travel intervals.

The cut-off spacecraft Lorentz factor would be:

$\{1/\{[1-[(v/c)^2]]^{1/2}\}\} = \{1/\{[1-[(\{c-\{\{1/[[8.521 \times (10^{60})]+e]\} c\}\}/c)^2]]^{1/2}\}\}$.

# Appendix 1. Prior Art In Staged Compression Gas Guns And Large Artillery.

There are many precedents for large barreled guns that were able to achieve a maximum chamber pressure of about 255 mega-newtons/$m^2$ and projectile accelerations of **a** = **F**/m = 27,000 m/$s^2$. Here, we conservatively assume a projectile mass of 2,700 pounds = 1,225 kg and a maximum chamber pressure of $2.55 \times 10^8$ N. The propulsive force on the round is assumed to be a maximum of $3.31 \times 10^7$ N. Note that these values are those achieved during World War 2 and which were reliably practiced in dozens of units as fielded on the 3rd generation battleships of that era.

From NASA's website:

https://www.nasa.gov/centers/wstf/site_tour/remote_hypervelocity_test_laboratory/two_stage_light_gas_guns.html

is the following summary herein quotations.

"1.0 caliber, .50 caliber, and two .17 caliber two stage light gas guns are housed in the Remote Hypervelocity Test Laboratory. These guns use gunpowder and highly compressed hydrogen to accelerate projectiles at speeds up to 27,500 feet per second to simulate impacts of particles on spacecraft and satellite materials and components.

Two Stage light gas guns use gunpowder, the first stage, and highly compressed hydrogen, the second stage, to accelerate projectiles at high velocities to simulate orbital debris impacts on spacecraft and satellite materials and components.

**First Stage**

The first stage uses conventional smokeless gunpowder as its propellant and works the same way as firing a bullet from a gun. The second-stage propellant uses a highly compressible light gas such as hydrogen. Since light gases have very low molecular weights, they are easily compressed to the high pressures needed to efficiently launch projectiles at hypervelocity speeds. The gun's breech contains the first-stage powder charge, set off by an electronic igniter. The ignition provides the explosion that drives a piston forward down the pump tube, rapidly compressing the hydrogen gas that provides the second-stage launch power. The front face of the piston is a hollow cone that forms a gas seal as it compresses the hydrogen at a speed of approximately 2,500 feet per second. The back end of the piston uses an O-ring to seal the expanding gases from the powder charge.

**Second Stage**

The taper-bored, high pressure (HP) section referred to as the high-pressure coupling halts the propelled piston at the end of its travel down the pump tube. In the HP section, rapid internal pressurization is followed by an extremely high level of impact caused by the halted piston. A petal-valve diaphragm retains the gas until it bursts and upon rupture, the launch package containing the sabot and the projectile is accelerated by the rapidly expanding light gas down the barrel into the expansion tank where a stripper plate separates the projectile from the sabot. The sphere enters the target tank (with air removed to replicate the vacuum of space) and hits the test article. The ASME-rated target tank is designed to contain the gas from the second-stage and the eruption of shrapnel and debris from the projectile's impact on the test article."

**FIGURE 1.** Looking down the 1.0 caliber two stage light gas gun from the open end of the breech toward the target tank just outside the building, 175 feet away. Projectiles achieve speeds of 23,000 feet per second in just 24 feet of barrel.
***Credits: NASA WSTF***
*Last Updated: Aug. 6, 2017*
*Editor: Judy Corbett*
Now, for a brief aside, consider the following excerpt from Chuck Hawks "U.S. Navy 16" Battleship Gun Facts" at https://www.chuckhawks.com/16-50_gun_facts.html

Here we provide a length quotation due to the large number of specific details provided therein.

"The US Navy's Iowa (four ships) class battleships carried a main armament of nine 16"/50 caliber guns in three triple turrets. The previous North Carolina (two ships) and South Dakota (four ships) classes carried a very similar main battery of nine 16"/45 caliber guns. These 10 ships, completed between 1941 and 1944, comprised the USN's "third generation" battleships and all saw service in WW II.

The designation 16"/50 means a 16" diameter shell and a barrel 50 calibers long. That would be 16x50=800 inches, or a barrel 66.66 feet long. The 16"/45 gun fired the same shells from a slightly shorter barrel 60 feet long.

The barrel had 96 rifling grooves (shades of Marlin's micro-groove type rifling, which is typical of cannons). The twist rate was one turn in 25 calibers, or 1:400". The maximum service pressure was 18.5 tons psi, or 370,000 pounds psi (corrected 37,000 psi).

Shells of different weights were fired, weighing from approximately 1,900 to 2,700 pounds. The heaviest shell was the AP (armor piercing) projectile, which had a maximum range 42,345 yards from the 16"/50 gun, or 39,000 yards from the 16"/45 gun (about 22 miles).

According to Jane's Fighting Ships (circa WW II), the muzzle velocity (MV) was up to 2800 fps with a 2100 pound shell and muzzle energy was 98,406 ft. tons. The rate of fire was about two rounds per minute."

So, there is precedent for large barreled guns able to achieve a maximum chamber pressure of about 255 mega-newtons/$m^2$ and projectile accelerations of **a** = **F**/m = 27,000 m/$s^2$. Here, we conservatively assume a projectile mass of 2,700 pounds = 1,225 kg and a maximum chamber pressure of 2.55 x $10^8$ N. The propulsive force on the round is assumed to be a maximum of 3.31 x $10^7$ N. Note that these values are those achieved during World War 2 and which were reliably practiced in dozens of units as fielded on the 3rd generation battleships of that era.

# Appendix 2. Enablement By Extreme Materials Made From Ordinary Elements.

A brief description of extreme materials composed of ordinary periodic table elements is given below. These materials although generally currently prohibitively expensive might be mass producible by nano-technology self-assembly or perhaps be widespread mining operations throughout the solar system.

**A) Carbon Nanotubes**

In 2000, a multiwall nanotube was observed to have a tensile strength of 63 gigapascals. This translates into a cable with a cross-sectional area of one square millimeter being able to hold 6,422 kg or 6.422 metric tons. A cable with a cross-section of one square centimeter could support 642 metric tons. A cable with a cross-section of one square inch could support 4,143 metric tons.

The table below was adapted from the same Wikipedia page and summarizes the relative mechanical strength properties of CNTs

Comparison of Mechanical Properties

| Material | Young's modulus (TPa) | Tensile strength (GPa) | Elongation at break (%) |
|---|---|---|---|
| SWNT | ~1 (from 1 to 5) | 13–53E | 16 |
| Armchair SWNT | 0.94T | 126.2T | 23.1 |
| Zigzag SWNT | 0.94T | 94.5T | 15.6–17.5 |
| Chiral SWNT | 0.92 | | |
| MWNT | 0.27E–0.8E–0.95E | 11E–63E–150E | |
| Stainless steel | 0.186E–0.214E | 0.38E–1.55E | 15–50 |
| Kevlar–29&149 | 0.06E–0.18E | 3.6E–3.8E | ~2 |

$^{E}$ = Experimental Observation; $^{T}$ = Theoretical Prediction
Chart adapted from Wikipedia at Courtesy of Wikipedia at: http://en.wikipedia.org/wiki/Carbon_nanotube

Metallic CNTs can carry an electric current density of [4 x ($10^9$)] amperes/($cm^2$). This is more than 1,000 times greater than metals such as copper (Wikipedia, 2011).

CNTs have theoretical temperature stability for temperatures as great as 2800° C in a vacuum and about 750° C in air.

This material may be of use in forward deployed starship shields as well as in metalized thin membrane like fabrics or grids for application in light sails. When charged, such sails can be used for Project Medusa style propulsion systems.

**B) Graphene**

Graphene consists of one atom-thick sheets of carbon atoms arranged in a repeating hexagonal pattern.

The electrical resistivity of graphene has been measured at ($10^{-6}$)[(Ohms)($cm^2$)] and thus has a conductivity that is higher than that of silver.

Graphene has been shown to have a bulk strength of 130 gigapascals or about (1,300,000)(14.7) pounds per square inch or 19.11 million PSI. This is a whopping 9,555 tons per square inch.

This material may be of use in forward deployed starship shields as well as in metalized thin membrane like fabrics or grids for application in light sails. When charged, such sails can be used for Project Medusa style propulsion systems.

**C) Graphene-Oxide Paper**

Graphene-oxide paper has been shown to have a tensile strength of 32 gigapascals or about (320,000)(14.7) pounds per square inch or 4.704 million PSI (Wikipedia, 2011). This material may also be of use in forward deployed starship shields as well as in metalized thin membrane like fabrics or grids for application in light sails. When charged, such sails can be used for Project Medusa style propulsion systems.

**D) Beta carbon nitride**

Beta carbon nitride (β-C3N4) is predicted to be harder than diamond and is crystalline in composition and formed by extremely short and strong chemical bonds between carbon and nitrogen atoms. Because the bonds are so strong, the possibility of producing ultra-strong threads composed of Beta carbon nitride becomes irresistible to researchers. So far, only nano-sized crystals of Beta carbon nitride have been produced (Wikipedia, 2010). This material may also be of use in forward deployed starship shields as well as in metalized thin membrane like fabrics or grids for application in light sails. When charged, such sails can be used for Project Medusa style propulsion systems.

**E) Heterodiamond**

Another material with extremely strong atomic bonds is heterodiamond. This material is composed of carbon, boron, and nitrogen atoms. The bulk modulus of heterodiamond is 282 Gpa and is surpassed only by that of diamond. Heterodiamond is extremely hard and possesses an excellent heat resistance. Heterodiamond may find application in starship forward shielding and also Project Orion style pusher plates. When fashioned into metalized thin membrane like fabrics or grids, the material can find application in light sails. When charged, such sails can be used for Project Medusa style propulsion systems.

**F) Diamond** in naturally occurring forms is the hardest known natural substance. Provided bulk portions of diamond can be fabricated, such a material would be useful for forward deployed starship shields and Project Orion pusher plates. Natural diamond is an excellent heat conductor and so it could be used to conduct heat from the outer surface of a Project Orion style pusher plate.

**G) Aggregated Diamond Nanorods**, or **ADNRs**, are a nanocrystalline form of diamond, also referred to as "nanodiamond" or hyperdiamond. In 2003, nanodiamond was produced by compression of graphite in 2003 and was found to be much harder than bulk diamond. Later produced by compression of fullerene, it was confirmed as the hardest and least compressible known material and had an isothermal bulk modulus of 491 gigapascals (GPa). Conventional diamond has a modulus of 442–446 Gpa. ADNRs are 0.3% denser than regular diamond. ADNRs were reported by the researchers performing the study as "having a hardness and Young's modulus comparable to that of natural diamond, but with "superior wear resistance". This "Bad Boy" material if somehow produced in large bulk quantities might make for an excellent forward deployed shield for starships and for Project Orion style pusher plates. Any potential issues of material brittleness would need to be addressed in the design of pusher plates primarily composed of ADNRs.

**H) Tantalum Hafnium Carbide**

**T**antalum hafnium carbide is sort of an old wonder material that is known for its extreme melting point. The material may conceivably be incorporated into star light and CMBR shielding for extremely relativistically blue-shifted background radiations. Other uses for the material can include monolithic or weave like grid light sails, charged Project Medusa type sails, and perhaps pusher plates for Project Orion style starships.

Tantalum hafnium carbide is defined by a general formula $Ta_xHf_{1-x}C_y$ which is a solid solution of tantalum carbide and hafnium carbide. The latter two carbides have the highest melting points among the binary compounds, 3,983 °C (7,201 °F) and 3,928 °C (7,102 °F), respectively. The composition $Ta_4HfC_5$ is believed to have a melting point of 4215 °C (7619 °F). Some tantalum-hafnium-carbide solutions are suggested to have melting points of 8,000 °F.

**I) Metallic Micro Lattice**

As of the time of this writing, TPM has reported on the development of the lightest known material, a form of metallic micro lattice, which is an aerated form of a nickel and phosphorus combination. The material is fabricated with micro-scale lattice cells and is 99.99 percent air by volume. The material exhibits good elasticity and strain properties and is much sturdier than aerogels.

The material has a density of 0.9 mg per cubic centimeter and thus may make an outstanding ultra-low weight thermal insulator of large balloon or membranous forms of fusion fuel tanks for starships. The new material may provide thermal insulation for layered cometary or ice-ball dispositions of fusion fuel such as hydrogen or deuterium ice.

Additionally, the material may find application in very thick but low mass density forward deployed shields for relativistic spacecraft. Such forward deployed dispositions can find application in the disruption of incoming dust particles where long paths lengths are desirable for disrupted dust particle debris for purposes of providing a mechanism to diffuse the highly concentrated massive kinetic energy jets formed by impacting dust particles. It is not inconceivable that such shields having a thickness of 10 kilometers to 100 kilometers but having the same mass as a water shield of the same cross-sectional dimensions but having a thickness of only 10 meters and 100 meters respectively could be augmented by magnetic and/or electric field emitters instilled within or alongside the shielding material. The electrodynamic fields could then differentially pull or push on charged particulate debris produced by the ionization of the incoming dust particles. The highly energetic plasma jets could be variably diffused as a function of kinetic energy to charge ratio, momentum to charge ratio, and relativistic mass to charge ratio.

According to TPM, '*"I think I actually had the idea to push for super-low density," said HRL's William Carter, one of the authors of the new paper, in an email to TPM.*

*"The main invention (the process to make microlattices) was done here at HRL by Dr. Alan Jacobsen in 2007. The process to get ultra-light materials was refined by Dr. Tobias Schaedler over the past year. We collaborated with Prof. Valdevit's group at UC Irvine to understand the mechanical properties, and Prof. Greer's group at Caltech to understand the base material properties."'*

The TMP report on this fascinating and potentially game changing material is available at:

**http://idealab.talkingpointsmemo.com/2011/11/making-the-worlds-lightest-material.php?ref=fpnewsfeed_beta**

**J) T-Carbon**

A theoretical form of carbon referred to as T-carbon is more like fluff than natural diamond but is 43 percent as dense and 65 percent as hard. The exotic material has been reported in Science News in **Diamond cousin proposed** at;

http://www.sciencenews.org/view/generic/id/70420/title/Diamond_cousin_proposed.

Gang Su, at the Graduate University of the Chinese Academy of Sciences in Beijing has co-authored a paper appearing in Physical Review Letters that describes the exotic material.

Several theoreticians have doubts about the stability of the proposed material and according to some researchers, the material would likely collapse due to instability based on its high energy state. Most stable materials settle into low energy states.

The good news is that there are many ways for carbon atoms to be configured into solids and so the possibility of finding just the right stable configurations, although difficult, has a degree of plausibility.

If such a material is possible to produce, its application for interstellar travel seems apparent. Such a material may provide ultra-low density but extremely hard shielding for a starship. Perhaps shielding several miles thick can be placed at the front of a spacecraft. Cosmic rays and plasma jets resulting from small dust particle impacts could be disrupted not only by the material's atomic nuclei, but also blunted by processes which include the breaking of the very strong molecular bonds among the material's constituent carbon atoms. The materials may be just hard enough to offer an electronic bonding based mechanism so as to absorb the cosmic rays and plasms jets produced by impacting microscale dust grains. Note that ionizing radiation is usually shielded by dense materials such as lead which ironically has a very low heat capacity and latent heat of fusion and vaporization. For such dense shielding materials, the disruptive effect on incident ionizing radiation is wrought by the materials density. For some ionization species and energy spectrums, the radiation absorption is affected by the particular properties of the shielding materials atomic nuclei.

The production of superhard low density materials may offer the advantage of providing electronic disruption of ionizing radiations while at the same time providing a long path for which to absorb the radiations. The absorption path-length scales in inverse proportion to the materials density for shields having the same cross-sectional area.

**K) Carbon Atom Chains**

Carbon atom chain based materials would be fabricated out of one carbon atom wide fibers. Such threads, if properly electronically configured, may have a tensile strength significantly higher than any other proposed molecular materials. The material's extremely high strength would be a consequence of the extremely strong bonds that can exist in certain electronic carbon configurations. Bulk carbon-based materials fabricated from carbon chains may have exotically high strength properties and thus might find use in metalized or otherwise highly reflective light sails including laser beam driven sails and sun-diver and other solar and stellar sails. Ultra-high mass specific area, microwave and radio-frequency beams sails, could be fabricated from any highly conductive versions of such materials. Micron scaled and nanoscale width metalized reflective threads consisting of aggregated carbon chains could be fashioned into grid-like stellar or laser beam sails and be used to very efficiently reflect infra-red, visible, soft ultra-violet light.

Nanotechnological or micro-robotic repair and reconfigurement mechanisms can optionally be included in spacecraft portions comprised of the above materials or any other materials. The shape and size of the portions of spacecraft comprised of any of the above listed materials or any other materials of construction may be chosen for specific mission criteria. The shape and size of said materials portions may also be adjustable in flight. Such additional morphological degrees of freedom can enable more appropriate interaction of the shield and reflector with the interstellar and intergalactic medium including but not limited to natural variations in: 1) plasma distribution by density, charge, and species; 2) neutral gas distribution by density and species; 3) ambient stellar and quasar light distributions according to energy spectrum and/or power flux density; 4) and interstellar and intergalactic magnetic field intensity and/or vector field orientation. Such considerations can be important for drag reduction, and for certain conditions, intentional increases in drag for re-routing and deceleration. Other considerations include adjustability for spacecraft propulsion system power outputs which may be required as a result of the above natural ambient variations in the interstellar and intergalactic medium.

# PART 2.

# A SECOND CONSIDERATION OF THE ELECTRON GAS GUN PELLET STREAM DISPENSING SYSTEM.

*Herein are proposed mechanisms for deploying principally nuclear fission and/or nuclear fusion fuel for remotely located spacecraft. The method of fuel placement includes as a primary options electron gas gun systems that launchs pellets to muzzle velocities of Keplerian to mildly relativistic scales. Accordingly, the fuel would be substantially prepositioned to be intercepted by relativistic spacecraft. Emphasis is placed on thermonuclear explosives that would be detonated at optimal locations proximate to a spacecraft pusher plate or charged chute system. The charged chutes would be attached to the spacecraft by spring-like or other elastic mechanism to dampen the electrodynamic shock of the explosions on the spacecraft and chute. Other various topics are addressed such as mechanisms for which a spacecraft would leverage captured explosive energy to energize propulsion apparatus that would collect background magnetic, electric, and ionic energy and convert portions of these energies to spacecraft kinetic energy.*

## THE BASIC SYSTEM.

## MORE ON LINEAR INDUCTION DRIVE MECHANISMS.

## SUM OF TWO VELOCITIES: GENERAL CASE.

## RELATIVISTIC PINWHEEL DYNAMO EFFECT.

## NEGATIVE ELECTROMAGNETIC REFRACTIVE INDEX PULLING LIGHT DRIVE.

## PRODUCTION OF ANTIMATTER FROM SUNLIGHT.

## VALIDATION OF SPECIFIC IMPULSES GREATER THAN ONE FOR PRINCIPALLY RELATIVISTIC ROCKET CRAFT.

CYCLER CONTROL AND PREDICTING BACKGROUND MAGNETIC FIELDS.

FORWARDLY, ORTHOGONALLY, OR OBLIQUELY INCIDENT PELLETS.

SHAPED CHARGE NUCLEAR FUEL PELLET.

OPTIMIZING NUCLEAR DEVICES FOR ROCKET PROPULSION SYSTEMS

FUSION ROCKET & BEAM RELAY PROPULSION.

INTERSTELLAR RAMJET & BEAM RELAY PROPULSION.

RADIATION PROTECTION AT HIGH LORENTZ FACTORS

BASIC SHIELDING II: ORDINARY ATOMS-BASED SPACE SHIELDS AND DRAG REDUCTION MECHANISMS FOR HIGH GAMMA FACTORS

ENABLEMENT BY EXTREME MATERIALS MADE FROM ORDINARY ELEMENTS

THE ELECTROSTATIC HAIRBRUSH REVISITED.

RELATIVISTIC PINWHEEL DYNAMO EFFECT AUGMENTED BY ELECTRODYNAMIC HAIR BRUSH.

# BUILDING BOMBS ENROUTE.

# GRAVITATIONAL ASSISTS TO VIRTUALLY LIGHT SPEED.

# SOME WHIMSICAL CLOSING ANECDOTES.

# SUMMARY.

# EPILOG

# REFERENCES:

I propose theoretical concept validation of a staged compression gun able to accelerate projectiles and spacecraft to Keplerian velocities and even to relativistic velocities. However, the pellet runway propulsion concepts provided below are not necessarily and end of the means itself as will become apparent later below.

For a gun with several stages, pods may be shot out of Earth's upper atmosphere and into interplanetary space where they may then use retro-rocket course correction to arrive on Mars. The pods may use direct rocket descent on Mars or aerobreaking methods followed by final descent by parachute and rockets.

The guns may be lofted into the high stratosphere by very large blimps or dirigibles and maintained on station by strong cables.

Explosive charges such as gun-powders, perhaps as simple as low cost primitive black powders, may be ferried to the balloon bases by cable-based elevators.

From NASA's website:

https://www.nasa.gov/centers/wstf/site_tour/remote_hypervelocity_test_laboratory/two_stage_light_gas_guns.html

is the following summary herein quotations.

"1.0 caliber, .50 caliber, and two .17 caliber two stage light gas guns are housed in the Remote Hypervelocity Test Laboratory. These guns use gunpowder and highly compressed hydrogen to accelerate projectiles at speeds up to 27,500 feet per second to simulate impacts of particles on spacecraft and satellite materials and components.

Two Stage light gas guns use gunpowder, the first stage, and highly compressed hydrogen, the second stage, to accelerate projectiles at high velocities to simulate orbital debris impacts on spacecraft and satellite materials and components.

**First Stage**

The first stage uses conventional smokeless gunpowder as its propellant and works the same way as firing a bullet from a gun. The second-stage propellant uses a highly compressible light gas such as hydrogen. Since light gases have very low molecular weights, they are easily compressed to the high pressures needed to efficiently launch projectiles at hypervelocity speeds. The gun's breech contains the first-stage powder charge, set off by an electronic igniter. The ignition provides the explosion that drives a piston forward down the pump tube, rapidly compressing the hydrogen gas that provides the second-stage launch power. The front face of the piston is a hollow cone that forms a gas seal as it compresses the hydrogen at a speed of approximately 2,500 feet per second. The back end of the piston uses an O-ring to seal the expanding gases from the powder charge.

**Second Stage**

The taper-bored, high pressure (HP) section referred to as the high-pressure coupling halts the propelled piston at the end of its travel down the pump tube. In the HP section, rapid internal pressurization is followed by an extremely high level of impact caused by the halted piston. A petal-valve diaphragm retains the gas until it bursts and upon rupture, the launch package containing the sabot and the projectile is accelerated by the rapidly expanding light gas down the barrel into the expansion tank where a stripper plate separates the projectile from the sabot. The sphere enters the target tank (with air removed to replicate the vacuum of space) and hits the test article. The ASME-rated target tank is designed to contain the gas from the second-stage and the eruption of shrapnel and debris from the projectile's impact on the test article."

**FIGURE 1.** Looking down the 1.0 caliber two stage light gas gun from the open end of the breech toward the target tank just outside the building, 175 feet away. Projectiles achieve speeds of 23,000 feet per second in just 24 feet of barrel.
***Credits: NASA WSTF***

*Last Updated: Aug. 6, 2017*
*Editor: Judy Corbett*

Now, for a brief aside, consider the following excerpt from Chuck Hawks "U.S. Navy 16" Battleship Gun Facts" at https://www.chuckhawks.com/16-50_gun_facts.html

Here we provide a length quotation due to the large number of specific details provided therein.

"The US Navy's Iowa (four ships) class battleships carried a main armament of nine 16"/50 caliber guns in three triple turrets. The previous North Carolina (two ships) and South Dakota (four ships) classes carried a very similar main battery of nine 16"/45 caliber guns. These 10 ships, completed between 1941 and 1944, comprised the USN's "third generation" battleships and all saw service in WW II.

The designation 16"/50 means a 16" diameter shell and a barrel 50 calibers long. That would be 16x50=800 inches, or a barrel 66.66 feet long. The 16"/45 gun fired the same shells from a slightly shorter barrel 60 feet long.

The barrel had 96 rifling grooves (shades of Marlin's micro-groove type rifling, which is typical of cannons). The twist rate was one turn in 25 calibers, or 1:400". The maximum service pressure was 18.5 tons psi, or 370,000 pounds psi (corrected 37,000 psi).

Shells of different weights were fired, weighing from approximately 1,900 to 2,700 pounds. The heaviest shell was the AP (armor piercing) projectile, which had a maximum range 42,345 yards from the 16"/50 gun, or 39,000 yards from the 16"/45 gun (about 22 miles).

According to Jane's Fighting Ships (circa WW II), the muzzle velocity (MV) was up to 2800 fps with a 2100 pound shell and muzzle energy was 98,406 ft. tons. The rate of fire was about two rounds per minute."

So, there is precedent for large barreled guns able to achieve a maximum chamber pressure of about 255 mega-newtons/m$^2$ and projectile accelerations of **a** = **F**/m = 27,000 m/s$^2$. Here, we conservatively assume a projectile mass of 2,700 pounds = 1,225 kg and a maximum chamber pressure of 2.55 x $10^8$ N. The propulsive force on the round is assumed to be a maximum of 3.31 x $10^7$ N. Note that these values are those achieved during World War 2 and which were reliably practiced in dozens of units as fielded on the 3rd generation battleships of that era.

Now, we consider in first order that quadrupling NASA's gas compression gun length and adding additional stages may enable projectile velocities as great as 55,000 feet per second or about 16,500 meters per second. This latter velocity is more than adequate for Mars orbit insertion and also into the main asteroid belt

Using an electron gas as final stage pressure embodiments, much greater velocities such as very mildly relativistic velocities could be obtained by guns of great length mounted on the Moon. Velocities on the order of at least roughly 750,000 meter/sec or 0.239 percent of the speed of light may be obtained Conceivably, higher velocities could be achieved with the electro-gas discharge. These extreme velocities when attributed to pellets on the scale of 40 kilograms/pellet would find great use in forwardly distributed nuclear fission or nuclear fusion fuel pellet runways to enable starships to reach large Lorentz factors with nearly no onboard nuclear fuels. For a 0.0001 light-year long distributed runway and spacecraft frame acceleration on the order of 0.5 earth g's or 4.905 m/s$^2$, a human crewed spacecraft could obtain a velocity of about 1 percent of light-speed and a Lorentz factor of 1.00005.

The above values for pellet stream velocities are in first order derived from the approximately 16.5 km/sec enabled by hydrogen compression multiplied by the square root of the ratio of the hydrogen atom mass and the electron mass which is 42.8623. This is so because the mass of the hydrogen atom is 1,837.17 times that of the electron. Since Newtonian momentum is proportional to velocity for a given mass, but assuming said square root of 1,837.17, we have said square root scaling of momentum but also said square root scaling of number of times on average a given electron will impart momentum to a gun barrel's wall per second. Here we consider pressure in the ideal gas law is a function of number of particle collisions per unit area per unit time multiplied by twice the average pre-incident momentum imparted to the barrel walls of the colliding particles while assuming elastic collisions. We thus obtain the approximately ¼ of a percent of the velocity of light for the exiting projectile.

Now, as the electron gas density increases, coulombic pressure will factor in thereby increasing the gas pressure beyond what it would be for a given average electron velocity. So, it is conceivable that an electron gas compression gun of many stages could accelerate a round to almost the speed of light. A gun barrel with a high electric flux topological reflector or insulator will make the electron-based coulombic force more concentrated thus increasing the potential velocity of the round to values even closer to the speed of light to attain extreme projectile Lorentz factors.

An anticipated method of providing electron gas is a powerful electrical discharge system such as by way of banks of super-capacitors. The charge may originate by powerful electrostatic discharge.

Other gun configurations would simply involve a long muzzle for which an electron discharge is timed to occur just as a round passes the locations of electrostatic discharge.

In order to accelerate a 10,000 metric ton craft to a Lorentz factor of 1.00005 and assuming that half of the pellet explosive energy is converted to spacecraft kinetic energy, the explosive energy embodied in the pellets would need to be approximately equivalent to 0.0001 the mass of the spacecraft or 1 metric ton. Assuming the yield of a nuclear explosive pellet is about 0.0025 of its invariant mass equivalent, the mass of the pellet runway would need to be about 400 metric tons.

A method of converting pellet explosive energy to spacecraft kinetic energy is the deployment of a large charge sail chute that would be pushed ahead of the spacecraft but which would be attached to the spacecraft by an elastic spring system. The recoiling chute would be pulled back to the spacecraft whereupon the next pellet would detonate.

A method of converting pellet explosive energy to spacecraft kinetic energy is the deployment of a large charge sail chute that would be pushed ahead of the spacecraft but which would be attached to the spacecraft by an elastic spring system. The recoiling chute would be pulled back to the spacecraft whereupon the next pellet would detonate.

For explosive charges each having a mass of 40 kilograms, 10,000 devices would be needed. Since each device would be standardized, assuming a unit can be mass produced at a cost of $100,000, the total cost of the fuel pellets would be $1 billion.

For the anticipated 0.0001 light-year long pellet stream, each pellet would need to be separated on average by $10^{-8}$ light-years or by 100 million meters.

Assuming gun muzzle velocity of 100,000 m/s, the average time between pellet launches would be 1,000 seconds.

Assuming a pellet shot mass of 40 kilograms, the kinetic energy of each pellet upon launch would be $2 \times 10^{11}$ joules. Thus, the time averaged power pulse of the gun per shot during the acceleration process would be approximately 200 gigawatts. This is roughly equivalent to the expansive gas power for large caliber artillery pieces.

More specifically, the average thermal loading on the electron gas gun barrel would be about 40 megawatts per 10 meter section of barrel length over a period of one second at 1,000 second intervals. Thus, the barrel would only need thermal sinking at a rate 1,000 times less than thermal loading. Thermal sinking can be obtained by radiative elements attached to the barrel with the option of employing liquid coolants.

For pellet accelerations of 100,000 meter/s$^2$, the propulsive force would be **F**= m**a** = 4 mega-newtons which works out to approximately 1 million pounds of force and of similar order of magnitude to modern artillery. The latter force value is equivalent to the impulse imparted on the gun and is the same for all subsequently provided examples below.

Assuming an orbital location of one AU from the sun, and an adjustable reflector which reflects about 1/3 of the incident sunlight but transmits the rest, we obtain the following relation.

$(4 \times 10^6 \text{ N}) = \langle S\rangle/c = \langle S\rangle/(3 \times 10^8 \text{ m/s})$. So, the required sunlight power to counteract the acceleration on the gun during the firing time would be $1.2 \times 10^{15}$ watts.

Assuming a barrel system mass of 5,000,000 kg, the net barrel acceleration during the firing operation would be $F/m = [(4 \times 10^6 \text{ N})/(5{,}000{,}000 \text{ kg})] = 0.8 \text{ m/s}^2$ and the displacement component of the barrel during the firing process would be $x = ½ \mathbf{a}t^2 = 0.25$ m/s.

Assuming a reverse acceleration time of 1,000 seconds, the barrel reverse acceleration need only be $8 \times 10^{-7}$ m/s$^2$ and thus the required sun-light power on a sail to counteract the firing acceleration effects is a mere $1.2 \times 10^9$ watts. The latter value can be obtained by a light-sail having collection area of a mere 1.2 km$^2$.

Assuming a projectile driving area of 0.01 square meters, the pressure in the barrel would be as high as $4 \times 10^8$ N/m$^2$.

Required gun barrel length is ½ **a**t$^2$ or 50 kilometers.

The total energy required to distribute the runway would be ½ mv$^2$ or $2 \times 10^{15}$ joules. The time averaged power input for the gun over a period of the required 10 million second runway construction process would be 200 megawatts which may in principle be provided by a space-based nuclear reactor.

The required electrical power input may also be achieve is a photo-voltaic array covering about 2/3 km$^2$ assuming a radiative flux density of 1,500 w/m$^2$ and a conversion efficiency of 20 percent.

The cost to install 10 kilowatts of solar PV panels on a house is $25,300 – $31,500 according to Home Guide at https://homeguide.com/costs/solar-panel-cost. So for a 200 megawatts of installed power at the above pricing for domestic home use, the cost would be ($25,300 – $31,500)(20,000) or $506,000,000 to $630,000,000.

Ordinary domestic use PV panels may be covered in a glass or plastic enclosure for use in space. These covering would help prevent oxidation by free radicals common in interplanetary space even though the gas concentration is extremely low compared to sea level atmospheric pressure.

Assuming the above 200 megawatt instillation can include the protective coverings for a manufacturing cost of $1 billion, and that the area specific mass of the system is 10 kilograms per square meter, a fully launched and assembled array of the panels would have mass of (667,000)(10) kilograms 6,670 metric tons. This value is about one order of magnitude greater than the cost of the International Space Station (ISS). However, the cost to place materials in space is projected to reduce by one to perhaps two orders of magnitude per unit mass installed. So, the cost to install the array may be roughly that of the current accrued cost by the ISS give or take one order of magnitude. The worst case scenario would have the rigid PV array installed and operating of about $1 Trillion but as low as $10 Billion.

Arrays operating on concentrated light can strongly reduce the capture area of the PV panels and thus the size of the array. There are some clever ways to greatly reduce the cost of concentrated sunlight systems which are discussed later below.

Turbo-electric systems powered by concentrated sunlight may also be of use. For example, a system of reflectors of capture area 1/3 $km^2$ could concentrate sunlight on a thermal mass to deliver 200 Mw usable electrical power provided a total system efficiency of 40 percent.

Note that most "solar panels are between 15% and 20% efficient, with outliers on either side of the range" according to energysage at https://news.energysage.com/what-are-the-most-efficient-solar-panels-on-the-market/

Domestic home use solar panels can easily work in space where the panels are enclosed in a low cost glass or uv-light resistant transparent plastic. The simply manufacturing process can greatly reduce the cost of the space-based PV arrays.

The thermal build up on the gas gun can be dissipated at similar rates as typical of smaller scale commercial electrical utility plants.

Assuming a barrel mass of 100 kilograms per meter of length, the total barrel mass would be 5 million kilograms or 5,000 metric tons. This is approximately one order of magnitude greater than that of the ISS. However, with anticipated launch to low Earth orbit (LEO) cost reductions by one order of magnitude or more and similarly also for placement of materials in interplanetary space for a cost of 10 million dollars per metric ton, the placement of the material composition of the gun in space would be accomplished for about $50 Billion at today's cost.

The cost of assembling the spacecraft in LEO would be roughly the same as the current instantiation of the ISS assuming a ten-fold cost reduction launch to LEO or about $100 Billion.

A nuclear reactor placed in space with the entire supporting infrastructure would have a cost similar to that accrued by the construction of the ISS at about $ 100 Billion.

So, the combined cost of the project components is estimated to be approximately $350 Billion at current dollar value. Over the life-time of the project, the cost of the project would be about $10 Billion per year.

The electron gas gun fueling of a spacecraft has merit in that relativistic spacecraft velocities can be achieved without the need for concern over otherwise storing a large quantity of nuclear explosives on the spacecraft and the lack of need of supporting architecture to store the devices on the spacecraft.

Each projectile may be fitted with a small low powered fine course correction rocket.

Needed data on the location of the pellets relative to the spacecraft and relative to each other would be provided by electronic transceiver apparatus installed within the pellets. The pellet transmissions can be tracked by Earth based systems and then sent to the exiting spacecraft.

The pellets would be detonated behind a charged space-craft chute having a collection area on the order of 100 square kilometers and attached to the spacecraft by an elastic spring like cable.

Assuming a chute mass-specific area density of 0.01 kilograms per square meter, the mass of the chute and attachment cables would be on the order of 1,000 metric tons.

In order to accelerate a 10,000 metric ton craft to a Lorentz factor of 1.00005 and assuming that 0.005 of the pellet explosive energy is converted to spacecraft kinetic energy, the explosive energy embodied in the pellets would need to be approximately equivalent to 0.01 the mass of the spacecraft or 100 metric tons. Assuming the yield of a nuclear explosive pellet is about 0.0025 of its invariant mass equivalent, the mass of the pellet runway would need to be about 40,000 metric tons.

A method of converting pellet explosive energy to spacecraft kinetic energy is the deployment of a large charge sail chute that would be pushed ahead of the spacecraft but which would be attached to the spacecraft by an elastic spring system. The recoiling chute would be pulled back to the spacecraft whereupon the next pellet would detonate.

For explosive charges each having a mass of 40 kilograms, one million devices would be needed. Since each device would be standardized, assuming a unit can be mass produced at a cost of $100,000, the total cost of the fuel pellets would be $100 billion.

For the anticipated 0.0001 light-year long pellet stream, each pellet would need to be separated on average by $10^{-10}$ light-years or by 1 million meters.

Assuming gun muzzle velocity of 100,000 m/s, the average time between pellet launches would be 10 seconds.

The time averaged thermal loading on the gun would be 100 times that of the previous example.

The total energy required to accelerate the pellets would be 100 times as great as that of the previous example.

The required time averaged electrical power required to operate the gun would be 20 gigawatts.

Assuming an orbital location of one AU from the sun, and an adjustable reflector which reflects about 1/3 of the incident sunlight but transmits the rest, we obtain the following relation.

$(4 \times 10^6 \text{ N}) = \langle S\rangle/c = \langle S\rangle/(3 \times 10^8 \text{ m/s})$. So, the required sunlight power to counteract the acceleration on the gun during the firing time would be $1.2 \times 10^{15}$ watts.

Assuming a barrel system mass of 5,000,000 kg, the net barrel acceleration during the firing operation would be $F/m = [(4 \times 10^6 \text{ N})/(5{,}000{,}000 \text{ kg})] = 0.8 \text{ m/s}^2$ and the displacement component of the barrel during the firing process would be $x = ½ \mathbf{a}t^2 = 0.25$ m.

Assuming a reverse acceleration time of 10 seconds, the barrel reverse acceleration need only be $8 \times 10^{-3} \text{ m/s}^2$ and thus the required sun-light power on a sail to counteract the firing acceleration effects is a mere $1.2 \times 10^{13}$ watts. The latter value can be obtained by a light-sail having collection area of a mere 12,000 $\text{km}^2$. The latter sail is rather large, but only the equivalent of 100 km by 120 km in size. Some solar sail concepts assume mass specific sail area of only one milligram per square meter thus enabling the entire reflective surface to have mass of 12,000 kilograms. A more robust sail would have an area density 100 times greater to yield a total sail mass of 1,200 metric tons.

Additionally, the total cost to assemble and launch the pellets would be about 100 times greater than in the previous example as would the cost of assembling and fielding the gun and its power sources.

For nuclear fission reactor power sources, the cost of launching and assembly would be very expensive as would the cost of any chosen rigid PV panel power supplies. However, as the space-travel and mining infrastructure becomes more highly developed, the total cost associated with these systems in current US dollars and as a fraction of GDP may and is expected to strongly reduce eventually to practical levels.

Here, we are considering gun placement in solar orbit but with adjustable orientation via retro-rockets, solar sails, and/or plasma sails. The orientation of the gun and its orbital radius can be re-adjusted as needed by sail tacking processes.

In order to archive spacecraft deceleration, electrodynamic breaking mechanisms may be employed.

For present consideration, a mechanism of choice would include one or more linear induction breaking mechanisms. Accordingly, one or more large highly conducting or super-conducting coils would be deployed. The coils would build up large currents which would then interact with the background magnetic fields to bring the spacecraft velocity down to Keplerian velocities; upon which small scale velocity correction rockets would be deployed. These rockets may include nuclear electric systems such as traditional ion thrusters or engines such as Ad Astra Rocket Company's VASIMIR Engine. See references for the VASIMIR Engine.

There is also an option for which a pre-deployed pellet stream displaced in the outer reaches of the solar system would be used to slow the craft. The breaking thermodynamics would be simply the reverse of the positive acceleration process.

To travel back to Earth, another pellet stream would be deployed to accelerate the spacecraft on a heading back to Earth. The thermodynamics of the return accelerating process would closely match that of the initial outbound acceleration.

Once again, yet another pellet stream would be used to slow the spacecraft before obtaining close proximity to Earth. The reason for the latter option is that nuclear explosions near the Earth may result in EMP effects which can damage satellites and electrical power grids on Earth.

The above proposed propulsion missions can be repeated many times and offer strong solutions to the problems of what to do in the near term with excess fissile plutonium and uranium stock piles

Once space-based mining of minerals becomes possible, the ability to find and refine uranium and thorium ores on the Moon, Mars, Mercury, the asteroids and the Kuiper Belt and Oort cloud object should enable enough fissile fuels or breeder reactor feedstocks to enable far more nuclear explosive charges to be produced than would be necessary for one or several missions.

The electron gas gun fueling of a spacecraft has merit in that relativistic spacecraft velocities can be achieved without the need for concern over otherwise storing a large quantity of nuclear explosives on the spacecraft and the lack of need of supporting architecture to store the devices on the spacecraft.

The spacecraft and the exiting fuel pellets would have fine course correction mechanisms and would inter-communicate to enable suitable temporal calibration of the spacecraft arrival in proximity to each pellet. Clock features in pellets can be based on robust quartz oscillations mechanisms since atomic clocks may not survive the launch forces.

For example, the pellets may have a charging feature that enables them to use the Lorentz turning force to adjust their flight paths as well as optional solar sails, small rockets, small explosive thrusting charges, and the like. Additionally, the pellets may have laser sails for which a beam of light from stations near Earth can induce a drive pressure on the pellets and/or spacecraft for finer course correction. The orientation of the sails can be adjusted for tacking style thrust to accelerate the pellets or spacecraft out alignment of the pellets and spacecraft velocity vectors.

These proposed methods entail a long term commitment and the development of suitable space travel industries. Once enabled, very great spacecraft Lorentz factors can be obtained with very small mass-ratios. Pellet launch rates and velocities can be up or down adjusted as needed. The above proposal is bold and costly however it is never too early to plan such missions. Once space-based mining and manufacturing develop, the feasibility will greatly increase.

In order to fabricate the number of needed nuclear charges, extremely low critical mass fissile fuels may be used for the nuclear primaries along with neutron reflectors to further reduce the critical mass of explosively compressed or combined fissile fuels during the detonation process.

The muzzle velocities of the guns can be adjusted so that the velocity, location, and timing of the ship arrival per pellet can be properly calibrated.

An interesting optimization would involve pellets fired in a direction tangential to Earth's orbit around the Sun and for pellets fired from Earth orbit, tangential to Earth orbit as well. This way, these two increased velocity components can enable reduced muzzle velocity.

Pellets may also be fired in such a flight pattern that they receive gravity assists from the gas giant planets of Jupiter and/or Saturn. Accordingly, the muzzle velocity of the pellets can be further reduced. These enhanced efficiencies can apply to many scenarios including but not limited to the ones presented in this book.

Launch rates of the pellets can be adjusted or made adjustable as can the yields of the pellets for optimal drive efficiency and abilities of the spacecraft systems to handle the nuclear blasts.

Additionally, optimized blends of fissile fuels can be used in the composition of the primaries thus offering research options to greatly reduce the required critical mass of the needed fissile fuels.

Additionally, U-238 based shots might replace conventional nuclear explosives for which neutron irradiation of the U-238 by a neutron beam originating from the ship would induce nuclear fission of the U-238 shots. The fissioning U-238 may set of a nuclear fusion fuel secondary which may have a yield 10 to 100 times that of the primary.

The Ronen fissile rule states that for a heavy element with $90 \leq Z \leq 100$, its isotopes with $2 \times Z - N = 43 \pm 2$, with few exceptions, are fissile (where N = number of neutrons and Z = number of protons

The fissile rule indicates 33 isotopes as likely fissile: Th-225, 227, 229; Pa-228, 230, 232; U-231, 233, 235; Np-234, 236, 238; Pu-237, 239, 241; Am-240, 242, 244; Cm-243, 245, 247; Bk-246, 248, 250; Cf-249, 251, 253; Es-252, 254, 256; Fm-255, 257, 259. Only fourteen (including a long-lived metastable nuclear isomer) have half-lives of at least a year: Th-229, U-233, U-235, Np-236, Pu-239, Pu-241, Am-242m, Cm-243, Cm-245, Cm-247, Bk-248, Cf-249, Cf-251 and Es-252. Of these, only U-235 is naturally occurring. It is possible to breed U-233 and Pu-239 from more common naturally occurring isotopes (Th-232 and U-238 respectively) by single neutron capture. The others are typically produced in smaller quantities through further neutron absorption.

Note that in nuclear fissile primaries, it is plausible that radioactive heavy isotopes may be installed that decay from a non-fissile isotope to a fissile isotope. This process may in principle extend the useful life-time of a nuclear fissile primary.

When adding tritium as a nuclear fission primary booster, the critical mass of a primary can be further reduced.

Assuming 33 known likely fissile isotopes, the number of subsets of these isotopes is [(2 EXP 33) – 1] Assuming the isotopic blends can be adjusted in a million different fractional combinations per set of two or more isotopes per primary, the number of testable combinations for optimization becomes almost equal to [(1,000,000 EXP [(2 EXP 33) – 1]].

Note that the shape of the chute can include elongated charged tubular configurations so as to more efficiently capture and back accelerate the plasma from the nuclear explosions. Moreover, the chute can have a pointed or tapered leading edge to reduce astrodynamics drag.

Exiting spacecraft may initially receive each shot at a rate of 1 per second to 1 every 100 seconds to enable timely spacecraft acceleration. Once the velocity of the spacecraft was brought up to high enough velocity, then the frequency of the shot in the background reference frame may be reduced.

The United States in reality can support the cost of the above projects since the country was able to fund the war efforts in the Middle East and central Asia during the first two decades of the 21st century with a total cost of several trillion dollars.

For a 0.01 light-year long distributed runway and spacecraft frame acceleration on the order of 0.5 earth g's or 4.905 $m/s^2$, a human crewed spacecraft could obtain a velocity of about 10 percent of light-speed and a Lorentz factor of 1.005.

In order to accelerate a 10,000 metric ton craft to a Lorentz factor of 1.005 and assuming that half of the pellet explosive energy is converted to spacecraft kinetic energy, the explosive energy embodied in the pellets would need to be approximately equivalent to 0.01 the mass of the spacecraft or 100 metric tons. Assuming the yield of a nuclear explosive pellet is about 0.0025 of its invariant mass equivalent, the mass of the pellet runway would need to be about 40,000 metric tons.

A method of converting pellet explosive energy to spacecraft kinetic energy is the deployment of a large charge sail chute that would be pushed ahead of the spacecraft but which would be attached to the spacecraft by an elastic spring system. The recoiling chute would be pulled back to the spacecraft whereupon the next pellet would detonate.

For explosive charges each having a mass of 40 kilograms, 1 million devices would be needed. Since each device would be standardized, assuming a unit can be mass produced at a cost of $100,000, the total cost of the fuel pellets would be $100 billion.

For the anticipated 0.01 light-year long pellet stream, each pellet would need to be separated on average by $10^{-8}$ light-years or by 100 million meters.

Assuming gun muzzle velocity of 100,000 m/s, the average time between pellet launches would be 1,000 seconds.

Assuming a pellet shot mass of 40 kilograms, the kinetic energy of each pellet upon launch would be $2 \times 10^{11}$ joules. Thus, the time averaged power pulse of the gun per shot during the acceleration process would be approximately 200 gigawatts. This is roughly equivalent to the expansive gas power for large caliber artillery pieces.

More specifically, the average thermal loading on the electron gas gun barrel would be about 40 megawatts per 10 meter section of barrel length over a period of one second at 1,000 second intervals. Thus, the barrel would only need thermal sinking at a rate 1,000 times less than thermal loading. Thermal sinking can be obtained by radiative elements attached to the barrel with the option of employing liquid coolants.

For pellet accelerations of 100,000 meter/s$^2$, the propulsive force would be **F**= m**a** = 4 mega-newtons which works out to approximately 1 million pounds of force and of similar order of magnitude to modern artillery.

Required gun barrel length is ½ **a**t$^2$ or 50 kilometers.

The total energy required to distribute the runway would be ½ mv$^2$ or $2 \times 10^{17}$ joules. The time averaged power input for the gun over a period of the required 1 billion second runway construction process would be 200 megawatts which may in principle be provided by a lunar-based nuclear reactor.

The thermal build up on the gas gun can be dissipated at similar rates as typical of smaller scale commercial electrical utility plants.

The time averaged power requirement to operate the gun and the cost and mass of the power generation stations would be the same as those for the first example considered in this book accept for cost components additionally accrued during the operation life-time of the gun.

Assuming a barrel mass of 100 kilograms per meter of length, the total barrel mass would be 5 million kilograms or 5,000 metric tons. This is approximately one order of magnitude greater than that of the ISS. However, with anticipated launch to low Earth orbit (LEO) cost reductions by one order of magnitude or more and similarly also for placement of materials on the lunar surface for a cost of 10 million dollars per metric ton, the placement of the material composition of the gun on the Moon would be accomplished for about $50 Billion at today's cost.

The cost of assembling the spacecraft in LEO would be roughly the same as the current instantiation of the ISS assuming a ten-fold cost reduction launch to LEO or about $100 Billion.

A nuclear reactor placed on the moon would have a cost similar to that accrued by the construction of the ISS at about $ 100 Billion.

So, the combined cost of the project components is estimated to be approximately $350 Billion at current dollar value. Over a thirty year project span, the cost of the project would be about $10 Billion per year.

In order to accelerate a 10,000 metric ton craft to a Lorentz factor of 1.005 and assuming that 0.05 of the pellet explosive energy is converted to spacecraft kinetic energy, the explosive energy embodied in the pellets would need to be approximately equivalent to 0.1 the mass of the spacecraft or 1,000 metric tons. Assuming the yield of a nuclear explosive pellet is about 0.0025 of its invariant mass equivalent, the mass of the pellet runway would need to be about 400,000 metric tons.

A method of converting pellet explosive energy to spacecraft kinetic energy is the deployment of a large charge sail chute that would be pushed ahead of the spacecraft but which would be attached to the spacecraft by an elastic spring system. The recoiling chute would be pulled back to the spacecraft whereupon the next pellet would detonate.

For explosive charges each having a mass of 40 kilograms, 10 million devices would be needed. Since each device would be standardized, assuming a unit can be mass produced at a cost of $100,000, the total cost of the fuel pellets would be $1 Trillion.

For the anticipated 0.01 light-year long pellet stream, each pellet would need to be separated on average by $10^{-9}$ light-years or by 10 million meters.

Assuming gun muzzle velocity of 100,000 m/s, the average time between pellet launches would be 100 seconds.

Assuming an orbital location of one AU from the sun, and an adjustable reflector which reflects about 1/3 of the incident sunlight but transmits the rest, we obtain the following relation.

($4 \times 10^6$ N) = <S>/c = <S>/($3 \times 10^8$ m/s). So, the required sunlight power to counteract the acceleration on the gun during the firing time would be $1.2 \times 10^{15}$ watts.

Assuming a barrel system mass of 5,000,000 kg, the net barrel acceleration during the firing operation would be F/m = [($4 \times 10^6$ N)/(5,000,000 kg)] = 0.8 m/s$^2$ and the displacement component of the barrel during the firing process would be x = ½ **a**t$^2$ = 0.25 m/s.

Assuming a reverse acceleration time of 100 seconds, the barrel reverse acceleration need only be $8 \times 10^{-5}$ m/s$^2$ and thus the required sun-light power on a sail to counteract the firing acceleration effects is a mere $1.2 \times 10^{11}$ watts. The latter value can be obtained by a light-sail having collection area of a mere 120 km$^2$. This is a mere equivalent of 12 km by 10 km or roughly the size of a modest city in land area. For a sail having capture area of 120 km and a mass specific area of one milligram per square meter, the total mass of the sail would be a mere 120 kilograms. A sail having area mass density of one gram per square meter would have mass of 120 metric tons.

The time averaged thermal loading on the gun would be 10 times that of the previous example.

The total energy required to accelerate the pellets would be 10 times as great as that of the previous example.

Additionally, the total cost to assemble and launch the pellets would be about 10 times greater than in the previous example.

The electron gas gun fueling of a spacecraft has merit in that relativistic spacecraft velocities can be achieved without the need for concern over otherwise storing a large quantity of nuclear explosives on the spacecraft and the lack of need of supporting architecture to store the devices on the spacecraft.

The nuclear fuel pellets for missions from earth to other stars positioned along the plane of the solar system would require that the projectiles be sprayed from the gun over a small time interval once approximately every complete lunar orbit around the Earth.

For mission to destinations orthogonally displaced from the plane of Earth's orbit around the sun, the projectiles can be continuously launched.

Each projectile may be fitted with a small, low powered, fine course correction rocket.

Needed data on the location of the pellets relative to the spacecraft and relative to each other would be provided by electronic transceiver apparatus installed within the pellets. The pellet transmissions can be tracked by Earth based systems and then sent to the exiting spacecraft.

The pellets would be detonated behind a charged space-craft chute having a collection area on the order of 100 square kilometers and attached to the spacecraft by an elastic spring-like cable.

Assuming a chute mass-specific area density of 0.01 kilograms per square meter, the mass of the chute and attachment cables would be on the order of 1,000 metric tons.

In order to archive relativistic spacecraft deceleration at the intended solar system, electrodynamic breaking mechanisms may be employed.

For present consideration, a mechanism of choice would include one or more linear induction breaking mechanisms. Accordingly, one or more large highly conducting or super-conducting coils would be deployed. The coils would build up large currents which would then interact with the background magnetic fields to bring the spacecraft velocity down to Keplerian velocities; upon which small scale velocity correction rockets would be deployed. These rockets may include nuclear electric systems such as traditional ion thrusters or engines such as Ad Astra Rocket Company's VASIMIR Engine.

Fine course correction of the spacecraft can also be enabled by a Stellar Cycler mechanism for which a spacecraft can attain a strong net charge and experience curvilinear acceleration by the relativistic Lorentz turning force.

For missions to destinations obliquely oriented with respect to the plane of the solar system, the Stellar Cycler method can re-route spacecraft on initial headings substantially orthogonal to the plane of the solar system. Thus, the problem of needing to fire pellets in short monthly intervals is avoided.

Once space-based mining of minerals becomes possible, the ability to find and refine uranium and thorium ores on the Moon, Mars, Mercury, the asteroids and the Kuiper Belt and Oort cloud object should enable enough fissile fuels or breeder reactor feedstocks to enable far more nuclear explosive charges to be produced than would be necessary for one or several missions. The electron gas gun fueling of a spacecraft has merit in that relativistic spacecraft velocities can be achieved without the need for concern over otherwise storing a large quantity of nuclear explosives on the spacecraft and the lack of need of supporting architecture to store the devices on the spacecraft.

These proposed methods entail a long term commitment and the development of suitable space travel industries. Once enabled, very great spacecraft Lorentz factors can be obtained with very small mass-ratios. Pellet launch rates and velocities can be up or down adjusted as needed. The above proposal is bold and costly however it is never too early to plan such missions. Once space-based mining and manufacturing develop, the feasibility will greatly increase.

In order to fabricate the number of needed nuclear charges, extremely low critical mass fissile fuels may be used for the nuclear primaries along with neutron reflectors to further reduce the critical mass of explosively compressed or combined fissile fuels during the detonation process.

The muzzle velocities of the guns can be adjusted so that the velocity, location, and timing of the ship arrival per pellet can be properly calibrated.

Launch rates of the pellets can be adjusted or made adjustable as can the yields of the pellets for optimal drive efficiency and abilities of the spacecraft systems to handle the nuclear blasts.

Exiting spacecraft may initially receive each shot at a rate of 1 per second to 1 every 100 seconds to enable timely acceleration. Once the velocity of the spacecraft was brought up to high enough velocity, then the frequency of the shot in the background reference frame may be reduced.

In short, the electron gas compression gun has merit in that it may enable relativistic spacecraft velocities without the need for large mass ratios and is simpler in design than traditional mass drivers.

Taken to extreme circumstances, spacecraft propelled by nuclear pellet streams may obtain velocities very close to the speed of light and extreme Lorentz factors. Tantalum-Hafnium- Carbide solutions in shielding material may in principle enable spacecraft to attain Lorentz factors at least as great as around 1,500 at which point the cosmic microwave background radiation will be relativistically blue-shifted and aberrated to an apparent black body temperature near the melting point of said carbide solutions.

For a 1,000 light-year long outwardly spiraling and distributed runway and tangential spacecraft frame acceleration on the order of 0.5 earth g's or 4.905 $m/s^2$, a human crewed spacecraft could obtain a velocity of about 99.9998 percent of light-speed and a Lorentz factor of 500.

In order to accelerate a 10,000 metric ton craft to a Lorentz factor of 500 and assuming that half of the pellets' explosive energy is converted to spacecraft kinetic energy, the explosive energy embodied in the pellets would need to be approximately equivalent to 1,000 times the mass of the spacecraft or 10,000,000 metric tons. Assuming the yield of a nuclear explosive pellet is about 0.0025 of its invariant mass equivalent, the mass of the pellet runway would need to be about 4 billion metric tons.

For explosive charges each having a mass of 40 kilograms, 100 billion devices would be needed. Since each device would be standardized, assuming a unit can be mass produced at a cost of $100, the total cost of the fuel pellets would be $10 trillion. We consider using fissile isotopes having small critical mass and nuclear fusion fuels having low activation energies. Neutron reflectors may further reduce the mass of the required fissile primaries. Innovative cost reductions for nuclear explosives are anticipated as new low mass fissile primaries are developed.

For the anticipated 1,000 light-year long pellet stream, each pellet would need to be separated on average by $10^{-8}$ light-years or by 100 million meters. (Note that it is fascinating to consider projects taking culturally deep future time periods for which one or more substantially linear or straight pellet runways would be constructed. Accordingly, the run-ways may far exceed 1,000 light-years in length. Such lengthy straight pellet streams may enable spacecraft to achieve ultra-high relativistic Lorentz factors without the need for the cycler mechanisms and associated very high charging of a spacecraft. Additionally, the spacecraft acceleration may be reduced to a comfortable acceleration, one that can substantially duplicate the invariant mass-specific gravitational force acting on objects on the surface of the Earth.)

Note that an interesting way to greatly reduce the required centripetal acceleration in the ship frame is to simply take the time to construct a 1,000 to 100,000 light-year radius pellet stream in spiraling form. A much larger spacecraft could harness the pellets yet with significantly reduced centripetal acceleration. The spacecraft may make several too many revolutions then disengage from the galaxy toward cosmically remote destinations.

Assuming a gun muzzle velocity of 100,000 m/s and average time of each pellet launch of one second, 100 guns in solar orbit would deploy the pellet stream in its initial position in 1 billion seconds.

The guns may have adjustable orientations so as to fire the pellets in the proper direction.

A spacecraft may begin its outwardly spirally journey with only a modest portion of the pellets being launched.

After 33 billion seconds or about 1,000 years, the portion of the spiral runway intake by the spacecraft would be about 1/3 of a light-year distant from Earth.

Now the relativistic Lorentz turning force:

$$|F_L| = |d[(\gamma)(M_0)(v)]/dt| = |(\gamma)(M_0)(dv/dt)| = |(\gamma)(M_0)(a)|,$$

where $M_0$ it the particle or craft rest mass, v is the particle or crafts velocity, a is the centripetal acceleration of the craft, and $\gamma = 1/\{[1 - [(v/C)^2]]^{1/2}\}$.

$|Fx| = |(q)(B_F)(V_y)| = |(\gamma)(M_0)[d(V_x)/dt]|$; $|Fy| = |-(q)(B)(V_x)| = |(gamma)(M_0)[d(V_y)/dt]|$; Fz = Zero N.

Note that $B_F = (500)(B)$

So assuming gamma of 500, $|\mathbf{a}| = |(F_L)/[(500)(10^7)]|$, for a value of centripetal acceleration or **a** = 10,000 m/s$^2$ in the spacecraft frame, the Lorentz turning force becomes 5 x $10^{13}$ newtons or about 5 x $10^9$ metric tons force in the background reference frame but only about 10 million metric tons or 1,000 g's in the spacecraft reference frame. The pellet distribution and uptake rate in the ship frame can be adjusted as needed. However, the spacecraft would require extreme electrical charging.

Now, the strength of the solar magnetic field at one AU from the Sun is about 6 nanotesla on average. So assuming Fx = (22,200)(5 x $10^9$ N), $V_x$ is essentially equal to almost light speed, and B = (500)(6 x $10^{-9}$ tesla), we compute q as 1.2333 x $10^{11}$ coulombs. Now, making a short hand approximation for the sake of simplicity that the electric charge consists of two portions, each of which is ½ the above electrical charge and located 10,000,000,000 meters away from each other, we obtain a tensile stress on the conduit of $F = k_e (q_1 q_2)/(r^2) = (8.99\times10^9 \text{ N}\cdot\text{m}^2\cdot\text{C}^{-2})\{[(1/2)(1.2333 \times 10^{11} \text{ C})]^2\}/(10^{10} \text{ m}^2) = 3.4186 \times 10^{11}$ N

Obviously, no known industrial material of ordinary charge configuration could handle the coulombic tension.

However, bulk element neutronium has a theoretical tensile strength of about $10^{34}$ N/m$^2$ and density of about $10^{18}$ kg/m$^3$. So, a drive sail made of neutron element that is one neutron thick or about $10^{-15}$ meters thick but also $10^{-6.5}$ meters wide and 10,000,000,000 meters long would have mass of 3,162,200 kg and a tensile strength of 3.1622 x $10^{12}$ N. Thus, the sail might be able to handle the strain provided the sail had a triangular or curved triangular shape that widens in the direction from the most distant portion of the sail from the ship to the ship. The mass of the ship and sail combined could be set at $10^7$ kg.

Now, the strength of the magnetic field of the Milky Way Galaxy is about 0.1 nanotesla on average. So assuming Fx = (5 x $10^9$ N), $V_x$ is essentially equal to almost light speed, and B = (500)(1 x $10^{-9}$ tesla), we compute q as 3.333 x $10^8$ coulombs. Now, making a short hand approximation for the sake of simplicity that the electric charge consists of two portions, each of which is ½ the above electrical charge and located 100,000 meters away from each other, we obtain a cpulombic tensile stress on the conduit of $F = k_e (q_1 q_2)/(r^2) = (8.99\times10^9 \text{ N}\cdot\text{m}^2\cdot\text{C}^{-2})\{[(1/2)(3.333 \times 10^8 \text{ C})]^2\}/(10^{10} \text{ m}^2) = 2.49717 \times 10^{16}$ N

Obviously, no known industrial material of ordinary charge configuration could handle the coulombic tension.

However, perhaps we could fabricate a thin drive membrane having an areal density of one milligram per square meter and comprised of nanometer scale electrically charged cells bound by topological isolators that result in no electrical field components along the plane of the membrane. If we assume the membrane has mass of one thousand metric tons, then the membrane would have a surface area of 1 million square kilometers and contain $10^{30}$ cells.

The topological insulator mechanism may also be applied to neutronium so as to enable much greater revolutionary velocities around the Sun.

The Sun has a magnetic field that varies across the surface. Its polar field is 0.0001T to 0.0002 T.

So let us assume that the solar magnetic field 0.01 AU from the solar center averages 0.0001 tesla.

So assuming gamma of 500, $|\mathbf{a}| = |(F_L)/[(500)(10^7)]|$, for a value of centripetal acceleration or **a** = 1,000,000 m/s$^2$ in the spacecraft frame, the Lorentz turning force becomes 5 x $10^{14}$ newtons or about 5 x $10^{11}$ metric tons force.

Again, the strength of the solar magnetic field at 0.01 AU from the Sun's center is about 0.0001 tesla on average.. So assuming Fx = (2,220,000)(5 x $10^9$ N), $V_x$ is essentially equal to almost light speed, and B = (500)(1 x $10^{-4}$ tesla), we compute q as 7.4 x $10^8$ coulombs. Now, making a short hand approximation for the sake of simplicity that the electric charge consists of two portions, each of which is ½ the above electrical charge and located 100,000,000 meters away from each other, we obtain a coulombic coulombic tensile stress on the conduit of $F = k_e (q_1 q_2)/(r^2) = (8.99\times10^9 \text{ N}\cdot\text{m}^2\cdot\text{C}^{-2})\{[(1/2)(7.4 \times 10^8 \text{ C})]^2\}/(10^{10} \text{ m}^2) = 1.2307 \times 10^7$ N

A matrix or lattice of properly orientated super-conducting electromagnets comprised of very small individual unit magnets can be properly distributed so as to balance the intra-coulombic cycle sail tension.

Accordingly, the coulombic sail can be a cone, done, a curved-cone or similar shape so that the electromagnets affixed to the sail have unblocked magnetic or uncancelled magnetic flux so as to efficiently relieve the tension on the sail.

At some point in time synchrotron radiation bleed off will become a problem. However, acceleration of protons in the Large Hadron Collider to a Lorentz factor of about 10,000 is done relatively efficiently. The mass specific synchrotron radiation bleed of decreases very strongly with increasing mass to charge ratios. So given the reality that the mass to charge ratio of protons is many orders of magnitude less than that of the proposed stellar cyclers and the radius of revolution of the stellar cyclers is many orders of magnitude greater than that of particles in the LHC gives hope that extreme solar cycling spacecraft can be enabled with suitably high-strength and refractory materials. Also note that the particle invariant mass specific synchrotron radiation power reduces rapidly with revolutional radius.

Another drive mechanism is plausible.

For example, consider a wire or conductive coil that is axially oriented in a direction parallel to the spacecraft velocity vector. Here, we assume that the wire or coil is made of neutronium, quarkonium, and the like extremely high tensile strength materials. Another fascinating option would involve more or less ordinary periodic table elemental composition but where the centripetal tension on the wire or cable is balanced by electrodynamic attraction of the rotating member to a central by electric and/or magnetic fields. The hub material would need to likely be made of neutronium or other similarly dense material. The linear induction coil generator will operate on more or less static background interstellar and intergalactic magnetic fields as well as magnetic fields in solar systems. For relativistic spacecraft, the background magnetic field components will be:

**Bx** = Bx

**By** = $\gamma\{B_y + \{[[v/[C^2]]E_z\}\} = \{1 - [(v/C)^2]\}^{-1/2}\{B_y + \{[[v/[C^2]]E_z\}\}$

and **Bz** = $\gamma\{B_z - \{[[v/[C^2]]E_y\}\} = \{1 - [(v/C)^2]\}^{-1/2}\{B_z - \{[[v/[C^2]]E_y\}\}$

Here $\mathbf{B_x}$ is the component of the background magnetic field that is parallel to the spacecraft velocity vector.

Now, in the spacecraft reference frame, the voltage developed within a conductive coil is equal to $E_{mf} = -N\Delta\phi/\Delta t = -N\Delta(BA)/\Delta t_{ship}$ where N is the number of cable turns, $\phi$ = BA is the magnetic flux, and $t_{ship}$ is the ship frame time.

Assuming $N_c$ coils where the $N_{c,gth,}$ coil has an plan form area $A_{c,gth,j}$, in the jth time step, and where the background magnetic field in the jth ship time step is $Bx_j + \gamma_j\{B_{y,j} + \{[[v_j/[C^2]]E_{z,j}\}\} + \gamma_j\{B_{z,j} - \{[[v_{,j}/[C^2]]E_{y,j}\}\} = Bx_j + \{1 - [(v_j/C)^2]\}^{-1/2}\{B_{y,j} + \{[[v_j/[C^2]]E_{z,j}\}\} + \{1 - [(v_j/C)^2]\}^{-1/2}\{B_{z,j} - \{[[v_j/[C^2]]E_{y,j}\}\}$.

The electric power generated in the cable having a resistance $R_j$ in the jth time step is equal to:

$P_j = V_j^2/R_j = E_{mf,j}{}^2/R_j = [-N_{c,gth}\,\Delta\phi j/\Delta t_{ship,j}]^2/R_j = \{-N_{c,gth}\,\Delta\{Bx_j + \gamma_j\{B_{y,j} + \{[[v_j/[C^2]]E_{z,j}\}\} + \gamma_j\{B_{z,j} - \{[[v_{,j}/[C^2]]E_{y,j}\}\}\}\,(A_{c,gth,j})/\Delta t_{ship,j}\}^2/R_j$

Another drive mechanism is plausible.

For example, consider the background magnetic in the spacecraft reference frame.

**Bx** = Bx

**By** = $\gamma\{B_y + \{[[v/[C^2]]E_z\}\} = \{1 - [(v/C)^2]\}^{-1/2}\{B_y + \{[[v/[C^2]]E_z\}\}$

and **Bz** = $\gamma\{B_z - \{[[v/[C^2]]E_y\}\} = \{1 - [(v/C)^2]\}^{-1/2}\{B_z - \{[[v/[C^2]]E_y\}\}$

Here $\mathbf{B_x}$ is the component of the background magnetic field that is parallel to the spacecraft velocity vector.

Here Bx, By, and Bz are the background frame magnetic field components.

The background magnetic field in the jth ship time step is $Bx_j + \gamma_j\{B_{y,j} + \{[[v_j/[C^2]]E_{z,j}\}\} + \gamma_j\{B_{z,j} - \{[[v_{,j}/[C^2]]E_{y,j}\}\} = Bx_j + \{1 - [(v_j/C)^2]\}^{-1/2}\{B_{y,j} + \{[[v_j/[C^2]]E_{z,j}\}\} + \{1 - [(v_j/C)^2]\}^{-1/2}\{B_{z,j} - \{[[v_j/[C^2]]E_{y,j}\}\}$.

Now, the magnetic energy stored in a volume of space is equal to:

$B_{volume} = (\text{Volume})(1/2)[B^2]/\mu$.

So the magnetic energy stored in a unit volume of space in the spacecraft reference frame for the jth time-step in the spacecraft reference frame will be:

$B_{volume,j} = (\text{Volume}_{shipframej})(1/2)[B_{spacecraft}{}^2]/\mu$

$= (\text{Volume}_j)(1/2)\{\{Bx_j + \{1 - [(v_j/C)^2]\}^{-1/2}\{B_{y,j} + \{[[v_j/[C^2]]E_{z,j}\}\} + \{1 - [(v_j/C)^2]\}^{-1/2}\{B_{z,j} - \{[[v_j/[C^2]]E_{y,j}\}\}\}^2\}/\mu$.

The total magnetic energy captured by the non-rotating pinwheels is:

$\Sigma(j = 1; j = m)[B_{volume,j}] = \Sigma(j = 1; j = m)\,\{(\text{Volume}_{shipframej})(1/2)[B_{spacecraft}{}^2]/\mu\}$

$= \Sigma(j = 1; j = m)\,\{(\text{Volume}_j)(1/2)\{\{Bx_j + \{\{1 - [(v_j/C)^2]\}^{-1/2}\{B_{y,j} + \{[[v_j/[C^2]]E_{z,j}\}\}\} + \{\{1 - [(v_j/C)^2]\}^{-1/2}\{B_{z,j} - \{[[v_j/[C^2]]E_{y,j}\}\}\}\}^2\}/\mu\}$.

Considering electric field analogue, we have the following analysis:

**Ex** = Ex

**Ey** = $\gamma[E_y - (vB_z)] = \{\{1 - [(v/C)^2]\}^{-1/2}\}[E_y - (vB_z)]$

and **Ez** = $\gamma[E_z + (vB_y)] = \{\{1 - [(v/C)^2]\}^{-1/2}\}[E_z + (vB_y)]$

Here $\mathbf{E_x}$ is the component of the background electric field that is parallel to the spacecraft velocity vector.

Here Ex, Ey, and Ez are the background magnetic field components.

The background electric field in the jth ship time step is $Ex_j + \gamma_j[E_{y,j} - (v_j B_{z,j})] + \gamma_j [E_{z,j} + (v_j B_{y,j})] = Ex_j + \{\{1 - [(v_j /C)^2]\}^{-1/2}\}[E_{y,j} - (v_j B_{z,j})] + \{\{1 - [(v_j /C)^2]\}^{-1/2}\}[E_{z,j} + (v_j B_{y,j})]$.

Now, the electric energy stored in a volume of space is equal to:

$E_{volume} = (\text{Volume})(1/2)(\varepsilon_0)[E^2]$.

So the electrical energy stored in a unit volume of space in the spacecraft reference frame for the jth time-step in the spacecraft reference frame will be:

$E_{volume,j} = (\text{Volume}_{shipframej})\ (1/2)(\varepsilon_0)[E^2]$

$= (\text{Volume}_j)\ (1/2)(\varepsilon_0)\{\{Ex_j + \{\{\{1 - [(v_j /C)^2]\}^{-1/2}\}[E_{y,j} - (v_j B_{z,j})]\} + \{\{\{1 - [(v_j /C)^2]\}^{-1/2}\}[E_{z,j} + (v_j B_{y,j})]\}\}^2\}$

Regarding linear induction coils, they may be augmented with relativistic rotational motion such as by way of induction coil pin-wheel rotors that rotate at velocities near the speed of light in the spacecraft reference frame.

For example, consider a wire or conductive coil that is axially oriented in a direction parallel to the spacecraft velocity vector. The linear induction coil generator will operate on more or less static background interstellar and intergalactic magnetic fields as well as magnetic fields in solar systems. For relativistic spacecraft, the background magnetic field components will be:

**Bx** = Bx

**By** = $\gamma\{B_y + \{[[v/[C^2]]E_z\}\} = \{1 - [(v/C)^2]\}^{-1/2}\{B_y + \{[[v/[C^2]]E_z\}\}$

and **Bz** = $\gamma\{B_z - \{[[v/[C^2]]E_y\}\} = \{1 - [(v/C)^2]\}^{-1/2}\{B_z - \{[[v/[C^2]]E_y\}\}$

Here $\mathbf{B_x}$ is the component of the background magnetic field that is parallel to the spacecraft velocity vector.

Now, in the spacecraft reference frame, the voltage developed within a conductive coil is equal to $E_{mf} = -N\Delta\phi/\Delta t = -N\Delta(BA)/\Delta t_{ship}$ where N is the number of cable turns, $\phi = BA$ is the magnetic flux, and $t_{ship}$ is the ship frame time.

Assuming $N_c$ coils where the $N_{c,gth,}$ coil has an plan form area $A_{c,gth,j}$, in the jth time step, and where the background magnetic field in the jth ship time step is $Bx_j + \gamma_j\{B_{y,j} + \{[[v_j/[C^2]]E_{z,j}\}\} + \gamma_j\{B_{z,j} - \{[[v_{,j}/[C^2]]E_{y,j}\}\} = Bx_j + \{1 - [(v_j/C)^2]\}^{-1/2}\{B_{y,j} + \{[[v_j/[C^2]]E_{z,j}\}\} + \{1 - [(v_j/C)^2]\}^{-1/2}\{B_{z,j} - \{[[v_j/[C^2]]E_{y,j}\}\}$.

The electric power generated in the cable having a resistance $R_j$ in the jth time step is equal to:

$P_j = V_j^2/R_j = E_{mf,j}{}^2/R_j = [-N_{c,gth}\ \Delta\phi j/\Delta t_{ship,j}]^2/R_j = \{-N_{c,gth}\ \Delta\{Bx_j + \gamma_j\{B_{y,j} + \{[[v_j/[C^2]]E_{z,j}\}\} + \gamma_j\{B_{z,j} - \{[[v_{,j}/[C^2]]E_{y,j}\}\}\}\ (A_{c,gth,j})/\Delta t_{ship,j}\}^2/R_j$

Another drive mechanism is plausible.

For example, consider the background magnetic in the spacecraft reference frame.

**Bx** = Bx

**By** = $\gamma\{B_y + \{[[v/[C^2]]E_z\}\} = \{1 - [(v/C)^2]\}^{1/2}\{B_y + \{[[v/[C^7]]E_z\}\}$

and **Bz** = $\gamma\{B_z - \{[[v/[C^2]]E_y\}\} = \{1 - [(v/C)^2]\}^{-1/2}\{B_z - \{[[v/[C^2]]E_y\}\}$

Here $\mathbf{B_x}$ is the component of the background magnetic field that is parallel to the spacecraft velocity vector.

Here Bx, By, and Bz are the background frame magnetic field components.

The background magnetic field in the jth ship time step is $Bx_j + \gamma_j\{B_{y,j} + \{[[v_j/[C^2]]E_{z,j}\}\} + \gamma_j\{B_{z,j} - \{[[v_{,j}/[C^2]]E_{y,j}\}\} = Bx_j + \{1 - [(v_j/C)^2]\}^{-1/2}\{B_{y,j} + \{[[v_j/[C^2]]E_{z,j}\}\} + \{1 - [(v_j/C)^2]\}^{-1/2}\{B_{z,j} - \{[[v_j/[C^2]]E_{y,j}\}\}$.

Now, the magnetic energy stored in a volume of space is equal to:

$B_{volume}$ = (Volume)(1/2)$[B^2]/\mu$.

So the magnetic energy stored in a unit volume of space in the spacecraft reference frame for the jth time-step in the spacecraft reference frame will be:

$B_{volume,j}$ = $(Volume_{shipframej})(1/2)[B_{spacecraft}{}^2]/\mu$

$= (Volume_j)(1/2)\{\{Bx_j + \{1 - [(v_j/C)^2]\}^{-1/2}\{B_{y,j} + \{[[v_j/[C^2]]E_{z,j}\}\} + \{1 - [(v_j/C)^2]\}^{-1/2}\{B_{z,j} - \{[[v_j/[C^2]]E_{y,j}\}\}\}^2\}/\mu$.

The total magnetic energy captured by the non-rotating pinwheels is:

$\Sigma(j = 1; j = m)[B_{volume,j}] = \Sigma(j = 1; j = m)\ \{(Volume_{shipframej})(1/2)[B_{spacecraft}{}^2]/\mu\}$

$= \Sigma(j = 1; j = m)\ \{(Volume_j)(1/2)\{\{Bx_j + \{\{1 - [(v_j/C)^2]\}^{-1/2}\{B_{y,j} + \{[[v_j/[C^2]]E_{z,j}\}\}\} + \{\{1 - [(v_j/C)^2]\}^{-1/2}\{B_{z,j} - \{[[v_j/[C^2]]E_{y,j}\}\}\}\}^2\}/\mu\}$.

Considering electric field analogue, we have the following analysis:

**Ex** = Ex

**Ey** = $\gamma[E_y - (vB_z)] = \{\{1 - [(v/C)^2]\}^{-1/2}\}[E_y - (vB_z)]$

and **Ez** = $\gamma[E_Z + (vB_y)] = \{\{1 - [(v/C)^2]\}^{-1/2}\}[E_z + (vB_y)]$

Here $\mathbf{E_x}$ is the component of the background electric field that is parallel to the spacecraft velocity vector.

Here Ex, Ey, and Ez are the background magnetic field components.

The background electric field in the jth ship time step is $Ex_j + \gamma_j[E_{y,j} - (v_j B_{z,j})] + \gamma_j [E_{z,j} + (v_jB_{y,j})] = Ex_j + \{\{1 - [(v_j /C)^2]\}^{-1/2}\}[E_{y,j} - (v_jB_{z,j})] + \{\{1 - [(v_j /C)^2]\}^{-1/2}\}[E_{z,j} + (v_j B_{y,j})]$.

Now, the electric energy stored in a volume of space is equal to:

$E_{volume}$ = (Volume)(1/2)($\varepsilon_0$)$[E^2]$.

So the electrical energy stored in a unit volume of space in the spacecraft reference frame for the jth time-step in the spacecraft reference frame will be:

$E_{volume,j} = (Volume_{shipframej})\ (1/2)(\varepsilon_0)[E^2]$

$= (Volume_j)\ (1/2)(\varepsilon_0)\{\{Ex_j + \{\{\{1 - [(v_j /C)^2]\}^{-1/2}\}[E_{y,j} - (v_jB_{z,j})]\} + \{\{\{1 - [(v_j /C)^2]\}^{-1/2}\}[E_{z,j} + (v_j B_{y,j})]\}\}^2\}$

The total electrical energy captured by the non-rotating pinwheels is:

$\Sigma(j = 1; j = m)\ \{(Volume_j)\ (1/2)(\varepsilon_0)\{\{Ex_j + \{\{\{1 - [(v_j /C)^2]\}^{-1/2}\}[E_{y,j} - (v_jB_{z,j})]\} + \{\{\{1 - [(v_j /C)^2]\}^{-1/2}\}[E_{z,j} + (v_j B_{y,j})]\}\}^2\}\}$

Considering relativistically rotating pinwheels, we develop a complicated scenario which includes modification by spacecraft translational gamma, pin-wheel rotational velocity in the spacecraft reference frame, orientation of pinwheel rotational axis with respect to the spacecraft translational heading, blade surface area, blade shape and blade pitch angle. However, the situation is even more complex because of the negative refraction index feature of the blades for magnetic and electric fields. We thus have a highly complex relativistic classical dynamics problem.

Essentially, a simple expression for both Standard Model and Mirror Matter Model magnetic and electric fields spacecraft energy gains taking into account the above complicating factors is:

$$\{\Sigma(j = 1;\ j = m)\ \{[\text{Standard Model and Mirror Matter Model}] : \{\{f[(\gamma_j),(v_{\text{rotation-ship-frame},j}),(\text{Axis}_{\text{pinwheel-relative},j}),(\text{Area}_{\text{blade},j}),(\text{Shape}_{\text{blade},j}),(\text{Pitch-angle}_{\text{blade},j}),(\text{Blade}_{\text{negative-refraction-index-magnetic-fields},j})]\}\ \{(\text{Volume}_j)(1/2)\{\{Bx_j + \{\{1 - [(v_j/C)^2]\}^{-1/2} \{B_{y,j} + \{[[v_j/[C^2]]E_{z,j}\}\}\} + \{\{1 - [(v_j/C)^2]\}^{-1/2} \{B_{z,j} - \{[[v_j/[C^2]]E_{y,j}\}\}\}\ ^2\}/\mu\}\}\}\}$$

$$+\ \{\Sigma(j = 1;\ j = m)\ \{[\text{Standard Model and Mirror Matter Model}] : \{\{f[(\gamma_j),(v_{\text{rotation-ship-frame},j}),(\text{Axis}_{\text{pinwheel-relative},j}),(\text{Area}_{\text{blade},j}),(\text{Shape}_{\text{blade},j}),(\text{Pitch-angle}_{\text{blade},j}),(\text{Blade}_{\text{negative-refraction-index-electric -fields},j})]\}\ \{(\text{Volume}_j)\ (1/2)(\varepsilon_0)\{\{Ex_j + \{\{\{1 - [(v_j/C)^2]\}^{-1/2}\}[E_{y,j} - (v_jB_{z,j})]\} + \{\{\{1 - [(v_j/C)^2]\}^{-1/2}\}[E_{z,j} + (v_j B_{y,j})]\}\}^2\}\}\}\}.$$

The pinwheels may optionally be plasma analogues which superconduct electrical current as they should and may rotate at any useful Keplarian or even relativistic velocities. A more complete set of formulas in details of the pinwheel mechanism is available in some of Essig's books in the list of suggested reading.

## More on linear induction drive mechanisms.

As for gravity and antigravity flux, that is an even more complicated beast which we will summarize by the following notation.

$$\{\{\Sigma(j = 1;\ j = m)\ [E_{\text{gravity},\ j}]\ \{f[(\gamma_j),(v_{\text{rotation-ship-frame},j}),(\text{Axis}_{\text{pinwheel-relative},j}),(\text{Area}_{\text{blade},j}),(\text{Shape}_{\text{blade},j}),(\text{Pitch-angle}_{\text{blade},j}),(\text{Blade}_{\text{negative-refraction-index-gravatic -fields},j})]\}$$

$$+\ \{\Sigma(j = 1;\ j = m)\ [E_{\text{gravity},\ j}]\ \{f[(\gamma_j),(v_{\text{rotation-ship-frame},j}),(\text{Axis}_{\text{pinwheel-relative},j}),(\text{Area}_{\text{blade},j}),(\text{Shape}_{\text{blade},j}),(\text{Pitch-angle}_{\text{blade},j}),(\text{Blade}_{\text{negative-refraction-index-antigravatic -fields},j})]\}\}$$

The combined static field energy terms for the six field species contemplated above is:

$$\{\{\{\Sigma(j = 1;\ j = m)\ \{[\text{Standard Model and Mirror Matter Model}] : \{\{f[(\gamma_j),(v_{\text{rotation-ship-frame},j}),(\text{Axis}_{\text{pinwheel-relative},j}),(\text{Area}_{\text{blade},j}),(\text{Shape}_{\text{blade},j}),(\text{Pitch-angle}_{\text{blade},j}),(\text{Blade}_{\text{negative-refraction-index-magnetic-fields},j})]\}\ \{(\text{Volume}_j)(1/2)\{\{Bx_j + \{\{1 - [(v_j/C)^2]\}^{-1/2} \{B_{y,j} + \{[[v_j/[C^2]]E_{z,j}\}\}\} + \{\{1 - [(v_j/C)^2]\}^{-1/2} \{B_{z,j} - \{[[v_j/[C^2]]E_{y,j}\}\}\}\ ^2\}/\mu\}\}\}\}$$

$$+\ \{\Sigma(j = 1;\ j = m)\ \{[\text{Standard Model and Mirror Matter Model}] : \{\{f[(\gamma_j),(v_{\text{rotation-ship-frame},j}),(\text{Axis}_{\text{pinwheel-relative},j}),(\text{Area}_{\text{blade},j}),(\text{Shape}_{\text{blade},j}),(\text{Pitch-angle}_{\text{blade},j}),(\text{Blade}_{\text{negative-refraction-index-electric -fields},j})]\}\ \{(\text{Volume}_j)\ (1/2)(\varepsilon_0)\{\{Ex_j + \{\{\{1 - [(v_j/C)^2]\}^{-1/2}\}[E_{y,j} - (v_jB_{z,j})]\} + \{\{\{1 - [(v_j/C)^2]\}^{-1/2}\}[E_{z,j} + (v_j B_{y,j})]\}\}^2\}\}\}\}\}$$

$$+\ \{\{\Sigma(j = 1;\ j = m)\ [E_{\text{gravity},\ j}]\ \{f[(\gamma_j),(v_{\text{rotation-ship-frame},j}),(\text{Axis}_{\text{pinwheel-relative},j}),(\text{Area}_{\text{blade},j}),(\text{Shape}_{\text{blade},j}),(\text{Pitch-angle}_{\text{blade},j}),(\text{Blade}_{\text{negative-refraction-index-gravatic -fields},j})]\}$$

$$+\ \{\Sigma(j = 1;\ j = m)\ [E_{\text{gravity},\ j}]\ \{f[(\gamma_j),(v_{\text{rotation-ship-frame},j}),(\text{Axis}_{\text{pinwheel-relative},j}),(\text{Area}_{\text{blade},j}),(\text{Shape}_{\text{blade},j}),(\text{Pitch-angle}_{\text{blade},j}),(\text{Blade}_{\text{negative-refraction-index-antigravatic -fields},j})]\}\}\}$$

**SUM OF TWO VELOCITIES: GENERAL CASE.**

Now, in the general case, the relativistic sum of two relativistic velocities V and U is

$$V +_{(\text{relativistic composition})} U = \{V + U_{\text{par}} + [(\text{alpha}_v)(U_{\text{perp}})]\}/\{1 + \{[v \text{ dot } u]/[C^2]\}\}$$

$$\text{Alpha}_v = 1/[\text{gamma}(V)] = \{1 - \{[IVI^2]/[C^2]\}\}^{1/2}.$$

As a result, the relativistic velocity of the crew members with respect to the back-ground for a rotating spacecraft with rotational velocity $V_r$ with respect to the spacecraft is given by

$$V_t +_{(\text{relativistic composition})} V_r = \{V_t + V_{\text{rpar}} + [(\text{alpha}_v)(V_{\text{rperp}})]\}/\{1 + \{[V_t \text{ dot } V_r]/[C^2]\}\}$$

$$= \{V_t + V_{\text{rpar}} + [[1/[\text{gamma}(V_t)]](V_{\text{rperp}})]\}/\{1 + \{[V_t \text{ dot } V_r]/[C^2]\}\}$$

$$= V_t +_{(\text{relativistic composition})} V_r = \{\{V_t + V_{\text{rpar}} + \{\{\{1 - \{[IV_tI^2]/[C^2]\}\}^{1/2}\}(V_{\text{rperp}})\}\}/\{1 + \{[V_t \text{ dot } V_r]/[C^2]\}\}\}$$

where $|V_t|$ is the magnitude of $V_t$.

Gamma as a function of $V_t$ and $V_r$ is plausibly equal to:

$$\text{gamma} = \{1/[1 - [[(V_t +_{(\text{relativistic composition})} V_r)/C]^{2}]]\}^{1/2}$$

$$= \{1/\{1 - \{\{\{\{V_t + V_{rpar} + [(\text{alpha}_v)(V_{rperp})]\}/\{1 + \{[V_t \text{ dot } V_r]/[C^{2}]\}\}\}/C\}^{2}\}\}\}^{1/2}$$

$$= \{1/\{1 - \{\{\{\{V_t + V_{rpar} + [[1/[\text{gamma}(V_t)]](V_{rperp})]\}/\{1 + \{[V_t \text{ dot } V_r]/[C^{2}]\}\}\}/C\}^{2}\}\}\}^{1/2}$$

$$= \{1/\{1 - \{\{\{\{V_t + V_{rpar} + \{\{\{1 - \{[|V_t|^{2}]/[C^{2}]\}\}^{1/2}\}(V_{rperp})\}\}/\{1 + \{[V_t \text{ dot } V_r]/[C^{2}]\}\}\}/C\}^{2}\}\}\}^{1/2}$$

$$= \{\{1/\{1 - \{\{\{\{V_t + V_{rpar} + \{\{\{1 - \{[|V_t|^{2}]/[C^{2}]\}\}^{1/2}\}(V_{rperp})\}\}/\{1 + \{[[(V_t)(V_r) \text{ cosine } (\Theta)]]/[C^{2}]\}\}\}/C\}^{2}\}\}\}^{½}\}$$

here $V_t$, $V_r$, $V_{rpar}$, $V_{rperp}$, and $\Theta$ are the space craft sail translational velocity, the space craft sail rotors' effective rotational velocity, the space craft sail rotors' velocity component that is parallel to the space craft translational velocity component, and the space craft sail rotors' velocity perpendicular to the space craft translational velocity.

**RELATIVISTIC PINWHEEL DYNAMO EFFECT.**

The non-relativistic Lorentz Force can be defined as:

$$F = q\,(E + v \times B)$$

Where q is the charge being acted on, E is the electric field, v is the velocity of the loop through the magnetic field, and B is the magnetic field and v x B is the cross-product of v and B.

For a non-relativistic dynamo, the electromotive force on a wire loop is:

$\varepsilon = (1/q) \int F \bullet dL = \int (E + v \times B) \bullet dL$ where $F \bullet dL$ is the scalar product or dot product of F and dL.

dL here is a differential or infinitesimal arc length along the loop.

Here, the line integral is evaluated along the length of the loop or the path of the loop.

For N loops of the same shape and size which is effectively N turns of a single loop, the electromotive force on a wire loop is:

$\varepsilon = N\,[(1/q) \int F \bullet dL] = N\,[\int (E + v \times B) \bullet dL]$ where $F \bullet dL$ is the scalar product or dot product of F and dL.

For N loops for which the size and shape of the loops can vary, the following approximation describes the voltage within the coil:

$$\varepsilon = \Sigma\,(n = 1, n = N_{coil})\,[(1/q) \int F \bullet dL]_n = \Sigma\,(n = 1, n = N_{coil})\,[\int (E + v \times B) \bullet dL]_n$$

Now, for the jth time step in the loop frame or corresponding travel interval, the relativistic Lorentz force may be expressed as:

$$F = q\,\{\{Ex_j + \{\{\{1 - [(v_j/C)^2]\}^{-1/2}\}[E_{y,j} - (v_j B_{z,j})]\} + \{\{\{1 - [(v_j/C)^2]\}^{-1/2}\}[E_{z,j} + (v_j B_{y,j})]\}\} + \{v \times \{\{Bx_j + \{\{1 - [(v_j/C)^2]\}^{-1/2}\{B_{y,j} + \{[[v_j/[C^2]]E_{z,j}\}\}\} + \{\{1 - [(v_j/C)^2]\}^{-1/2}\{B_{z,j} - \{[[v_j/[C^2]]E_{y,j}\}\}\}\}\}\}\}$$

For a relativistic dynamo, the electromotive force on a wire loop in the loop frame is:

$$\varepsilon = (1/q) \int \bullet dL = \int \{\{Ex_j + \{\{\{1 - [(v_j/C)^2]\}^{-1/2}\}[E_{y,j} - (v_jB_{z,j})]\} + \{\{\{1 - [(v_j/C)^2]\}^{-1/2}\}[E_{z,j} + (v_j B_{y,j})]]\}\} + \{v \times \{\{Bx_j + \{\{1 - [(v_j/C)^2]\}^{-1/2}\{B_{y,j} + \{[[v_j/[C^2]]E_{z,j}\}\}\} + \{\{1 - [(v_j/C)^2]\}^{-1/2}\{B_{z,j} - \{[[v_j/[C^2]]E_{y,j}\}\}\}\}\}\}\} \bullet dL$$

For N loops of the same shape and size which is effectively N turns of a single loop, the electromotive force on the coil in the coil frame is:

$$\varepsilon = N\,[(1/q) \int F \bullet dL] = N\,\{\int\{\{Ex_j + \{\{\{1 - [(v_j/C)^2]\}^{-1/2}\}[E_{y,j} - (v_jB_{z,j})]\} + \{\{\{1 - [(v_j/C)^2]\}^{-1/2}\}[E_{z,j} + (v_j B_{y,j})]]\}\} + \{v \times \{\{Bx_j + \{\{1 - [(v_j/C)^2]\}^{-1/2}\{B_{y,j} + \{[[v_j/[C^2]]E_{z,j}\}\}\} + \{\{1 - [(v_j/C)^2]\}^{-1/2}\{B_{z,j} - \{[[v_j/[C^2]]E_{y,j}\}\}\}\}\}\}\} \bullet dL\}$$ where $F \bullet dL$ is the scalar product or dot product of F and dL.

For N loops for which the size and shape of the loops can vary, the following approximation describes the voltage within the coil in the coil frame:

$$\varepsilon = \Sigma\,(n = 1, n = N_{coil})\,[(1/q) \int F \bullet dL]_n = \Sigma\,(n = 1, n = N_{coil})\,\{\int\{\{Ex_j + \{\{\{1 - [(v_j/C)^2]\}^{-1/2}\}[E_{y,j} - (v_jB_{z,j})]\} + \{\{\{1 - [(v_j/C)^2]\}^{-1/2}\}[E_{z,j} + (v_j B_{y,j})]]\}\} + \{v \times \{\{Bx_j + \{\{1 - [(v_j/C)^2]\}^{-1/2}\{B_{y,j} + \{[[v_j/[C^2]]E_{z,j}\}\}\} + \{\{1 - [(v_j/C)^2]\}^{-1/2}\{B_{z,j} - \{[[v_j/[C^2]]E_{y,j}\}\}\}\}\}\}\} \bullet dL\}_n.$$

Taking into account the composite velocity of the rotor blades and coils contained within the rotor blades, we obtain an approximate but more detailed description of the electromotive force generated within the coils.

Accordingly; the EMF for N uniformly sized and shaped coils the coil in the coil frame becomes:

$$\varepsilon = N\,[(1/q) \int F \bullet dL] = N\,\{\int\{\{Ex_j + \{\{\{1 - [((V_{t,j} +_{(\text{relativistic composition})} V_{r,j})/C)^2]\}^{-1/2}\}[E_{y,j} - (v_jB_{z,j})]\} +_{(\text{relativistic composition})} \{\{\{1 - [((V_{t,j} +_{(\text{relativistic composition})} V_{r,j})_j/C)^2]\}^{-1/2}\}[E_{z,j} + ((V_{t,j} +_{(\text{relativistic composition})} V_{r,j})B_{y,j})]]\}\} + \{(V_{t,j} +_{(\text{relativistic composition})} V_{r,j}) \times \{\{Bx_j + \{\{1 - [((V_{t,j} +_{(\text{relativistic composition})} V_{r,j})/C)^2]\}^{-1/2}\{B_{y,j} + \{[[(V_{t,j} +_{(\text{relativistic composition})} V_{r,j})/[C^2]]E_{z,j}\}\}\} + \{\{1 - [((V_{t,j} +_{(\text{relativistic composition})} V_{r,j})/C)^2]\}^{-1/2}\{B_{z,j} - \{[[(V_{t,j} +_{(\text{relativistic composition})} V_{r,j})/[C^2]]E_{y,j}\}\}\}\}\}\}\} \bullet dL\};$$

where $F \bullet dL$ is the scalar product or dot product of F and dL.

For N loops for which the size and shape of the loops can vary, the following approximation describes the voltage within the coil in the coil in the coil frame:

$$\varepsilon = \Sigma\,(n = 1, n = N_{coil})\,[(1/q) \int F \bullet dL]_n = \Sigma\,(n = 1, n = N_{coil})\,\{\int\{\{Ex_j + \{\{\{1 - [((V_{t,j} +_{(\text{relativistic composition})} V_{r,j})/C)^2]\}^{-1/2}\}[E_{y,j} - (v_jB_{z,j})]\} + \{\{\{1 - [((V_{t,j} +_{(\text{relativistic composition})} V_{r,j})_j/C)^2]\}^{-1/2}\}[E_{z,j} + ((V_{t,j} +_{(\text{relativistic composition})} V_{r,j})B_{y,j})]]\}\} + \{(V_{t,j} +_{(\text{relativistic composition})} V_{r,j}) \times \{\{Bx_j + \{\{1 - [((V_{t,j} +_{(\text{relativistic composition})} V_{r,j})/C)^2]\}^{-1/2}\{B_{y,j} + \{[[(V_{t,j} +_{(\text{relativistic composition})} V_{r,j})/[C^2]]E_{z,j}\}\}\} + \{\{1 - [((V_{t,j} +_{(\text{relativistic composition})} V_{r,j})/C)^2]\}^{-1/2}\{B_{z,j} - \{[[(V_{t,j} +_{(\text{relativistic composition})} V_{r,j})/[C^2]]E_{y,j}\}\}\}\}\}\}\} \bullet dL\}_n;$$

where $F \bullet dL$ is the scalar product or dot product of F and dL.

Formulating explicitly for the rotor velocity components in the spacecraft reference frame, we obtain;

Accordingly; the EMF for N uniformly sized and shaped coils the coil in the coil frame becomes:

$$\varepsilon = N\,[(1/q) \int F \bullet dL] = N\,\{\int\{\{Ex_j + \{\{\{1 - [(((\{\{V_{t,j} + V_{rpar,j} + \{\{\{1 - \{[|V_{t,j}|^2]/[C^2]\}\}^{1/2}\}(V_{rperp,j})\}\}/\{1 + \{[V_{t,j} \text{ dot } V_{r,j}]/[C^2]\}\}\})/C)^2]\}^{-1/2}\}[E_{y,j} - (v_jB_{z,j})]\} + \{\{\{1 - [(((\{\{V_{t,j} + V_{rpar,j} + \{\{\{1 - \{[|V_{t,j}|^2]/[C^2]\}\}^{1/2}\}(V_{rperp,j})\}\}/\{1 + \{[V_{t,j} \text{ dot } V_{r,j}]/[C^2]\}\}\})/C)^2]\}^{-1/2}\}[E_{z,j} + (((\{\{V_{t,j} + V_{rpar,j} + \{\{\{1 - \{[|V_{t,j}|^2]/[C^2]\}\}^{1/2}\}(V_{rperp,j})\}\}/\{1 + \{[V_{t,j} \text{ dot } V_{r,j}]/[C^2]\}\}\}\})B_{y,j})]]\}\} + \{((\{\{V_{t,j} + V_{rpar,j} + \{\{\{1 - \{[|V_{t,j}|^2]/[C^2]\}\}^{1/2}\}(V_{rperp,j})\}\}/\{1 + \{[V_{t,j} \text{ dot } V_{r,j}]/[C^2]\}\}\}\}) \times \{\{Bx_j + \{\{1 - [(((\{\{V_{t,j} + V_{rpar,j} + \{\{\{1 - \{[|V_{t,j}|^2]/[C^2]\}\}^{1/2}\}(V_{rperp,j})\}\}/\{1 + \{[V_{t,j} \text{ dot } V_{r,j}]/[C^2]\}\}\}\})/C)^2]\}^{-1/2}\{B_{y,j} + \{[[((\{\{V_{t,j} + V_{rpar,j} + \{\{\{1 - \{[|V_{t,j}|^2]/[C^2]\}\}^{1/2}\}(V_{rperp,j})\}\}/\{1 + \{[V_{t,j} \text{ dot } V_{r,j}]/[C^2]\}\}\}\})/[C^2]]E_{z,j}\}\}\} + \{\{1 - [(((\{\{V_{t,j} + V_{rpar,j} + \{\{\{1 - \{[|V_{t,j}|^2]/[C^2]\}\}^{1/2}\}(V_{rperp,j})\}\}/\{1 + \{[V_{t,j} \text{ dot } V_{r,j}]/[C^2]\}\}\}\})/C)^2]\}^{-1/2}\{B_{z,j} - \{[[((\{\{V_{t,j} + V_{rpar,j} + \{\{\{1 - \{[|V_{t,j}|^2]/[C^2]\}\}^{1/2}\}(V_{rperp,j})\}\}/\{1 + \{[V_{t,j} \text{ dot } V_{r,j}]/[C^2]\}\}\}\})/[C^2]]E_{y,j}\}\}\}\}\}\}\} \bullet dL\};$$

For N loops for which the size and shape of the loops can vary, the following approximation describes the voltage within the coil in the coil in the coil frame:

$$\varepsilon = \Sigma (n = 1, n = N_{coil}) [(1/q) \int F \bullet dL]_n = \Sigma (n = 1, n = N_{coil}) \{\int\{\{Ex_j + \{\{\{1 - [(((\{\{V_{t,j} + V_{rpar,j} + \{\{\{1 - \{[|V_{t,j}|^{2}]/[C^{2}]\}\}^{1/2}\}(V_{rperp,j})\}\}/\{1 + \{[V_{t,j} \text{ dot } V_{r,j}]/[C^{2}]\}\}\}) /C)^{2}]\}^{-1/2}\}[E_{y,j} - (v_jB_{z,j})]\} + \{\{\{1 - [(((\{\{V_{t,j} + V_{rpar,j} + \{\{\{1 - \{[|V_{t,j}|^{2}]/[C^{2}]\}\}^{1/2}\}(V_{rperp,j})\}\}/\{1 + \{[V_{t,j} \text{ dot } V_{r,j}]/[C^{2}]\}\}\}) /C)^{2}]\}^{-1/2}\}[E_{z,j} + (((\{\{V_{t,j} + V_{rpar,j} + \{\{\{1 - \{[|V_{t,j}|^{2}]/[C^{2}]\}\}^{1/2}\}(V_{rperp,j})\}\}/\{1 + \{[V_{t,j} \text{ dot } V_{r,j}]/[C^{2}]\}\}\})B_{y,j})]\}\} + \{((\{\{V_{t,j} + V_{rpar,j} + \{\{\{1 - \{[|V_{t,j}|^{2}]/[C^{2}]\}\}^{1/2}\}(V_{rperp,j})\}\}/\{1 + \{[V_{t,j} \text{ dot } V_{r,j}]/[C^{2}]\}\}\}) \times \{\{Bx_j + \{\{1 - [(((\{\{V_{t,j} + V_{rpar,j} + \{\{\{1 - \{[|V_{t,j}|^{2}]/[C^{2}]\}\}^{1/2}\}(V_{rperp,j})\}\}/\{1 + \{[V_{t,j} \text{ dot } V_{r,j}]/[C^{2}]\}\}\})/C)^{2}]\}^{-1/2} \{B_{y,j} + \{[[((\{\{V_{t,j} + V_{rpar,j} + \{\{\{1 - \{[|V_{t,j}|^{2}]/[C^{2}]\}\}^{1/2}\}(V_{rperp,j})\}\}/\{1 + \{[V_{t,j} \text{ dot } V_{r,j}]/[C^{2}]\}\}\})/[C^{2}]]E_{z,j}\}\}\} + \{\{1 - [(((\{\{V_{t,j} + V_{rpar,j} + \{\{\{1 - \{[|V_{t,j}|^{2}]/[C^{2}]\}\}^{1/2}\}(V_{rperp,j})\}\}/\{1 + \{[V_{t,j} \text{ dot } V_{r,j}]/[C^{2}]\}\}\})/C)^{2}]\}^{-1/2} \{B_{z,j} - \{[[((\{\{V_{t,j} + V_{rpar,j} + \{\{\{1 - \{[|V_{t,j}|^{2}]/[C^{2}]\}\}^{1/2}\}(V_{rperp,j})\}\}/\{1 + \{[V_{t,j} \text{ dot } V_{r,j}]/[C^{2}]\}\}\})/[C^{2}]]E_{y,j}\}\}\}\}\}\} \bullet dL\}_n;$$

where $F \bullet dL$ is the scalar product or dot product of F and dL.

Now, electrical power though a conductive loop can be expressed as:

$$P = QV/t = (I)(V);$$

where P is the work done per unit of time.

Accordingly; the power for N uniformly sized and shaped coils the coil in the coil frame becomes:

$$P_{electric} = (I_{electric})\{N [(1/q) \int F \bullet dL]\} = (I_{electric})\{N\{\int\{\{Ex_j + \{\{\{1 - [(((\{\{V_{t,j} + V_{rpar,j} + \{\{\{1 - \{[|V_{t,j}|^{2}]/[C^{2}]\}\}^{1/2}\}(V_{rperp,j})\}\}/\{1 + \{[V_{t,j} \text{ dot } V_{r,j}]/[C^{2}]\}\}\}) /C)^{2}]\}^{-1/2}\}[E_{y,j} - (v_jB_{z,j})]\} + \{\{\{1 - [(((\{\{V_{t,j} + V_{rpar,j} + \{\{\{1 - \{[|V_{t,j}|^{2}]/[C^{2}]\}\}^{1/2}\}(V_{rperp,j})\}\}/\{1 + \{[V_{t,j} \text{ dot } V_{r,j}]/[C^{2}]\}\}\}) /C)^{2}]\}^{-1/2}\}[E_{z,j} + (((\{\{V_{t,j} + V_{rpar,j} + \{\{\{1 - \{[|V_{t,j}|^{2}]/[C^{2}]\}\}^{1/2}\}(V_{rperp,j})\}\}/\{1 + \{[V_{t,j} \text{ dot } V_{r,j}]/[C^{2}]\}\}\})B_{y,j})]\}\} + \{((\{\{V_{t,j} + V_{rpar,j} + \{\{\{1 - \{[|V_{t,j}|^{2}]/[C^{2}]\}\}^{1/2}\}(V_{rperp,j})\}\}/\{1 + \{[V_{t,j} \text{ dot } V_{r,j}]/[C^{2}]\}\}\}) \times \{\{Bx_j + \{\{1 - [(((\{\{V_{t,j} + V_{rpar,j} + \{\{\{1 - \{[|V_{t,j}|^{2}]/[C^{2}]\}\}^{1/2}\}(V_{rperp,j})\}\}/\{1 + \{[V_{t,j} \text{ dot } V_{r,j}]/[C^{2}]\}\}\})/C)^{2}]\}^{-1/2} \{B_{y,j} + \{[[((\{\{V_{t,j} + V_{rpar,j} + \{\{\{1 - \{[|V_{t,j}|^{2}]/[C^{2}]\}\}^{1/2}\}(V_{rperp,j})\}\}/\{1 + \{[V_{t,j} \text{ dot } V_{r,j}]/[C^{2}]\}\}\})/[C^{2}]]E_{z,j}\}\}\} + \{\{1 - [(((\{\{V_{t,j} + V_{rpar,j} + \{\{\{1 - \{[|V_{t,j}|^{2}]/[C^{2}]\}\}^{1/2}\}(V_{rperp,j})\}\}/\{1 + \{[V_{t,j} \text{ dot } V_{r,j}]/[C^{2}]\}\}\})/C)^{2}]\}^{-1/2} \{B_{z,j} - \{[[((\{\{V_{t,j} + V_{rpar,j} + \{\{\{1 - \{[|V_{t,j}|^{2}]/[C^{2}]\}\}^{1/2}\}(V_{rperp,j})\}\}/\{1 + \{[V_{t,j} \text{ dot } V_{r,j}]/[C^{2}]\}\}\})/[C^{2}]]E_{y,j}\}\}\}\}\}\} \bullet dL\}\};$$

where $F \bullet dL$ is the scalar product or dot product of F and dL.

For N loops for which the size and shape of the loops can vary, the following approximation describes the power generated by the coil in the coil in the coil frame:

$$P_{electric,j} = (I_{electric,j})\{\Sigma(n = 1, n = N_{coil})[(1/q) \int F_j \bullet dL]_n\} = (I_{electric,j})\{\Sigma(n = 1, n = N_{coil})\{\int\{\{Ex_j + \{\{\{1 - [(((\{\{V_{t,j} + V_{rpar,j} + \{\{\{1 - \{[|V_{t,j}|^{2}]/[C^{2}]\}\}^{1/2}\}(V_{rperp,j})\}\}/\{1 + \{[V_{t,j} \text{ dot } V_{r,j}]/[C^{2}]\}\}\}) /C)^{2}]\}^{-1/2}\}[E_{y,j} - (v_jB_{z,j})]\} + \{\{\{1 - [(((\{\{V_{t,j} + V_{rpar,j} + \{\{\{1 - \{[|V_{t,j}|^{2}]/[C^{2}]\}\}^{1/2}\}(V_{rperp,j})\}\}/\{1 + \{[V_{t,j} \text{ dot } V_{r,j}]/[C^{2}]\}\}\})_j /C)^{2}]\}^{-1/2}\}[E_{z,j} + (((\{\{V_{t,j} + V_{rpar,j} + \{\{\{1 - \{[|V_{t,j}|^{2}]/[C^{2}]\}\}^{1/2}\}(V_{rperp,j})\}\}/\{1 + \{[V_{t,j} \text{ dot } V_{r,j}]/[C^{2}]\}\}\})B_{y,j})]\}\} + \{((\{\{V_{t,j} + V_{rpar,j} + \{\{\{1 - \{[|V_{t,j}|^{2}]/[C^{2}]\}\}^{1/2}\}(V_{rperp,j})\}\}/\{1 + \{[V_{t,j} \text{ dot } V_{r,j}]/[C^{2}]\}\}\}) \times \{\{Bx_j + \{\{1 - [(((\{\{V_{t,j} + V_{rpar,j} + \{\{\{1 - \{[|V_{t,j}|^{2}]/[C^{2}]\}\}^{1/2}\}(V_{rperp,j})\}\}/\{1 + \{[V_{t,j} \text{ dot } V_{r,j}]/[C^{2}]\}\}\})/C)^{2}]\}^{-1/2} \{B_{y,j} + \{[[((\{\{V_{t,j} + V_{rpar,j} + \{\{\{1 - \{[|V_{t,j}|^{2}]/[C^{2}]\}\}^{1/2}\}(V_{rperp,j})\}\}/\{1 + \{[V_{t,j} \text{ dot } V_{r,j}]/[C^{2}]\}\}\})/[C^{2}]]E_{z,j}\}\}\} + \{\{1 - [(((\{\{V_{t,j} + V_{rpar,j} + \{\{\{1 - \{[|V_{t,j}|^{2}]/[C^{2}]\}\}^{1/2}\}(V_{rperp,j})\}\}/\{1 + \{[V_{t,j} \text{ dot } V_{r,j}]/[C^{2}]\}\}\})/C)^{2}]\}^{-1/2} \{B_{z,j} - \{[[((\{\{V_{t,j} + V_{rpar,j} + \{\{\{1 - \{[|V_{t,j}|^{2}]/[C^{2}]\}\}^{1/2}\}(V_{rperp,j})\}\}/\{1 + \{[V_{t,j} \text{ dot } V_{r,j}]/[C^{2}]\}\}\})/[C^{2}]]E_{y,j}\}\}\}\}\}\}_n \bullet dL\}\}$$

where $F \bullet dL$ is the scalar product or dot product of F and dL.

The electrical energy generated over the jth time step is thus equal to:

$$E_{coil} = [d\, P_{electric,j}/dt_{coils}] \{1/\{\{1 - [[(V_t +_{(\text{relativistic composition})} V_r)/C]^2]\}^{1/2}\}\}$$

$$= \{d\{(I_{electric,j})\{\Sigma(n = 1, n = N_{coil})[(1/q) \int F_j \bullet dL]_n\}\}\} = \{d\{(I_{electric,j})\{\Sigma(n = 1, n = N_{coil})\{\int\{\{Ex_j + \{\{\{1 - [(((\{\{V_{t,j} + V_{rpar,j} + \{\{\{1 - \{[|V_{t,j}|^{2}]/[C^{2}]\}\}^{1/2}\}(V_{rperp,j})\}\}/\{1 + \{[V_{t,j} \text{ dot } V_{r,j}]/[C^{2}]\}\}\}) /C)^{2}]\}^{-1/2}\}[E_{y,j} - (v_jB_{z,j})]\} + \{\{\{1 - [(((\{\{V_{t,j} + V_{rpar,j} + \{\{\{1 - \{[|V_{t,j}|^{2}]/[C^{2}]\}\}^{1/2}\}(V_{rperp,j})\}\}/\{1 + \{[V_{t,j} \text{ dot } V_{r,j}]/[C^{2}]\}\}\})_j /C)^{2}]\}^{-1/2}\}[E_{z,j} + (((\{\{V_{t,j} + V_{rpar,j} + \{\{\{1 - \{[|V_{t,j}|^{2}]/[C^{2}]\}\}^{1/2}\}(V_{rperp,j})\}\}/\{1 + \{[V_{t,j} \text{ dot } V_{r,j}]/[C^{2}]\}\}\})B_{y,j})]\}\} + \{((\{\{V_{t,j} + V_{rpar,j} + \{\{\{1 - \{[|V_{t,j}|^{2}]/[C^{2}]\}\}^{1/2}\}(V_{rperp,j})\}\}/\{1 + \{[V_{t,j} \text{ dot } V_{r,j}]/[C^{2}]\}\}\}) \times \{\{Bx_j + \{\{1 - [(((\{\{V_{t,j} + V_{rpar,j} + \{\{\{1 - \{[|V_{t,j}|^{2}]/[C^{2}]\}\}^{1/2}\}(V_{rperp,j})\}\}/\{1 + \{[V_{t,j} \text{ dot } V_{r,j}]/[C^{2}]\}\}\})/C)^{2}]\}^{-1/2} \{B_{y,j} + \{[[((\{\{V_{t,j} + V_{rpar,j} + \{\{\{1 - \{[|V_{t,j}|^{2}]/[C^{2}]\}\}^{1/2}\}(V_{rperp,j})\}\}/\{1 + \{[V_{t,j} \text{ dot } V_{r,j}]/[C^{2}]\}\}\})/[C^{2}]]E_{z,j}\}\}\} + \{\{1 - [(((\{\{V_{t,j} + V_{rpar,j} + \{\{\{1 - \{[|V_{t,j}|^{2}]/[C^{2}]\}\}^{1/2}\}(V_{rperp,j})\}\}/\{1 + \{[V_{t,j} \text{ dot } V_{r,j}]/[C^{2}]\}\}\})/C)^{2}]\}^{-1/2} \{B_{z,j} - \{[[((\{\{V_{t,j} + V_{rpar,j} + \{\{\{1 - \{[|V_{t,j}|^{2}]/[C^{2}]\}\}^{1/2}\}(V_{rperp,j})\}\}/\{1 + \{[V_{t,j} \text{ dot } V_{r,j}]/[C^{2}]\}\}\})/[C^{2}]]E_{y,j}\}\}\}\}\}\}_n \bullet dL\}\}\}/dt_{coil(s)}\} \{1/\{\{1 - [[(V_{t,j} +_{(\text{relativistic composition})} V_{r,j})/C]^2]\}^{1/2}\}\}$$

= {d{($I_{electric,j}$)}{Σ(n = 1, n = $N_{coil}$)[(1/q) ∫ $F_j$ • dL]$_{n}$}}} = {d{($I_{electric,j}$)}{Σ(n = 1, n = $N_{coil}$)}{∫{{$Ex_j$ + {{{1 – [(({{$V_{t,j}$ + $V_{rpar,j}$ + {{{1 – {[I$V_{t,j}$ I $^{2}$]/[C $^{2}$]}} $^{1/2}$}($V_{rperp,j}$)}}/{1 + {[$V_{t,j}$ dot $V_{r,j}$]/[C $^{2}$]}}}) /C)$^{2}$]}$^{-1/2}$ }[$E_{y,j}$ - ($v_jB_{z,j}$)]} + {{{1 – [(({{$V_{t,j}$ + $V_{rpar,j}$ + {{{1 – {[I$V_{t,j}$ I $^{2}$]/[C $^{2}$]}} $^{1/2}$}($V_{rperp,j}$)}}/{1 + {[$V_{t,j}$ dot $V_{r,j}$]/[C $^{2}$]}}})$_{j}$ /C)$^{2}$]}$^{-1/2}$ }[$E_{z,j}$ + (({{$V_{t,j}$ + $V_{rpar,j}$ + {{{1 – {[I$V_{t,j}$ I $^{2}$]/[C $^{2}$]}} $^{1/2}$}($V_{rperp,j}$)}}/{1 + {[$V_{t,j}$ dot $V_{r,j}$]/[C $^{2}$]}}})$B_{y,j}$)]}} + {({{$V_{t,j}$ + $V_{rpar,j}$ + {{{1 – {[I$V_{t,j}$ I $^{2}$]/[C $^{2}$]}} $^{1/2}$}($V_{rperp,j}$)}}/{1 + {[$V_{t,j}$ dot $V_{r,j}$]/[C $^{2}$]}}}) x {{$Bx_j$ + {{1 – [(({{$V_{t,j}$ + $V_{rpar,j}$ + {{{1 – {[I$V_{t,j}$ I $^{2}$]/[C $^{2}$]}} $^{1/2}$}($V_{rperp,j}$)}}/{1 + {[$V_{t,j}$ dot $V_{r,j}$]/[C $^{2}$]}}})/C)$^{2}$]}$^{-1/2}$ {$B_{y,j}$ + {[[({{$V_{t,j}$ + $V_{rpar,j}$ + {{{1 – {[I$V_{t,j}$ I $^{2}$]/[C $^{2}$]}} $^{1/2}$}($V_{rperp,j}$)}}/{1 + {[$V_{t,j}$ dot $V_{r,j}$]/[C $^{2}$]}}})/[C$^{2}$]]$E_{z,j}$}}} + {{1 – [(({{$V_{t,j}$ + $V_{rpar,j}$ + {{{1 – {[I$V_{t,j}$ I $^{2}$]/[C $^{2}$]}} $^{1/2}$}($V_{rperp,j}$)}}/{1 + {[$V_{t,j}$ dot $V_{r,j}$]/[C $^{2}$]}}})/C)$^{2}$]}$^{-1/2}$ {$B_{z,j}$ - {[[({{$V_{t,j}$ + $V_{rpar,j}$ + {{{1 – {[I$V_{t,j}$ I $^{2}$]/[C $^{2}$]}} $^{1/2}$}($V_{rperp,j}$)}}/{1 + {[$V_{t,j}$ dot $V_{r,j}$]/[C $^{2}$]}}})/[C$^{2}$]]$E_{y,j}$}}}}}}}$_{n}$ • dL}}}/$dt_{coil(s)}$} {1/{{1 - [[($V_{t,j}$ $+_{(relativistic\ composition)}$ $V_{r,j}$)/C]$^{2}$]}$^{1/2}$}{{1/{1 – {{{{$V_{t,j}$ + $V_{rpar,j}$ + {{{1 – {[I$V_{t,j}$I $^{2}$]/[C $^{2}$]}} $^{1/2}$}($V_{rperp,j}$)}}/{1 + {[[($V_{t,j}$)($V_{r,j}$) cosine (Θ)]]/[C $^{2}$]}}}}/C} $^{2}$}}} $^{½}$}.

Now, the rotor blade power intake and generation is affected by blade and coil pitch angle, surface area, coil radial extension width and length profiles, coil conductivity, and other factors. Thus, we will modify the latter formula to yield;

$E_{coil}$ = [d $P_{electric,j}$/$dt_{coils}$] {1/{{1 - [[($V_t$ $+_{(relativistic\ composition)}$ $V_r$)/C]$^{2}$]}$^{1/2}$}}{f[(Pitch angle),(Surface area),(Coil radial extension width and length profile),(Coil conductivity),[Σ Other factors]]}

= {d{($I_{electric,j}$)}{Σ(n = 1, n = $N_{coil}$)[(1/q) ∫ $F_j$ • dL]$_{n}$}}}{f[(Pitch angle),(Surface area),(Coil radial extension width and length profile),(Coil conductivity),[Σ Other factors]]} = {{d{($I_{electric,j}$)}{Σ(n = 1, n = $N_{coil}$)}{∫{{$Ex_j$ + {{{1 – [(({{$V_{t,j}$ + $V_{rpar,j}$ + {{{1 – {[I$V_{t,j}$ I $^{2}$]/[C $^{2}$]}} $^{1/2}$}($V_{rperp,j}$)}}/{1 + {[$V_{t,j}$ dot $V_{r,j}$]/[C $^{2}$]}}}) /C)$^{2}$]}$^{-1/2}$ }[$E_{y,j}$ - ($v_jB_{z,j}$)]} + {{{1 – [(({{$V_{t,j}$ + $V_{rpar,j}$ + {{{1 – {[I$V_{t,j}$ I $^{2}$]/[C $^{2}$]}} $^{1/2}$}($V_{rperp,j}$)}}/{1 + {[$V_{t,j}$ dot $V_{r,j}$]/[C $^{2}$]}}})$_{j}$ /C)$^{2}$]}$^{-1/2}$ }[$E_{z,j}$ + (({{$V_{t,j}$ + $V_{rpar,j}$ + {{{1 – {[I$V_{t,j}$ I $^{2}$]/[C $^{2}$]}} $^{1/2}$}($V_{rperp,j}$)}}/{1 + {[$V_{t,j}$ dot $V_{r,j}$]/[C $^{2}$]}}})$B_{y,j}$)]}} + {({{$V_{t,j}$ + $V_{rpar,j}$ + {{{1 – {[I$V_{t,j}$ I $^{2}$]/[C $^{2}$]}} $^{1/2}$}($V_{rperp,j}$)}}/{1 + {[$V_{t,j}$ dot $V_{r,j}$]/[C $^{2}$]}}}) x {{$Bx_j$ + {{1 – [(({{$V_{t,j}$ + $V_{rpar,j}$ + {{{1 – {[I$V_{t,j}$ I $^{2}$]/[C $^{2}$]}} $^{1/2}$}($V_{rperp,j}$)}}/{1 + {[$V_{t,j}$ dot $V_{r,j}$]/[C $^{2}$]}}})/C)$^{2}$]}$^{-1/2}$ {$B_{y,j}$ + {[[({{$V_{t,j}$ + $V_{rpar,j}$ + {{{1 – {[I$V_{t,j}$ I $^{2}$]/[C $^{2}$]}} $^{1/2}$}($V_{rperp,j}$)}}/{1 + {[$V_{t,j}$ dot $V_{r,j}$]/[C $^{2}$]}}})/[C$^{2}$]]$E_{z,j}$}}} + {{1 – [(({{$V_{t,j}$ + $V_{rpar,j}$ + {{{1 – {[I$V_{t,j}$ I $^{2}$]/[C $^{2}$]}} $^{1/2}$}($V_{rperp,j}$)}}/{1 + {[$V_{t,j}$ dot $V_{r,j}$]/[C $^{2}$]}}})/C)$^{2}$]}$^{-1/2}$ {$B_{z,j}$ - {[[({{$V_{t,j}$ + $V_{rpar,j}$ + {{{1 – {[I$V_{t,j}$ I $^{2}$]/[C $^{2}$]}} $^{1/2}$}($V_{rperp,j}$)}}/{1 + {[$V_{t,j}$ dot $V_{r,j}$]/[C $^{2}$]}}})/[C$^{2}$]]$E_{y,j}$}}}}}}}$_{n}$ • dL}}}/$dt_{coil(s)}$} {1/{{1 - [[($V_{t,j}$ $+_{(relativistic\ composition)}$ $V_{r,j}$)/C]$^{2}$]}$^{1/2}$}{f[(Pitch angle),(Surface area),(Coil radial extension width and length profile),(Coil conductivity),[Σ Other factors]]}}

= {{d{($I_{electric,j}$)}{Σ(n = 1, n = $N_{coil}$)[(1/q) ∫ $F_j$ • dL]$_{n}$}}} = {d{($I_{electric,j}$)}{Σ(n = 1, n = $N_{coil}$)}{∫{{$Ex_j$ + {{{1 – [(({{$V_{t,j}$ + $V_{rpar,j}$ + {{{1 – {[I$V_{t,j}$ I $^{2}$]/[C $^{2}$]}} $^{1/2}$}($V_{rperp,j}$)}}/{1 + {[$V_{t,j}$ dot $V_{r,j}$]/[C $^{2}$]}}}) /C)$^{2}$]}$^{-1/2}$ }[$E_{y,j}$ - ($v_jB_{z,j}$)]} + {{{1 – [(({{$V_{t,j}$ + $V_{rpar,j}$ + {{{1 – {[I$V_{t,j}$ I $^{2}$]/[C $^{2}$]}} $^{1/2}$}($V_{rperp,j}$)}}/{1 + {[$V_{t,j}$ dot $V_{r,j}$]/[C $^{2}$]}}})$_{j}$ /C)$^{2}$]}$^{-1/2}$ }[$E_{z,j}$ + (({{$V_{t,j}$ + $V_{rpar,j}$ + {{{1 – {[I$V_{t,j}$ I $^{2}$]/[C $^{2}$]}} $^{1/2}$}($V_{rperp,j}$)}}/{1 + {[$V_{t,j}$ dot $V_{r,j}$]/[C $^{2}$]}}})$B_{y,j}$)]}} + {({{$V_{t,j}$ + $V_{rpar,j}$ + {{{1 – {[I$V_{t,j}$ I $^{2}$]/[C $^{2}$]}} $^{1/2}$}($V_{rperp,j}$)}}/{1 + {[$V_{t,j}$ dot $V_{r,j}$]/[C $^{2}$]}}}) x {{$Bx_j$ + {{1 – [(({{$V_{t,j}$ + $V_{rpar,j}$ + {{{1 – {[I$V_{t,j}$ I $^{2}$]/[C $^{2}$]}} $^{1/2}$}($V_{rperp,j}$)}}/{1 + {[$V_{t,j}$ dot $V_{r,j}$]/[C $^{2}$]}}})/C)$^{2}$]}$^{-1/2}$ {$B_{y,j}$ + {[[({{$V_{t,j}$ + $V_{rpar,j}$ + {{{1 – {[I$V_{t,j}$ I $^{2}$]/[C $^{2}$]}} $^{1/2}$}($V_{rperp,j}$)}}/{1 + {[$V_{t,j}$ dot $V_{r,j}$]/[C $^{2}$]}}})/[C$^{2}$]]$E_{z,j}$}}} + {{1 – [(({{$V_{t,j}$ + $V_{rpar,j}$ + {{{1 – {[I$V_{t,j}$ I $^{2}$]/[C $^{2}$]}} $^{1/2}$}($V_{rperp,j}$)}}/{1 + {[$V_{t,j}$ dot $V_{r,j}$]/[C $^{2}$]}}})/C)$^{2}$]}$^{-1/2}$ {$B_{z,j}$ - {[[({{$V_{t,j}$ + $V_{rpar,j}$ + {{{1 – {[I$V_{t,j}$ I $^{2}$]/[C $^{2}$]}} $^{1/2}$}($V_{rperp,j}$)}}/{1 + {[$V_{t,j}$ dot $V_{r,j}$]/[C $^{2}$]}}})/[C$^{2}$]]$E_{y,j}$}}}}}}}$_{n}$ • dL}}}/$dt_{coil(s)}$} {1/{{1 - [[($V_{t,j}$ $+_{(relativistic\ composition)}$ $V_{r,j}$)/C]$^{2}$]}$^{1/2}$}{{1/{1 – {{{{$V_{t,j}$ + $V_{rpar,j}$ + {{{1 – {[I$V_{t,j}$I $^{2}$]/[C $^{2}$]}} $^{1/2}$}($V_{rperp,j}$)}}/{1 + {[[($V_{t,j}$)($V_{r,j}$) cosine (Θ)]]/[C $^{2}$]}}}}/C} $^{2}$}}} $^{½}$}{f[(Pitch angle),(Surface area),(Coil radial extension width and length profile),(Coil conductivity),[Σ Other factors]]}}.

Now F = $\mathbf{F_1} + \mathbf{F_2} + \ldots = Q\mathbf{E}$.

Therefore, the electric force exerted on a given differential surface area charge element of the central axial charge member, Q, by a single intake particle of positive charge, q, is initially equal to:

$F = Q \{[1/(4\pi\varepsilon_0)][\gamma q \mathbf{R}]/\{\{[\gamma^2 R^2 \cos^2\theta] + [R^2 \sin^2\theta]\}^{3/2}\}\}$.

Here, q is in integer multiples of positronic or protonic electric charge units expressed in units of Coulomb.

The incremental step-wise series of forces exerted on the charged particle as it is pulled into the charged axial surface by a given differential surface area charge element of the charged axial surface, Q is:

$\Sigma F_h\ (h = 1, h = n_{axial}) = Q \{\Sigma\ (h = 1, h = n_{axial})\ \{[1/(4\pi\varepsilon_0)][\gamma_h q \mathbf{R_h}]/\{\{[\gamma_h^2 R_h^2 \cos^2\theta_h] + [R_h^2 \sin^2\theta_h]\}^{3/2}\}\}\}$.

For a stream of m positively charged intake particles entering the cone where the particles are distributed in the background frame over a unit spatial interval in the background frame; the electric force exerted by the intake particles on the differential axial charge element is equal to:

For a stream of k positively charged intake particles entering the cone where the particles are distributed in the background frame over a unit spatial interval in the background frame; the electric force exerted by the particles on the differential axial charge element is equal to:

$F_{elect} = Q \{\Sigma(k = 1, k = o) \{\Sigma (h =1, h = n_{axial}) \{[1/(4\pi\varepsilon_0)][\gamma_{h,k} (sq_{h,k}) \mathbf{R_{h,k}}]/\{\{[\gamma_{h,k}{}^2 R_{h,k}{}^2 \cos^2 \theta_{h,k}] + [R_{h,k}{}^2 \sin^2 \theta_{h,k}]\}^{3/2}\}\}\}\}(\gamma)(v/C)$,

where s is a positive integer representing the number of net electronic charge units.

The total electric force exerted on the axial charged member by the stream of particles where the particles are distributed in the background frame over a unit spatial interval in the background frame is equal to:

The total electric force exerted on the axial charged member by the stream of k charged particles where the particles are distributed in the background frame over a unit spatial interval in the background frame is equal to.

$F_{elect} = \{\Sigma (u = 1, u = n_{axial}) [Q_{u,axial}(x,y,z,t)]\{\Sigma(k = 1, k = m) \{\Sigma (h =1, h = n) \{[1/(4\pi\varepsilon_0)][\gamma_{h,k} (sq_{h,k}) \mathbf{R}_{h,k}]/\{\{[\gamma_{h,k}{}^2 R_{h,k}{}^2 \cos^2 \theta_{h,k}] + [R_{h,k}{}^2 \sin^2 \theta_{h,k}]\}^{3/2}\}\}\}\}\}(\gamma)(v/C)$

The electric force exerted on a given differential surface area charge element of the charged cone, Q, by a single intake particle of negative charge, q, is equal to:

$F = Q \{[1/(4\pi\varepsilon_0)][\gamma q \mathbf{R}]/\{\{[\gamma^2 R^2 \cos^2 \theta] + [R^2 \sin^2 \theta]\}^{3/2}\}\}$.

Here, q is in integer multiples of electronic charge in electric charge units expressed in units of Coulomb.

The incremental step-wise series of forces exerted on the charged particle as it is pulled into the charged conical surface by a given differential surface area charge element of the charged conical surface, Q is:

$\Sigma F_h (h =1, h = n_{cone}) = Q \{\Sigma (h =1, h = n_{cone}) \{[1/(4\pi\varepsilon_0)][\gamma_h q \mathbf{R}_h]/\{\{[\gamma_h{}^2 R_h{}^2 \cos^2 \theta_h] + [R_h{}^2 \sin^2 \theta_h]\}^{3/2}\}\}\}$.

For a stream of k positively charged intake particles entering the cone where the particles are distributed in the background frame over a unit spatial interval in the background frame; the electric force exerted by the particles on the differential conical charge element is equal to:

$F_{elect} = Q \{\Sigma(k = 1, k = o) \{\Sigma (h =1, h = n_{cone}) \{[1/(4\pi\varepsilon_0)][\gamma_{h,k} (sq_{h,k}) \mathbf{R_{h,k}}]/\{\{[\gamma_{h,k}{}^2 R_{h,k}{}^2 \cos^2 \theta_{h,k}] + [R_{h,k}{}^2 \sin^2 \theta_{h,k}]\}^{3/2}\}\}\}\}(\gamma)(v/C)$

where s is a positive integer representing the number of net electronic charge units.

The total electric force exerted on the charged conical member by the stream of k charged particles where the particles are distributed in the background frame over a unit spatial interval in the background frame is equal to:

$F_{elect} = \{\Sigma (u = 1, u = n_{cone}) [Q_{u,cone}(x,y,z,t)]\{\Sigma(k = 1, k = m) \{\Sigma (h =1, h = n) \{[1/(4\pi\varepsilon_0)][\gamma_{h,k} (sq_{h,k}) \mathbf{R}_{h,k}]/\{\{[\gamma_{h,k}{}^2 R_{h,k}{}^2 \cos^2 \theta_{h,k}] + [R_{h,k}{}^2 \sin^2 \theta_{h,k}]\}^{3/2}\}\}\}\}\}(\gamma)(v/C)$

The following factors or terms are non-dimensional scalar quantities.

When the bristles' electric flux density enhancements are taken into account, each of the above equations for overall axial member pulling effects and/or overall conical pulling effects on the intake chargons is modified by the factoral suffix;

$\{\Sigma (i = 0, i = n) \{\Sigma (r = 1, r = N) [f(\text{bristle shape}_{r,i}, \text{intake chargon } \gamma_{r,i}, \text{bristle electric-field density dist}_{r,i}, \text{total bristle electric field energy}_{r,i})]/N\}\}$.

For the axial bristles, we obtain:

$\{\Sigma (i = 0, i = n) \{\Sigma (r = 1, r = N) [f(\text{bristle shape}_{r,i}, \text{intake chargon } \gamma_{r,i}, \text{bristle electric-field density dist}_{r,i}, \text{total bristle electric field energy}_{r,i})]/N_{axial}\}\}$.

For the cone bristles, we obtain:

$\{\Sigma (i = 0, i = n) \{\Sigma (r = 1, r = N) [f(\text{bristle shape}_{r,i}, \text{intake chargon } \gamma_{r,i}, \text{bristle electric-field density dist}_{r,i}, \text{total bristle electric field energy}_{r,i})]/N_{conical}\}\}$.

Here, $N_{axial}$ and $N_{conical}$ are the numbers of axial and conical bristles. Thus, the above sigma functions are averages which take into account the collective electric flux concentration effects of the entire respective sets of bristles.

The factors, $\{\Sigma\ (i = 0, i = n)\ \ \{\Sigma\ (r = 1, r = N)\ [f(\text{magnetic}_{r,i})]/N_{axial}\}\}$, and $\{\Sigma\ (i = 0, i = n)\ \ \{\Sigma\ (r = 1, r = N)\ [f(\text{magnetic}_{r,i})]/N_{conical}\}\}$ take into account the average collective effects of magnetic field elements associated with the focusing of the intake chargons onto the bristle tips.

Taken together, the two factors describing the bristle effects and operations are for axial bristles:

$$\{\{\Sigma\ (i = 0, i = n)\ \ \{\Sigma\ (r = 1, r = N)\ [f(\text{magnetic}_{r,i})]/N_{axial}\}\}\{\Sigma\ (i = 0, i = n)\ \ \{\Sigma\ (r = 1, r = N)\ [f(\text{bristle shape}_{r,i}, \text{intake chargon } \gamma_{r,i}, \text{bristle electric-field density dist}_{r,i}, \text{total bristle electric field energy}_{r,i})]/N_{axial}\}\}\}$$

and for the cone bristles,

$$\{\{\Sigma\ (i = 0, i = n)\ \ \{\Sigma\ (r = 1, r = N)\ [f(\text{bristle shape}_{r,i}, \text{intake chargon } \gamma_{r,i}, \text{bristle electric-field density dist}_{r,i}, \text{total bristle electric field energy}_{r,i})]/N_{conical}\}\}\{\Sigma\ (i = 0, i = n)\ \ \{\Sigma\ (r = 1, r = N)\ [f(\text{magnetic}_{r,i})]/N_{conical}\}\}\}.$$

The sub-functions bristle $\text{shape}_{r,i}$, intake chargon $\gamma_{r,i}$, bristle electric-field density $\text{dist}_{r,i}$, total bristle electric field $\text{energy}_{r,I}$ account for each bristle: the bristle shape, the Lorentz factor of the intake chargons as modified by the enhanced electric flux densities near the tips of the bristles, the bristle electric field density distribution, and total bristle electric field based energy.

For cases where no bristles are included in the conical design, i.e., the bristle mechanism is neglected, the above modification factors for the bristles may be set to one thus indicating a non-operative function. Setting the bristle terms equal to zero would be in error because then the energy gain expressions which include the bristle factors would become equal to zero, thus wrongly implying that the associated terms for chargon drive yield zero drive energy input for chargon drive modes.

All of the above mentioned intake chargon drives as also plausibly include mirror matter model analogues. We use the prefix operator [SM, MMM] in the formulations below for gamma and velocity to indicate both Standard Model and Mirror Matter Model charge mechanisms.

The charged intake matter that enters the cone is assumed to be completely absorbed into either the cone surface or the central axial member and converted to energy under the condition where astrodynamic drag energy is completely absorbed and recycled so that the net astrodynamic drag imposed by such intake matter is completely neutralized. The energy of conversion of the intake charged mass is assumed to be converted to a virtually 100 percent efficient light speed exhaust stream such as a photonic exhaust, neutrino exhaust, and/or gravitational radiation based exhaust that is close to perfectly collimated.

Therefore, the force produced by the beamed exhaust produced by the energy conversion of the charged mass absorbed into the axial member is equal to;

$$\langle S\rangle/C = \langle\{[SM, MMM]:\ d\ \{\Sigma\ (u = 1, u = n_{axial})\ [M_{ua}C^2]\ (\gamma)(v/C)\}/dt\}\rangle/C,$$

where $\langle S\rangle$ is the ship time averaged light speed exhaust thrust based power, $n_{axial}$ is the number of charged intake particles per unit background spatial interval in the background reference frame, $M_{ua}$ is the slowed mass of the uth intake particle in the ship's reference frame or the invariant mass of the particle applied in the space-craft's reference frame for particles being absorbed into the axial member, and t is the background frame time .

The force produced by the beamed exhaust produced by the energy conversion of the charged mass absorbed into the conical surface is equal to;

$$\langle S\rangle/C = \langle\{[SM, MMM]:\ d\ \{\Sigma\ (u = 1, u = n_{cone})\ [M_{uc}C^2]\ (\gamma)(v/C)\}/dt\}\rangle/C,$$

where $\langle S\rangle$ is the ship time averaged light speed exhaust thrust based power, $n_{conel}$ is the number of charged intake particles per unit background spatial interval in the background reference frame, $M_{uc}$ is the slowed mass of the uth intake particle in the ship's reference frame or the invariant mass of the particle applied in the space-craft's reference frame for particles being absorbed into the conical surface, and t is the background frame time .

The above mentioned axial and conical intake chargon conversion to light speed exhaust mechanisms also plausibly include mirror matter model analogues. We use the prefix operator [SM, MMM] in the formulations below for gamma and velocity to indicate both Standard Model and Mirror Matter Model charge mechanisms.

The total power with respect to the sail's reference frame for a given Lorentz factor is therefore:

$$P = \{\{\int(y_1,y_2) \{\int(x_1,x_2) \{\int (0, \pi) \{\{\{(T_{cmbr}) /\{\gamma [1 + [(v/C) \cos \theta]]\}\}^4\} \sigma e\} d\theta\}dx\}dy\}(2)\} + \{[\text{Standard Model and Mirror Matter Model}]d\{\int < \{\Sigma (u = 1, u = n_{axial}) (2)[Q_{u,axial}(x,y,z,t)]\{\Sigma(k = 1, k = m) \{\Sigma (h =1, h = n) \{[1/(4\pi\varepsilon_0)][\gamma_{h,k} (sq_{h,k}) \mathbf{R}_{h,k}]/\{\{[\gamma_{h,k}^2 R_{h,k}^2 \cos^2 \theta_{h,k}] + [R_{h,k}^2 \sin^2 \theta_{h,k}]\}^{3/2}\}\}\}\}\}(\gamma)(v/C)> \cdot dx\}/dt\} + \{[\text{Standard Model and Mirror Matter Model}]d\{\int < \{\Sigma (u = 1, u = n_{cone}) (2)[Q_{u,cone}(x,y,z,t)]\{\Sigma(k = 1, k = m) \{\Sigma (h =1, h = n) \{[1/(4\pi\varepsilon_0)][\gamma_{h,k} (sq_{h,k}) \mathbf{R}_{h,k}]/\{\{[\gamma_{h,k}^2 R_{h,k}^2 \cos^2 \theta_{h,k}] + [R_{h,k}^2 \sin^2 \theta_{h,k}]\}^{3/2}\}\}\}\}\}(\gamma)(v/C)> \cdot dx\}/dt\} + \{[SM, MMM]: d \{\Sigma (u = 1, u = n_{axial}) [M_{ua}C^2] (\gamma)(v/C)\}/dt\} + \{[SM, MMM]: d \{\Sigma (u = 1, u = n_{cone}) [M_{uc}C^2] (\gamma)(v/C)\}/dt\}$$

$$= \{\{\int(y_1,y_2) \{\int(x_1,x_2) \{\int (0, \pi) \{\{\{(T_{cmbr}) /\{\{1/\{[1 + [(v/C)^2 ]]^{1/2}\}\} [1 + [(v/C) \cos \theta]]\}\}^4\} \sigma e\} d\theta\}dx\}dy\}(2)\} + \{[\text{Standard Model and Mirror Matter Model}] d\{\int < \{\Sigma (u = 1, u = n_{axial}) (2)[Q_{u,axial}(x,y,z,t)]\{\Sigma(k = 1, k = m) \{\Sigma (h =1, h = n) \{[1/(4\pi\varepsilon_0)]\{ \{1/\{1 - [(v_{h,k}/C)^2 ]\}^{1/2}\} (sq_{h,k}) \mathbf{R}_{h,k}\}/\{\{\{ \{\{1/\{1 - [(v_{h,k}/C)^2 ]\}^{1/2}\}^2\} R_{h,k}^2 \cos^2 \theta_{h,k}\} + [R_{h,k}^2 \sin^2 \theta_{h,k}]\}^{3/2}\}\}\}\}\}\{1/\{1 -[(v/C)^2 ]\}^{1/2}\} (v/C)> \cdot dx\} /dt\} + \{[\text{Standard Model and Mirror Matter Model}] d\{\int < \{\Sigma (u = 1, u = n_{cone}) (2)[Q_{u,cone}(x,y,z,t)]\{\Sigma(k = 1, k = m) \{\Sigma (h =1, h = n) \{[1/(4\pi\varepsilon_0)]\{ \{1/\{1 - [(v_{h,k}/C)^2 ]\}^{1/2}\} (sq_{h,k}) \mathbf{R}_{h,k}\}/\{\{\{\{1/\{1 - [(v_{h,k}/C)^2 ]\}^{1/2}\}^2\} R_{h,k}^2 \cos^2 \theta_{h,k}\} + [R_{h,k}^2 \sin^2 \theta_{h,k}]\}^{3/2}\}\}\}\}\}\{1/\{1 - [(v/C)^2 ]\}^{1/2}\} (v/C)> \cdot dx\}/dt\} + \{[SM, MMM]: d \{\Sigma (u = 1, u = n_{axial}) [M_{ua}C^2] \{1/\{1 - [(v/C)^2 ]\}^{1/2}\} (v/C)\}/dt\} + \{[SM, MMM]: d \{\Sigma (u = 1, u = n_{cone}) [M_{uc}C^2] \{1/\{1 - [(v/C)^2 ]\}^{1/2}\} (v/C)\}/dt\}$$

Now, a fascinating electrically charged rotor system would consist of a rotor or pin-wheel of plasma rotating at near the speed of light. The plasma may be electrons, protons, ions, and the like and may be magnetically affixed to the spacecraft. Alternatively, the plasma may be electrically affixed to the spacecraft. The near light-speed rotors of plasma would greatly increased the relativistic Lorentz turing force compared to a non-rotating solid sail of the same mass, eventhough the mass of the plasma rotor could be ultra-relativistic in the spacecraft reference frame. The radius of the plasma rotor may be as great as 10,000 km for greatly reduced synchron radiation and associated losses.

Another mechanism would utilize explosions off-set to the star-bound side of the craft. Accordingly, the craft would be turned by explosions set off properly behind the craft and the thrust would be captured by one or more specially dedicated sails. Note that relativistic aberration would require that the explosions be adjusted in location with respect to the craft.

## Negative electromagnetic refractive index pulling light drive.

For extreme Lorentz factors, the CMBR and starlight will be highly blue-shifted and will be relativistically aberrated to what would approach a point source in front of the spacecraft at $\gamma = \infty$. A sail parallel to the spacecraft velocity vector made of a suitable negative electromagnetic refraction index material will be pulled forward even by light incident on the sail at a very shallow angle from in front of the spacecraft.

To enhance the negative refraction index sails capture of EM energy, the sails may have negative index hairs or cilia distributed along its length.

Negative refraction index materials have actually been measured to be pulled on by incident light. Duke University and other academic and government labs are researching the various aspects of negative refraction index materials.

I have no problem with spacecraft being pulled forward by forward incident light. The Big Bang may have been the most recent free lunch. There is no reason why the Big Bang could not have started with miniscule quantities of mass-energy.

A good abstract for a great paper on negative super-pressure of light acting on a negative refractive index material is

Henri Lezec
(Center for Nanoscale Science and Technology, NIST)

Forty years ago, V. Veselago derived the electromagnetic properties of a hypothetical material having simultaneously-negative values of electric permittivity and magnetic permeability [1]. Such a material, denominated "left-handed", was predicted to exhibit a negative index of refraction, as well as a number of other counter-intuitive optical properties. For example, it was hypothesized that a perfect mirror illuminated with a plane wave would experience a negative radiation pressure (pull) when immersed in a left-handed medium, as opposed to the usual positive radiation pressure experienced when facing a dielectric medium such as air or glass. Since left-handed materials are not available in nature, considerable efforts are currently under way to implement them under the form of artificial "metamaterials" — composite media with tailored bulk optical characteristics resulting from constituent structures which are smaller in both size and density than the effective wavelength in the medium. Here we show how surface-plasmon modes propagating in a stacked array of metal-insulator-metal (MIM) waveguides can be harnessed to yield a volumetric left-handed metamaterial characterized by an in-plane-isotropic negative index of refraction over a broad frequency range spanning the blue and green. By sculpting this material with a focused-ion beam we realize prisms and micro-cantilevers which we use to demonstrate, for the first time, (a) in-plane isotropic negative-refraction at optical frequencies, and (b) negative radiation pressure. We predict and experimentally verify a negative "superpressure", the magnitude of which exceeds the photon pressure experienced by a perfect mirror by more than a factor of two. 1) V. Veselago, \textit{ Sov. Phys. Usp. }10, p.509 (1968).

Available at:

http://meetings.aps.org/Meeting/MAR09/Event/93172

The sail might not need to be held by guy lines. A strong magnetic field based coupling or electrical charged based connection might work.

Another option is to fabricate the sail guy lines out of graphene, carbon nanotubes, boron nitride nanotubes, graphene oxide paper, and the like. A cable constructed from such materials could stretch for about 20 to 50 kilometers yet still handle tens to hundreds of Earth G's. The tensile strength of graphene is close to 18 million PSI for perfect forms.

Sun light may be negatively refracted all the while the craft is thrust forward by tangential explosive thrust.

To cancel out extreme acceleration based g-forces for the crew, the following mechanisms might be employed; 1) electric charge dispositions within the interior of the ship; 2) electric fields set up within the interior of the ship; and/or 3) magnetic fields set up within the ship. Sufficiently strong magnetic fields would magnetize any solid, liquid, or gaseous matter within the ship and thus can be used for cancelling out g forces.

A field effect mechanism used to cancel out ship based g-forces would likely need one or more extremely sturdy anchors that are appropriately placed within the spacecraft. The anchors would need to be suitably robust so as not to be destroyed by the stresses, strains, and/or compressional forces produced within them.

Alternatively, the crew members' bodies could be either frozen in a suspended animation state, or be placed alive and awake within hydrostatically sealed pressure vessels containing breathable oxygenated liquids. The pressure vessel mechanism could be augmented with any or all of the field effect mechanisms described in the previous paragraph.

Assuming a pellet shot mass of 40 kilograms, the kinetic energy of each pellet upon launch would be 2 x $10^{11}$ joules. Thus, the time averaged power pulse of the gun per shot during the acceleration process would be approximately 200 gigawatts. This is roughly equivalent to the expansive gas power for large caliber artillery pieces.

More specifically, the average thermal loading on the electron gas gun barrel would be about 40 megawatts per 10 meter section of barrel length. This would require a cooling affluent and large surface area radiator.

The two hundred giga-watts power per gun would be supplied by photovoltaic membranes having capture areas of 1,000 $km^2$ and areal density of 10 grams per $m^2$. Thus, the membranes would have mass of 10,000 metric tons. The membranes may be used as light-sails to assist in tacking processes by solar radiation to assist on-station maintenance. Of course, the membrane conversion efficiency would be 20 percent.

Eventually, a significant increase in velocity will build-up on each gun placement. To compensate, the gun can have reverse firing patterns so that two or more arcuate segments of the pellet runway can be launched from each gun.

Assuming light-sail methods are used to maintain orbital position with respect to the Sun and an orbital location of one AU from the sun, and an adjustable reflector which reflects about 1/3 of the incident sunlight but transmits the rest, we obtain the following relation.

(4 x $10^6$ N) = <S>/c = <S>/(3 x $10^8$ m/s). So, the required sunlight power to counteract the acceleration on the gun during the firing time would be 1.2 x $10^{15}$ watts. This would require a sail capture area of 1,200,000 $km^2$ for each of the 100 guns and thus would require a vast increase in available space flight and resource infrastructure relative to current levels.

For pellet accelerations of 100,000 $m/s^2$, the propulsive force would be **F**= m**a** = 4 mega-newtons which works out to approximately 1 million pounds of force and of similar order of magnitude to modern artillery. Required gun barrel length is ½ **a**$t^2$ or 50 kilometers.

Herein, we neglect the gravitational attraction of the pellets and the Sun. This attraction would reduce the effective exiting velocity of the pellets from the solar system by a factor of about 1/3 of the launch velocity where the pellets are launched from a 1 AU radius solar orbit. This condition applies to all specific examples provided in this book.

Assuming a barrel mass of 100 kilograms per meter of length, the total barrel mass per gun would be 5 million kilograms or 5,000 metric tons. This is approximately one order of magnitude greater than that of the ISS. One hundred guns as anticipated would have mass of 500,000 metric tons. However, with anticipated launch to low Earth orbit (LEO) cost reductions by two orders of magnitude or more and similarly also for placement of materials in solar orbit at $10,000/metric ton, the placement of the material composition of the guns and PV membranes would be accomplished for about $15 Billion at today's cost.

The cost of assembling the spacecraft in LEO would be roughly the same as the current instantiation of the ISS assuming a one hundred-fold cost reduction launch to LEO or about $100 Billion. The greater portion of the cost would be the fabrication of spacecraft modules instead of the greatly reduced cost of LEO insertion.

So, the combined cost of the project components is estimated to be approximately $50 Trillion at current dollar value. Over a thirty year project span, the cost of the project would be about $1.66 Trillion per year. The bulk of the cost would be orbiting of the nuclear explosives.

Additionally, optimized blends of fissile fuels can be used in the composition of the primaries thus offering research options to greatly reduce the required critical mass of the needed fissile fuels.

Note that the shape of the chute can include elongated charged tubular configurations so as to more efficiently capture and back accelerate the plasma from the nuclear explosions. Moreover, the chute can have a pointed or tapered leading edge to reduce astrodynamics drag.

Here, we are considering gun placement in solar orbit but with adjustable orientation via retro-rockets, solar sails, and/or plasma sails. The orientation of the gun and its orbital radius can be re-adjusted as needed by sail tacking processes.

## Production of antimatter from sunlight.

Regarding powerful sun-light reflectors for production of antimatter and/or for maintaining stable solar orbits, the reflectors may be oriented at 45 degree or less relative to the incident sunlight. The reflectors may then concentrate the sun-light on to a central combination solar energy harvesting and electrical power generator while the remaining sunlight is reflected directly in parallel with the direction of the sunlight away from the sun and incident on the apparatus.

Alternatively, a more complex system would have additional reflector mechanism that are used only for orbit position maintenance similar to the ones mention above while another set of reflectors concentrates sunlight for and use antimatter production.

Solar energy conversion apparatus may include highly efficient solar panel, thermoelectric generators, and turbo-electric generators.

A fascinating long term program would include the pre-manufacture of hydrogen fusion pellets throughout the entire galaxy and greater cosmic light-cone.

Accordingly, manufacturing packages would be launch to manufacture the pellets and which would have self-replicating functions so that eventually a cosmic light-cone distribution of pellets is deployed.

Nano-technology self-assembly aspects can be included in the pellet manufacturing industry as well as optional macroscopic manufacturing equipment.

Provided the universe will continue to expand more or less at its current rate, it is plausible that the pellet manufacturing system would extend into adjacent light-cones and thus become a web of Milky Way Galaxy centered super-light-cone enterprise.

A fascinating prospect would include an advanced future industry for which either zero point field energy or perhaps dark energy is converted to hydrogen and helium in matter and antimatter forms. As such, the density of our universe portion of our universe can be maintained so as to prevent a thermodynamic heat death.

As an interesting aside, we may consider growing super-light-cone distributions of fuel pellets.

Accordingly, we may set the pellets on mutual collision courses but from outside our currently light-cone. Ideally, six near cosmic-light-cone mass pellets would be incident into our current cosmic light-cone in a six sided cube like pattern. Thus, there would be close to six light-cone mass units of fuel incident into our light-cone.

A fascinating prospect would entail all cosmic light-cone massed pellets impinging into our cosmic light-cone before gravitation had a chance to "catch up" with the about six cosmic light-cone mass units before a black-hole state could materialize for the entire seven light-cone system.

Given that locations outside of our cosmic light-cone are screened by superluminal recession velocities, in some sense, these extra-cosmic light-cones would not be well defined thermodynamically and causally with respect to our cosmic light-cone. An implication of this lack of definition may be such that more than six cosmic light-cone mass systems may impinge on our cosmic light-cone simultaneously. Perhaps a very large finite number or even an infinite number of such cosmic light-cone massed units may impinge onto our cosmic light-cone before an enveloping black hole state could manifest. This result may require that each impinging cosmic light-cone mass system is naturally and/or artificially screened from the adjacent ones.

Returning to the main topic here on pellet runways, we can consider matter-antimatter fuel pellets as well.

For matter-antimatter reactor powered, manned, interstellar spacecraft or perhaps manned matter-antimatter rocket spacecraft, we will first assume a mass ratio of 10. Using the relativistic rocket equation:

$$\Delta v = C\, tanh\, [(I_{sp}/C)\, ln\, (M_0/M_1)] = 0.98\ C.$$

For $M_0/M_1 = 100$:

$$\Delta v = C\, tanh\, [(I_{sp}/C)\, ln\, (100)] = 0.9998\ C.$$

The Lorentz factor of 0.98 C is:

$$1/\{1 - [(v/C)^2]\}^{1/2} = 1/\{1 - [(0.98\ C/C)^2]\}^{1/2} = 5.0252.$$

For v = 0.9998 C, the Lorentz factor is:

$$1/\{1 - [(0.9998\ C/C)^2]\}^{1/2} = 50.0025.$$

These above examples assume that both the matter and antimatter portions of the fuel are carried along from the start of the mission and that the efficiency of the system is very close to 100 percent. Such an assumption is perhaps a tall order with current matter-antimatter rocket concepts.

We can go to further extremes on the relativistic rocket themes by speculating that antimatter rockets might derive their normal matter reactants from the interstellar medium where the propulsion system efficiency would approach 100 percent, exactly, and where the $I_{sp}$ therefore is greater than C.

Although the exhaust velocity cannot be greater than C, for the case where normal matter can be extracted from space in a manner such that the drag energy can be recycled, the momentum delivered to the spacecraft per unit of utilized onboard fuel is twice that obtainable for cases were both components are carried on board from the start. Specific impulse can be viewed as the quantity of momentum delivered to the spacecraft per unit of onboard fuel.

About $10^{15}$ watts could be collected in the case where 100 million membranous solar concentrators are deployed in a 1 AU solar orbit and where each solar concentrator has a capture area of 10,000 square meters. Each of these concentrators could have a high power density PV cell that may be as much as 40 or more percent efficient. This would result in 400 TW of electrical power being generated. In one year, $[4 \times (10^{14})][3 \times (10^7)]$ joules of electrical power may be produced or the equivalent of about 120 metric tons of matter converted into energy. Twelve metric tons of antimatter per year could be produced by the above systems if the antimatter can be produced with 20 percent efficiency. At nearly 100 percent efficiency, 60 metric tons per year could be produced. The antimatter generators would always produce an equal amount of normal matter except for reactions that violate CPT invariance since antimatter is always produced along with normal matter. However, the details of CPT invariance are another story and so are not included here.

Twelve thousand metric tons of antimatter could be generated per year at 20 percent efficiency from 1,000 such stations. Sixty-thousand metric tons per year could be produced at nearly 100 percent efficiency. The antimatter generation could be ramped up several more orders of magnitude in order to produce millions if not hundreds of millions of tons of antimatter per year, provided we can develop a workable infrastructure. In 100 years, this would amount to tens of billions of tons.

One caveat is the production of very low cost and durable reflectors with current technology and cheap abundant materials or technology and materials to be developed that are suitably light-weight and robust in the environment of outer space.

My brother John and I have invented patented apparatuses that include, but are not limited to, very low cost, high mass specific power output inflatable reflectors, made of durable, high modulus, reflective, membranous materials. We managed to produce reflectors that have a mass specific power output on Earth's surface of up to 10 kilowatts/kilogram using 0.5 mil metalized Mylar or 0.5 mil metalized nylon. The method of manufacture involves efficient flat sheet manufacturing patterns using mainly 4, 6, or 8 sheets of thermally bonded, adhesively bonded, or otherwise, bonded materials. For our first prototypes, we used a clothes iron to thermally bond metalized polymer film based toy balloon cut-outs of various inner and outer radii. We were more than able to cook hot dogs to a char even in intermittent sunlight using the devices in a total of about 15 minutes or with about 7 minutes of sunlight.

The mass specific power yield of our reflectors will increase as thin film materials of greater strength are developed.

Some potential exists for using exotic, super-high strength materials, for making the devices such as carbon nanotube membranous sheets of anywhere from a few nanometers in thickness to tens of nanometers in thickness thus increasing the mass specific power yield of the reflectors by 4 orders of magnitude. Other potential super high strength materials of construction include:

1) graphene oxide paper;
2) boron-nitride nanotubes;
3) graphene;
4) carbon atom chains;
5) diamond fibers;
6) Beta carbon nitride fibers.

Such apparatuses can also be useful as solar sails and beam sails. The thinner the sail, the lower the mass per unit area of sail, and therefore, thinner sails can be used to capture more energy because they can be made larger than thicker sails of the same mass. There are some lower limits to sail thickness because sails that are too thin or which are not made of the proper materials would allow too much sunlight to pass through and thus result in a strong loss in efficiency.

All that would be required from the reflector material standpoint to collect $10^{15}$ watts with our current technology is 100 billion kilograms of material or 100 million metric tons. Ten billion metric tons would suffice for 1,000 such stations. Perhaps the building and deployment of such reflectors would provide a carbon sink due to the carbon requirement for the production of any carbonaceous high strength polymeric materials used to fabricate the concentrators. This could result in reduced atmospheric carbon and thus a mitigated problem of global warming.

For reflectors made of carbon nanotube materials or perhaps the even stronger carbon graphene, the combined mass required for the reflector portions of the 1,000 conjectured stations is only 10 million metric tons. A mere 10 million metric tons could be used to collect $10^{18}$ watts of solar energy.

Note that some mechanism for allowing the collection stations to remain in steady positions around the sun would be required. Perhaps some sort of angled adjustment procedure, ion or electron rocket thrust, electrodynamic-hydrodynamic-plasma-drive thrust, or other means can be used to keep the antimatter generation stations in stable orbit about the Sun. A very interesting mechanism would entail the deployment of a negative index of refraction type of material affixed to the solar radiation concentrating stations. A negative refraction index material has a strange property by which such materials are pulled forward by impinging light instead of being pushed by the light. More will be said about negative refraction index materials again in later sections of this book.

Note that we achieved full stable deployment of our devices which were usually only about one meter in diameter with a relative pressure of about 0.1 PSI or less. The larger the device, the lower the internal pressure can be to deploy the devices.

Another caveat is the ability to launch the collection stations. I think the problem is tractable this very century if we can get the hardware in solar orbit at 1 AU. At 0.1 AU, the required mass of the reflective materials drops by 100 fold. However, low cost and efficient access to solar orbit is needed in order launch and deploy the systems at 1 AU.

What could be done with 60,000 metric tons of matter-antimatter fuel that would somehow be utilized with almost 100 percent efficiency? Assume a spacecraft with a final payload mass of 6,000 metric tons. Using the relativistic rocket equation,

$\Delta v = C\ tanh\ [(I_{sp}/C)\ ln\ (M_0/M_1)] = 0.98\ C$, where the mass ratio is equal to 10. This corresponds to a relativistic Lorentz factor of about 5.

For a mass ratio of 100, e.g., a fully fueled mass to dry weight of the vehicle of (60,000 metric tons)/(600 metric tons), $\Delta v$ is equal to 0.9998 C. This corresponds to a relativistic Lorentz factor of 50.

Consider the case where only the antimatter fuel component is taken along from the start, perhaps in the form of anti-hydrogen ice. Consequently, the effective $I_{sp}$ is greater than C for systems that operate at near 100 percent efficiency where the matter fuel is collected in route.

Assume that the human life expectancy can be augmented to 1,100 years in duration. A Lorentz factor of 5 would permit roughly 5,000 light-year trips for the original living crew: a Lorentz factor of 50, 50,000 light-year trips; a Lorentz factor of 500, 500,000 light-year trips; and a Lorentz factor of 5,000, 5 million light-year trips. A life expectancy of 1,100 years would permit the travelers to live out the remaining 100 years of their lives in relaxation and style on the beaches, lake-sides, and mountain resorts on any of an un-told number of beautiful worlds yet to be discovered.

I feel that provided an expansive and bold funding initiative could be established, we could launch such missions within the next two centuries in droves, and for missions that are limited to terminal Lorentz factors of say between 5 and 50, this very century, perhaps within the lifetimes of some of our children.

We need not use rocket thrust with the need to carry extra fuel in order to arrive safely at the points of destination. As indicated before, the spacecraft could be slowed by any of the previously listed electro-dynamic breaking mechanisms.

Magnetic breaking could be accomplished by deploying a large superconducting coil which would build up extremely high current as it passed through the interstellar or intergalactic magnetic fields thus producing a magnetic field to react against the space-based fields in a drag inducing manner. Electrodynamic-hydrodynamic-plasma breaking could be accomplished by a reverse interstellar ramjet type of mechanism. Magnetic sail-based breaking could be accomplished through the deployment of a magnetic bottle consisting of plasma deployed around the ship and held fixed by electrodynamic fields.

As with most relativistic rocket concepts, the above electrodynamic deceleration mechanisms can be augmented if needed by reverse rocket thrust of the same form(s) as that used to accelerate the spacecraft.

I think of the meager infrastructure that the New World settlers had here in what would become the U.S. and now we have super highways, 100 story buildings, hundreds of airports, a roughly 300 advanced ship Navy, dozens of large cities, and the list goes on and on.

I have a gut feeling that we can produce vast quantities of antimatter from the Sun and we still do not know what the properties of bulk quantities of antimatter are due to CPT violation which occurs in certain particle pair creations. As of the time of this writtingm we have not yet determined whether antimatter possesses an antigravity component as a deviation from its otherwise ordinary gravity as Frank Close speculates in his book entitled **Antimatter** (Close, 2009) . New research has not detected any gravatic difference however deviations smaller than the scale of dectability cannot yet be ruled out.

One way or another, I feel that antimatter sequestration from natural sources such as within the magnetosphere of the gas giant planets within our Solar System, or its artificial production in bulk quantities, will prove extremely useful for our manned interstellar missions, within the next two centuries and even bolder missions beyond.

Again, note the following relativistic rocket equations for constant acceleration:

$$t = (C/a)\ \sinh(aT/C) = [[(d/C)^2] + (2d/a)]^{1/2}$$

$$d = [(C^2)/a]\ [[\cosh(aT/C)] - 1] = [C^2)/a]\ \{\{[1 + [(at/C)^2]]^{1/2}\} - 1\}$$

$$v = C\ \tanh(aT/C) = (at) / \{[1 + [(at/C)^2]]^{1/2}\}$$

$$T = (C/a)\ \text{inversesinh}\ (at/C) = (C/a)\ \text{inversecosh}\ [[ad/(C^2)] + 1]$$

$$\gamma = \cosh(aT/c) = [1 + [(at/C)\ ^2]]\ ^{1/2} = [ad/(C^2)] + 1$$

Consider $t = (C/a)\sinh(aT/C) = [[(d/C)^2] + (2d/a)]^{1/2}$

For a constant 1 G acceleration: and for $d = 9.461 \times 10^{17}$ meters = $10^2$ light-years:

$t = \{[[[9.461 \times (10^{17})]/C]^2] + [(2)[9.461 \times (10^{17})]/(9.81)]\}^{1/2}$ seconds

= $3.186263 \times 10^9$ seconds = 100.97 years background time.

For $d = 10^3$ light-years:

$t = \{[[[9.461 \times (10^{18})]/C]^2] + [(2)[9.461 \times (10^{18})]/(9.81)]\}^{1/2}$ seconds

= $3.1589 \times 10^{10}$ seconds = 1,001.012 years.

For $d = 10^4$ light-years:

$t = \{[[[9.461 \times (10^{19})]/C]^2] + [(2)[9.461 \times (10^{19})]/(9.81)]\}^{1/2}$ seconds

= $3.1561555 \times 10^{11}$ seconds = 10,001.46 years.

For $d = 10^5$ light-years:

$t = \{[[[9.461 \times (10^{20})]/C]^2] + [(2)[9.461 \times (10^{20})]/(9.81)]\}^{1/2}$ seconds

= $3.15588 \times 10^{12}$ seconds = 100,005.95 years:

For $d = 10^6$ light-years:

$t = \{[[[9.461 \times (10^{21})]/C]^2] + [(2)[9.461 \times (10^{21})]/(9.81)]\}^{1/2}$ seconds

= $3.155853 \times 10^{13}$ seconds = 1,000,050.8 years.

For a constant 0.1 G acceleration: and for $d = 9.461 \times 10^{17}$ meters = $10^2$ light-years:

$t = \{[[[9.461 \times (10^{17})]C]^2] + [(2)[9.461 \times (10^{17})]/(0.981)]\}^{1/2}$ seconds

= $3.44793 \times 10^9$ seconds = 109.2607 years background time.

For $d = 10^3$ light-years:

$t = \{[[[9.461 \times (10^{18})]/C]^2] + [(2)[9.461 \times (10^{18})]/(0.981)]\}^{1/2}$ seconds

= $3.18626 \times 10^{10}$ seconds = 1,009.69 years.

For $d = 10^4$ light-years:

$t = \{[[[9.461 \times (10^{19})]/C]^2] + [(2)[9.461 \times (10^{19})]/(0.981)]\}^{1/2}$ seconds

= $3.15890 \times 10^{11}$ seconds = 10,010.2 years.

For $d = 10^5$ light-years:

$t = \{[[[9.461 \times (10^{20})]/C]^2] + [(2)[9.461 \times (10^{20})]/(0.981)]\}^{1/2}$ seconds

= $3.156155 \times 10^{12}$ seconds = 100,014.67 years.

For $d = 10^6$ light-years:

$t = \{[[[9.461 \times (10^{21})]/C]^2] + [(2)\ [9.461 \times (10^{21})]/(0.981)]\}^{1/2}$ seconds

$= 3.15588 \times 10^{13}$ seconds = 1,000,059.5 years.

For a constant 10 G acceleration and for $d = 9.461 \times 10^{17}$ meters = $10^2$ light-years:

$t = \{[[[9.461 \times (10^{17})]C]^2] + [(2)\ [9.461 \times (10^{17})]/(98.1)]\}^{1/2}$ seconds

$= 3.158904 \times 10^9$ seconds = 109.2607 years background time.

For $d = 10^3$ light-years:

$t = \{[[[9.461 \times (10^{18})]/C]^2] + [(2)\ [9.461 \times (10^{18})]/(\ 98.1)]\}^{1/2}$ seconds

$= 3.156155 \times 10^{10}$ seconds = 1,000.1467 years.

For $d = 10^4$ light-years:

$t = \{[[[9.461 \times (10^{19})]/C]^2] + [(2)\ [9.461 \times (10^{19})]/(98.1)]\}^{1/2}$ seconds

$= 3.155880 \times 10^{11}$ seconds = 10,000.595 years.

For $d = 10^5$ light-years:

$t = \{[[[9.461 \times (10^{20})]/C]^2] + [(2)\ [9.461 \times (10^{20})]/(98.1)]\}^{1/2}$ seconds

$= 3.155853 \times 10^{12}$ seconds = 100,005,08 years.

For $d = 10^6$ light-years:

$t = \{[[[9.461 \times (10^{21})]/C]^2] + [(2)\ [9.461 \times (10^{21})]/(98.1)]\}^{1/2}$ seconds

$= 3.1558502 \times 10^{13}$ seconds = 1,000,049.95 years.

Notice how the number of years background time seems to converge on the value of $d/C$ as the distance, acceleration, or both the distance and the acceleration increase. The background time, $t$, approaches $d/C$ exactly for $d = \infty$, finite $a$, or for finite $d$ and infinite $a$. Technically speaking, the background time will approach $[d/C] + e$ where $e = (1/\infty)[d/C]$ as $d$ approaches infinity and where $e$ is thus finite.

Thus, the following limits hold:

Lim $t = d/C$

$d \rightarrow \infty$, for finite a.

Lim $t = d/C$

$a \rightarrow \infty$, for finite $d$.

Yet once again, consider in general nuclear fission powered, nuclear fusion powered, and matter-antimatter reaction powered starships that use an enclosed thermal mass type of reaction energy collection mechanism. Either the heat generated would need to be rapidly conducted away from the thermal mass and converted to propulsion energy or the temperature capacity of the refractories utilized would need to be very high to enable accelerations of one Earth G or more. Perhaps one way heat conductors or diodic heat conductors that operate in a manner similar to fiber optics or even to that of electrical super-conductors might be needed to transfer the extreme heat loading resulting from the high power to spacecraft mass ratios commensurate with accelerations of one G or more.

## Validation of specific impulses greater than one for principally relativistic rocket craft.

The velocities attainable by the relativistic matter-antimatter rocket are also obtainable with matter-antimatter pellet streams.

However, a perhaps more effective mechanism would involve a spacecraft carrying pure antimatter, perhaps in the form of anti-protium while extracting the normal hydrogen from space.

Accordingly, the anti-protium would be combined with ordinary protium to perhaps as much as double the effective specific impulse.

2 C is twice that for the case of $I_{sp}$ = 1 C.

Accordingly, we will assume that the factoral increase in Lorentz factor of the spacecraft for each said additional stage burn-through for the case of $I_{sp}$ = 2 C is twice that of the corresponding stages for the case of $I_{sp}$ = 1 C. Thus, for each additional said stage, the spacecraft Lorentz factor is increased by a factor of 4.

Considered from another perspective, imagine a spacecraft traveling at a Lorentz factor of 1,000. Consider that the mass-energy of the spacecraft at γ = 1,000 is equal to 1,000 units. Further consider that the spacecraft has a final purely antimatter stage equal in invariant mass to the final payload beginning at γ = 1,000. Now, consider that the spacecraft stage is 1/10 spent where the matter half is taken from interstellar or intergalactic space in a lossless manner. The spacecraft's kinetic energy increase is 50 units or 5 percent thus yielding a spacecraft kinetic energy of 1,050 units. However, the invariant mass of the spacecraft goes down to 95 percent from its starting invariant mass at γ = 1,000. The Lorentz factor of the spacecraft thus goes up to [1,050/(0.95)] = 1,105.263158.

Now consider that the next 1/10 of the spacecraft stage is consumed where the matter half is taken from interstellar or intergalactic space in a lossless manner. The relativistic mass-energy of the second 1/10 of the stage is now equal to (1,105.263158)(0.050) units = 55.26315789 units. The spacecraft's kinetic energy increase is 55.26315789 units thus yielding a spacecraft kinetic energy of 1,160.526316 units. However, the invariant mass of the spacecraft goes down to 90 percent from its starting invariant mass at γ = 1,000. The Lorentz factor of the spacecraft thus goes up to [1,160.526316 /(0.9)] = 1,289.473684.

Now consider that the next 1/10 of the spacecraft stage is consumed where the matter half is taken from interstellar or intergalactic space in a lossless manner. The relativistic mass-energy of the third 1/10 of the stage is now equal to (1,289.473684)(0.050) units = 64.47368422 units just before its burning begins. The spacecraft's kinetic energy increase is 64.47368422 units thus yielding a spacecraft kinetic energy of 1,225 units. However, the invariant mass of the spacecraft goes down to 85 percent from its starting invariant mass at γ = 1,000. The Lorentz factor of the spacecraft thus goes up to [1,225/(0.85)] = 1,441.176471.

Now consider that the next 1/10 of the spacecraft stage is consumed where the matter half is taken from interstellar or intergalactic space in a lossless manner. The relativistic mass-energy of the third 1/10 of the stage is now equal to (1,441.176471)(0.050) units = 72.05882354 units just before its burning begins. The spacecraft's kinetic energy increase is 72.05882354 units thus yielding a spacecraft kinetic energy of 1,297.058824 units. However, the invariant mass of the spacecraft goes down to 80 percent from its starting invariant mass at γ = 1,000. The Lorentz factor of the spacecraft thus goes up to [1,297.058824/(0.80)] = 1,621,323529.

Now consider that the next 1/10 of the spacecraft stage is consumed where the matter half is taken from interstellar or intergalactic space in a lossless manner. The relativistic mass-energy of the third 1/10 of the stage is now equal to (1,621,323529)(0.050) units = 81.06617645 units just before its burning begins. The spacecraft's kinetic energy increase is 81.06617645 units thus yielding a spacecraft kinetic energy of 1,378.12500 units. However, the Invariant mass of the spacecraft goes down to 0.75 percent from its starting invariant mass at γ = 1,000. The Lorentz factor of the spacecraft thus goes up to [1,378.12500/(0.75)] = 1,837.50000.

Now consider that the next 1/10 of the spacecraft stage is consumed where the matter half is taken from interstellar or intergalactic space in a lossless manner. The relativistic mass-energy of the third 1/10 of the stage is now equal to (1,837.50000)(0.050) units = 91.8750000 units just before its burning begins. The spacecraft's kinetic energy increase is 91.8750000 units thus yielding a spacecraft kinetic energy of 1,470 units. However, the invariant mass of the spacecraft goes down to 70 percent from its starting invariant mass at γ = 1,000. The Lorentz factor of the spacecraft thus goes up to [1,470/(0.70)] = 2,100.

Now consider that the next 1/10 of the spacecraft stage is consumed where the matter half is taken from interstellar or intergalactic space in a lossless manner. The relativistic mass-energy of the third 1/10 of the stage is now equal to (2,100)(0.050) units = 105 units just before its burning begins. The spacecraft's kinetic energy increase is 105 units thus yielding a spacecraft kinetic energy of 1,575 units. However, the invariant mass of the spacecraft goes down to 65 percent from its starting invariant mass at γ = 1,000. The Lorentz factor of the spacecraft thus goes up to [1,575 /(0.65)] = 2,423.076923.

Now consider that the next 1/10 of the spacecraft stage is consumed where the matter half is taken from interstellar or intergalactic space in a lossless manner. The relativistic mass-energy of the third 1/10 of the stage is now equal to (2,423.076923)(0.050) units = 121.1538462 units just before its burning begins. The spacecraft's kinetic energy increase is 121.1538462 units thus yielding a spacecraft kinetic energy of 1,696.153846 units. However, the invariant mass of the spacecraft goes down to percent from its starting invariant mass at γ = 1,000. The Lorentz factor of the spacecraft thus goes up to [1,696.153846 /(0.60)] = 2,826.923077.

Now consider that the next 1/10 of the spacecraft stage is consumed where the matter half is taken from interstellar or intergalactic space in a lossless manner. The relativistic mass-energy of the third 1/10 of the stage is now equal to (2,826.923077)(0.05) units = 141.3461539 units just before its burning begins. The spacecraft's kinetic energy increase is 135.0641026 units thus yielding a spacecraft kinetic energy of 1,837.5 units. However, the invariant mass of the spacecraft goes down to 55 percent from its starting invariant mass at γ = 1,000. The Lorentz factor of the spacecraft thus goes up to [1,837.5/(0.55)] = 3,340.909091.

Now consider that the next 1/10 of the spacecraft stage is consumed where the matter half is taken from interstellar or intergalactic space in a lossless manner. The relativistic mass-energy of the third 1/10 of the stage is now equal to (3,340.909091)(0.050) units = 167.0454545 units just before its burning begins. The spacecraft's kinetic energy increase is 167.0454545 units thus yielding a spacecraft kinetic energy of 2,00454545 units. However, the invariant mass of the spacecraft goes down to 50 percent from its starting invariant mass at γ = 1,000. The Lorentz factor of the spacecraft thus goes up to [2,00454545/(0.50)] = 4,009.090909 ≈ 4.

The ratio of 4,009.090909 and 4 is equal to 1.002272727. The two values are within about 1/500 of each other and so the discrepancy can be interpreted from both the compounded error resulting from repeated calculations and the fact that not all of the thrust energy is converted to spacecraft kinetic energy.

Consider the verification of the correctness of an assumed specific impulse of 2 for the basic relativistic rocket equation where the rocket fuel is purely antimatter and where the normal matter portion is extracted from interstellar or intergalactic space in a lossless manner for the case of a perfectly efficient spacecraft propulsion system. Recurrence relations can be used to compute spacecraft Lorentz factors for cases where the specific impulse is increased in increments of 2 C and also for cases where any of the effectively superluminal specific impulses are otherwise inefficiently manifested. From such considerations, we can prove the mass ratio specific Lorentz factor for arbitrarily effectively super-luminal specific impulses with or without arbitrary losses in otherwise ideally efficient propulsion systems.

For spacecraft having an effective fuel specific impulse of 2 C, after the first stage is burnt, the relativistic Lorentz factor of the craft will be equal to 3. We assume a value of 3 because of inefficiencies in propulsion at modest fractions of C.

After the second stage is burnt, γ = 3(4) = 12.

After the third stage is burnt, γ = 3(4)(4) = 48.

After the fourth stage is burnt, γ = 3(4)(4)(4) = 192.

After the fifth stage is burnt, γ = 3(4)(4)(4)(4) = 768.

After the sixth stage is burnt, γ = 3(4)(4)(4)(4)(4) = 3,072.

After the seventh stage is burnt, γ = 3(4)(4)(4)(4)(4)(4) = 12,288.

After the eighth stage is burnt, γ = 3(4)(4)(4)(4)(4)(4)(4) = 49,152.

After the ninth stage is burnt, γ = 3(4)(4)(4)(4)(4)(4)(4)(4) = 196,608.

After the tenth stage is burnt, γ = 3(4)(4)(4)(4)(4)(4)(4)(4)(4) = 786,432.

After the eleventh stage is burnt, $\gamma = 3(4)(4)(4)(4)(4)(4)(4)(4)(4) = 3{,}145{,}728$.

Our final result from the series of 11 steps is 3,145,728/2,094,086.734 times or 1.50219 times that obtained through the relativistic rocket equation. For the 11 step method, we obviously assumed that all of the fuel energy would be converted to ship based kinetic energy. In reality, since the spacecraft spends much of its first stage well below the speed of light, and even spends the second stage at significantly below the speed of light, the conversion of the light speed thrust energy to kinetic energy is inefficient.

If we obtain more potential energy from the reactive interstellar medium than from the antimatter fuel, the potential Lorentz factor for the craft is much higher. Any cloaking mechanism would need to operate in such a manner that the propulsion system would be functional. Thus, the collision energy or potential collision energy resulting from the incoming interstellar and intergalactic particle stream would need to be completely recycled or nearly so, or be cloaked, but not so in a manner such that interstellar or intergalactic plasma could not be reacted against by electrodynamic fields produced by the craft such as magnetic fields, electric fields, and electromagnetic fields for the case where electrodynamic-hydrodynamic-plasma-drives operate at extreme Lorentz factors. Some sort of electroweak mechanism may also be of use in repelling neutral matter that is nonetheless reactive to electrodynamic fields by some unspecified electroweak interaction propulsion mechanism.

A huge caveat here is the assumption that the drag and friction energy could be nearly completely recycled in a nearly lossless manner.

In addition, another even bigger caveat is that the above method can actually work given the conservative energy and momentum laws. In the event that these conservative laws would ordinarily prevent such hyper-relativistic specific-impulse gains, perhaps the effect can be made to work on classically borrowed energy, information conversion to energy, or be interpretable as some sort of potential energy extraction from the vacuum energy state of empty space.

I feel like a bit of a naïve in describing this concept. However, if it can be made to work, perhaps even if the workability is interpretable as vacuum state potential energy extraction, we may have one heck of a way to obtain stupendously high relativistic Lorentz factors.

Thus, perhaps the energy imparted to the spacecraft for $I_{sp}$ = 2 C fuel can be viewed in terms of energy depletion from the interstellar and intergalactic medium at least with respect to the spacecraft's local environment. The energy might in a sense be classically borrowed and returned as the spacecraft decelerated to a halt. Alternatively, it may be the case that a unified field of information-mass-energy is conserved where information is converted into energy by some sort of conservative "info-dynamics" laws.

I have not be able to determine that the common relativistic rocket equation used frequently in this text would not apply for the case above case where only the antimatter fuel component would be carried on board from the start of the mission, under the condition that drag energy is completely recycled and used for virtually 100 percent efficient thrust streams. We seem to be running into energy conservation issues with the $I_{sp}$ = 2C notion, but heck, if our universe was one big free lunch, then why not look for special relativistic free lunches in our highly evolved universe.

It might be the case that net information is conserved with or without its unification with mass-energy. A conservation of information or potential information might be shown to supersede energy conservation principles and may thus permit $I_{sp}$ values of 2 C for a rocket that carries only its antimatter fuel component along from the start of its journey.

Science News recently reported on the laboratory production of a minuscule, quantum level of energy, from information. The article describes how information was converted into energy but nonetheless suggests that the total energy involved in producing the effect was many, many, orders of magnitude greater than the energy so produced. Quite simply, the energy that went into operating the apparatus which was typical in scale of laboratory grade set-ups was many orders of magnitude greater than the energy produced from the information. The production of the energy may merely get lost in the accounting books by the production process. In fact, the situation was compared to the production of a single nuclear reaction amidst a set up within a full scale nuclear power plant. The potential energy required to cause the nuclear reaction is already present and harnessable within the reactor system including the fuel rods and control systems. Nonetheless, the fact that information was in a sense converted into energy in the former laboratory experiment shows that a new paradigm involving information-mass-energy interconvertability may someday arrive.

We can consider cases where the propulsion system efficiency is less than if not significantly less than two. Such systems would imply various levels of astrodynamic drag, other propulsion system inefficiencies, and the like.

Now consider situations where drag energy is completely recycled and where the propulsion system is non-optimally efficient, say having an efficiency of 3/4, that is, only 3/4 of the matter antimatter reaction energy is transformed into an optimized light speed exhaust stream.

Consider the series of masses below for the total rocket fuel invariant mass starting with the final portion of rocket fuel equal in invariant mass to $M_1$ and ending with the mass of the first rocket stage, where the rocket fuel takes the form of pure antimatter and where the thrusting process proceeds with an efficiency of 3/4.

Consider that the nth stage for the case of specific impulse equal to 1.5 C yields 1.5 times the energy than that for the case of specific impulse equal to 1 C under the condition that in both scenarios, the starting Lorentz factor would be the same for the beginning of nth stage. We will assume that the increase in relativistic kinetic energy brought about through first stage burn-through for the case of $I_{sp}$ = 1.5 C is 1.5 times that for the case of $I_{sp}$ = 1 C. Accordingly, we will assume that the factoral increase in Lorentz factor of the spacecraft for each said additional stage burn-through for the case of $I_{sp}$ = 1.5 C is 1.5 times that of the corresponding stages for the case of $I_{sp}$ = 1 C thus yielding an increase in Lorentz factor by a factor of 3 for each said stage.

(1)M1, 2(M1), 4(M1), 8(M1), 16(M1), 32(M1), 64(M1), 128(M1), 256(M1), 512(M1), 1,024(M1).

Using the relativistic rocket equation:

$$\Delta v = C \tanh [(I_{sp}/C) \ln (M_0/M_1)]$$

$$= C \tanh [(1¾\ C/C) \ln (M_0/M_1)]$$

$$= C \tanh [(1¾) \ln (2{,}048)] = 0.99999999976674.$$

This corresponds to a Lorentz factor of 46,340.96106.

After the first stage is burnt, $\gamma$ = 2.5.

After the second stage is burnt, $\gamma$ = (2.5)(3) = 7.5.

After the third stage is burnt, $\gamma$ = (2..5)(3)(3) = 22.5.

After the fourth stage is burnt, $\gamma$ = (2.5) (3)(3)(3) = 67.5

After the fifth stage is burnt, $\gamma$ = (2.5) (3)(3)(3)(3) = 202.5.

After the sixth stage is burnt, $\gamma$ = (2.5) (3)(3)(3)(3)(3) = 607.5.

After the seventh stage is burnt, $\gamma$ = (2.5) (3)(3)(3)(3)(3)(3) = 1.822.5.

After the eighth stage is burnt, $\gamma$ = (2.5) (3)(3)(3)(3)(3)(3)(3) = 5,467.5.

After the ninth stage is burnt, $\gamma$ = (2.5) (3)(3)(3)(3)(3)(3)(3)(3) = 16,402.5.

After the tenth stage is burnt, $\gamma$ = (2.5) (3)(3)(3)(3)(3)(3)(3)(3)(3) = 49,207.5.

After the eleventh stage is burnt, $\gamma$ = (2.5) (3)(3)(3)(3)(3)(3)(3)(3)(3)(3) = 147,622.5.

Our final result from the series of 11 steps is 147,622.5/46,340.96106 times or 3.18557 times that obtained through the relativistic rocket equation. For the 11 step method, we obviously assumed that all of the fuel energy would be converted to ship based kinetic energy. In reality, since the spacecraft spends much of its first stage well below the speed of light, and even spends the second stage at significantly below the speed of light, the conversion of the light speed thrust energy to kinetic energy is inefficient.

Now consider situations where drag energy is completely recycled and where the propulsion system is non-optimally efficient, say having an efficiency of 2/3, that is, only 2/3 of the matter antimatter reaction energy is transformed into an optimized light speed exhaust stream.

Consider the series of masses below for the total rocket fuel invariant mass starting with the final portion of rocket fuel equal in invariant mass to $M_1$ and ending with the mass of the first rocket stage, where the rocket fuel takes the form of pure antimatter and where the thrusting process proceeds with an efficiency of 2/3.

Consider that the nth stage for the case of specific impulse equal to 1⅓ C yields 1⅓ times the energy than for the case of specific impulse equal to 1 C under the condition that in both scenarios, the starting Lorentz factor would be the same for the beginning of nth stage. We will assume that the increase in relativistic kinetic energy brought about through first stage burn-through for the case of $I_{sp}$ = 1⅓ C is 1⅓ times that for the case of $I_{sp}$ = 1 C. Accordingly, we will assume that the factoral increase in Lorentz factor of the spacecraft for each additional stage burn-through for the case of $I_{sp}$ = 1⅓ C is 1⅓ times that of the corresponding stages for the case of $I_{sp}$ = 1 C. Thus, we will assume that the increase in spacecraft Lorentz factor for each said additional burnt stage is 2 ⅔.

(1)M1, 2(M1), 4(M1), 8(M1), 16(M1), 32(M1), 64(M1), 128(M1), 256(M1), 512(M1), 1,024(M1).

Using the relativistic rocket equation:

$\Delta v = C \tanh [(I_{sp}/C) \ln (M_0/M_1)]$

$= C \tanh [(1⅓\ C/C) \ln (M_0/M_1)]$

$= C \tanh [(1⅓) \ln (2{,}048)] = 0.99999999704324.$

This corresponds to a Lorentz factor of 13,003.65919.

After the first stage is burnt, γ = 2.333333333.

After the second stage is burnt, γ = (2.33333333)(2.66666667) = 6.222222222.

After the third stage is burnt, γ = (2.33333333)(2.66666667)(2.66666667) = 16.59259256.

After the fourth stage is burnt, γ = (2.33333333) (2.66666667)(2.66666667) (2.66666667) = 44.24691369.

After the fifth stage is burnt, γ = (2.33333333) (2.66666667)(2.66666667) (2.66666667) (2.66666667) = 117.9917679.

After the sixth stage is burnt, γ = (2.33333333) (2.66666667)(2.66666667) (2.66666667) (2.66666667) (2.66666667) = 314.6447185.

After the seventh stage is burnt, γ = (2.33333333) (2.66666667)(2.66666667) (2.66666667) (2.66666667) (2.66666667) (2.66666667) = 839.052581.

After the eighth stage is burnt, γ = (2.33333333) (2.66666667)(2.66666667) (2.66666667) (2.66666667) (2.66666667) (2.66666667) (2.66666667) = 2,237.473552.

After the ninth stage is burnt, γ = (2.33333333) (2.66666667)(2.66666667) (2.66666667) (2.66666667) (2.66666667) (2.66666667) (2.66666667) (2.66666667) = 5,966.596146.

After the tenth stage is burnt, γ = (2.33333333) (2.66666667)(2.66666667) (2.66666667) (2.66666667) (2.66666667) (2.66666667) (2.66666667) (2.66666667) (2.66666667) = 15,910.92303.

After the eleventh stage is burnt, γ = (2.33333333) (2.66666667)(2.66666667) (2.66666667) (2.66666667) (2.66666667) (2.66666667) (2.66666667) (2.66666667) (2.66666667) (2.66666667) = 42,429.12806.

Our final result from the series of 11 steps is 42,429.12806/13,003.65919 times or 3.26286 times that obtained through the relativistic rocket equation. For the 11 step method, we obviously assumed that all of the fuel energy would be converted to ship based kinetic energy. In reality, since the spacecraft spends much of its first stage well below the speed of light, and even spends the second stage at significantly below the speed of light, the conversion of the light speed thrust energy to kinetic energy is inefficient.

Notice how the ratio of the gamma values obtained by the 11 step processes and those obtained through the relativistic rocket equation for specific impulse values 2 C, 1¾ C, and 1⅓ C increase. It seems to be the case that the ratios will reach a maximum because for the case of specific impulse equal to 1 C, the ratio is lower than that for the cases of specific impulses equal to 1¾ C, and 1⅓ C. Part of the difference between the specific impulses obtained by the 11 step methods and the corresponding values obtained through the relativistic rocket equation can be explained away by the fact that the spacecraft spends more time accelerating inefficiently while burning its first few stages than while burning its latter stages. The thrusting process approaches perfect efficiency for Lorentz factors equal to infinity or for velocities equal to C, exactly.

Nonetheless, there must exist a maximum in the plot of the ratios bounded by the conditions of specific impulse equal to 1 C and specific impulse equal to 2 C. Thus, I conclude that there is a fudge factor that modifies my computations at least for the specific process where I used the eleven step methods.

The above methods of calculating Lorentz factors for super-relativistic specific impulse fueling mechanisms can also be applied to cases where the energy conversion process and thrusting is 100 percent efficient but where the ratio of antimatter to matter for the reactions is greater than one but not so high as to cause a drop in specific impulse to values less than or equal to 1 C. A combination of fueling mechanisms having a maximum theoretical specific impulse of less than 2 C combined with arbitrary inefficiencies in rocket thrusting mechanisms and non-zero astrodynamic drag can also yield effectively super-relativistic specific impulse values that are less than the maximum value of 2 C.

We now move on to consider the case where a relativistic rocket with a specific impulse for pure carried along antimatter fuel is ordinarily equal to 2 C. but where the rocket thrust power is matched by remotely beamed power.

The terminal velocity of such a system is plausibly equal to:

$$\Delta v = C \text{ Tanh } \{[I_{sp}/C]\ln(M_0/M_1)]\}$$

$$= C \text{ Tanh } \{[4C/C]\ln(M_0/M_1)]\}$$

$$= C \text{ Tanh } \{[4]\ln(M_0/M_1)]\}.$$

Consider again, the staged series;

(1)M1, 2(M1), 4(M1), 8(M1), 16(M1), 32(M1), 64(M1), 128(M1), 256(M1), 512(M1), 1,024(M1).

Each stage for the scenario of effective specific impulse equal to 4 C is associated with a spent rocket fuel mass specific integrated thrust power output component for fuel that is double what it would be otherwise for the $I_{sp}$ = 2 scenario for the case of an nth stage where, in both scenarios, the starting Lorentz factor of the spacecraft at the beginning of nth stage burning is the same. However, the rocket thrust component is matched by the beam drive component thus yielding a 16 fold increase in spacecraft Lorentz factor for each consumed stage. This is four times the effective stage specific Lorentz factor increase for the case of a specific impulse of 2 C and 8 times the effective stage specific Lorentz factor increase for the case of a specific impulse of 1 C.

After the first stage is burnt, $\gamma$ =17 or 1 + ($4^2$). We will assume a value of 17 here because the inefficiencies of propulsion at modest fractions of C will be much shorter in used fuel mass specific duration than in the cases where the specific impulse is equal to 2 C or less.

After the second stage is burnt, $\gamma = 17 (4^2) = 272$.

After the third stage is burnt, $\gamma = 17(4^2) (4^2) = 4{,}352$.

After the fourth stage is burnt, $\gamma = 17(4^2) (4^2) (4^2) = 69{,}632$.

After the fifth stage is burnt, $\gamma = 17 (4^2) (4^2) (4^2) (4^2) = 1{,}114{,}112$.

After the sixth stage is burnt, $\gamma = (17)(4^2) (4^2)(4^2) (4^2) (4^2) = 17{,}825{,}792$.

After the seventh stage is burnt, $\gamma = (17)(4^2) (4^2) (4^2) (4^2) (4^2) (4^2) = 285{,}212{,}672$.

After the eighth stage is burnt, $\gamma = (17)(4^2) (4^2) (4^2) (4^2) (4^2) (4^2) (4^2) = 4{,}563{,}402{,}752$.

After the ninth stage is burnt, $\gamma = (17)(4^2) (4^2)(4^2) (4^2) (4^2) (4^2) (4^2) (4^2) = 73{,}014{,}444{,}030{,}000$

After the tenth stage is burnt, $\gamma = (17)(4^2)(4^2)(4^2)(4^2)(4^2)(4^2)(4^2)(4^2)(4^2)$ = 1,168,231,105,000.

After the eleventh stage is burnt, $\gamma = (17)(4^2)(4^2)(4^2)(4^2)(4^2)(4^2)(4^2)(4^2)(4^2)(4^2)$ = 18,691,697,670,000.

Our final result from the series of the 5 considered steps is 1,114,112/524,288 times or 2.125 times that obtained through the relativistic rocket equation. For the 11 step method, we obviously assumed that all of the fuel energy would be converted to ship based kinetic energy. In reality, since the spacecraft spends much of its first stage well below the speed of light, and even spends the second stage at significantly below the speed of light, the conversion of the light speed thrust energy to kinetic energy is inefficient.

Since my EXCEL spread sheet has limited numbers of decimal place computation, I will run the relativistic rocket equation with values of mass ratio associated with a 5 stage rocket for which the relativistic Lorentz factor of the craft was computed to be $17 (4^2)(4^2)(4^2)(4^2)$ = 1,114,112. The associated mass ratio is (1 + 1 + 2 + 4 + 8 + 16) = 32. Thus, I will use the following analytic expression:

$$\Delta v = C \text{ Tanh } \{[I_{sp}/C]\ln(M_0/M_1)]\}$$

$$= C \text{ Tanh } \{[4C/C]\ln(M_0/M_1)]\}$$

$$= C \text{ Tanh } \{[4]\ln(64)]\} = 524{,}288.$$

Note that the ratio of gamma obtained by the five steps method and that obtained by the relativistic rocket equation is equal to 1,114,112/524,288 = 2.125 which is about equal to the respective ratios for the previously considered super-relativistic fuels. The latter ratio would be even closer to unity if we had assumed a lower first stage burn-through Lorentz factor to compensate for the fact that the spacecraft does not achieve near light speed until after a significant portion of the first stage is consumed.

We can also consider the case of a thrust optimized spacecraft carrying both components of fuel along from the start of the journey in full thus resulting in a maximum specific impulse equal to 1 C and where the thrust power is matched by beam drive power. In this case, we can see from simple inspection that the effective specific impulse will be equal to 2 C. The stage specific Lorentz factors will be the same as for the $I_{sp}$ = 2 C scenario previously covered.

The above methods of calculating Lorentz factors for super-relativistic specific impulse fueling mechanisms can also be applied to cases where the energy conversion process and thrusting is 100 percent efficient but where the ratio of antimatter to matter for the reactions is greater than one but not so high as to cause a drop in specific impulse to values less than or equal to 1 C. A combination of fueling mechanisms having a maximum theoretical specific impulse of less than 2 C combined with arbitrary inefficiencies in rocket thrusting mechanisms and non-zero astrodynamic drag can also yield effectively super-relativistic specific impulse values that are less than the maximum value of 2 C.

We now consider the case of a rocket that carries only its antimatter fuel along from the beginning and where a beam drive mechanism imparts twice as much energy into the rocket per unit matter-antimatter fuel used and where the system is virtually 100 percent efficient.

Consider the scenario where the spacecraft is traveling at a Lorentz factor of say, 1,000, and where the spacecraft is about to burn its final stage which has an invariant mass equal to its payload.

Now, the specific impulse of a beam rocket that obtains all of its normal matter fuel from the background and which has a per stage energy generation that is matched by the beam mode is 4 C according to the previous derivation. However, for the case where the beam energy per stage is twice that of the former case for a specific impulse of 4 C, we have 1.5 times more per stage propulsion power than in the case of a specific impulse of 4 C due to an extra 50 percent increase in stage specific propulsion power for a given Lorentz factor. Therefore, we will assume that the generated specific impulse of the rocket is effectively equal to 6 C.

Consider again the scenario where the spacecraft is traveling at a Lorentz factor of say, 1,000, and where the spacecraft is about to burn its final stage which has an invariant mass equal to its payload. We go further to consider the case where the entire stage is burned under the above scenario of specific impulse equal to 6 C. The factoral component of the increase in spacecraft relativistic Lorentz factor due to the stage burning for the case of $I_{sp}$ = 4 C is 16 which is a factor of 4 greater than in the case of $I_{sp}$ = 2. Therefore, we will assume by symmetrical exponential growth that the factoral increase in spacecraft Lorentz factor per stage burnt is equal to (4)(16) = 64 for the case of specific impulse equal to 6 C. This is plausible because the effective fuel energy delivered to the craft for the $I_{sp}$ = 6 C case goes up by a factor of two relative to the $I_{sp}$ = 4 C case for a given stage starting Lorentz factor and the stage specific beam energy delivered to the craft for the $I_{sp}$ = 6 C goes up by a factor of two relative to the case of $I_{sp}$ = 4 C where the nth stage for both specific impulse values would start at the same Lorentz factor. Since the fuel is doubly stretched and the beam power is twice that of the rocket thrust mechanism, the total stage specific Lorentz factor increase is (4)(2 + 2)(4) = 64.

Consider again, the staged series;

(1)M1, 2(M1), 4(M1), 8(M1), 16(M1), 32(M1), 64(M1), 128(M1), 256(M1), 512(M1), 1,024(M1).

After the first stage is burnt, γ =65 + (64). We will assume a value of 65 here because the inefficiencies of propulsion at modest fractions of C will be much shorter in used fuel mass specific duration than in the cases where the specific impulse is equal to 2 C or less.

After the second stage is burnt, γ = (65)(64) = 4,160.

After the third stage is burnt, γ = (65)(64)(64) = 266,240.

After the fourth stage is burnt, γ = (65)(64)(64)(64) = 17,039,360.

After the fifth stage is burnt, γ = 17 (65)(64)(64)(64)(64) = 1,090,519,040.

After the sixth stage is burnt, γ = (65)(64)(64)(64)(64)(64) = 69,793,218,560.

After the seventh stage is burnt, γ = (65)(64)(64)(64)(64)(64)(64) = 4,466,765,988,000.

After the eighth stage is burnt, γ = (65)(64)(64)(64)(64)(64)(64)(64) = 285,873,023,200,000.

After the ninth stage is burnt, γ = (65)(64)(64)(64)(64)(64)(64)(64)(64) = 18,295,873,490,000,000.

After the tenth stage is burnt, γ = (65)(64)(64)(64)(64)(64)(64)(64)(64)(64) = 1,170,935,903,000,000,000.

After the eleventh stage is burnt, γ =(65)(64)(64)(64)(64)(64)(64)(64)(64)(64)(64) = 74,939,897,800,000,000,000.

Let us do an inductive sanity check for values derived through the first four stages.

The ratio of gamma obtained by the multistep discreet method and that obtained by the relativistic rocket equation for the first stage is 65/32.0078125 = 2.03075421.

The ratio of gamma obtained by the multistep discreet method and that obtained by the relativistic rocket equation for the second stage is 4,160/2048.000123 = 2.031249878.

The ratio of gamma obtained by the multistep discreet method and that obtained by the relativistic rocket equation for the third stage is 266,240/131,072 = 2.03125.

The ratio of gamma obtained by the multistep discreet method and that obtained by the relativistic rocket equation for the fourth stage is 17,039,360/8,323,830.135 = 2.047057631.

These ratios are all virtually equal to each other and are about equal to the analogous ratios computed for all of the previous greater than 1 C specific impulse scenarios covered. The ratios increase slightly per step because of compounding errors due to the assumption that all propulsion energy is converted to spacecraft kinetic energy whereas in reality, a relativistic rocket converts all propulsion energy to spacecraft kinetic energy only at infinite Lorentz factors.

Our final result from the series of the five 5 considered steps is 17,039,360/8,323,830.135 = 2.047057631 times that obtained through the relativistic rocket equation. For the 11 step method, we obviously assumed that all of the fuel energy would be converted to ship based kinetic energy. In reality, since the spacecraft spends much of its first stage well below the speed of light, and even spends the second stage at significantly below the speed of light, the conversion of the light speed thrust energy to kinetic energy is inefficient.

The specific impulse by definition will drop to 4 C for scenarios where the ideal craft with a fuel specific impulse = 6 C otherwise operates with an efficiency of only 2/3. In such a case, one simply notes by inspection the equality of the stage specific factoral

Lorentz factor increase for the inefficient $I_{sp}$ = 6 C case with that of the 100 percent efficient $I_{sp}$ = 4 C case in order to obtain the stage specific factoral Lorentz factor increase.

The specific impulse by definition will drop to 3 C for scenarios where the ideal specific impulse = 6 C and where craft otherwise operates with an efficiency of only 1/2. In such a specific case, one simply notes by inspection the equality of the stage specific factoral Lorentz factor increase for the inefficient $I_{sp}$ = 6 C case with that of the 100 percent efficient $I_{sp}$ = 4 C case in order to obtain the stage specific factoral Lorentz factor increase.

Consider the case where both matter and antimatter fuel components are carried along from the start of the mission and where the beamed energy provides twice the stage specific propulsion power of the rocket thrusting mechanism at a given Lorentz factor. The specific impulse will equal to

$I_{sptotal} = I_{sprocket} + I_{spbeam} = I_{sprocket} + [(2)(I_{sprocket})] = C + [2\ C] = 3\ C.$

We now consider the case of a rocket that carries only its antimatter fuel along from the beginning and where a beam drive mechanism imparts three times as much power into the rocket per unit matter-antimatter fuel used and where the system is virtually 100 percent efficient.

Consider the scenario where the spacecraft is traveling at a Lorentz factor of say, 1,000, and where the spacecraft is about to burn its final stage which has an invariant mass equal to its payload. The specific impulse of a beam rocket that obtains all of its normal matter fuel from the background and which has a per stage rocket thrust energy generation that is 1/2 of that by the beam mode, is 6 C according to the previous derivation. However, for the case where the fractional beam power per stage is 3/4 for case for a specific impulse of 8 C, we have 1⅓ times more per stage propulsion power at a given Lorentz factor than in the case of a specific impulse of 6 C due to an extra ⅓ increase in stage specific propulsion power, all else remaining the same. Therefore, we will assume that the generated specific impulse of the rocket is effectively equal to 8 C.

Consider again the scenario where the spacecraft is traveling at a Lorentz factor of say, 1,000, and where the spacecraft is about to burn its final stage which has an invariant mass equal to its payload. We go further to consider the case where the entire stage is burned under the above scenario of specific impulse equal to 8 C. Note that the factoral component of the increase in spacecraft relativistic Lorentz factor due to the stage burning for the case of $I_{sp}$ = 6 C is 64 which is a factor of 4 greater than in the case of $I_{sp}$ = 4. Thus, we will assume by symmetrical exponential growth that the factoral increase in spacecraft Lorentz factor per stage burnt is equal to [(4)(16)](4) = (64)(4) = 256 for the case of specific impulse equal to 8 C. This is plausible because the effective instantaneous fuel energy delivered to the craft for the $I_{sp}$ = 8 C case goes up by a factor of four relative to the $I_{sp}$ = 2 C case and the total instantaneous propulsion system power goes up by a factor of 4 relative to the case of $I_{sp}$ = 2C. Furthermore, the fuel supply is instantaneously stretched by a factor 4. Thus, the stage specific spacecraft factoral Lorentz factor increase for the $I_{sp}$ = 8 C case is (4)(4)(4)(4) = 256.

Consider again, the staged series;

(1)M1, 2(M1), 4(M1), 8(M1), 16(M1), 32(M1), 64(M1), 128(M1), 256(M1), 512(M1), 1,024(M1).

After the first stage is burnt, $\gamma$ =257 = 256 + 1. We will assume a value of 257 here because the inefficiencies of propulsion at modest fractions of C will be much shorter in used fuel mass specific duration than in the cases where the specific impulse is equal to 2 C or less.

After the second stage is burnt, $\gamma$ = (257)(256) = 65,792.

After the third stage is burnt, $\gamma$ = (257)(256)(256) = 16,842,752.

After the fourth stage is burnt, $\gamma$ = (257)(256)(256)(256) = 4,311,744,512.

After the fifth stage is burnt, $\gamma$ = (257)(256)(256)(256)(256) = $1.103806595 \times 10^{12}$.

After the sixth stage is burnt, $\gamma$ = (257)(256)(256)(256)(256)(256) = $2.825744883 \times 10^{14}$.

After the seventh stage is burnt, $\gamma$ = (257)(256)(256)(256)(256)(256)(256) = $7.233906901 \times 10^{16}$.

After the eighth stage is burnt, $\gamma$ = (257)(256)(256)(256)(256)(256)(256)(256) = $1.851880167 \times 10^{19}$.

After the ninth stage is burnt, $\gamma$ = (257)(256)(256)(256)(256)(256)(256)(256)(256) = $4.740813227 \times 10^{21}$.

After the tenth stage is burnt, γ = (257)(256)(256)(256)(256)(256)(256)(256)(256)(256) = 1.213648186 x $10^{24}$.

After the eleventh stage is burnt, γ = (257)(256)(256)(256)(256)(256)(256)(256)(256)(256)(256) = 3.106939356 x $10^{26}$.

Let us do an inductive sanity check for values derived through the first three stages.

The ratio of gamma obtained by the multistep discreet method and that obtained by the relativistic rocket equation for the first stage is 257/128.0019531 = 2.007781864.

The ratio of gamma obtained by the multistep discreet method and that obtained by the relativistic rocket equation for the second stage is 65,792/32,768.00391 = 2.00781226.

The ratio of gamma obtained by the multistep discreet method and that obtained by the relativistic rocket equation for the third stage is 16,842,752 /8,323,830.135 = 2.023437736.

Thus, a Lorentz factor of about 1.5 x $10^{26}$ is plausible for an $I_{sp}$ = 8 spacecraft having a starting mass ratio of only 2,048.

These ratios are all virtually equal to each other and are about equal to the analogous ratios computed for all of the previous greater than 1 C specific impulse scenarios covered. The ratios increase slightly per step because of compounding errors due to the assumption that all propulsion energy is converted to spacecraft kinetic energy. In reality, a relativistic rocket converts all of its propulsion energy to spacecraft kinetic energy only at infinite Lorentz factors.

Our final result from the series of 11 steps is 3,145,728/2,094,086.734 times or 1.50219 times that obtained through the relativistic rocket equation. For the 11 step method, we obviously assumed that all of the fuel energy would be converted to ship based kinetic energy. In reality, since the spacecraft spends much of its first stage well below the speed of light, and even spends the second stage at significantly below the speed of light, the conversion of the light speed thrust energy to kinetic energy is inefficient.

Consider scenarios where the ideal specific impulse = 8 C and where the craft otherwise operates with an efficiency of only ¾. The specific impulse by definition will drop to 6 C. In such a case, in order to obtain the stage specific factoral Lorentz factor increase, one simply notes by inspection the equality of the stage specific factoral Lorentz factor increase for the inefficient $I_{sp}$ = 8 C case with that of the 100 percent efficient $I_{sp}$ = 6 C case.

Consider scenarios where the ideal specific impulse = 8 C and where the craft otherwise operates with an efficiency of only ½. The specific impulse by definition will drop to 4 C. In such a case, in order to obtain the stage specific factoral Lorentz factor increase, one simply notes by inspection the equality of the stage specific factoral Lorentz factor increase for the inefficient $I_{sp}$ = 8 C case with that of the 100 percent efficient $I_{sp}$ = 4 C case.

Consider scenarios where the ideal specific impulse = 8 C and where craft otherwise operates with an efficiency of only ¼. The specific impulse by definition will drop to 2 C. In such a case, in order to obtain the stage specific factoral Lorentz factor increase, one simply notes by inspection the equality of the stage specific factoral Lorentz factor increase for the inefficient $I_{sp}$ = 8 C case with that of the 100 percent efficient $I_{sp}$ = 2 C case.

Consider the case where both matter and antimatter fuel components are carried along from the start of the mission and where the beamed energy provides three times the stage specific propulsion energy of the rocket thrusting mechanism. The specific impulse will simply equal:

$I_{sptotal} = I_{sprocket} + I_{spbeam} = I_{sprocket} + [(3)(I_{sprocket})] = C + [3\ C] = 4\ C.$

We will consider just one additional specific case and that is where a rocket carries only its antimatter fuel along from the beginning and where a beam drive mechanism imparts four times as much power into the rocket per unit matter-antimatter fuel used and where the system is virtually 100 percent efficient.

Consider the scenario where the spacecraft is traveling at a Lorentz factor of say, 1,000, and where the spacecraft is about to burn its final stage which has an invariant mass equal to its payload.

Now, consider the case where the fractional beam power per stage is 4/5. We have 1.25 times more per stage propulsion power at a given Lorentz factor than in the case of a specific impulse of 8 C due to an extra ¼ increase in stage specific propulsion power, i.e., due to a 25 percent increase in stage propulsion power. Therefore, we will assume that the effective generated specific impulse of the rocket is equal to 10 C.

Consider again the scenario where the spacecraft is traveling at a Lorentz factor of say, 1,000, and where the spacecraft is about to burn its final stage which has an invariant mass equal to its payload. We go further to consider the case where the entire stage is burned under the above scenario of specific impulse equal to 10 C. Note again that the factoral component of the increase in spacecraft relativistic Lorentz factor due to the stage burning for the case of $I_{sp}$ = 8 C is 256 which is a factor of 4 greater than in the case of an $I_{sp}$ = 6. Thus, we will assume by symmetrical exponential growth that the factoral increase in spacecraft Lorentz factor per stage burnt is equal to [(4)(16)](4)(4) = (64)(4)(4) = (256)(4) = 1,024 for the case of specific impulse equal to 10 C. This fact can be intuitively grasped from the reality that by quadrupling the beam energy component of the propulsion system relative to the rocket thrust mechanism, the rocket fuel is used (2)(2) = 4 times more stage specific instantaneously efficiently than in the $I_{sp}$ = 4 C and thus the fuel supply is stretched by another factor of 4. However, the beam drive power grows by a factor of 4 relative to that for the case of $I_{sp}$ = 4 C thus resulting in yet another factor of 4 increase in stage specific Lorentz factor growth thereby resulting in the stage specific spacecraft Lorentz factor increase equal to (16)(4)(4)(4) = 1,024.

Consider again, the staged series;

(1)M1, 2(M1), 4(M1), 8(M1), 16(M1), 32(M1), 64(M1), 128(M1), 256(M1), 512(M1), 1,024(M1).

After the first stage is burnt, γ = 1,025 = 1,024 + 1. We will assume a value of 1,025 here because the inefficiencies of propulsion at modest fractions of C will be much shorter in used fuel mass specific duration than in the cases where the specific impulse is equal to 2 C or less.

After the second stage is burnt, γ = (1,025)(1,024) = 1,049,600.

After the third stage is burnt, γ = (1,025)(1,024)(1,024) = 1,074,790,400.

After the fourth stage is burnt, γ = (1,025)(1,024)(1,024)(1,024) = 1.10058537 x $10^{12}$

After the fifth stage is burnt, γ = (1,025)(1,024)(1,024)(1,024)(1,024) = 1.126999418 x $10^{15}$.

After the sixth stage is burnt, γ = (1,025)(1,024)(1,024)(1,024)(1,024)(1,024) = 1.154047405 x $10^{18}$.

After the seventh stage is burnt, γ = (1,025)(1,024)(1,024)(1,024)(1,024)(1,024)(1,024) = 1.181744542 x $10^{21}$.

After the eighth stage is burnt, γ = (1,025)(1,024)(1,024)(1,024)(1,024)(1,024)(1,024)(1,024) = 1.210106411 x $10^{24}$..

After the ninth stage is burnt, γ = (1,025)(1,024)(1,024)(1,024)(1,024)(1,024)(1,024)(1,024)(1,024) 1.239148965 x $10^{27}$.

After the tenth stage is burnt, γ = (1,025)(1,024)(1,024)(1,024)(1,024)(1,024)(1,024)(1,024)(1.024)(1,024) = 1.26888854 x $10^{30}$.

After the eleventh stage is burnt, γ=(1,025)(1,024)(1,024)(1,024)(1,024)(1,024)(1,024)(1,024)(1.024)(1,024)(1,024) = 1.299341865 x $10^{33}$.

Let us do an inductive sanity check for values derived through the first two stages.

The ratio of gamma obtained by the multistep discreet method and that obtained by the relativistic rocket equation for the first stage is 1,025/512.0004883 = 2.001951216.

The ratio of gamma obtained by the multistep discreet method and that obtained by the relativistic rocket equation for the second stage is 1,049,600/524,288 = 2.001953125.

These ratios are all virtually equal to each other and are about equal to the analogous ratios computed for all of the previous greater than 1 C specific impulse scenarios covered. The ratios increase slightly per step because of compounding errors due to the assumption that all propulsion energy is converted to spacecraft kinetic energy. In reality, a relativistic rocket converts all propulsion energy to spacecraft kinetic energy only at infinite Lorentz factors.

Thus, the Lorentz factor of a spacecraft with a mass ratio of 2,048 having an effective specific impulse of 10 C is equal to about 6.5 x $10^{32}$.

The specific impulse by definition will drop to 8 C for scenarios where the specific impulse = 10 C and where the craft otherwise operates with an efficiency of only 4/5.

The effective specific impulse by definition will drop to 6 C for scenarios where the specific impulse otherwise equals 10 C and where the craft otherwise operates with an efficiency of only 3/5. In such a specific case, in order to obtain the stage specific factoral Lorentz factor increase, one simply notes by inspection the equality of the stage specific factoral Lorentz factor increase for the inefficient $I_{sp}$ = 10 C case with that of the 100 percent efficient $I_{sp}$ = 6 C case.

The effective specific impulse will drop to 4 C for scenarios where the specific impulse otherwise is equal to 10 C and where the craft otherwise operates with an efficiency of only 2/5. In such a specific case, in order to obtain the stage specific factoral Lorentz factor increase, one simply notes by inspection the equality of the stage specific factoral Lorentz factor increase for the inefficient $I_{sp}$ = 10 C case with that of the 100 percent efficient $I_{sp}$ = 4 C case.

Consider the case where both matter and antimatter fuel components are carried along from the start of the mission and where the beamed energy provides four times the stage specific propulsion energy of the rocket thrusting mechanism. The specific impulse will simply equal:

$I_{sptotal} = I_{sprocket} + I_{spbeam} = I_{sprocket} + [(4)(I_{sprocket})] = C + [4C] = 5\ C.$

A pattern is easily discernable here. For cases where a spacecraft obtains all of its normal matter components from the medium of space but where the spacecraft is partially driven by a beam mode for which the beam drive power is a positive integer factor, n, greater than the rocket thrust, the specific impulse of the rocket will be equal to [(2)(1 + n)] C. The factoral increase in Lorentz factor of the spacecraft will be close to $4^{(n+1)}$ for spacecraft having the following series of stages:

(1)M1, 2(M1), 4(M1), 8(M1), 16(M1), 32(M1), 64(M1), 128(M1), 256(M1), 512(M1), 1,024(M1), ... .

Thus, the relativistic rocket equation yields:

$$\Delta v = C \operatorname{Tanh} \{\{[(2)(1 + n)]\ C/C\}\ln(M_0/M_1)\},$$

for such systems. Terminal gamma therefore equals:

$$\gamma = \{1 - \{\{\{ C \operatorname{Tanh} \{\{[(2)(1 + n)]\ C/C\}\ln(M_0/M_1)\}\}/C\}^2\}\ \}^{-1/2}$$

For the case where the efficiency drops to a fractional value of (a/b) where a and b are positive and real, the relativistic rocket equation yields:

$$\Delta v = C \operatorname{Tanh} \{\{[(2)(1 + n)](a/b)\ C/C\}\ln(M_0/M_1)\}$$

for such systems. Terminal gamma therefore equals:

$$\gamma = \{1 - \{\{\{ C \operatorname{Tanh} \{\{[(2)(1 + n)](a/b)\ C/C\}\ln(M_0/M_1)\}\}/C\}^2\}\ \}^{-1/2}.$$

We can also consider cases where other propulsion modes such as magnetic sails, negative refraction index pull sails, hydrogen burning interstellar ramjets, and the like augment the relativistic rocket or the beam sail relativistic rocket.

Consider cases where the mth mode of propulsion contributes an additional fraction of propulsion power, $f_m$, of the rocket thrusting mechanism where the specific impulse of the rocket itself is equal to 1 C. The modified relativistic rocket equation yields:

$$\Delta v = C \operatorname{Tanh} \{\{[1C + (f_m C)]/C\}\ln(M_0/M_1)\}$$

$$= C \operatorname{Tanh} \{\{[(1 + f_m)]\ C/C\}\ln(M_0/M_1)\}.$$

The associated Lorentz factor is equal to:

$$\gamma = \{1- \{\{\{C \operatorname{Tanh} \{\{[1C + (f_m C)]/C\}\ln(M_0/M_1)\}\}\ /C\}^2\}\}^{-1/2}$$

$$= \{1- \{\{\{ C \operatorname{Tanh} \{\{[(1 + f_m)]\ C/C\}\ln(M_0/M_1)\}\}/C\}^2\}\}^{-1/2}.$$

Consider cases where the total propulsion energy stream is otherwise ideal but where the efficiency otherwise drops from 100 percent to a fractional value of (a/b) where both a and b are positive and real. The relativistic rocket equation yields:

$\Delta v = C \text{ Tanh } \{\{[1C + (f_m C)](a/b)/C\}\ln(M_0/M_1)\}$

$= C \text{ Tanh } \{\{[(1 + f_m)] (a/b)C/C\}\ln(M_0/M_1)\}$.

The associated Lorentz factor is equal to:

$\gamma = \{1- \{\{\{C \text{ Tanh } \{\{[1C + (f_m C)]/C\}\ln(M_0/M_1)\}\}(a/b)/C\}^2\}\}^{-1/2}$

$= \{1- \{\{\{ C \text{ Tanh } \{\{[(1 + f_m)] C/C\}\ln(M_0/M_1)\}\}(a/b)/C\}^2\}\}^{-1/2}$.

Consider cases where there are multiple modes of non-rocket propulsion methods assisting the relativistic rocket propulsion system where each non-rocket mode, $m_i$, provides an additional fraction of propulsion power, $f_{mi}$, of the rocket thrusting mechanism to the spacecraft where the specific impulse of the rocket itself is equal to 1 C. The modified relativistic rocket equation yields:

$\Delta v = C \text{ Tanh } \{\{[1C + (f_{m1}C) + (f_{m2}C) + (f_{m3}C) + \ldots + (f_{mp}C)] /C\}\ln(M_0/M_1)\}$

$= C \text{ Tanh } \{\{(1 + f_{m1} + f_{m2} + f_{m3} + \ldots + f_{mp}) C/C\}\ln(M_0/M_1)\}$.

The associated Lorentz factor is equal to:

$\gamma = \{1- \{\{\{ C \text{ Tanh } \{\{[1C + (f_{m1}C) + (f_{m2}C) + (f_{m3}C) + \ldots + (f_{mp}C)] /C\}\ln(M_0/M_1)\}\} /C\}^2\}\}^{-1/2}$

$= \{1- \{\{\{ C \text{ Tanh } \{\{(1 + f_{m1} + f_{m2} + f_{m3} + \ldots + f_{mp}) C/C\}\ln(M_0/M_1)\}\} /C\}^2\}\}^{-1/2}$.

Consider cases where the total propulsion energy stream is otherwise ideal but where the overall propulsion system efficiency otherwise drops from 100 percent to a fractional value of (a/b) where both a and b are positive and real. The relativistic rocket equation yields:

$\Delta v = C \text{ Tanh } \{\{[1C + (f_{m1}C) + (f_{m2}C) + (f_{m3}C) + \ldots + (f_{mp}C)] (a/b)/C\}\ln(M_0/M_1)\}$

$= C \text{ Tanh } \{\{(1 + f_{m1} + f_{m2} + f_{m3} + \ldots + f_{mp}) (a/b)C/C\}\ln(M_0/M_1)\}$.

Alternatively,

$\Delta v = C \text{ Tanh } \{\{[1C + [\Sigma (1,p) (f_{mi}C)]] (a/b)/C\}\ln(M_0/M_1)\}$

$= C \text{ Tanh } \{\{\{1 + [[\Sigma (1,p) f_{mi}](a/b) C]\}/C\}\ln(M_0/M_1)\}$.

The associated Lorentz factor is equal to:

$\gamma = \{1- \{\{\{ C \text{ Tanh } \{\{[1C + (f_{m1}C) + (f_{m2}C) + (f_{m3}C) + \ldots + (f_{mp}C)](a/b)/C\}\ln(M_0/M_1)\}\} /C\}^2\}\}^{-1/2}$

$= \{1- \{\{\{ C \text{ Tanh } \{\{(1 + f_{m1} + f_{m2} + f_{m3} + \ldots + f_{mp})(a/b) C/C\}\ln(M_0/M_1)\}\} /C\}^2\}\}^{-1/2}$

$= \{1- \{\{\{ C \text{ Tanh } \{\{[1C + [\Sigma (1,p) (f_{mi}C)]] (a/b)/C\}\ln(M_0/M_1)\}\} /C\}^2\}\}^{-1/2}$

$= \{1- \{\{\{ C \text{ Tanh } \{\{\{1 + [[\Sigma (1,p) f_{mi}](a/b) C]\}/C\}\ln(M_0/M_1)\}\} /C\}^2\}\}^{-1/2}$.

Consider cases where the mth mode of propulsion contributes an additional fraction of propulsion power, $f_m$, of the rocket thrusting mechanism where the specific impulse of the rocket itself is equal to 2 C such as for matter-antimatter rockets that obtain only and all of its normal matter reactants from space. The modified relativistic rocket equation yields:

$\Delta v = C \text{ Tanh } \{\{[2C + [2(f_m C)]]/C\}\ln(M_0/M_1)\}$

$= C \text{ Tanh } \{\{[2 + (2f_m)] C/C\}\ln(M_0/M_1)\}$.

The associated Lorentz factor is equal to:

$\gamma = \{1- \{\{\{C \text{ Tanh } \{\{[2C + [2(f_m C)]]/C\}\ln(M_0/M_1)\}\} /C\}^2\}\}^{-1/2}$

$= \{1- \{\{\{ C \text{ Tanh } \{\{[2 + [2(f_m)]] C/C\}ln(M_0/M_1)\}\}/C\}^2\}\}^{-1/2}.$

Consider cases where the total energy drive stream is otherwise ideal but where the efficiency otherwise drops from 100 percent to a fractional value of (a/b) where a and b are positive and real, and where the ideal specific impulse of the rocket mechanism would be equal to 2 C. The relativistic rocket equation yields:

$\Delta v = C \text{ Tanh } \{\{[2C + [2(f_mC)]]/ (a/b)/C\}ln(M_0/M_1)\}$

$= C \text{ Tanh } \{\{[2 + [2(f_m)]](a/b)C/C\}ln(M_0/M_1)\}.$

The associated Lorentz factor is equal to:

$\gamma = \{1- \{\{\{C \text{ Tanh } \{\{[(2C) + [2(f_mC)]]/C\}ln(M_0/M_1)\}\}(a/b)/C\}^2\}\}^{-1/2}$

$= \{1- \{\{\{ C \text{ Tanh } \{\{[2 + [2(f_m)]] C/C\}ln(M_0/M_1)\}\}(a/b)/C\}^2\}\}^{-1/2}.$

Consider cases where there are multiple modes of non-rocket propulsion methods assisting the relativistic rocket propulsion system where each non-rocket mode, $m_i$, provides an additional fraction of propulsion power, $f_{mi}$, of the rocket thrusting mechanism to the spacecraft where the specific impulse of the rocket itself is equal to 2 C. The modified relativistic rocket equation yields:

$\Delta v = C \text{ Tanh } \{\{[(2C) + [2(f_{m1}C)]+ [2(f_{m2}C)] + [2(f_{m3}C)] + ... + [2(f_{mp}C)]] /C\}ln(M_0/M_1)\}$

$= C \text{ Tanh } \{\{[2 + 2f_{m1} + 2f_{m2} + 2f_{m3} + ... + 2f_{mp}] C/C\}ln(M_0/M_1)\}.$

The associated Lorentz factor is equal to:

$\gamma = \{1- \{\{\{ C \text{ Tanh } \{\{[(2C) + [2(f_{m1}C)] + [2(f_{m2}C)] + [2(f_{m3}C)] + ... + [2(f_{mp}C)]] /C\}ln(M_0/M_1)\}\} /C\}^2\}\}^{-1/2}$

$= \{1- \{\{\{ C \text{ Tanh } \{\{[(2C) + 2f_{m1} + 2f_{m2} + 2f_{m3} + ... + 2f_{mp}] C/C\}ln(M_0/M_1)\}\} /C\}^2\}\}^{-1/2}.$

Consider cases where the total propulsion energy stream is otherwise ideal but where the efficiency otherwise drops from 100 percent to a fractional value of (a/b) where a and b are positive and real, and where the ideal specific impulse of the rocket mechanism would be equal to 2 C. The relativistic rocket equation yields:

$\Delta v = C \text{ Tanh } \{\{[(2C) + [2(f_{m1}C)]+ [2(f_{m2}C)] + [2(f_{m3}C)] + ... + [2(f_{mp}C)]] (a/b)/C\}ln(M_0/M_1)\}$

$= C \text{ Tanh } \{\{[2 + 2f_{m1} + 2f_{m2} + 2f_{m3} + ... + 2f_{mp}](a/b) C/C\}ln(M_0/M_1)\}.$

The associated Lorentz factor is equal to:

$\gamma = \{1- \{\{\{ C \text{ Tanh } \{\{[(2C) + [2(f_{m1}C)] + [2(f_{m2}C)] + [2(f_{m3}C)] + ... + [2(f_{mp}C)]] (a/b)/C\}ln(M_0/M_1)\}\} /C\}^2\}\}^{-1/2}$

$= \{1- \{\{\{ C \text{ Tanh } \{\{[(2C) + 2f_{m1} + 2f_{m2} + 2f_{m3} + ... + 2f_{mp}] (a/b) C/C\}ln(M_0/M_1)\}\} /C\}^2\}\}^{-1/2}.$

Alternatively:

$\Delta v = C \text{ Tanh } \{\{[(2C) + [\Sigma (1,p) [2(f_{mi}C)]]] (a/b)/C\}ln(M_0/M_1)\}$

$= C \text{ Tanh } \{\{\{2 + [[\Sigma (1,p) (2f_{mi})](a/b) C]\}/C\}ln(M_0/M_1)\}.$

The associated Lorentz factor is equal to:

$\gamma = \{1- \{\{\{ C \text{ Tanh } \{\{[(2C) + [\Sigma (1,p) [2(f_{mi}C)]]] (a/b)/C\}ln(M_0/M_1)\}\} /C\}^2\}\}^{-1/2}$

$= \{1- \{\{\{ C \text{ Tanh } \{\{\{2 + [[\Sigma (1,p) [2f_{mi}]](a/b) C]\}/C\}ln(M_0/M_1)\}\} /C\}^2\}\}^{-1/2}.$

Consider cases where an otherwise ideal mth mode of propulsion would contribute an additional fraction of propulsion power, $f_m$, of the rocket thrusting mechanism where the specific impulse of the rocket itself is equal to 1 C, but where the efficiency of the mth mode is reduced to $e_m$. The modified relativistic rocket equation yields:

$$\Delta v = C\ \text{Tanh}\ \{\{[1C + [(e_m)(f_m C)]]/C\}\ln(M_0/M_1)\}$$

$$= C\ \text{Tanh}\ \{\{[1 + [(e_m)(f_m)]]\ C/C\}\ln(M_0/M_1)\}.$$

The associated Lorentz factor is equal to:

$$\gamma = \{1- \{\{\{C\ \text{Tanh}\ \{\{[1C + [(e_m)(f_m C)]]/C\}\ln(M_0/M_1)\}\}\ /C\}^2\}\}^{-1/2}$$

$$= \{1- \{\{\{\ C\ \text{Tanh}\ \{\{[1 + [(e_m)(f_m)]]\ C/C\}\ln(M_0/M_1)\}\}/C\}^2\}\}^{-1/2}.$$

Consider cases where the rocket propulsion energy stream is otherwise ideal but where the efficiency drops from 100 percent to a fractional value of (a/b) where both a and b are positive and real, and where the efficiency of an otherwise ideal mth additional mode of propulsion having a power fraction of $f_m$ of the otherwise ideal relativistic rocket thrust is reduced to $e_m$. The relativistic rocket equation yields:

$$\Delta v = C\ \text{Tanh}\ \{\{[[(a/b)C] + [(e_m)(f_m C)]]/C\}\ln(M_0/M_1)\}$$

$$= C\ \text{Tanh}\ \{\{[(a/b) + [(e_m)(f_m)]]\ C/C\}\ln(M_0/M_1)\}.$$

The associated Lorentz factor is equal to:

$$\gamma = \{1- \{\{\{C\ \text{Tanh}\ \{\{[[(a/b)C] + [(e_m)(f_m C)]]/C\}\ln(M_0/M_1)\}\}/C\}^2\}\}^{-1/2}$$

$$= \{1- \{\{\{\ C\ \text{Tanh}\ \{\{[(a/b) + [(e_m)(f_m)]]\ C\ /C\}\ln(M_0/M_1)\}\}(a/b)/C\}^2\}\}^{-1/2}.$$

Consider cases where there are multiple modes of non-rocket propulsion methods assisting the relativistic rocket propulsion system where each non-rocket mode, $m_i$, would provide an additional fraction of propulsion power, $f_{mi}$, of an ideal rocket thrusting mechanism but where the efficiency of the mith mode is reduced to $e_{mi}$ and where the specific impulse of the rocket itself is equal to 1 C. The modified relativistic rocket equation yields:

$$\Delta v = C\ \text{Tanh}\ \{\{[1C + [(e_{m1})(f_{m1}C)] + [(e_{m2})(f_{m2}C)] + [(e_{m3})(f_{m3}C)] + \ldots + [(e_{mp})(f_{mp}C)]]\ /C\}\ln(M_0/M_1)\}$$

$$= C\ \text{Tanh}\ \{\{[1 + [e_{m1}\ f_{m1}] + [e_{m2}\ f_{m2}] + [e_{m3}\ f_{m3}] + \ldots + [e_{mp}\ f_{mp}]]\ C/C\}\ln(M_0/M_1)\}.$$

The associated Lorentz factor is equal to:

$$\gamma = \{1- \{\{\{\ C\ \text{Tanh}\ \{\{[1C + [(e_{m1})(f_{m1}C)]+ [(e_{m2})(f_{m2}C)] + [(e_{m3})(f_{m3}C)] + \ldots + [(e_{mp})(f_{mp}C)]]\ /C\}\ln(M_0/M_1)\}\}\ /C\}^2\}\}^{-1/2}$$

$$= \{1- \{\{\{\ C\ \text{Tanh}\ \{\{[1 + [e_{m1}\ f_{m1}] + [e_{m2}\ f_{m2}] + [e_{m3}\ f_{m3}] + \ldots + [e_{mp}\ f_{mp}]]\ C/C\}\ln(M_0/M_1)\}\}\ /C\}^2\}\}^{-1/2}.$$

Consider cases where there are multiple modes of non-rocket propulsion assisting the relativistic rocket propulsion system where each non-rocket mode, $m_i$, would provide an additional fraction of propulsion power, $f_{mi}$, of an ideal rocket thrusting mechanism but where the efficiency of the mith additional mode is reduced to $e_{mi}$ and where the specific impulse of the rocket itself ideally equal to 1 C, is reduced in efficiency to yield an efficiency a/b. The modified relativistic rocket equation yields:

$$\Delta v = C\ \text{Tanh}\ \{\{[[(a/b)C] + [(e_{m1})(f_{m1}C)] + [(e_{m2})(f_{m2}C)] + [(e_{m3})(f_{m3}C)] + \ldots + [(e_{mp})(f_{mp}C)]]\ /C\}\ln(M_0/M_1)\}$$

$$= C\ \text{Tanh}\ \{\{[(a/b) + [e_{m1}\ f_{m1}] + [e_{m2}\ f_{m2}] + [e_{m3}\ f_{m3}] + \ldots + [e_{mp}\ f_{mp}]]\ C/C\}\ln(M_0/M_1)\}.$$

The associated Lorentz factor is equal to:

$$\gamma = \{1- \{\{\{\ C\ \text{Tanh}\ \{\{[[(a/b)C] + [(e_{m1})(f_{m1}C)] + [(e_{m2})(f_{m2}C)] + [(e_{m3})(f_{m3}C)] + \ldots + [(e_{mp})(f_{mp}C)]]\ /C\}\ln(M_0/M_1)\}\}\ /C\}^2\}\}^{-1/2}$$

$$= \{1- \{\{\{\ C\ \text{Tanh}\ \{\{[(a/b) + [e_{m1}\ f_{m1}] + [e_{m2}\ f_{m2}] + [e_{m3}\ f_{m3}] + \ldots + [e_{mp}\ f_{mp}]]\ C/C\}\ln(M_0/M_1)\}\}\ /C\}^2\}\}^{-1/2}.$$

Alternatively:

$$\Delta v = C\ \text{Tanh}\ \{\{[[(a/b)C] + [\Sigma\ (1,p)\ (e_{mi})(f_{mi}C)]]\ /C\}\ln(M_0/M_1)\}$$

$$= C\ \text{Tanh}\ \{\{\{[(a/b) + [\Sigma\ (1,p)(e_{mi})(f_{mi})]]\ C\}/C\}\ln(M_0/M_1)\}.$$

The associated Lorentz factor is equal to:

$$\gamma = \{1- \{\{\{ C \text{ Tanh } \{\{[[(a/b)C] + [\Sigma (1,p) (e_{mi})(f_{mi}C)]] /C\}\ln(M_0/M_1)\}\} /C\}^2\}\}^{-1/2}$$

$$= \{1- \{\{\{ C \text{ Tanh } \{\{\{[(a/b) + [\Sigma (1,p)(e_{mi})(f_{mi})]] C\}/C\}\ln(M_0/M_1)\}\} /C\}^2\}\}^{-1/2} .$$

Consider cases where an otherwise ideal mth mode of propulsion would contribute an additional fraction of propulsion power, $f_m$, of the rocket thrusting mechanism where the specific impulse of the rocket itself is equal to 2 C such as for a rocket that obtains only and all of its normal matter fuel from interstellar or intergalactic space, but where the efficiency of the mth mode is reduced to $e_m$. The modified relativistic rocket equation yields:

$$\Delta v = C \text{ Tanh } \{\{[2C + [2(e_m)(f_mC)]]/C\}\ln(M_0/M_1)\}$$

$$= C \text{ Tanh } \{\{[2 + [2(e_m)(f_m)]] C/C\}\ln(M_0/M_1)\}.$$

The associated Lorentz factor is equal to:

$$\gamma = \{1- \{\{\{C \text{ Tanh } \{\{[2C + [2(e_m)(f_mC)]]/C\}\ln(M_0/M_1)\}\} /C\}^2\}\}^{-1/2}$$

$$= \{1- \{\{\{ C \text{ Tanh } \{\{[2 + [2(e_m)(f_m)]] C/C\}\ln(M_0/M_1)\}\}/C\}^2\}\}^{-1/2}.$$

Consider cases where the rocket propulsion energy stream is otherwise ideal but where the efficiency drops from 100 percent to a fractional value of (a/b) where a and b are positive and real, where the specific impulse of the rocket itself is ideally equal to 2 C such as for a rocket that obtains only and all of its normal matter fuel from interstellar or intergalactic space, and where the efficiency of an otherwise ideal mth additional mode of propulsion having a power fraction of $f_m$ of the otherwise ideal relativistic rocket thrust is reduced to $e_m$. The relativistic rocket equation yields:

$$\Delta v = C \text{ Tanh } \{\{[[2(a/b)C] + [2(e_m)(f_mC)]]/C\}\ln(M_0/M_1)\}$$

$$= C \text{ Tanh } \{\{[[2(a/b)] + [2(e_m)(f_m)]] C/C\}\ln(M_0/M_1)\}.$$

The associated Lorentz factor is equal to:

$$\gamma = \{1- \{\{\{C \text{ Tanh } \{\{[[2(a/b)C] + [2(e_m)(f_mC)]]/C\}\ln(M_0/M_1)\}\}/C\}^2\}\}^{-1/2}$$

$$= \{1- \{\{\{ C \text{ Tanh } \{\{[[2(a/b)] + [2(e_m)(f_m)]] C /C\}\ln(M_0/M_1)\}\}/C\}^2\}\}^{-1/2}.$$

Consider cases where there are multiple modes of non-rocket propulsion methods assisting the relativistic rocket propulsion system where each non-rocket mode, $m_i$, would provide an additional fraction of propulsion power, $f_{mi}$, of an ideal rocket thrusting mechanism but where the efficiency of the mith mode is reduced to $e_{mi}$ and where the specific impulse of the rocket itself is equal to 2 C such as for a rocket that obtains only and all of its normal matter fuel from interstellar or intergalactic space. The modified relativistic rocket equation yields:

$$\Delta v = C \text{ Tanh } \{\{[2C + [2(e_{m1})(f_{m1}C)] + [2(e_{m2})(f_{m2}C)] + [2(e_{m3})(f_{m3}C)] + \ldots + [2(e_{mp})(f_{mp}C)]] /C\}\ln(M_0/M_1)\}$$

$$= C \text{ Tanh } \{\{[2 + [2(e_{m1} f_{m1})] + [2(e_{m2} f_{m2})] + [2(e_{m3} f_{m3})] + \ldots + [2(e_{mp} f_{mp})]] C/C\}\ln(M_0/M_1)\}.$$

The associated Lorentz factor is equal to:

$$\gamma = \{1- \{\{\{ C \text{ Tanh } \{\{[2C + [2(e_{m1})(f_{m1}C)] + [2(e_{m2})(f_{m2}C)] + [2(e_{m3})(f_{m3}C)] + \ldots + [2(e_{mp})(f_{mp}C)]] /C\}\ln(M_0/M_1)\}\} /C\}^2\}\}^{-1/2}$$

$$= \{1- \{\{\{ C \text{ Tanh } \{\{[2 + [2(e_{m1} f_{m1})] + [2(e_{m2} f_{m2})] + [2(e_{m3} f_{m3})] + \ldots + [2(e_{mp} f_{mp})]] C/C\}\ln(M_0/M_1)\}\} /C\}^2\}\}^{-1/2}.$$

Consider cases where there are multiple modes of non-rocket propulsion assisting the relativistic rocket propulsion system where each non-rocket mode, $m_i$, would provide an additional fraction of propulsion power, $f_{mi}$, of an ideal rocket thrusting mechanism but where the efficiency of the mith additional mode is reduced to $e_{mi}$ and where the specific impulse of the rocket itself ideally equal to 2 C such as for a rocket that obtains only and all of its normal matter fuel from interstellar or intergalactic space, is reduced in efficiency to yield an efficiency (a/b). The modified relativistic rocket equation yields:

$$\Delta v = C \text{ Tanh } \{\{[[2(a/b)C] + [2(e_{m1})(f_{m1}C)] + [2(e_{m2})(f_{m2}C)] + [2(e_{m3})(f_{m3}C)] + \ldots + [2(e_{mp})(f_{mp}C)]] /C\}\ln(M_0/M_1)\}$$

$= C \text{ Tanh } \{\{[[2(a/b)] + [2(e_{m1} f_{m1})] + [2(e_{m2} f_{m2})] + [2(e_{m3} f_{m3})] + \ldots + [2(e_{mp} f_{mp})]] C/C\}\ln(M_0/M_1)\}.$

The associated Lorentz factor is equal to:

$$\gamma = \{1- \{\{\{ C \text{ Tanh } \{\{[[2(a/b)C] + [2(e_{m1})(f_{m1}C)] + [2(e_{m2})(f_{m2}C)] + [2(e_{m3})(f_{m3}C)] + \ldots + [2(e_{mp})(f_{mp}C)]] /C\}\ln(M_0/M_1)\}\} /C\}^2\}\}^{-1/2}$$

$$= \{1- \{\{\{ C \text{ Tanh } \{\{[[2(a/b)] + [2(e_{m1} f_{m1})] + [2(e_{m2} f_{m2})] + [2(e_{m3} f_{m3})] + \ldots + [2(e_{mp} f_{mp})]] C/C\}\ln(M_0/M_1)\}\} /C\}^2\}\}^{-1/2}.$$

Alternatively:

$$\Delta v = C \text{ Tanh } \{\{[[2(a/b)C] + [\Sigma (1,p) (2)(e_{mi})(f_{mi}C)]] /C\}\ln(M_0/M_1)\}$$

$$= C \text{ Tanh } \{\{\{[[2(a/b)] + [\Sigma (1,p)(2)(e_{mi})(f_{mi})]] C\}/C\}\ln(M_0/M_1)\}.$$

The associated Lorentz factor is equal to:

$$\gamma = \{1- \{\{\{ C \text{ Tanh } \{\{[[2(a/b)C] + [\Sigma (1,p) (2)(e_{mi})(f_{mi}C)]] /C\}\ln(M_0/M_1)\}\} /C\}^2\}\}^{-1/2}$$

$$= \{1- \{\{\{ C \text{ Tanh } \{\{\{[[2(a/b)] + [\Sigma (1,p)(2)(e_{mi})(f_{mi})]] C\}/C\}\ln(M_0/M_1)\}\} /C\}^2\}\}^{-1/2}.$$

Consider cases where the mth mode of propulsion contributes an additional fraction of propulsion power, $f_m$, of the rocket thrusting mechanism where the specific impulse of the rocket itself is equal to 2 C such as for matter-antimatter rockets that obtain only and all of its normal matter reactants from space. The modified relativistic rocket equation yields:

$$\Delta v = C \text{ Tanh } \{\{[2C + [2(f_m C)]]/C\} \ln(M_0/M_1)\}$$

$$= C \text{ Tanh } \{\{[2 + (2f_m)] C/C\}\ln(M_0/M_1)\}.$$

The associated Lorentz factor is equal to:

$$\gamma = \{1- \{\{\{C \text{ Tanh } \{\{[2C + [2(f_m C)]]/C\}\ln(M_0/M_1)\}\} /C\}^2\}\}^{-1/2}$$

$$= \{1- \{\{\{ C \text{ Tanh } \{\{[2 + [2(f_m)]] C/C\}\ln(M_0/M_1)\}\}/C\}^2\}\}^{-1/2}.$$

Consider cases where the total energy drive stream is ideal but where the efficiency otherwise drops from 100 percent to a fractional value of (a/b) where both a and b are positive and real. The relativistic rocket equation yields:

$$\Delta v = C \text{ Tanh } \{\{[2C + [2(f_m C)]]/ (a/b)/C\}\ln(M_0/M_1)\}$$

$$= C \text{ Tanh } \{\{[2 + [2(f_m)]](a/b)C/C\}\ln(M_0/M_1)\}.$$

The associated Lorentz factor is equal to:

$$\gamma = \{1- \{\{\{C \text{ Tanh } \{\{[(2C) + [2(f_m C)]]/C\}\ln(M_0/M_1)\}\}(a/b)/C\}^2\}\}^{-1/2}$$

$$= \{1- \{\{\{ C \text{ Tanh } \{\{[2 + [2(f_m)]] C/C\}\ln(M_0/M_1)\}\}(a/b)/C\}^2\}\}^{-1/2}.$$

Consider cases where there are multiple modes of non-rocket propulsion methods assisting the relativistic rocket propulsion system where each non-rocket mode, $m_i$, provides an additional fraction of propulsion power, $f_{mi}$, of the rocket thrusting mechanism to the spacecraft where the specific impulse of the rocket itself is equal to 1 C. The modified relativistic rocket equation yields:

$$\Delta v = C \text{ Tanh } \{\{[(2C) + [2(f_{m1}C)]+ [2(f_{m2}C)] + [2(f_{m3}C)] + \ldots + [2(f_{mp}C)]] /C\}\ln(M_0/M_1)\}$$

$$= C \text{ Tanh } \{\{[2 + 2f_{m1} + 2f_{m2} + 2f_{m3} + \ldots + 2f_{mp}] C/C\}\ln(M_0/M_1)\}.$$

The associated Lorentz factor is equal to:

$$\gamma = \{1- \{\{\{ C \text{ Tanh } \{\{[(2C) + [2(f_{m1}C)] + [2(f_{m2}C)] + [2(f_{m3}C)] + \ldots + [2(f_{mp}C)]] /C\}\ln(M_0/M_1)\}\} /C\}^2\}\}^{-1/2}$$

$$= \{1- \{\{\{ C \text{ Tanh } \{\{[(2C) + 2f_{m1} + 2f_{m2} + 2f_{m3} + \ldots + 2f_{mp}] C/C\}\ln(M_0/M_1)\}\} /C\}^2\}\}^{-1/2}.$$

Consider cases where the total propulsion energy stream is ideal but where the efficiency otherwise drops from 100 percent to a fractional value of (a/b) where both a and b are positive and real. The relativistic rocket equation yields:

$$\Delta v = C\ \mathrm{Tanh}\ \{\{[(2C) + [2(f_{m1}C)] + [2(f_{m2}C)] + [2(f_{m3}C)] + \ldots + [2(f_{mp}C)]]\ (a/b)/C\}\ln(M_0/M_1)\}$$

$$= C\ \mathrm{Tanh}\ \{\{[2 + 2f_{m1} + 2f_{m2} + 2f_{m3} + \ldots + 2f_{mp}](a/b)\ C/C\}\ln(M_0/M_1)\}.$$

The associated Lorentz factor is equal to:

$$\gamma = \{1- \{\{\{ C\ \mathrm{Tanh}\ \{\{[(2C) + [2(f_{m1}C)] + [2(f_{m2}C)] + [2(f_{m3}C)] + \ldots + [2(f_{mp}C)]]\ (a/b)/C\}\ln(M_0/M_1)\}\}\ /C\}^2\}\}^{-1/2}$$

$$= \{1- \{\{\{ C\ \mathrm{Tanh}\ \{\{[(2C) + 2f_{m1} + 2f_{m2} + 2f_{m3} + \ldots + 2f_{mp}]\ (a/b)\ C/C\}\ln(M_0/M_1)\}\}\ /C\}^2\}\}^{-1/2}.$$

Alternatively:

$$\Delta v = C\ \mathrm{Tanh}\ \{\{[(2C) + [\Sigma\ (1,p)\ [2(f_{mi}C)]]]\ (a/b)/C\}\ln(M_0/M_1)\}$$

$$= C\ \mathrm{Tanh}\ \{\{\{2 + [[\Sigma\ (1,p)\ (2f_{mi})](a/b)\ C]\}/C\}\ln(M_0/M_1)\}.$$

The associated Lorentz factor is equal to:

$$\gamma = \{1- \{\{\{ C\ \mathrm{Tanh}\ \{\{[(2C) + [\Sigma\ (1,p)\ [2(f_{mi}C)]]]\ (a/b)/C\}\ln(M_0/M_1)\}\}\ /C\}^2\}\}^{-1/2}$$

$$= \{1- \{\{\{ C\ \mathrm{Tanh}\ \{\{\{2 + [[\Sigma\ (1,p)\ [2f_{mi}]](a/b)\ C]\}/C\}\ln(M_0/M_1)\}\}\ /C\}^2\}\}^{-1/2}.$$

In general, the relative power contribution per mode of multimode propulsion matter-antimatter rockets can be judiciously chosen and adjusted for specific mission criteria. The relative power contribution of each mode can optionally be variable in flight. Such additional degrees of freedom can enable more appropriate interaction of the spacecraft and its propulsion systems with the interstellar and intergalactic medium including but not limited to natural variations in: 1) plasma distribution by density, charge, and species; 2) neutral gas distribution by density and species; 3) ambient stellar and quasar light distributions according to energy spectrum and power flux density; and 4) interstellar and intergalactic magnetic field intensity and vector field orientation. Such considerations can be important for drag reduction, and for certain conditions, intentional increases in drag for re-routing and deceleration. Another important consideration is relative power contribution adjustability as may be desired or necessary based on the above background interstellar and intergalactic medium variations.

Once again, there is one vitally important caveat in order for many of the above super-luminal specific impulse scenarios to be realized. A means of collecting background matters in a drag-less and otherwise essentially lossless manner is needed. For such systems to operate, the effective drag inducing momentum exchange between the massive interstellar or intergalactic species and the spacecraft would need to at the very least, be mostly eliminated. Perhaps some form of matter wave negative refraction index material based cloak and collection mechanism may be deployed. Currently, there is no known method of reducing the subject momentum exchange as such processes would violate current momentum conservation laws. However, since perhaps the birth of our universe in a Big Bang may have been an ultimate form of free lunch, perhaps we can analogously duplicate nature's ability for self-evolution by someday realizing the dream of gainful super-luminal specific impulse for rocket vehicle applications.

## Cycler control and predicting background magnetic fields.

Solar cycling spacecraft will need to be able to predict background magnetic field flux patters.

To facilitate stable and safe propulsion, a need to remove or aquire electric charge and enabling magnetic field emissions is necessary.

Accordingly, the charge on a towline or other charged member(s) will need to be adjustable. So a safe and reliable means to capture and off-load electrons, protons, and ions will be required.

To facilitate these requirements, a series of orbiting satellites are needed which sample background magnetic flux and orientations can be of use.

For example, bands of solar orbiting satellites located at various optimized radial coordinates from the sun can be of use. The number of bands is to be determined but should be chosen with satellite spacing on the order of the typical times solar wind fluxuations take place over the distances of satellite separation. Actually, the scale of satellite seperations should likely be at least half an order of magnitude less than typical or likely plasma flux wave cycles passing through the intermediary space between adjacent satellites.

The sattelites would neeed broadcast methods that are reliable and of band widths required for safe and full information aquisiation by the cycler spacecraft. Carrier waves may include any frequencies ranging from low frequency rf to visible light laser systems. Other systems may include flash-bulb style rf to visible light wave-fronts.

The spacecraft may also include insitu magnetic flux measuring equipment so as to have self-reported information and ability to immeadiately adjust tow line electric charge and other propulsion systems including but not limited to thrust vectoring systems.

Large plasma fluxuations need monitoring as well. These include corona mass ejections and other similar adverse space weather events. Here, satellite systems can play the role of early warming systems to advice spacecraft crew and auto-pilot mechanisms.

The communications beacons as such would have emissions that are highly relativistically Doppler shifted with respect to the spacecraft. The orientations of the Poynting vectors of the signals intercepted by the starship will be highly rotated and in most cases either highly blue-shifted or highly red-shifted. All these aberrations will be needed to be taken into account in the spacecraft design and flight protocol.

An interesting emergency feature would include the storage of multiple tow line systems so that in the event that a line or other charged member needs to be detached from the spacecraft during a propulsive emergency, other tow-lines can be quickly deploved to maintain the spacecraft on its desired heading. Additionally, tow-lines and the like could have fractional breakaway mechansims so that only a limited portion of a charged member need be detached.

A fascinating prospect would include the growth of tow lines perhaps as stored replacements by incorporation of background interplanetary gas into tow lines by nano-technology self-assembly mechanisms. The required elements can be capture by a spacecraft by suitable electronic chargon breaking mechanism whereupon the low ship frame velocity ions can be neutralized and incorporated into the tow lines.

Radar ranging and predction of thr paths asteroids, comets, pebble, and other sizable debris is also needed. Forwardly directed impactors or nuclear explosives can be useful for ionizing large desbris so that the debris can be converted to a form that can be electrodynamically diverted.

## Forwardly, orthogonally, or obliquely incident pellets.

The pellet runways may have predeployed fuel pellets that are forward or otherwise not backwardly incident on the craft.

For example, these latter pellets can be intaked via electrodynamic means to slow them down to a substantially stationary state with respect to the craft.

Accordingly, the drag energy associated with these pellets which would be electrically and/or magnetically charged would be captured and recycled into propulsion energy ideally with virtually one hundred percent efficiency. The ship-frame deceleration of the pellets would be accomplished by field effect, or linear induction mechanism.

The pellets would be detonated at optimized locations and velocities with respect to the spacecraft and any chutes used to capture at least a limited portion of the explosive energy.

We now provide a rather advanced technical digression for purposes that will be made clear in a summarizing paragraph afterward.

## Shaped charge nuclear fuel pellet.

The possibility of producing a shaped-charge type of nuclear or thermonuclear device that can concentrate the explosive energy flux upon its detonation by 6 orders of magnitude above that which is possible for a standard spherically symmetric nuclear device of the same yield has been proposed in the literature. According to some sources, a concentric pattern of simultaneous detonation or a radially stepped concentric pattern of detonation of the material of either a thermonuclear and/or a nuclear fission device, where the nuclear fuel charge has a relatively simple disk like configuration, might lead to the concentration of the explosive energy in the disk's center or axis of rotation to a level as much as 6 orders of magnitude greater than that possible with a standard spherically detonating device of the same overall yield (source not locatable).

In each of the following examples, opposing jets produced by two or more of the devices could provide the center-of-mass frame to enable the production of slower moving and more controllable strangelets, charmedlets, bottomlets, and/or toplets.

A) The first device might simply have one primary wherein a nuclear explosive disk undergoes explosive reaction simultaneously throughout. Alternatively, the reaction can be produced within differential radial portions of the nuclear explosive in a precisely timed manner so as to compound the pressure of any compressed plasma at the very center of the disk, thereby increasing the central temperature and forcing the mass-energy to exit the central axis of the detonating disk in the form of an extreme temperature and pressure jet.

In order to temporarily force the jet in one axial direction instead of producing a symmetrical bipolar jet that would otherwise result, some sort of very dense tamper or cladding might be incorporated into one face of the disk. In this manner, the super-hot jet that would otherwise be projected outward and perpendicularly to the uncladded side would be reflected back to amplify the heat and pressure of the jet formed in the opposite direction. The cladding might include a plate or ring radially extending out a distance from the center; or the cladding may consist of multiple dense materials, where the radial distribution of the cladding materials may optionally differ in density, material type, or quantity to optimize the formation of a more unipolar jet. The cladding might take the form of a very dense, perhaps neutron reflective, material. Alternatively, the jet reflection might be facilitated by the introduction of a secondary explosive jet produced by similar means as the first jet. Here, the collision and rebounding of the primary jet would act to force the back in the direction from which it originated. Note that the cladding would be reduced to plasma at essentially the speed of light and therefore the cladding would need to act on the primary jet very quickly.

Another configuration of a similar device might include cladding on both sides with an optional opening at the very center on one cladded face so that the blast pressure and temperature can be amplified, thus increasing the pressure, temperature, and density of the unipolar jet. Yet another configuration of this same basic design would entail a ring of dense material around the perimeter of a double-sided cladded device.

B) A second major configuration of large shaped charged atomic devices might include a stacked series of primary disks where each disk optionally has any of the characteristics mentioned above. Optionally, disks with dissimilar characteristics chosen from the examples given above may be incorporated within a stacked arrangement and may each be detonated simultaneously or in series or other temporal patterns to maximize the pressure, temperature, density, power density, and total jet energy.

An optional but non-limiting form of similar devices could include a long cylinder in which the progression of the nuclear reactions throughout the cylinder in a radial and/or length-wise detonation pattern would be optimized to yield the highest temperatures, pressures, densities, power densities, and/or overall energy contained within the jet. Dense cladding of appropriate materials may be included at optimal positions along the outer surface of the cylinder, or cladding in the form of plates or disks which may be distributed in a lengthwise order along the axis of the cylinder or otherwise arranged to achieve the desired thermodynamic properties of the jet.

C) In a third general configuration, multiple atomic disks would produce inwardly directed coplanar jets. The same arrangement might be used with multiple matter-antimatter primaries each having a cylindrical form such as those mentioned under section B) above. The multiplicity of devices could be detonated simultaneously so that its jets converge at the same time, producing a bipolar or unipolar jet of increased temperature, pressure, density, power density, and overall total energy relative to the single primary based configurations described under sections A) and B) above. A tamper plate or mass may be incorporated on one face within the ring like distribution of primaries so that the jet formed is preferentially concentrated in one direction.

In another form of the third basic configuration, each cylinder or each stack of plates, of which there would be several, would be wedge-like or truncated-wedge-like where the space between the cylinders or wedges would be minimized or eliminated to prevent or mitigate the divergence of the pressure pulse before it could be amplified into a unipolar jet.

D) In a fourth general configuration, a series of cylindrical arrangements where each arrangement optionally has any of the configurations described above in section C) would be stacked, one on top of the other with optionally very little or no space between each of the cylindrical arrangements, thus permitting amplified; pressures, temperatures, densities, power densities, and total energy content of the unipolar jet.

E) In a fifth general configuration, each of the substantially stacked cylindrical arrangements described in section D) could be included in multiplicity where these cylindrical components would be arranged in a manner similar to that described in section C) above, with opposing jets functioning similarly.

Note that additional hierarchies similar to those described under sections C) thru E) can be included, but configurations are not described to avoid unnecessary redundancy. However, practically any number of hierarchies can be incorporated into a single device.

F) A sixth basic configuration would include a conical wedge shaped region filled with any form of appropriate nuclear explosive which may optionally have optimized energy absorbing properties. This configuration may include any of the configurations in sections B) through E) above. The resulting devices would resemble in form, a shaped charged conventional warhead such as those commonly used in heavy armor defeating projectiles. Note that each of the substantially cylindrical arrangements described in sections C) through E) can include a similar plasma or neutral matter squeezing mechanism to enhance the shaped charge effect.

Any of the configurations described under sections B) through F) and/or some or all of the subcomponents thereof may be configured to produce a conical arrangement of jets. Therefore, the devices or sub-components can be configured to form jets that are not coplanar but converge in a conical pattern. In some designs, such an arrangement may be necessary in order to permit the jets formed from canceling out.

Shaped charge devices can more effectively propel starships that are traveling at extremely relativistic velocities compared to spherically symmetrically exploding nuclear devices. To the extent that the columniation of the jets produced by the exploding devices can be improved, the width of the jets decreased, and the temperature of the jets increased, virtually unlimited spacecraft Lorentz factors are possible for spacecraft propelled by shaped charge nuclear fission and/or nuclear fusion pellets.

Future generations can establish linear, circular or spiraling fusion and/or fission pellet runways as humankind establishes itself in the cosmos.

Linear fusion pellet runways may extend the radius of the currently observable universe, and perhaps further, in consideration of the forward progression of a highly relativistic spacecraft's light cone as the spacecraft travels away from Earth. Of course, a gradual reduction in the rate of universal space-time expansion would aid in this by helping maintain the linear mass density of the pellet stream at a more uniform level, thus enabling higher spacecraft accelerations and higher terminal Lorentz factors. However, to the extent that the spacecraft could forever utilize such nuclear pellets, the Lorentz factors obtainable by the spacecraft in the depths of eternity are maximally bounded by values at least equal to Omega which is the smallest infinite ordinal.

For extreme Lorentz factor spacecraft, relativistic aberrational effects would need to be considered. However, such effects can at least initially be dealt with by the deployment of sails with a very large surface area that would capture most or all of the bomb blast energy by blast reflection and/or absorption. Robust sails made of stabilized neutronic fibers, quarkonium fibers.

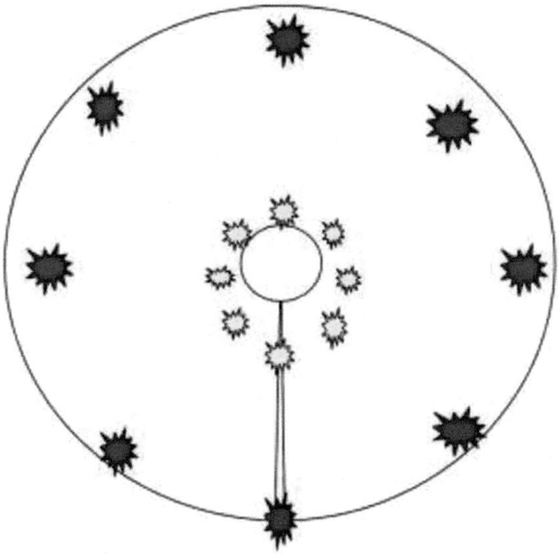

**FIGURE 2.** is a top plan view of a simple shaped charged nuclear or subnuclear explosive disk. Explosion denoted by the same color occur simultaneously. The two sets of explosions can occur simultaneously or sequentially in either of the two orders.

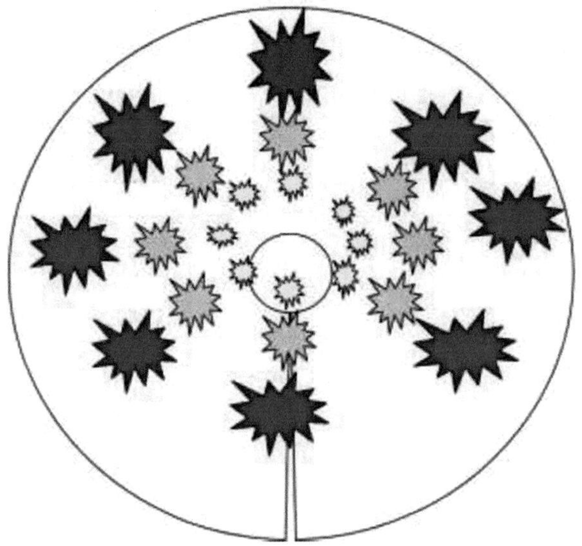

**FIGURE 3** is a top plan view of a simple shaped charged nuclear or subnuclear explosive disk. Explosion denoted by the same color occur simultaneously. The three sets of explosions can occur simultaneously or sequentially in any of the possible serial orders.

**FIGURE 4** depicts another configuration of a shaped charge nuclear or subnuclear explosive device. Note the trapezoidal partitions represent side-views of explosive disks, or partitioned wedges. The order of detonation of the partitions can be simultaneous or optionally in sequence in a manner suitable for production of maximum central temperatures and pressures.

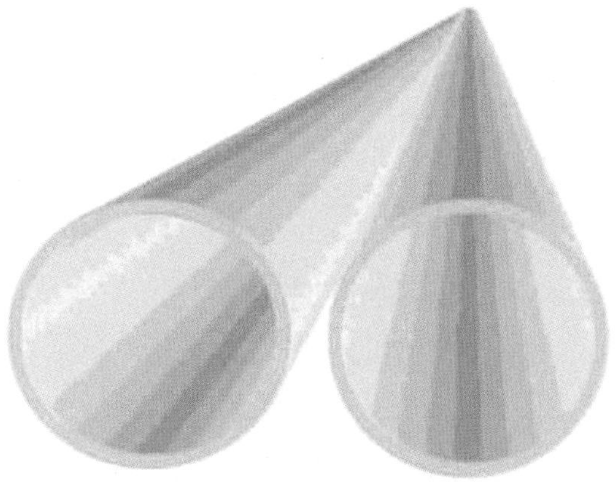

**FIGURE 5** depicts a side elevational perspective view of an azimuthal portion of yet another configuration of shaped charged explosive device. The cylinders would be packed with radially serial arrangements of explosive disks, sub-compositions of shorter disks, or optionally be filled with a non-serially partitioned explosive mixture. The detonation pattern for each cylinder can be radially or lengthwise simultaneous throughout the cylinder; or optionally, the composition(s) can be detonated in an arbitrarily radial and/or length-wise pattern.

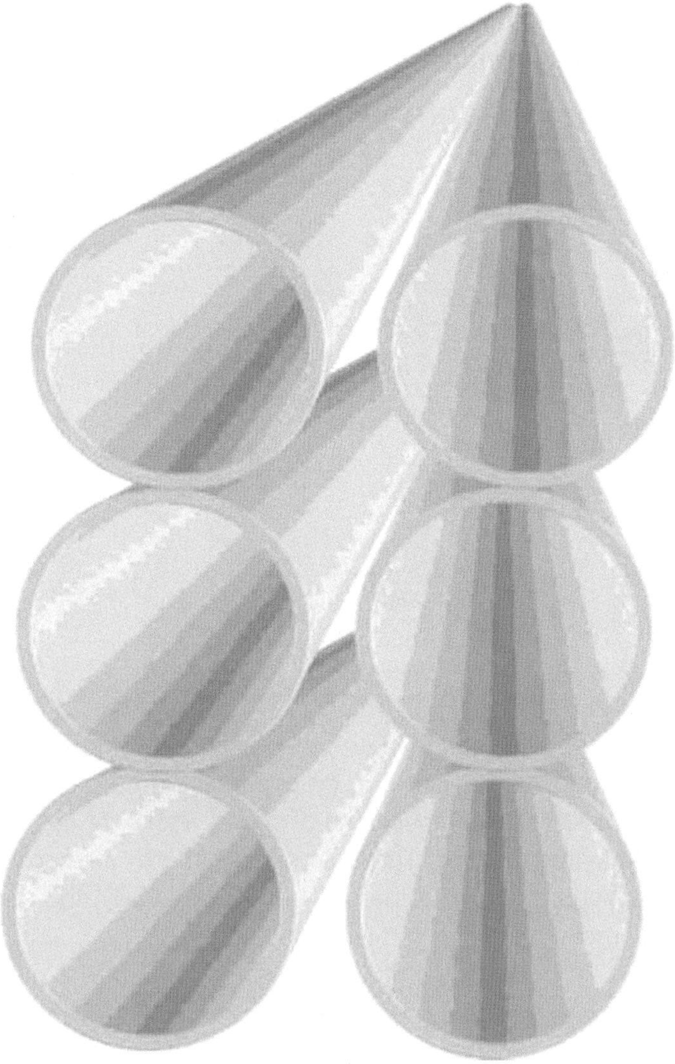

**FIGURE 6** depicts a side elevational perspective view of an azimuthal portion of yet another configuration of a shaped charged explosive device. The overall device consists of a linear stacked arrangement of multiple devices such as that depicted in the previous figure. The cylinders would be packed with radially serial arrangements of explosive disks, sub-compositions of shorter cylinders, or optionally be filled with a non-serially partitioned explosive mixture. The detonation pattern for each cylinder can be radially or lengthwise simultaneous throughout the cylinder; or optionally, the composition(s) can be detonated in an arbitrarily radial and/or length-wise pattern. The relative detonation pattern of each of assemblage of cylinders can be judiciously arbitrary with respect to the other assemblages. Three layers are shown but the number of layers in a functioning device is somewhat arbitrary.

**FIGURE 7** depicts converging jets from shaped charged nuclear or subnuclear devices. The arrangement shown is planar but the relative orientation of the devices is somewhat arbitrary. A conically converging series of jets may produce higher temperatures and pressures in some cases.

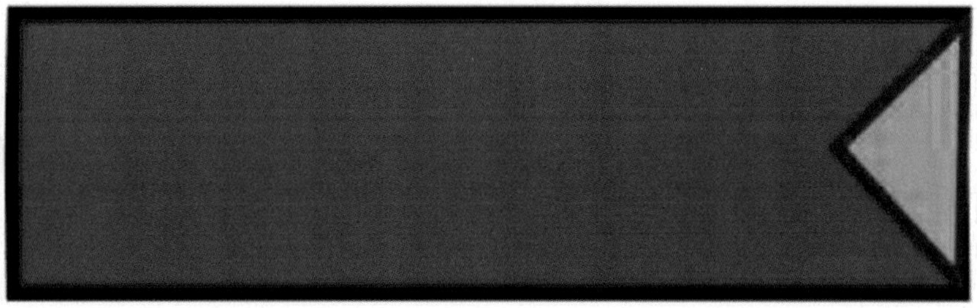

**FIGURE** 8 depicts a shaped charge nuclear or sub-nuclear explosive device. The red shading indicates an arbitrary species of nuclear or sub-nuclear explosive. The blue shading or wedge shows a dense material that is squeezed by the exploding red material to produce a concentrated jet pointing to the right. The thick black lines indicate dense cladding to temporarily provide inertial and/or mechanical resistance to the exploding material. The right-triangle shape of the blue material is arbitrary since the aspect ratio of the expellant material can be chosen over a wide range of shapes and sizes.

## Optimizing Nuclear Devices for Rocket Propulsion Systems

There are several candidates for mechanisms to perform nuclear fusion to propel spacecraft of a pulsed fusion reaction driven form and many of the associated nuclear explosive types that have not yet been developed. The first one involves an ion or electron beam system that would project intense beams into nuclear fusion fuel pellets. Upon fusion of the pellets, the energy released would drive a rocket spacecraft forward in a manner similar to the mechanism contemplated in the original Project Orion style designs.

A second system would operate in a manner similar to the plasma beam fusion mechanism, except that the heat and pressure required to induce nuclear fusion would be obtained from super intense laser pulses aimed symmetrically around a given fusion fuel pellet. Note that the National Ignition Facility at Lawrence Livermore National Laboratory is currently attempting to achieve this. The problem has been keeping the "pellet" perfectly spherical when the lasers hit it.. All attempts to date have resulted in deformation of the fusionable material and failure of the experiment. Again, upon fusion of the pellets, the energy released would drive a rocket spacecraft forward in a manner similar to that of the original Project Orion style designs.

In yet another system, a mechanism called a Z-pinch might be used to heat portions of fusion pellets to fusing temperatures and pressures. The fusion reaction would be self-propagating throughout the remainder of the pellet provided requisite temperatures and pressures could be generated by the initial fusion reactions. The Z-pinch may in theory enable arbitrary forms of nuclear bomb stagings. In a U.S.-based laboratory such Z-pinches are an assemblage of very thin wires that surround a very small fuel pellet. When an intense surge of electrical current passes through the Z-pinch's wires, the wires are reduced to plasma with a temperature on the order of a couple of million Kelvin. These efforts are being undertaken to study the feasibility of future versions of Z-pinches for possible use in power nuclear fusion reactors for commercial electrical power production. The devices thus far have not yielded significant releases of fusion energy from the pellets because the generated temperatures and pressures are currently insufficient to induce significant levels of nuclear fusion. Physicists, however, remain hopeful that such efforts will eventually lead to fuel pellet fusion.

Antimatter might be stored in a neutral form such as an anti-hydrogen ice, either in continuous bulk form or in the form of ready to annihilate tiny pellets that would be fired into or otherwise mixed with the normal matter hydrogen fuel, which may take the form of hydrogen ice. Alternatively, the anti-hydrogen ice could be ionized whereupon beams of anti-protons would be fired into portions of hydrogen fuel to be fused thereby causing mini-thermonuclear explosions. The heat generated by the matter-antimatter annihilation would be used to fuse hydrogen within the fuel pellets.

Another mechanism would use antimatter to induce nuclear fission. It may be possible for a nuclear fissile fuel such as U-233, U-235, Np-237, Pu-239, Am-241 or similar materials to be fashioned into small pebbles where anti-protons would be instilled in internal molecular cages. The antimatter would be unlocked and released at an appropriate time thus allowing it to intermingle with the fissionable isotopes. A cascade of fission reactions would ensue releasing a large flux of neutrons causing a pellet to become immediately super-critical. Note that U-233, U-235, and Pu-239 have a fission yield of 81.95 TJ/kg, 83.14 TJ/kg, and 83.61 TJ/kg respectively and are thus about equally useful for the associated specific impulses. However, it is possible that some unknown isotopes or yet to be discovered super-heavy elements and associated isotopes may have significantly greater mass specific fission yields. Designs of pure thermonuclear fuel pellets might include antimatter catalyzed fission pellets encased or cladded in a fusionable fuel such as hydrogen ice, solid metallic hydrogen, lithium deuteride, liquid Helium-3 or Helium-4 containing micro-vessels, and similar high yield fusion fuels. The density of solid metallic hydrogen is currently being researched - see: http://iopscience.iop.org/1742-6596/244/4/042020.

For larger pure thermonuclear pellets, a three stage mechanism such as a fission-fusion-fission sequence could be used. Here, the thermonuclear two stage pellet would be clad in a U-238 tamper. The fusion neutrons generated by the two stage device would be captured by the U-238 tamper thus causing the U-238 to fission.

A four stage pellet would involve a much larger supply of fusion fuel such as hydrogen ice, solid metallic hydrogen, lithium deuteride, liquid Helium-3 and similar high yield fusion fuels. The fuels would be heated or heated and compressed to fusion temperatures or fusion temperatures and pressures. The required temperatures can be reduced for cases where pressure and fuel density is increased. The required pressures can be reduced for cases where the densities are increased.

Another option might entail an encasement of a pure fusion fuel pellet within a cladding of fissile material such as U-233, U-235, Np-237, Pu-239, Am-241 or perhaps even U-238. The large neutron flux would stimulate the formation of a supercritical fissile stage. The fissile stage could then heat or heat and compress a much larger deposition of fusion fuel. Provided sufficient temperatures are attained from the fissile stages, perhaps non- traditional fusion fuels may be utilized such as ordinary water, heavy water, carbon, etc. The anti-protons used to start the initial fusion reactions might be either encased in molecular cages within the subject fuels or injected into the pellets at the appropriate times. The injected antiprotons may either be generated enroot or be stored onboard the spacecraft in either an anti-proton trap or in the form of an anti-hydrogen ice. A cryogenic anti-helium liquid is also a plausible form of fusion inducing antimatter.

One can imagine an interstellar ramjet being powered by antimatter catalyzed fusion, however, this may be highly dependent on capturing interstellar hydrogen or helium-3 and slowing it to a standstill with respect to the spacecraft while recycling the drag energy and friction induced heat to reduce or eliminate interstellar drag. Alternatively, the interstellar fusion fuel could be concentrated while it passes through a reaction chamber with little or no slowing while at the same time being mixed with or bombarded with anti-protons or anti-helium-3 nuclei.

The U.S. Air Force has been funding antimatter research for a number of years, as have other DoD entities such as DARPA, not to mention the Dept. of Energy. They do not direct the research objectives, but DoD funders do consider the application of successful research in terms of defense (which does not always mean weaponization). In any case, to date the cost of antimatter production for such purposes has been prohibitive.

For the above fusion neutron sources, the Z-pinch fusion stage can in principle be substituted for the antimatter catalyzed fusion stage for fusion-fission, and fusion-fission-fusion pellets.

Positronium initiated nuclear fusion might also provide a viable mechanism to ignite fusion fuel pellets. Such positronium catalyzed nuclear fusion stages might thereby provide neutrons for fusion-fission, and fusion-fission-fusion, pellets. The advantage of pure fusion, fission-fusion, fusion-fission, fusion-fission-fusion, and fission-fusion-fission pellets is that the mass specific yield of these proposed micro-bombs can be much higher in principle than that of currently fielded nuclear weapons. A modern nuclear warhead has a yield limited to about 2.2 megatons per metric ton. Pure fissile fuels top this by one order of magnitude at 22.5 megatons per metric ton, and pure fusion fuels go even higher at a maximum of about 160 megatons per metric ton. Thus, when used responsibly, pure fission or thermonuclear pellets can drive spacecraft to relativistic velocities. Project Orion Style spacecraft could obtain velocities easily within the range of 0.3 C to 0.5 C with large mass-ratios using such devices. Fission or thermonuclear pellet runways laid out in front of the spacecraft would enable very high Lorentz factors to be obtained in principle.

Other nuclear explosives may include mechanisms for generating powerful EMP bursts.

For example, the device may be magnetized and upon exploding, production of magnetic flux compression. It is even plausible that the device may have highly conducting or materials that become highly conducting when reduced to a plasma to produce magnetic flux compression thereby converting blast energy to EMP energy.

The EMP energy may be back-reflected by a magnetic plasma bottle sail.

Additionally, the pellets may be lobed with opposing electrostatic charge so as to induce a partition of the lobe-based plasma upon detonation. The charge differentiation may conceivably lead to negative and positive charged plasma portions to differentiate under the influence of positive and negative electric fields per two or more chutes of opposite net electrical charge. The plasma may be back-reflected by magnetic plasma bottle sails of appropriately signed charge.

Other methods of providing acceleration of the spacecraft include so-called negative electromagnetic refractive index pulling sails. Accordingly, these optics would be pulled forward by incident light in the direction from which the light-comes.

Nuclear fusion ramjets are another option as are electrodynamic-hydrodynamic-plasma-drives. Magnetic plasma bottle sails may also be employed.

A fascinating but perhaps remote possibility would involve electronic subcomposition for which particles much lower in invariant mass would be liberated and used as a gas plasma for multistage compression guns. Thus, the general conceptualizations presented in this book can have far-ranging significance into the deep future as our civilization's knowledge of particles and fields continues to develop.

Such subelectron plasma guns might be able to accelerate pellets to ultra-relativistic velocities where the pellets could catch up to spacecraft already traveling at highly relativistic velocities. Moreover, a predeployed ultra-relativistic pellet stream could be captured by an ultra-relativistic spacecraft that previously leveraged pellet intake energy supplies by capturing potential energy from the background.

Fuel pellets may themselves include propulsion mechanisms so that their kinetic energies with respect to the background can may many times greater than their muzzle energies.

What can be accomplished using the Sun has bearing on the use of other stars and quasars for stellar cycler type spacecraft. Here we assume that similar gas guns could be set up around these bodies and used to boost the velocity of spacecraft by fueling. Especially luminous stars along the main sequence and quasars would permit greater cycling radii thus reducing the loading on the craft due to centripetal acceleration. Included below are several diagrams illustrating the negative electromagnetic refractive index effect and stellar cycler pellet runways. **Note that Karl Schroeder was the originator of the Stellar Cycler concept and has no doubt already thought of similar applications before I was even aware of the general concept.**

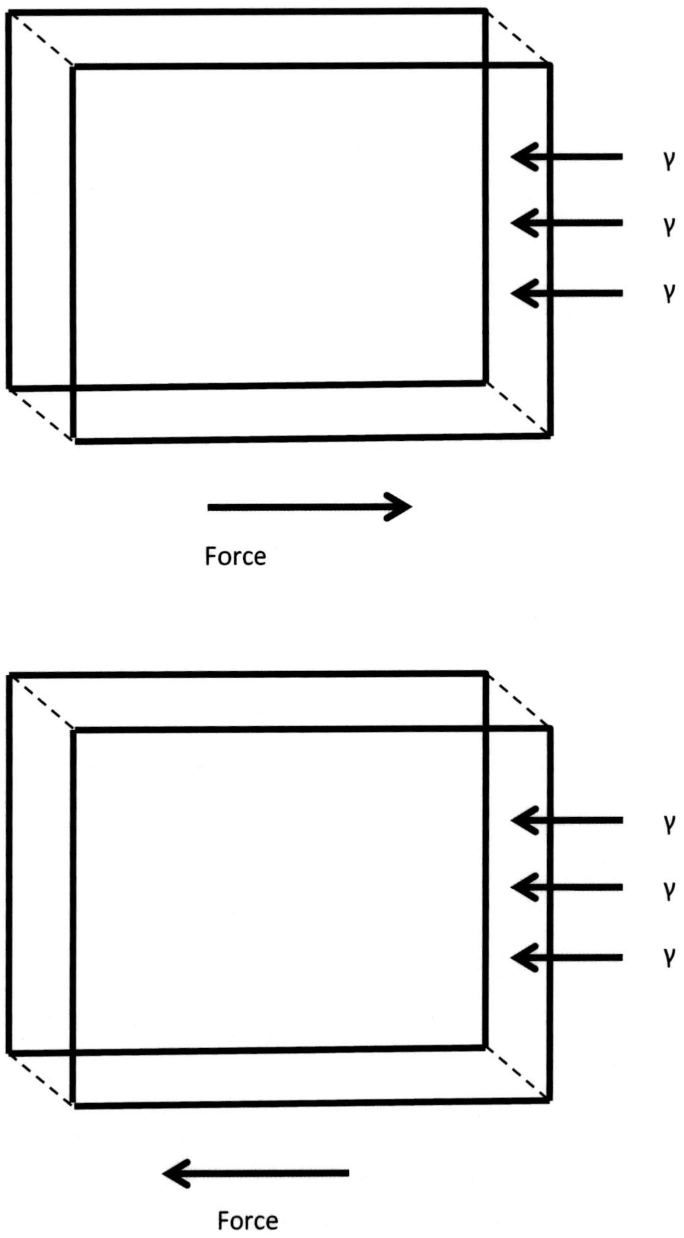


**FIGURES 9 A, B**: FIGURE 9 A depicts the pulling effect due to the negative refraction index mechanism. FIGURE 9 B contrasts the positive refraction index aspect of ordinary materials. The incident electromagnetic radiation is indicated by γ. The force exerted by impinging radiation is indicated by F. It is now common knowledge that impinging light of reactive frequencies can have a pulling component on negative refractive index materials. In at least one case, net negative pressure was obtained by light incident on a negative refractive index material.

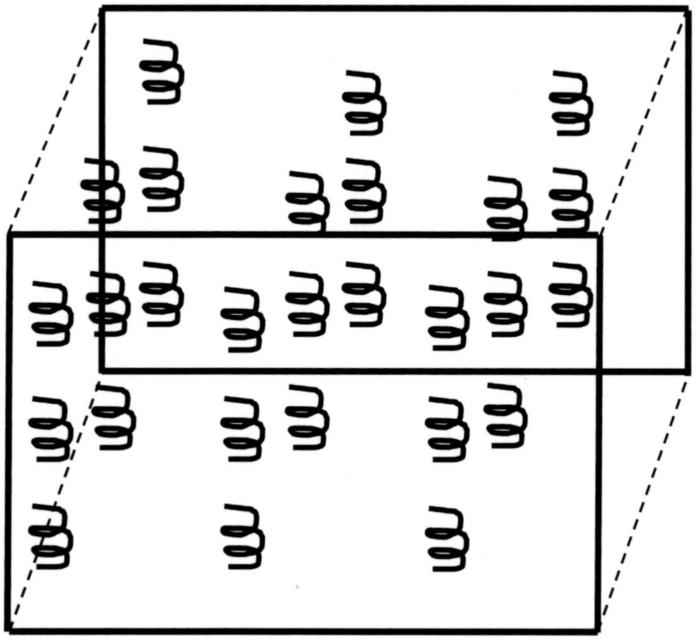

**FIGURE 10** depicts an electromagnetic negative refraction index metamaterial. Note the inductor like coils which serve as the material's active cells.

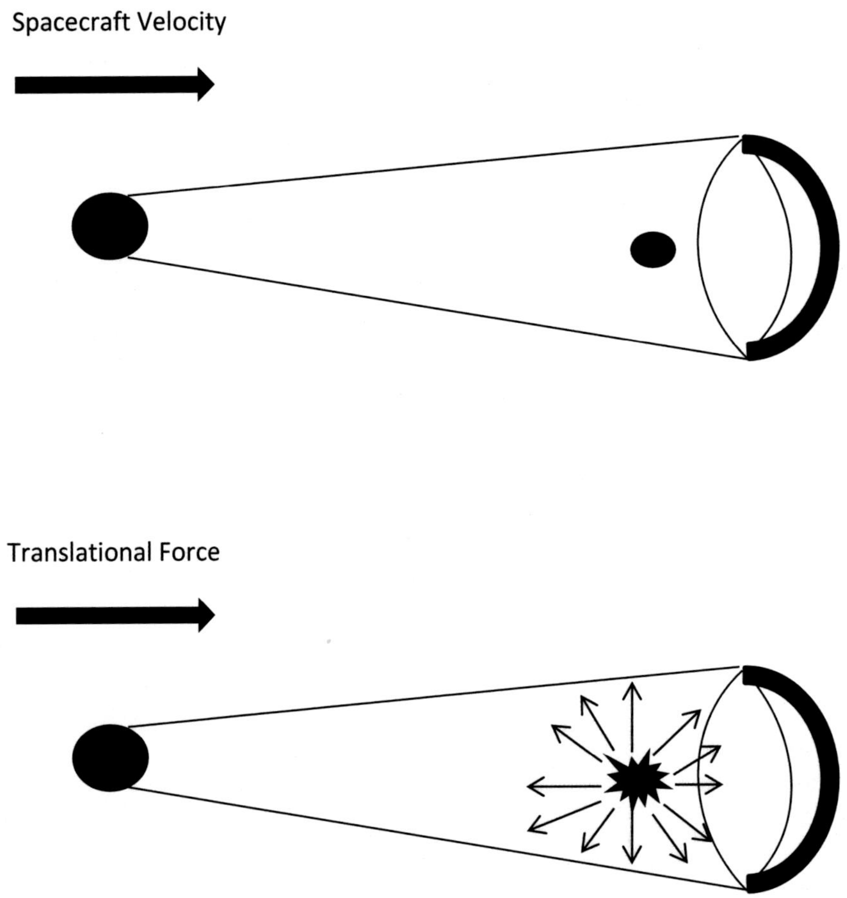


**FIGURES 11 A, B**: FIGURE 11 A depicts a Project Medusa style deployed nuclear bomb blast sail. In both figures, only two tow lines are shown. In practice, numerous tow lines would be required so as to distribute the loading of the blasts. Tow-lines at various radial locations along the sail with respect to the center of the sail will also likely be necessary for many specific sail configurations and performance criteria. FIGURE 11 B depicts an exploding nuclear device behind the sail. Project Medusa was a later version of the atomic bomb pulse drive studied by the U.S. Government under the name Project Orion. Project Medusa anticipates a very large electrically charged chute propelled by nuclear explosions. Accordingly, the chute would be electrically charged and would divert the plasma incident on the sail backward to obtain efficient thrust. Note that these chutes may conceivably be made of negative gamma ray, x-ray, and u-v light negative refractive index materials. Thus, detonations in front of the sail-chutes instead of behind may enable powerful drive modes. A sail in some cases for which the incident photons would be too penetrating may be configured as a thick plate or block. Such a thick absorber made be fabricated out of negative refractive index material and also have turbo-electric power generation mechanisms so as to convert waste heat into electrical power for propelling the craft forward.

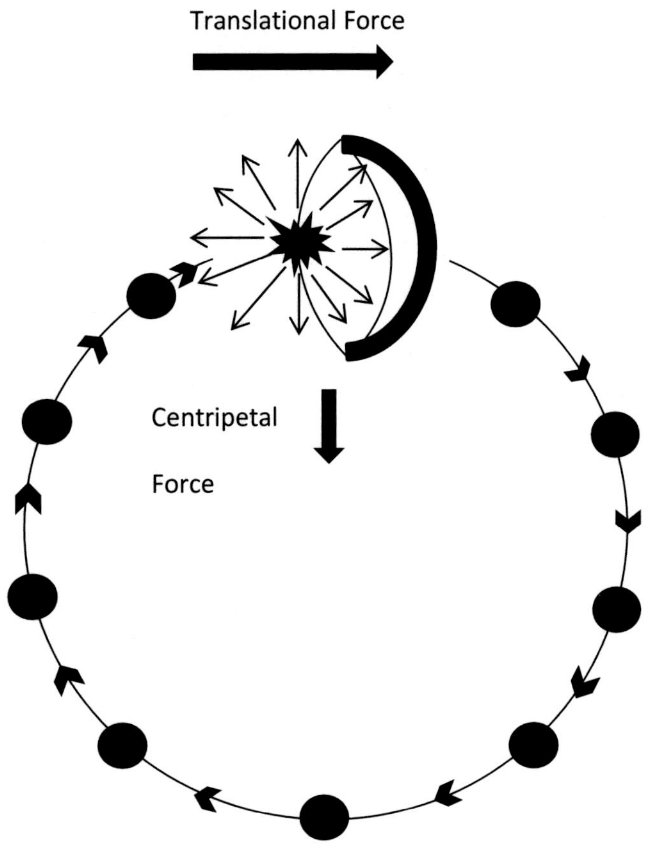


**FIGURE 12.** depicts a curvilinear pellet stream such as for a stellar cycler spacecraft.

Note that the shape and size of the bomb sail(s) can be chosen for specific mission criteria. The shape and size of the sail(s) may also be adjustable in flight. Such additional morphological degrees of freedom can enable more appropriate interaction of the sail(s) with the interstellar and intergalactic medium including but not limited to natural variations in: 1) plasma distribution by density, charge, and species; 2) neutral gas distribution by density and species; 3) ambient stellar and quasar light distributions according to energy spectrum and power flux density; 4) and interstellar and intergalactic magnetic field intensity and vector field orientation. Such considerations can be important for drag reduction, and in certain conditions, intentional increases in drag such as for re-routing and deceleration. Such natural variations can enhance or degrade spacecraft performances including but not limited to propulsion system efficiency.

A fascinating class of pellets would include steel or iron pellets that may incident into a spacecraft kinetic energy extractor from any direction, even in directions anti-parallel to the spacecraft velocity vectors. These pellets would either have their kinetic energy extracted or when forwardly incident, be further accelerated by a magnetic mass driver.

Other pellet stream schemes would include initially charged pellets or pellets that acquire an electrical charge in flight which would be incident from behind or any other direction of the spacecraft.

Still other pellets may be magnetized so that the pellets have greater electrodynamic interaction force with a field effect propulsion mechanism affixed to the ship.

Still other fuel pellet systems can include multi-level systems for which initially laid down pellet streams would serve as a fuel source for still other pellets such as for a two level pellet system. Optionally but not necessarily, a second level pellet stream can serve as a fuel source for a third level pellet stream and so-on. Multi-level pellet streams can optionally rely on the same secondary propulsion and deployment schemes, or any other scheme similarly or dissimilarly from pellet to pellet in a given hierarchy or amongst different hierarchies.

The kinetic energy of the pellets in the background could be completely depleted by the spacecraft and converted to spacecraft kinetic energy. After the pellets kinetic energy was depleted, the pellets could then be detonated to produce additional propulsion energy, of course, for cases where the pellets contain fuel for exothermic yields.

## Fusion Rocket & Beam Relay Propulsion.

**The following digression provides what might otherwise seem as overly redundant examples. However, these examples are provided to mathematically and computationally support the assertions made below. Extreme mathematicalized scenarios require extraordinary mathematical evidence and clarity thus the reason for including so many examples.**

Imagine a fusion rocket having a mass-ratio of $10^4$ and a fuel specific impulse 0.119 C. Furthermore, assume that the payload has mass ratio of $10^4$ and the same fuel specific impulse. Assume four additional stages of the same proportion.

Now assume the first stage starts out and burns enough fuel to yield an effective first stage mass-ratio of $10^3$.

The terminal velocity and Lorentz factor of the third stage will thus be 0.267349984 c and 1.037775555 respectively.

Now, assume that the remainder of the fusion fuel in the first stage is processed in an efficient fusion reactor to power a laser beam or microwave beam but where the fusion waste products are held and thus serve as a way to conserve first stage momentum. So we will assume for the sake of current argument that the Lorentz factor of the first stage remains at 1.037775555. The lasing energy is thus equal to $(0.007)(0.1)[M_1/M_0](M_0)[C^2](1.037775555) = 7.2644288\ (M_0)[C^2]$.

Now, assume that all of this lasing energy is converted to second stage spacecraft kinetic energy such as via an ultra-efficient ion rocket magneto-hydrodynamic-plasma-drive, electro-hydrodynamic-plasma-drive, or electromagneto-hydrodynamic-plasma-drive. Thus the starting Lorentz factor of the second stage once fully accelerated is [7.2644288 + 1] = 8.2644288.

Now, assume that the second stage similarly lases or beams on the third stage, once again, this time using all initial second stage fusion fuel. Thus the energy generated will be equal to $(0.007)(0.9999)[M_1/M_0](M_0)[C^2](8.2644288) = 578.452165\ (M_0)[C^2]$.

Now, assume that all of this lasing energy is converted to third stage spacecraft kinetic energy such as via an ultra-efficient ion rocket magneto-hydrodynamic-plasma-drive, electro-hydrodynamic-plasma-drive, or electromagneto-hydrodynamic-plasma-drive. Thus the starting Lorentz factor of the third stage once fully accelerated is [578.452165 + 1] = 579.452165.

Now, assume that the third stage similarly lases or beam on the fourth stage, once again, this time using all initial third stage fusion fuel. Thus the energy generated will be equal to $(0.007)(0.9999)[M_1/M_0](M_0)[C^2](579.452165) = 40{,}557.595\ (M_0)[C^2]$.

Now, assume that all of this lasing energy is converted to fourth stage spacecraft kinetic energy such as via an ultra-efficient ion rocket magneto-hydrodynamic-plasma-drive, electro-hydrodynamic-plasma-drive, or electromagneto-hydrodynamic-plasma-drive. Thus the starting Lorentz factor of the fourth stage once fully accelerated is [40,557.595 + 1] = 579.452165.

Now, assume that the fourth stage similarly lases or beams on the fifth stage, once again, this time using all initial fourth stage fusion fuel. Thus the energy generated will be equal to $(0.007)(0.9999)[M_1/M_0](M_0)[C^2](40{,}558.595) = 2{,}838{,}817.767\ (M_0)[C^2]$.

Now, assume that all of this lasing energy is converted to fifth stage spacecraft kinetic energy such as via an ultra-efficient ion rocket magneto-hydrodynamic-plasma-drive, electro-hydrodynamic-plasma-drive, or electromagneto-hydrodynamic-plasma-drive. Thus the starting Lorentz factor of the fifth stage once fully accelerated is [2,838,817.767 + 1] = 2,838,818.767.

Now, assume that the fifth stage similarly lases or beams on the sixth stage, once again, this time using all initial fifth stage fusion fuel. Thus the energy generated will be equal to $(0.007)(0.9999)[M_1/M_0](M_0)[C^2](\ 2{,}838{,}818.767) = 198{,}697{,}441.9\ (M_0)[C^2]$.

Now, assume that all of this lasing energy is converted to sixth stage spacecraft kinetic energy such as via an ultra-efficient ion rocket magneto-hydrodynamic-plasma-drive, electro-hydrodynamic-plasma-drive, or electromagneto-hydrodynamic-plasma-drive. Thus the starting Lorentz factor of the sixth stage once fully accelerated is [198,697,441.9 + 1] = 198,697,442.9.

For a spacecraft having a starting mass of $10^{24}$ metric tons, we derive a final state initial Lorentz factor of 198,697,442.9.

Now, imagine a fusion rocket having a mass-ratio of $10^5$ and a fuel specific impulse 0.119 C. Furthermore, assume that the payload has mass ratio of $10^5$ and the same fuel specific impulse. Assume four additional stages of the same proportion.

Now assume the first stage starts out and burns enough fuel to yield an effective first stage mass-ratio of $10^4$.

The terminal Lorentz factor and velocity of the third stage will thus be 0.267349984 and 1.037775555 respectively.

Now, assume that the remainder of the fusion fuel in the first stage is processed in an efficient fusion reactor to power a laser beam or microwave beam but where the fusion waste products are held and thus serve as a way to conserve first stage momentum. So we will assume for the sake of current argument that the Lorentz factor of the first stage remains at 1.037775555. The lasing energy is thus equal to $(0.007)(0.1)[M_1/M_0](M_0)[C^2](1.037775555) = 72.644288\ (M_0)[C^2]$.

Now, assume that all of this lasing energy is converted to second stage spacecraft kinetic energy such as via an ultra-efficient ion rocket magneto-hydrodynamic-plasma-drive, electro-hydrodynamic-plasma-drive, or electromagneto-hydrodynamic-plasma-drive. Thus the starting Lorentz factor of the second stage once fully accelerated is [72.644288 + 1] = 73.644288.

Now, assume that the second stage similarly lases or beams on the third stage, once again, this time using all initial second stage fusion fuel. Thus the energy generated will be equal to $(0.007)(0.99999)[M_1/M_0](M_0)[C^2](73.644288) = 51{,}550.48\ (M_0)[C^2]$.

Now, assume that all of this lasing energy is converted to third stage spacecraft kinetic energy such as via an ultra-efficient ion rocket magneto-hydrodynamic-plasma-drive, electro-hydrodynamic-plasma-drive, or electromagneto-hydrodynamic-plasma-drive. Thus the starting Lorentz factor of the third stage once fully accelerated is [51,550.48 + 1] =51,551.48.

Now, assume that the third stage similarly lases or beams on the fourth stage, once again, this time using all initial third stage fusion fuel. Thus, the energy generated will be equal to $(0.007)(0.99999)[M_1/M_0](M_0)[C^2](51{,}551.48) = 36{,}085{,}675\ (M_0)[C^2]$.

Now, assume that all of this lasing energy is converted to fourth stage spacecraft kinetic energy such as via an ultra-efficient ion rocket magneto-hydrodynamic-plasma-drive, electro-hydrodynamic-plasma-drive, or electromagneto-hydrodynamic-plasma-drive. Thus the starting Lorentz factor of the fourth stage once fully accelerated is [36,085,675 + 1] =36,085,676.

Now, assume that the fourth stage similarly lases or beams on the fifth stage, once again, this time using all of the initial fourth stage fusion fuel. Thus, the energy generated will be equal to $(0.007)(0.99999)[M_1/M_0](M_0)[C^2](36{,}085{,}676) = 2.52597206 \times 10^{10}\ (M_0)[C^2]$.

Now, assume that all of this lasing energy is converted to fifth stage spacecraft kinetic energy such as via an ultra-efficient ion rocket magneto-hydrodynamic-plasma-drive, electro-hydrodynamic-plasma-drive, or electromagneto-hydrodynamic-plasma-drive. Thus, the starting Lorentz factor of the fifth stage once fully accelerated is $[2.52597206 \times 10^{10}] + 1$.

Now, assume that the fifth stage similarly lases or beams on the sixth stage, once again, this time using all initial fifth stage fusion fuel. Thus the energy generated will be equal to $(0.007)(0.99999)[M_1/M_0](M_0)[C^2]\ [2.52597206 \times 10^{10}] = 1.76816276 \times 10^{13}\ (M_0)[C^2]$.

Now, assume that all of this lasing energy is converted to sixth stage spacecraft kinetic energy such as via an ultra-efficient ion rocket magneto-hydrodynamic-plasma-drive, electro-hydrodynamic-plasma-drive, or electromagneto-hydrodynamic-plasma-drive. Thus, the starting Lorentz factor of the sixth stage once fully accelerated is $[1.76816276 \ldots \times 10^{13}] + 1 = 198{,}697{,}442.9$.

For a spacecraft having a starting mass of $10^{30}$ metric tons, we derive a final state initial Lorentz factor of $[1.76816276 \times 10^{13}] + 1$

Now, imagine a fusion rocket having a mass-ratio of $10^6$ and a fuel specific impulse 0.119 C. Furthermore, assume that the payload has mass ratio of 106 and the same fuel specific impulse. Assume four additional stages of the same proportion.

Now assume the first stage starts out and burns enough fuel to yield an effective first stage mass-ratio of $10^5$.

The terminal Lorentz factor and velocity of the first stage will thus be 0.267349984 and 1.037775555 respectively.

Now, assume that the remainder of the fusion fuel in the first stage is processed in an efficient fusion reactor to power a laser beam or microwave beam but where the fusion waste products are held and thus serve as a way to conserve first stage momentum. So we will assume for the sake of current argument that the Lorentz factor of the first stage remains at 1.037775555. The lasing energy is thus equal to $(0.007)(0.1)[M_1/M_0](M_0)[C^2](1.037775555) = 726.44288\ (M_0)[C^2]$.

Now, assume that all of this lasing energy is converted to second stage spacecraft kinetic energy such as via an ultra-efficient ion rocket magneto-hydrodynamic-plasma-drive, electro-hydrodynamic-plasma-drive, or electromagneto-hydrodynamic-plasma-drive. Thus the starting Lorentz factor of the second stage once fully accelerated is [726.44288 + 1] = 727.44288.

Now, assume that the second stage similarly lases or beams on the third stage, once again, this time using all initial second stage fusion fuel. Thus the energy generated will be equal to $(0.007)(0.999999)[M_1/M_0](M_0)[C^2](727.44288) = 5{,}092{,}095(M_0)[C^2]$.

Now, assume that all of this lasing energy is converted to third stage spacecraft kinetic energy such as via an ultra-efficient ion rocket magneto-hydrodynamic-plasma-drive, electro-hydrodynamic-plasma-drive, or electromagneto-hydrodynamic-plasma-drive. Thus the starting Lorentz factor of the third stage once fully accelerated is [5,092,095 + 1] = 5,092,096.

Now, assume that the third stage similarly lases or beams on the fourth stage, once again, this time using all initial third stage fusion fuel. Thus, the energy generated will be equal to $(0.007)(0.999999)[M_1/M_0](M_0)[C^2](5{,}092{,}096) = 3.564463 \ldots \times 10^{10}\ (M_0)[C^2]$.

Now, assume that all of this lasing energy is converted to fourth stage spacecraft kinetic energy such as via an ultra-efficient ion rocket magneto-hydrodynamic-plasma-drive, electro-hydrodynamic-plasma-drive, or electromagneto-hydrodynamic-plasma-drive. Thus the starting Lorentz factor of the fourth stage once fully accelerated is [3.564463 ... x $10^{10}$] + 1.

Now, assume that the fourth stage similarly lases or beams on the fifth stage, once again, this time using all of the initial fourth stage fusion fuel. Thus, the energy generated will be equal to (0.007)(0.999999)[$M_1/M_0$]($M_0$)[$C^2$] {[3.564463 ... x $10^{10}$] + 1} = 2.495122 ... x $10^{14}$ ($M_0$)[$C^2$].

Now, assume that all of this lasing energy is converted to fifth stage spacecraft kinetic energy such as via an ultra-efficient ion rocket magneto-hydrodynamic-plasma-drive, electro-hydrodynamic-plasma-drive, or electromagneto-hydrodynamic-plasma-drive. Thus, the starting Lorentz factor of the fifth stage once fully accelerated is [2.495122 ... x $10^{14}$] + 1.

Now, assume that the fifth stage similarly lases or beams on the sixth stage, once again, this time using all initial fifth stage fusion fuel. Thus the energy generated will be equal to (0.007)(0.999999)[$M_1/M_0$]($M_0$)[$C^2$] {[2.495122 ... x $10^{14}$] + 1} = 1.74658 x $10^{18}$ ($M_0$)[$C^2$].

Now, assume that all of this lasing energy is converted to sixth stage spacecraft kinetic energy such as via an ultra-efficient ion rocket magneto-hydrodynamic-plasma-drive, electro-hydrodynamic-plasma-drive, or electromagneto-hydrodynamic-plasma-drive. Thus, the starting Lorentz factor of the sixth stage once fully accelerated is [1.74658 ... x $10^{18}$] + 1.

For a spacecraft having a starting mass of $10^{36}$ metric tons, we derive a final state initial Lorentz factor of

[1.74658 ... x $10^{18}$] + 1.

Now, imagine a fusion rocket having a mass-ratio of $10^7$ and a fuel specific impulse 0.119 C. Furthermore, assume that the payload has mass ratio of $10^7$ and the same fuel specific impulse. Assume four additional stages of the same proportion.

Now assume the first stage starts out and burns enough fuel to yield an effective first stage mass-ratio of $10^7$.

The terminal Lorentz factor and velocity of the first stage will thus be 0.267349984 and 1.037775555 respectively.

Now, assume that the remainder of the fusion fuel in the first stage is processed in an efficient fusion reactor to power a laser beam or microwave beam but where the fusion waste products are held and thus serve as a way to conserve first stage momentum. So we will assume for the sake of current argument that the Lorentz factor of the first stage remains at 1.037775555. The lasing energy is thus equal to (0.007)(0.1)[$M_1/M_0$]($M_0$)[$C^2$](1.037775555) = 7,264.4288 ($M_0$)[$C^2$].

Now, assume that all of this lasing energy is converted to second stage spacecraft kinetic energy such as via an ultra-efficient ion rocket magneto-hydrodynamic-plasma-drive, electro-hydrodynamic-plasma-drive, or electromagneto-hydrodynamic-plasma-drive. Thus the starting Lorentz factor of the second stage once fully accelerated is [7,264.4288 + 1] =7,265.4288.

Now, assume that the second stage similarly lases or beams on the third stage, once again, this time using all initial second stage fusion fuel. Thus the energy generated will be equal to (0.007)(0.9999999)[$M_1/M_0$]($M_0$)[$C^2$]( 7,265.4288) = 508,579,965 ($M_0$)[$C^2$].

Now, assume that all of this lasing energy is converted to third stage spacecraft kinetic energy such as via an ultra-efficient ion rocket magneto-hydrodynamic-plasma-drive, electro-hydrodynamic-plasma-drive, or electromagneto-hydrodynamic-plasma-drive. Thus the starting Lorentz factor of the third stage once fully accelerated is [508,579,965 + 1] = 508,579,966.

Now, assume that the third stage similarly lases or beams on the fourth stage, once again, this time using all initial third stage fusion fuel. Thus, the energy generated will be equal to (0.007)(0.9999999)[$M_1/M_0$]($M_0$)[$C^2$](508,579,966) = 3.560059 ... x $10^{13}$ ($M_0$)[$C^2$].

Now, assume that all of this lasing energy is converted to fourth stage spacecraft kinetic energy such as via an ultra-efficient ion rocket magneto-hydrodynamic-plasma-drive, electro-hydrodynamic-plasma-drive, or electromagneto-hydrodynamic-plasma-drive. Thus the starting Lorentz factor of the fourth stage once fully accelerated is [3.560059 ... x $10^{13}$] + 1.

Now, assume that the fourth stage similarly lases or beams on the fifth stage, once again, this time using all of the initial fourth stage fusion fuel. Thus, the energy generated will be equal to (0.007)(0.9999999)[$M_1/M_0$]($M_0$)[$C^2$] {[3.560059 ... x $10^{10}$] + 1} = 2.49204 ... x $10^{18}$ ($M_0$)[$C^2$].

Now, assume that all of this lasing energy is converted to fifth stage spacecraft kinetic energy such as via an ultra-efficient ion rocket magneto-hydrodynamic-plasma-drive, electro-hydrodynamic-plasma-drive, or electromagneto-hydrodynamic-plasma-drive. Thus, the starting Lorentz factor of the fifth stage once fully accelerated is [2.49204 ... x $10^{18}$] + 1.

Now, assume that the fifth stage similarly lases or beams on the sixth stage, once again, this time using all initial fifth stage fusion fuel. Thus the energy generated will be equal to (0.007)(0.9999999)[$M_1/M_0$]($M_0$)[$C^2$] {[2.49204 ... x $10^{18}$] + 1} = 1.74658 x $10^{18}$ ($M_0$)[$C^2$].

Now, assume that all of this lasing energy is converted to sixth stage spacecraft kinetic energy such as via an ultra-efficient ion rocket magneto-hydrodynamic-plasma-drive, electro-hydrodynamic-plasma-drive, or electromagneto-hydrodynamic-plasma-drive. Thus, the starting Lorentz factor of the sixth stage once fully accelerated is [1.744428 ... x $10^{23}$] + 1.

For a spacecraft having a starting mass of $10^{42}$ metric tons, we derive a final state initial Lorentz factor of

[1.74658 ... x $10^{23}$] + 1.

Now, imagine a fusion rocket having a mass-ratio of $10^8$ and a fuel specific impulse 0.119 C. Furthermore, assume that the payload has mass ratio of $10^8$ and the same fuel specific impulse. Assume four additional stages of the same proportion.

Now assume the first stage starts out and burns enough fuel to yield an effective first stage mass-ratio of $10^8$.

The terminal Lorentz factor and velocity of the first stage will thus be 0.267349984 and 1.037775555 respectively.

Now, assume that the remainder of the fusion fuel in the first stage is processed in an efficient fusion reactor to power a laser beam or microwave beam but where the fusion waste products are held and thus serve as a way to conserve first stage momentum. So we will assume for the sake of current argument that the Lorentz factor of the first stage remains at 1.037775555. The lasing energy is thus equal to (0.007)(0.1)[$M_1/M_0$]($M_0$)[$C^2$](1.037775555) = 72,644.288 ($M_0$)[$C^2$].

Now, assume that all of this lasing energy is converted to second stage spacecraft kinetic energy such as via an ultra-efficient ion rocket magneto-hydrodynamic-plasma-drive, electro-hydrodynamic-plasma-drive, or electromagneto-hydrodynamic-plasma-drive. Thus the starting Lorentz factor of the second stage once fully accelerated is [72,644.288 + 1] =72,645.288.

Now, assume that the second stage similarly lases or beams on the third stage, once again, this time using all initial second stage fusion fuel. Thus the energy generated will be equal to (0.007)(0.99999999)[$M_1/M_0$]($M_0$)[$C^2$](72,645.288) = 5.08517 ... x $10^{10}$ ($M_0$)[$C^2$].

Now, assume that all of this lasing energy is converted to third stage spacecraft kinetic energy such as via an ultra-efficient ion rocket magneto-hydrodynamic-plasma-drive, electro-hydrodynamic-plasma-drive, or electromagneto-hydrodynamic-plasma-drive. Thus the starting Lorentz factor of the third stage once fully accelerated is [5.08517 ... x $10^{10}$] + 1.

Now, assume that the third stage similarly lases or beams on the fourth stage, once again, this time using all initial third stage fusion fuel. Thus, the energy generated will be equal to (0.007)(0.99999999)[$M_1/M_0$]($M_0$)[$C^2$] {[5.08517 ... x $10^{10}$] + 1} = 3.5596 ... x $10^{16}$ ($M_0$)[$C^2$].

Now, assume that all of this lasing energy is converted to fourth stage spacecraft kinetic energy such as via an ultra-efficient ion rocket magneto-hydrodynamic-plasma-drive, electro-hydrodynamic-plasma-drive, or electromagneto-hydrodynamic-plasma-drive. Thus the starting Lorentz factor of the fourth stage once fully accelerated is [3.5596 ... x $10^{16}$] + 1.

Now, assume that the fourth stage similarly lases or beams on the fifth stage, once again, this time using all of the initial fourth stage fusion fuel. Thus, the energy generated will be equal to (0.007)(0.99999999)[$M_1/M_0$]($M_0$)[$C^2$] {[2.49173 ... x $10^{22}$] + 1} = {[2.49173 ... x $10^{22}$] + 1} ($M_0$)[$C^2$].

Now, assume that all of this lasing energy is converted to fifth stage spacecraft kinetic energy such as via an ultra-efficient ion rocket magneto-hydrodynamic-plasma-drive, electro-hydrodynamic-plasma-drive, or electromagneto-hydrodynamic-plasma-drive. Thus, the starting Lorentz factor of the fifth stage once fully accelerated is [2.49173 ... x $10^{22}$] + 1.

Now, assume that the fifth stage similarly lases or beams on the sixth stage, once again, this time using all initial fifth stage fusion fuel. Thus the energy generated will be equal to (0.007)(0.99999999)[$M_1/M_0$]($M_0$)[$C^2$] {[2.49173 ... x $10^{22}$] + 1} = 1.7442 ... x $10^{28}$ ($M_0$)[$C^2$].

Now, assume that all of this lasing energy is converted to sixth stage spacecraft kinetic energy such as via an ultra-efficient ion rocket magneto-hydrodynamic-plasma-drive, electro-hydrodynamic-plasma-drive, or electromagneto-hydrodynamic-plasma-drive. Thus, the starting Lorentz factor of the sixth stage once fully accelerated is [1.7442 ... x $10^{28}$] + 1.

For a spacecraft having a starting mass of $10^{48}$ metric tons, we derive a final state initial Lorentz factor of

[1.74658 ... x $10^{28}$] + 1.

Now, imagine a fusion rocket having a mass-ratio of $10^9$ and a fuel specific impulse 0.119 C. Furthermore, assume that the payload has mass ratio of $10^9$ and the same fuel specific impulse. Assume four additional stages of the same proportion.

Now assume the first stage starts out and burns enough fuel to yield an effective first stage mass-ratio of $10^9$.

The terminal Lorentz factor and velocity of the first stage will thus be 0.267349984 and 1.037775555 respectively.

Now, assume that the remainder of the fusion fuel in the first stage is processed in an efficient fusion reactor to power a laser beam or microwave beam but where the fusion waste products are held and thus serve as a way to conserve first stage momentum. So we will assume for the sake of current argument that the Lorentz factor of the first stage remains at 1.037775555. The lasing energy is thus equal to $(0.007)(0.1)[M_1/M_0](M_0)[C^2](1.037775555) = 726{,}442.88\ (M_0)[C^2]$.

Now, assume that all of this lasing energy is converted to second stage spacecraft kinetic energy such as via an ultra-efficient ion rocket magneto-hydrodynamic-plasma-drive, electro-hydrodynamic-plasma-drive, or electromagneto-hydrodynamic-plasma-drive. Thus the starting Lorentz factor of the second stage once fully accelerated is [726,442.88 + 1] =726,442.88.

Now, assume that the second stage similarly lases or beams on the third stage, once again, this time using all initial second stage fusion fuel. Thus the energy generated will be equal to $(0.007)(0.999999999)[M_1/M_0](M_0)[C^2](726{,}442.88) = 5.08510 \ldots \times 10^{12}\ (M_0)[C^2]$.

Now, assume that all of this lasing energy is converted to third stage spacecraft kinetic energy such as via an ultra-efficient ion rocket magneto-hydrodynamic-plasma-drive, electro-hydrodynamic-plasma-drive, or electromagneto-hydrodynamic-plasma-drive. Thus the starting Lorentz factor of the third stage once fully accelerated is $[5.08510 \ldots \times 10^{12}] + 1$.

Now, assume that the third stage similarly lases or beams on the fourth stage, once again, this time using all initial third stage fusion fuel. Thus, the energy generated will be equal to $(0.007)(0.999999999)[M_1/M_0](M_0)[C^2]\ \{[5.08517 \ldots \times 10^{12}] + 1\} = 3.55957 \ldots \times 10^{19}\ (M_0)[C^2]$.

Now, assume that all of this lasing energy is converted to fourth stage spacecraft kinetic energy such as via an ultra-efficient ion rocket magneto-hydrodynamic-plasma-drive, electro-hydrodynamic-plasma-drive, or electromagneto-hydrodynamic-plasma-drive. Thus the starting Lorentz factor of the fourth stage once fully accelerated is $[3.55957 \ldots \times 10^{19}] + 1$.

Now, assume that the fourth stage similarly lases or beams on the fifth stage, once again, this time using all of the initial fourth stage fusion fuel. Thus, the energy generated will be equal to $(0.007)(0.999999999)[M_1/M_0](M_0)[C^2]\ \{[3.55957 \ldots \times 10^{19}] + 1\} = \{[2.491699 \ldots \times 10^{26}] + 1\}\ (M_0)[C^2]$.

Now, assume that all of this lasing energy is converted to fifth stage spacecraft kinetic energy such as via an ultra-efficient ion rocket magneto-hydrodynamic-plasma-drive, electro-hydrodynamic-plasma-drive, or electromagneto-hydrodynamic-plasma-drive. Thus, the starting Lorentz factor of the fifth stage once fully accelerated is $[2.491699 \ldots \times 10^{26}] + 1$.

Now, imagine a fusion rocket having a mass-ratio of $10^{10}$ and a fuel specific impulse 0.119 C. Furthermore, assume that the payload has mass ratio of $10^{10}$ and the same fuel specific impulse. Assume four additional stages of the same proportion.

Now assume the first stage starts out and burns enough fuel to yield an effective first stage mass-ratio of $10^{10}$.

The terminal Lorentz factor and velocity of the first stage will thus be 0.267349984 and 1.037775555 respectively.

Now, assume that the remainder of the fusion fuel in the first stage is processed in an efficient fusion reactor to power a laser beam or microwave beam but where the fusion waste products are held and thus serve as a way to conserve first stage momentum. So we will assume for the sake of current argument that the Lorentz factor of the first stage remains at 1.037775555. The lasing energy is thus equal to $(0.007)(0.1)[M_1/M_0](M_0)[C^2](1.037775555) = 7{,}264{,}428.8\ (M_0)[C^2]$.

Now, assume that all of this lasing energy is converted to second stage spacecraft kinetic energy such as via an ultra-efficient ion rocket magneto-hydrodynamic-plasma-drive, electro-hydrodynamic-plasma-drive, or electromagneto-hydrodynamic-plasma-drive. Thus the starting Lorentz factor of the second stage once fully accelerated is [7,264,428.8 + 1] =7,264,429.8.

Now, assume that the second stage similarly lases or beams on the third stage, once again, this time using all initial second stage fusion fuel. Thus the energy generated will be equal to $(0.007)(0.9999999999)[M_1/M_0](M_0)[C^2](7{,}264{,}429.8) = 5.08510 \ldots \times 10^{14}\ (M_0)[C^2]$.

Now, assume that all of this lasing energy is converted to third stage spacecraft kinetic energy such as via an ultra-efficient ion rocket magneto-hydrodynamic-plasma-drive, electro-hydrodynamic-plasma-drive, or electromagneto-hydrodynamic-plasma-drive. Thus the starting Lorentz factor of the third stage once fully accelerated is $[5.08510 \ldots \times 10^{14}] + 1$.

Now, assume that the third stage similarly lases or beams on the fourth stage, once again, this time using all initial third stage fusion fuel. Thus, the energy generated will be equal to $(0.007)(0.99999999999)[M_1/M_0](M_0)[C^2]\ \{[5.08510 \ldots \times 10^{14}] + 1\} = 3.55957 \ldots \times 10^{22}\ (M_0)[C^2]$.

Now, assume that all of this lasing energy is converted to fourth stage spacecraft kinetic energy such as via an ultra-efficient ion rocket magneto-hydrodynamic-plasma-drive, electro-hydrodynamic-plasma-drive, or electromagneto-hydrodynamic-plasma-drive. Thus the starting Lorentz factor of the fourth stage once fully accelerated is $[3.55957 \ldots \times 10^{22}] + 1$.

Now, assume that the fourth stage similarly lases or beams on the fifth stage, once again, this time using all of the initial fourth stage fusion fuel. Thus, the energy generated will be equal to $(0.007)(0.999999999)[M_1/M_0](M_0)[C^2]\ \{[3.55957 \ldots \times 10^{22}] + 1\} = \{[2.491699 \ldots \times 10^{30}] + 1\}\ (M_0)[C^2]$.

Now, assume that all of this lasing energy is converted to fifth stage spacecraft kinetic energy such as via an ultra-efficient ion rocket magneto-hydrodynamic-plasma-drive, electro-hydrodynamic-plasma-drive, or electromagneto-hydrodynamic-plasma-drive. Thus, the starting Lorentz factor of the fifth stage once fully accelerated is $[2.491699 \ldots \times 10^{30}] + 1$.

Now, imagine a fusion rocket having a mass-ratio of $10^{12}$ and a fuel specific impulse 0.119 C. Furthermore, assume that the payload has mass ratio of $10^{12}$ and the same fuel specific impulse. Assume two additional stages of the same proportion.

Now assume the first stage starts out and burns enough fuel to yield an effective first stage mass-ratio of $10^{12}$.

The terminal Lorentz factor and velocity of the first stage will thus be 0.267349984 and 1.037775555 respectively.

Now, assume that the remainder of the fusion fuel in the first stage is processed in an efficient fusion reactor to power a laser beam or microwave beam but where the fusion waste products are held and thus serve as a way to conserve first stage momentum. So we will assume for the sake of current argument that the Lorentz factor of the first stage remains at 1.037775555. The lasing energy is thus equal to $(0.007)(0.1)[M_1/M_O](M_O)[C^2](1.037775555) = 726{,}442{,}880\ (M_O)[C^2]$.

Now, assume that all of this lasing energy is converted to second stage spacecraft kinetic energy such as via an ultra-efficient ion rocket magneto-hydrodynamic-plasma-drive, electro-hydrodynamic-plasma-drive, or electromagneto-hydrodynamic-plasma-drive. Thus the starting Lorentz factor of the second stage once fully accelerated is [726,442,880 + 1] = 726,442,881.

Now, assume that the second stage similarly lases or beams on the third stage, once again, this time using all initial second stage fusion fuel. Thus the energy generated will be equal to $(0.007)(0.99999999999)[M_1/M_O](M_O)[C^2](726{,}442{,}881) = 5.08510 \ldots \times 10^{18}\ (M_O)[C^2]$.

Now, assume that all of this lasing energy is converted to third stage spacecraft kinetic energy such as via an ultra-efficient ion rocket magneto-hydrodynamic-plasma-drive, electro-hydrodynamic-plasma-drive, or electromagneto-hydrodynamic-plasma-drive. Thus the starting Lorentz factor of the third stage once fully accelerated is $[5.08510 \ldots \times 10^{18}] + 1$.

Now, assume that the third stage similarly lases or beams on the fourth stage, once again, this time using all initial third stage fusion fuel. Thus, the energy generated will be equal to $(0.007)(0.99999999999)[M_1/M_O](M_O)[C^2]\ \{[5.08510 \ldots \times 10^{18}] + 1\} = 3.55957 \ldots \times 10^{28}\ (M_O)[C^2]$.

Now, assume that all of this lasing energy is converted to fourth stage spacecraft kinetic energy such as via an ultra-efficient ion rocket magneto-hydrodynamic-plasma-drive, electro-hydrodynamic-plasma-drive, or electromagneto-hydrodynamic-plasma-drive. Thus the starting Lorentz factor of the fourth stage once fully accelerated is $[3.55957 \ldots \times 10^{28}] + 1$.

Now, let us assume that the final of fifth stage has a mass that is $10^{-10}$ of that of the fourth stage and that the fourth stage processes all of its fusion fuel into lasing or beam energy which is then converted to fifth stage kinetic energy. Thus, the factoral increase in Lorentz factor will be $(0.007)(10^{12}) = 7 \times 10^9$.

The final spacecraft Lorentz factor will be $\{[3.55957 \ldots \times 10^{28}] + 1\}[7 \times 10^9] = 2.49 \times 10^{38}$.

Obviously, the mass of the initial system would need to be roughly the same as that of the observable universe. However, the universe may be of infinite extent and perhaps instead of experiencing a big rip in another inflationary epoch, perhaps the rate of expansion will begin to significantly slow many eons from now. It is possible that the effects of Dark Energy which appears to be accelerating the rate of expansion will eventually reduce its expansionary effects.

For the proposed extreme systems to work, a given stage would need to remain in the same cosmic light-cone as the preceding and proceeding stage. Otherwise, the beam chain would be broken. A slowing of universal expansion would be helpful in this regard.

Numerous smaller stages may be used to replace each large stage.

Other materials may be used as a mean of energy storage in each stage including extremely long lived nuclear isomers, exotic QCD fuels such as those that rely on conversion of ordinary baryonic matter to strange matter, charmonium, bottomonium, and matter-antimatter fuels.

Moreover explosive charges may be used instead of beaming apparatuses. These charges may include shaped atomic or thermonuclear charges, pure fusion charges, long lived nuclear isomeric charges, exotic QCD charges and matter-antimatter charges.

What is even more interesting is that any of the stages used may be accelerated by a mass-driver or by other means to be pre-placed along the anticipated heading of the spacecraft. Moreover, the charges may be preplaced in a stationary or non-stationary relation with respect to the background from differing locations along the initial length of the spacecraft anticipated itinerary.

Imagine again a fusion rocket having a mass-ratio of $10^4$ and a fuel specific impulse 0.119 C. Furthermore, assume that the payload has mass ratio of $10^4$ and the same fuel specific impulse. Assume two additional stages of the same proportion.

Now assume the first stage starts out and burns enough fuel to yield an effective first stage mass-ratio of $10^3$.

The terminal Lorentz factor and velocity of the third stage will thus be 0.267349984 and 1.037775555 respectively.

Now, assume that the remainder of the fusion fuel in the first stage is processed in an efficient fusion reactor to power a laser beam or microwave beam but where the fusion waste products are held and thus serve as a way to conserve first stage momentum. So we will assume for the sake of current argument that the Lorentz factor of the first stage remains at 1.037775555. The lasing energy is thus equal to $(0.007)(0.1)[M_1/M_0](M_0)[C^2](1.037775555) = 7.2644288\ (M_0)[C^2]$.

Now, assume that all of this lasing energy is converted to second stage spacecraft kinetic energy such as via an ultra-efficient ion rocket magneto-hydrodynamic-plasma-drive, electro-hydrodynamic-plasma-drive, or electromagneto-hydrodynamic-plasma-drive. Thus the starting Lorentz factor of the second stage once fully accelerated is [7.2644288 + 1] = 8.2644288.

Now, assume that the second stage similarly lases or beam on the third stage, once again, this time using all initial second stage fusion fuel. Thus the energy generated will be equal to $(0.007)(0.9999)[M_1/M_0](M_0)[C^2](8.2644288) = 578.452165\ (M_0)[C^2]$.

Now, assume that all of this lasing energy is converted to third stage spacecraft kinetic energy such as via an ultra-efficient ion rocket magneto-hydrodynamic-plasma-drive, electro-hydrodynamic-plasma-drive, or electromagneto-hydrodynamic-plasma-drive. Thus the starting Lorentz factor of the third stage once fully accelerated is [578.452165 + 1] = 579.452165.

Now, assume that the third stage similarly lases or beam on the fourth stage, once again, this time using all initial third stage fusion fuel. Thus the energy generated will be equal to $(0.007)(0.9999)[M_1/M_0](M_0)[C^2](579.452165) = 40{,}557.595\ (M_0)[C^2]$.

Now, assume that all of this lasing energy is converted to fourth stage spacecraft kinetic energy such as via an ultra-efficient ion rocket magneto-hydrodynamic-plasma-drive, electro-hydrodynamic-plasma-drive, or electromagneto-hydrodynamic-plasma-drive. Thus the starting Lorentz factor of the fourth stage once fully accelerated is [40,557.595 + 1] =40,558.595.

Not bad for a $10^{16}$ metric ton starting mass. Even a three stage craft and a $10^{12}$ metric tons starting mass will get you to a Lorentz factor of 579.452165 which is enough for manned missions anywhere in the Galaxy.

We can also conjecture four stage craft where the initial mass is held to $10^{12}$ metric tons.

Imagine again a fusion rocket having a mass-ratio of $10^3$ and a fuel specific impulse 0.119 C. Furthermore, assume that the payload has mass ratio of $10^3$ and the same fuel specific impulse. Assume two additional stages of the same proportion.

Now assume the first stage starts out and burns enough fuel to yield an effective first stage mass-ratio of $10^2$.

The terminal Lorentz factor and velocity of the third stage will thus be 0.267349984 and 1.037775555 respectively.

Now, assume that the remainder of the fusion fuel in the first stage is processed in an efficient fusion reactor to power a laser beam or microwave beam but where the fusion waste products are held and thus serve as a way to conserve first stage momentum. So we will assume for the sake of current argument that the Lorentz factor of the first stage remains at 1.037775555. The lasing energy is thus equal to $(0.007)(0.1)[M_1/M_0](M_0)[C^2](1.037775555) = 0.72644288\ (M_0)[C^2]$.

Now, assume that all of this lasing energy is converted to second stage spacecraft kinetic energy such as via an ultra-efficient ion rocket magneto-hydrodynamic-plasma-drive, electro-hydrodynamic-plasma-drive, or electromagneto-hydrodynamic-plasma-drive. Thus the starting Lorentz factor of the second stage once fully accelerated is [0.072644288 + 1] = 1.072644288

Now, assume that the second stage similarly lases or beam on the third stage, once again, this time using all initial second stage fusion fuel. Thus the energy generated will be equal to $(0.007)(0.999)[M_1/M_0](M_0)[C^2](1.072644288) = 7.5010015\ (M_0)[C^2]$.

Now, assume that all of this lasing energy is converted to third stage spacecraft kinetic energy such as via an ultra-efficient ion rocket magneto-hydrodynamic-plasma-drive, electro-hydrodynamic-plasma-drive, or electromagneto-hydrodynamic-plasma-drive. Thus the starting Lorentz factor of the third stage once fully accelerated is [7.5010015 + 1] = 8.5010015

Now, assume that the third stage similarly lases or beam on the fourth stage, once again, this time using all initial third stage fusion fuel. Thus the energy generated will be equal to $(0.007)(0.999)[M_1/M_0](M_0)[C^2](8.5010015) = 59.4475\ (M_0)[C^2]$.

Now, assume that all of this lasing energy is converted to fourth stage spacecraft kinetic energy such as via an ultra-efficient ion rocket magneto-hydrodynamic-plasma-drive, electro-hydrodynamic-plasma-drive, or electromagneto-hydrodynamic-plasma-drive. Thus the starting Lorentz factor of the fourth stage once fully accelerated is [59.4475 + 1] =60.4475.

Not bad for a $10^{12}$ metric ton starting mass. Now if the final stage had a matter-antimatter fuel mass-ratio of 10, for which the final payload would be a 100 metric ton spacecraft commensurate with the dubiously reported typical space-alien crewed UFOs, the final Lorentz factor after antimatter burn-through will be about 604. For cases where the craft would extract its normal matter components from interstellar space in a drag energy recycling manner, the maximum Lorentz factor would be about 6,040.

Thus, we now have the tools to build general formulas describing the proposed method as follows.

Stage 2: $\{\{[(0.007)(F_{mr,remaining,stage1})[M_{1,1}/M_{0,1}](M_{0,1})[C^2](\gamma_1)]\ /[(M_{0,1})[C^2]]\} + 1\}$

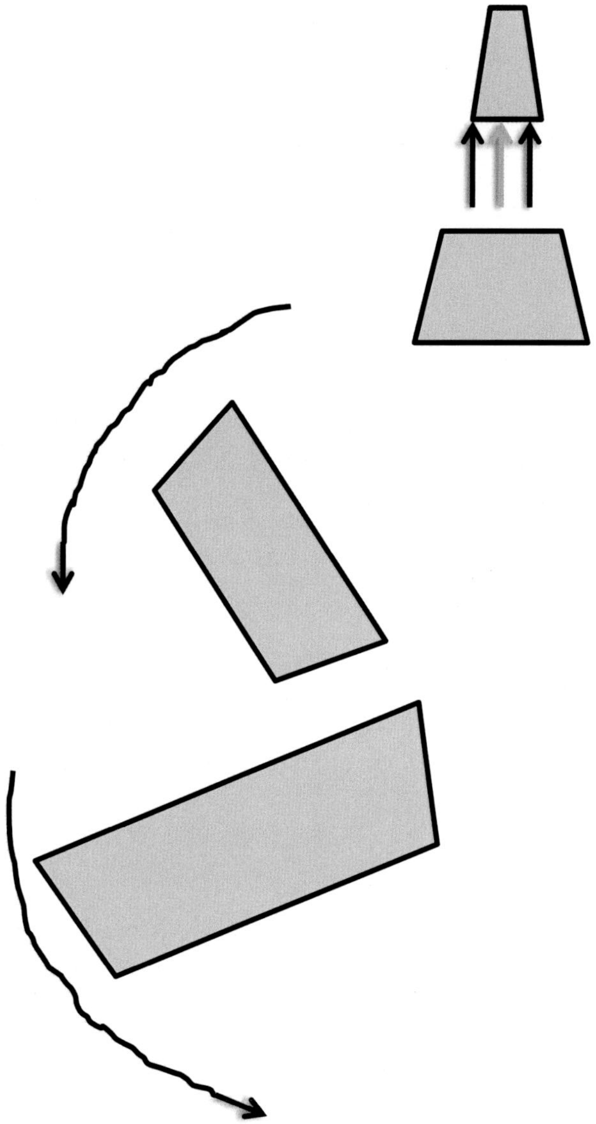

**FIGURE 15** depicts the beamed energy stage propulsion mechanisms with self-steering discharged stages.

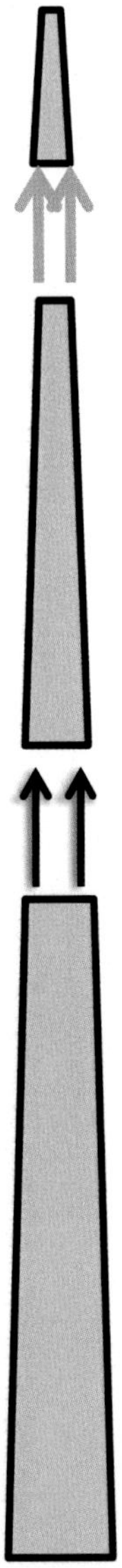

**FIGURE 16** depicts the astrodynamic, train-like, staged, beaming mechanism. Red arrows indicate beamed energy.

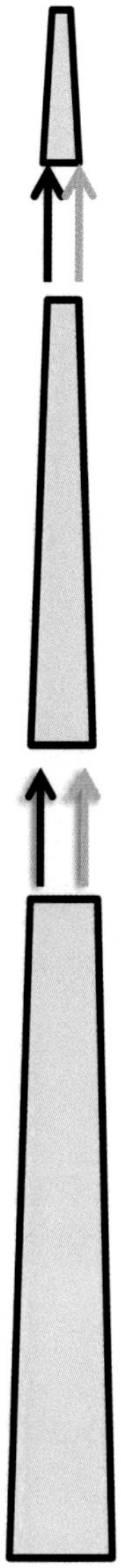

**FIGURE 17** depicts the astrodynamic, train-like, staged, beaming mechanism. Red arrows and black arrows indicate multiple species of beamed energy.

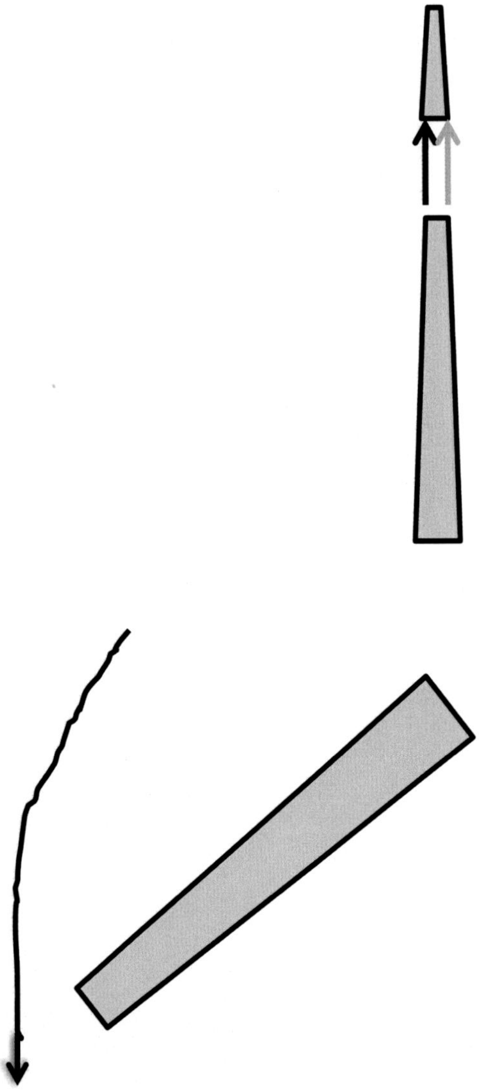

**FIGURE 18** depicts the train-like beamed energy stage propulsion mechanisms with self-steering discharged stages.

More than one species of beamed energy may in principle be used to accelerate the non-basic stages.

For example, a beam of oppositely charged particles may be used to transfer energy from one stage to another. The particles may include protons, electrons, antiprotons, positrons, or heavier normal matter and antimatter atomic nuclei.

Neutral particles such as extremely relativistic neutrons, and also neutral atoms and molecules may be used as the energy transmission means.

Macroscopic and mesoscale pellets may also be employed for the same purposes. The pellets may be charged or uncharged, magnetized or non-magnetic. Where magnetic, the means of magnetization can include permanent magnets, or superconductive current carrying coils.

It is even possible that gravitational waves may transmit energy from stage to stage.

Another plausible mechanism is neutrino based energy transmission.

The more extreme systems describe previously require a system starting mass roughly equal to the hydrogenic mass of the observable universe. This would be a bounding limit for initial localized hydrogen supplies, but the spacecraft energy can grow to relatively enormously greater levels for cases where the fusion derived energy is used to react against background media in such a way that net energy is extracted from the background in addition to the fusion derived energy per differential space intervals traveled.

Now, sub-canonical-ensemble Lorentz factor spacecraft must be completely cloaked but I will leave that problem for future materials researchers and thermodynamicists.

We can indeed in principle obtain the above referenced Lorentz factors without recourse to matter-antimatter fuels.

Regarding extreme spacecraft fuel depositions, one option is to construct a linear fuel stage oriented orthogonally to the anticipated spacecraft velocity vector. Another solution would be a circular or toroidal ring shaped fuel and stage configuration. As other options, the stage may include regular polygonal or irregular polygonal shapes or non-circular curvilinear shapes. Such staging can be applied for masses within an order of magnitude of that of the observable universe where the extensity of the first stage is on the order of one cosmic light cone radius for current era space-time expansion rates.

However, it is conceivable that the stages could be confined to well below the volume of our solar system for which gravitational field cancellation materials or features would be included in the staging. Such measures might include antigravity field generators and/or negative mass.

Regardless, the propositions of this chapter are made as upper boundary value conditions rather than as "QED" style proofs that the proposed and formulated scenarios will ever be attained.

## Interstellar Ramjet & Beam Relay Propulsion.

Another aspect of relay stations would include those accelerated by interstellar ramjets. The first stage would be accelerated to mildly relativistic velocities by fusion rocket methods. The first stage would then beam on a second stage which would be both accelerated by the beamed energy and ISR mode. The second stage would beam upon the third stage which would be accelerated by both the ISR mode and the beamed energy received from the second stage. Like-wise, the fourth stage would be so accelerated as well as the fifth, and sixth, and so-on.

An interesting situation may arise here. The final stage of the series or the final payload may end up receding from the first stage or home base at a velocity greater than C due to universal expansionary effects.

For example, a system may evolve so that the relative spatial expansion velocity between the first stage and the second stage is 0.4 C, as would be the velocity between the second stage and the third stage, and the third stage and the fourth stage and so-on. Such a system could evolve so that the recessional velocity of each state with respect to the preceding stage is $C - \epsilon$ where $\epsilon$ is a very small fraction of one.

It is possible that the timed launching of one stage relative to another as well as the travel velocity of each stage through space could be preprogrammed to enable the at least temporary maintenance of such a system.

A suitable decreasing of the space-interval specific rate of expansion would help in such regards.

Another option is to seed the interval between a separated stage and the preceding stage with nano-technology self-replicating or self-growing seed which would grow into fully functioning interstellar ramjets and beaming stages.

For example, a seed may germinate upon internally timed spouting with or without additional growth signals from the stage of deployment and/or its preceding stages. The sprouts may harness background star-light, quasar-light, blue-shifted CMBR, or interstellar fusion fuel or other background fuel sources.

Alternatively, the spouts may obtain growth energy from mass or energy beams from a preceding stage and/or the stage of release. Such beams can include fusion fuel, fission fuel, matter-antimatter fuel, antimatter fuel, particle beams, fuel pellets, kinetic energy shot, laser beams, phase array or directed microwave or radio-frequency beams, gravitational radiation beams, electron beams, and plausibly neutrino beams.

Another alternative would include energy derived from an interstellar ramjet feature pre-installed or self-grown into the sprouts for which a limited portion of the ISR derived fusion energy would be employed for powering further stage self-growth.

Regardless of the mechanism for powering self-growth of stage sprouts, the materials of manufacture can be completely supplied with a given seed from its release and/or captured in route from natural and/or artificial sources.

For example, carbon, hydrogen, carbonaceous super-materials, dense heavy element based refractory materials and the like can be included in reservoirs or pod like storage locations within a given seed. Such materials may optionally include fully assembled beams, trusses, motors, gears, conduits, and the like out of which a fully functioning interstellar ramjet would be manufactured.

Alternatively, the elemental composition used to grow a sprout could be largely included in a regular or irregularly shaped; powdered, granular, corded, membranous, sheet, plate, or block-like portions. Nanotechnology means would then be used to extract the raw materials for growth of the sprouts.

In cases where the materials of ISR craft composition are largely not including within the initial seed state, the materials may be scooped up from the vacuum of interstellar and/or intergalactic space by electrodynamic field-effect methods. Spouts fed with hydrogen, carbon, oxygen, and nitrogen are especially intriguing in this regard because these four elements are highly reactive and can relatively easily be fashioned into super materials which are extremely refractory.

As an example, background hydrogen may be fashioned into hydrogen metal, which as yet, still unobserved in stable forms within a laboratory environment. Hydrogen metals have been proposed as an extremely high strength but low density material which may either have superconducting properties or almost perfect conventional conductivity. Carbon may be fashioned into graphene, carbon nanotubes whether single walled or multi-walled, graphite fiber, single atom-wide carbon chain-based fabrics, diamond fiber fabric, and bulk diamond and the like. When mixed with oxygen, graphene-oxide paper can be manufactured. Nitrogen can be combined with oxygen to form Beta-Carbon-Nitride which in some forms has been implicated as being harder than diamond. Carbon-nitride can include bulk continuous samples, or carbon nitride thread.

Construction materials for growing spouts can be beamed from proximate stages and can include any of the ordinary elemental species found naturally in interstellar and intergalactic space, but can be further enriched with heavier elements. Alternatively, heavy metallic elements can be beamed as fully fabricated sub-components, pelletized beams, or charged and/or neutral particle beams.

Once fully mature, sprout stages can produce a secondary series of seeds which can grow by any suitable mechanisms described above. Like-wise, tertiary, quaternary, pentary, and so-on series of seeds and spouts may germinate and grow into fully functioning ISR stages.

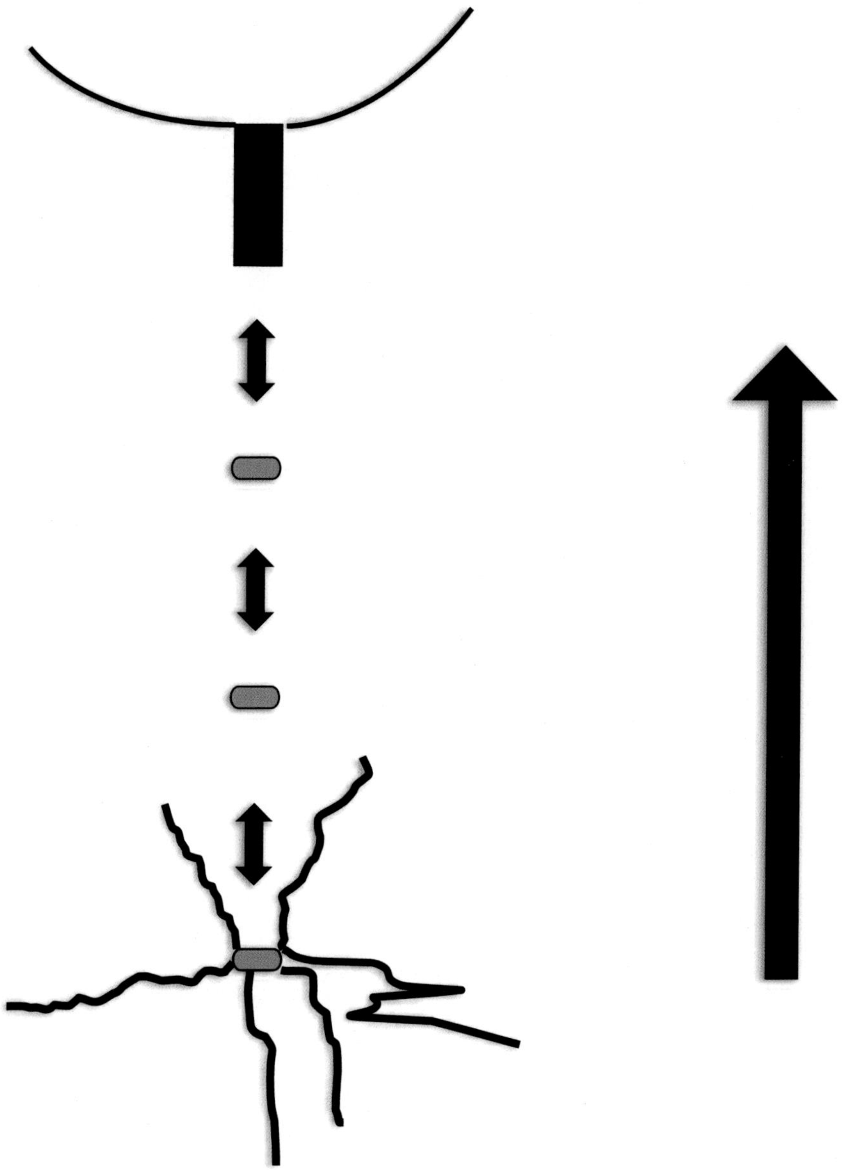

**FIGURE 19** depicts propulsive seeds (ovals) and growing seed (spider shaped lower portion) and interstellar ramjet (upper portion of diagram).

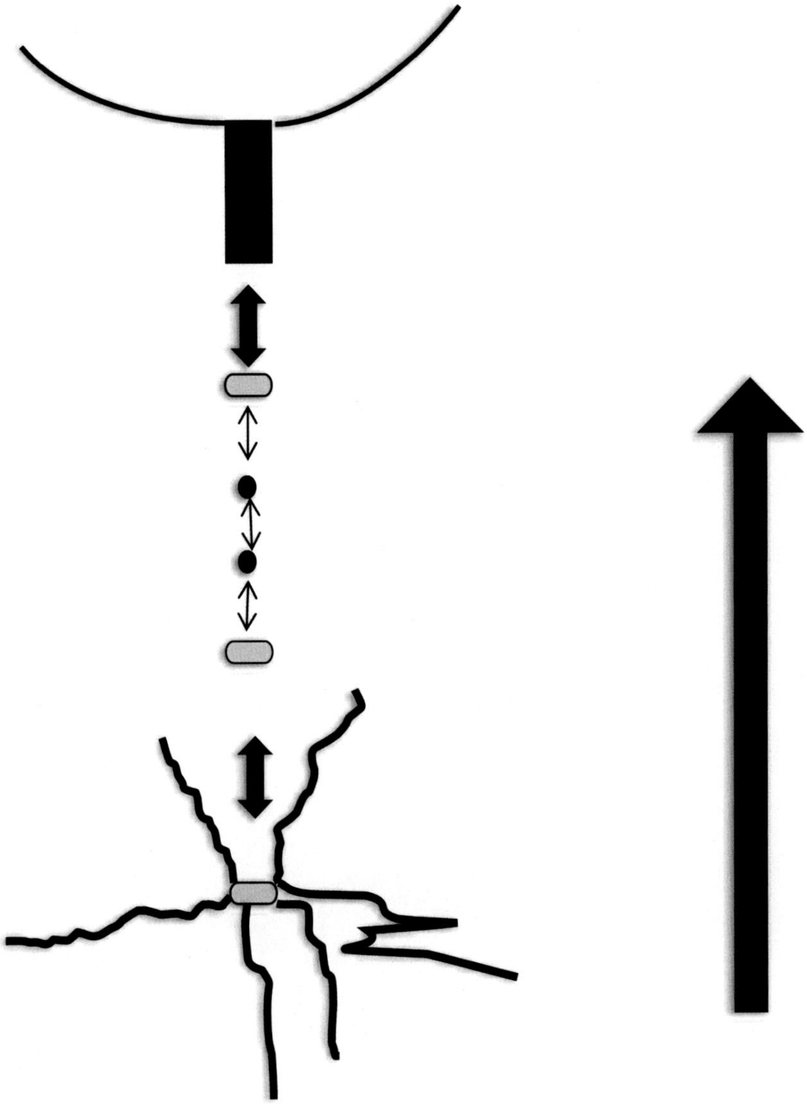

**FIGURE 20** depicts propulsive seeds (gray ovals), growing seed (spider shaped lower portion), and secondary embryos (ellipses), and interstellar ramjet (upper black portion).

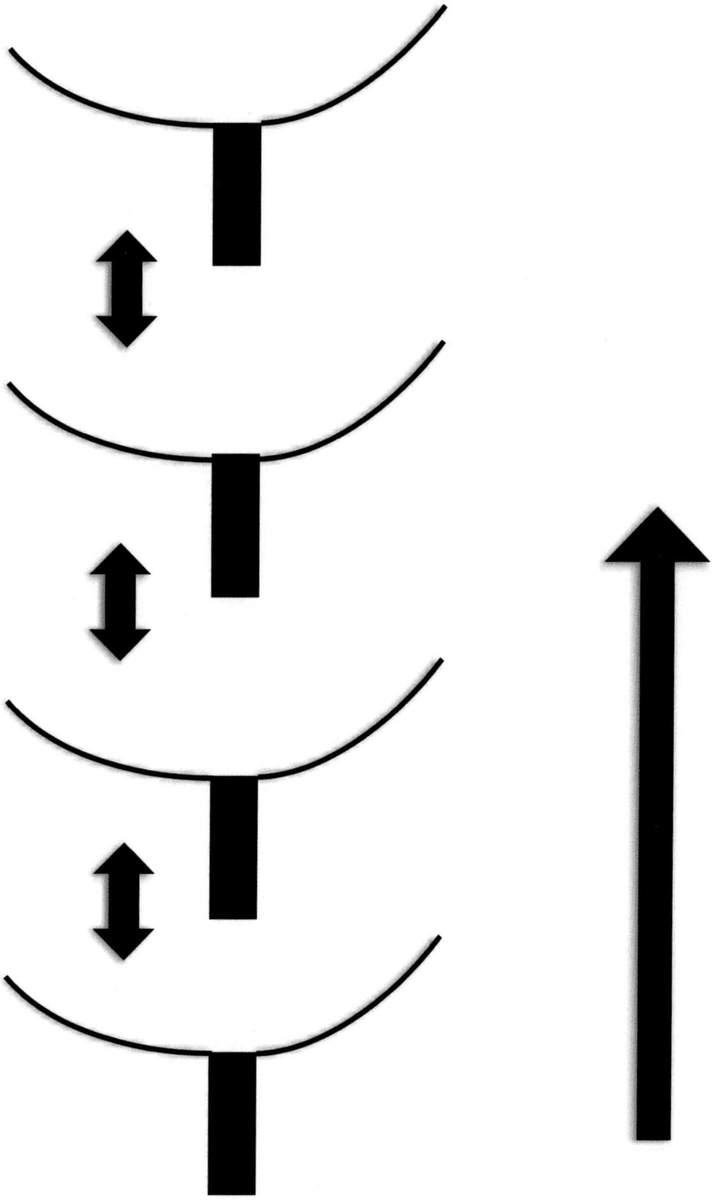

**FIGURE 21** depicts series of  propulsive interstellar ramjets.

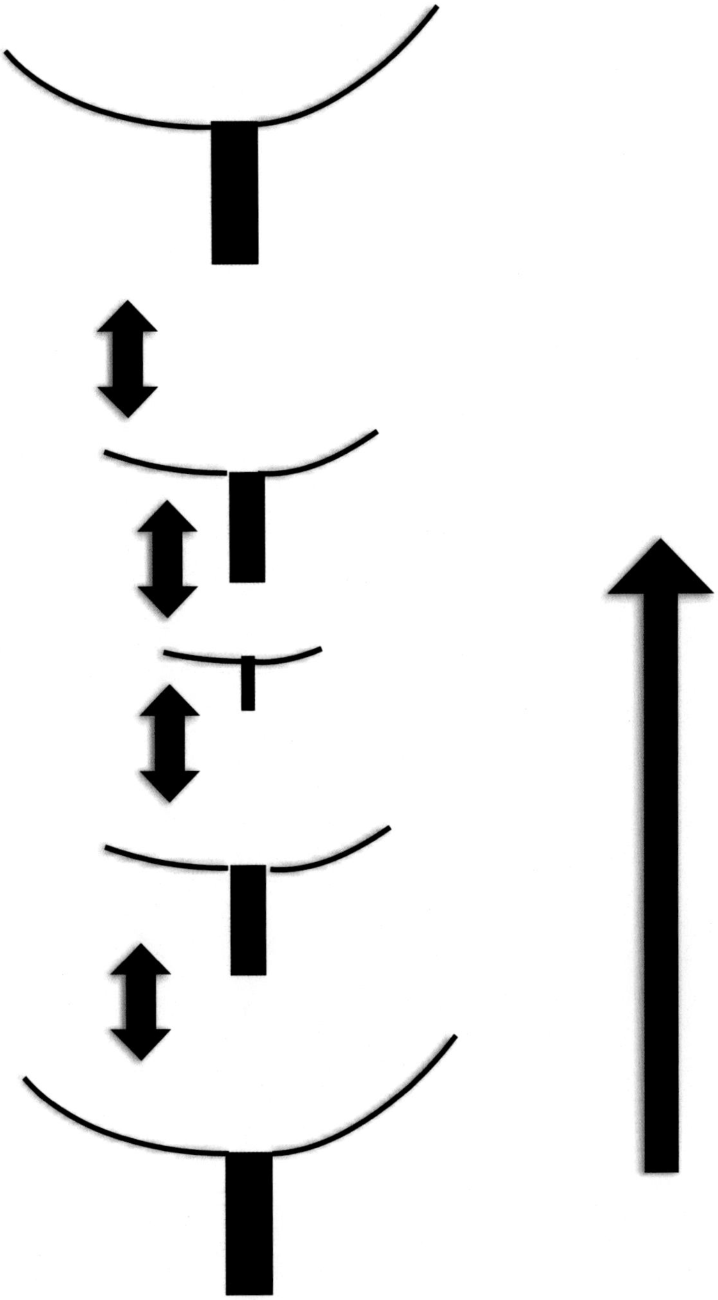

**FIGURE 22** depicts propulsive primary, secondary, and trinary interstellar ramjets.

Regardless of how the above pseudo-causality chain could be maintained, the concept perhaps has relevance to quantum mechanical matter and energy teleportation as well as information teleportation and signaling. Such a mechanism might conceivably be used for superluminal travel and/or communication.

Superluminal methods would at the very least involve an initial conveyance of entangled matter or energy from one stage to another. This would require that entangled states could be cloned or transferred to any seeds deployed between the primary stages. Cloning entangled states is an anathema per the No Cloning Theorem.

So, nuclear fusion fuel pellet streams acting as relativistic spacecraft can in principle enable what the author refers to as sub-canonical ensemble Lorentz factors.

Relativistic craft and pellets can include any and all of the following: 1) Fusion Rockets, 2) Fission Rockets, 3) Fission Fragment Drives, 4) Fusion powered ion, electron, photon, and/or neutrino rockets, 5) Fission powered ion, electron, photon, and/or neutrino rockets, 6) matter antimatter rockets that carry both components of fuel on board from the start of the mission, 7) matter antimatter rockets that carry only their antimatter fuel component(s) along from the start of the mission, 8] matter antimatter reactor powered ion, electron, photon, and/or neutrino rockets, 9) fusion fuel pellet linear runway powered craft, 10) fission fuel pellet linear runway powered craft, 11) fission-fusion fuel pellet linear runway powered craft, 12) nuclear isomer fuel pellet linear runway powered craft, 13) matter antimatter fuel pellet linear runway powered craft, 14) antimatter fuel pellet linear runway powered craft, 15) fusion fuel pellet circulinear runway powered craft, 16) fission fuel pellet circulinear runway powered craft, 17) fission-fusion fuel pellet circulinear runway powered craft, 18) nuclear isomer fuel pellet circulinear runway powered craft, 19) matter antimatter fuel pellet circulinear runway powered craft, 20) antimatter fuel pellet circulinear runway powered craft, 21) nuclear fission powered electro-hydrodynamic-plasma drive craft, 22) nuclear fusion powered electro-hydrodynamic-plasma drive craft, 23) matter antimatter reaction powered electro-hydrodynamic-plasma drive craft, 24) nuclear fission powered magneto-hydrodynamic-plasma drive craft, 25) nuclear fusion powered magneto-hydrodynamic-plasma drive craft, 26) matter antimatter reaction powered electro-hydrodynamic-plasma drive craft, 27) nuclear fission powered electro-magneto-hydrodynamic-plasma drive craft, 28) nuclear fusion powered electro-magneto-hydrodynamic-plasma drive craft, 29) matter-antimatter powered electro-magneto-hydrodynamic-plasma drive craft, 30) fusion powered magnetic field effect drive, 31) fission powered magnetic field effect drive, 32) matter antimatter reaction powered field effect drive, 33) single pass solar dive and fry sail driven craft, 34) single pass stellar dive and fry sail driven craft, 35) single pass quasar dive and fry sail driven craft, 36) multi-pass stellar cycler solar dive and fry sail driven craft, 37) multi-pass stellar cycler dive and fry sail driven craft, 38) multi-pass cycler quasar dive and fry sail driven craft, 38) laser beam driven relativistic sail craft, 40) microwave beam driven relativistic sail craft, 41) radio-frequency beam driven relativistic sail craft, 42) massive neutral particle beam driven sail craft, 43) massive charged particle beam driven sail craft, 44) massive particle beam fission fuel powered craft, 45) massive particle beam fusion fuel powered craft, 46) massive particle beam matter antimatter beam fuel powered craft, 47) antimatter beam fuel powered craft, 48) nuclear bomb pulse driven propulsion of the original Project Orion forms, 49) pure nuclear fusion bomb pulse driven propulsion analogous to the original Project Orion forms, 50) matter antimatter bomb driven propulsion analogous to the original Project Orion forms, 51) one side reflective cosmic microwave background radiation sails, 52) multiple beam bounce propulsion methods, 53) any improved interstellar ramjet craft, 54) fusion rocket or fusion powered electron, ion, photon, or neutrino rockets utilizing single body or serially multiple body powered gravitational assists, 55) fission rocket or fission powered electron, ion, photon, or neutrino rockets utilizing single body or serially multiple body powered gravitational assists, 56) matter antimatter rocket or matter antimatter reaction electron, ion, photon, or neutrino rockets utilizing single body or serially multiple body powered gravitational assists and others. If the propulsion system is multi-modal, then even if only one stage is used for each mode, the number of combinations and thus the number of possible propulsion systems is at least equal to (2 EXP 56) – 1 = 7.205 x 10 EXP 16 = 72.05 Quadrillion = 72,050 trillion!

In each of these 56 categories, several sub-methods have been proposed and so the number of possible multi-mode/multi-stage propulsion systems is many, many orders of magnitude greater yet. If say, each category permits 10 sub-categories, which I can reasonably assure you is likely a conservative estimate, then the total number of possible propulsion systems, all else being the same is equal to about (2 EXP 560) - 1 which is approximately equal to 10 EXP 160. This is roughly 10 EXP 70 times the number of atoms, electrons, photons, and neutrinos in the observable universe. Most of these methods permit very high Lorentz factors, in many cases, virtually unlimited relativistic Lorentz factors given the virtually if not actually unlimited future time periods available to sequester ever greater resources of fuel via an ever expanding human space based resource collection infrastructure.

All of the above craft propulsion mechanisms may be facilitated by initial boosts to high Kelperian to mildly relativistic velocities by the fuel pellet mechanisms proposed herein and/or as may become known in the future.

Interesting closing consideration include concepts for which a mass driver or gas gun system would be assembled over a period roughly equal to the age of our universe but for which the atomic distances of the gun's barrel composition would not increase over the same period. At current expansion rates, we would expect that over another current universal age limit, the space at the far end of the gun would become twice our current cosmic light-cone radius units distant from the Milky Way Galaxy. However, before that would even happen, it is conceivable that a projectile could be loaded into the gun at the far end on a projection back toward our galaxy. The projectile even after traveling a mere few light-years or so, might end up traveling faster than light with respect to the space-time that was proximate to the far end of the barrel for which said space-time would have been displaced by universal expansion. The same result would apply to free background low velocity matter such as background gas, dust, and plasma. A big caveat is that the barrel of the gun would be required to have near to zero astrodynamic coefficient of drag. The barrel otherwise would snap under tensile stress. Additionally, the barrel would need to be sufficiently refractory so as to withstand the relativistically blue-shifted and aberrated background light-sources.

Another facilitating concept is the development and distribution of several to numerous guns as useful for step-wise acceleration of the pellets. These guns may themselves be accelerated to relativistic velocities so that incident fuel pellets can achieve longer acceleration drive times and path-lengths with respect to the gun and the background.

Another fascinating scenario includes nano-technology molecular assembly and larger unit scale macro-assembly of pellet runways that would grow by self-assembly to the scale of a future cosmic light-cone radius in our universe. Runways might even grow to arbitrary multiples of cosmic light-cone radii bounded by infinite numbers of cosmic light-cone radii in future temporal infinity. Sensitivity to initial conditions and other chaotic manifestations such as strange att4ractors might be of use in perturbing interstellar and intergalactic hydrogen and/or other less common elements in the background to efficiently clump together to enhance the gun construction efficiency.

High velocity assembly pods can be fired or otherwise transported to distant locations to begin the construction of portions of the gun in remote locations.

# Radiation Protection At High Lorentz Factors

The following excerpt in quotations is from an article that was fairly recently posted at *New Scientist* (Jamieson, 2010**)**. I must vehemently disagree with the conclusions made therein. I present my arguments after the excerpt below.

*"Star Trek fans, prepare to be disappointed. Kirk, Spock and the rest of the crew would die within a second of the USS Enterprise approaching the speed of light.*

*The problem lies with Einstein's special theory of relativity. It transforms the thin wisp of hydrogen gas that permeates interstellar space into an intense radiation beam that would kill humans within seconds and destroy the spacecraft's electronic instruments.*

*Interstellar space is an empty place. For every cubic centimeter, there are fewer than two hydrogen atoms, on average, compared with 30 billion billion atoms of air here on Earth. But according to William Edelstein of the Johns Hopkins University School of Medicine in Baltimore, Maryland, that sparse interstellar gas should worry the crew of a spaceship travelling close to the speed of light even more than Romulans decloaking off the starboard bow."*

Nonetheless, Edelstein has provided much food for thought by discussing the radiation hazards of high gamma factor space travel. His analysis refutes the need to consider very significant shielding at high gamma factors.

Perhaps magnetic fields of strengths on the order of 100 tesla to 10,000 tesla can be set up around the craft by exotic superconducting electromagnets. Alternatively, perhaps the magnetic fields could be produced by some exotic form of permanent magnet made of atomistic materials which are not of the periodic table types.

If a strong enough magnetic field can be deployed, then the interstellar hydrogen atoms will take on a dipole moment and can be directed away from the spacecraft or into a safe repository where the atoms can then be used as hydrogen fusion fuel to assist spacecraft acceleration. Charged matter could be diverted by a magscoop or an electric field scoop type of mechanism.

It is possible that the ship could be cloaked in a matter wave cloak which would be the matter wave analogue of a negative electromagnetic index of refraction, meta-material cloak. However, note that negative EM refraction index metamaterials are only operative on wavelengths of EM radiation that are on the order of or greater in size than the active cell size of the materials. Each cell consists of a conductive inductor coil-like loop embedded in a dielectric base material.

Now, EM negative refraction index materials have the peculiar property such that they are pulled forward by the impinging EM radiation. Consequently, such materials might make an excellent pull sail that would be gainful amidst the incoming CMBR and star light instead of being pushed by the light. Note that the full scope of classical electromagnetic theory ramifications of negative EM refraction index materials has yet to be worked-out. Thus, the subject of classical electrodynamics has received a whole new territorial field of unknowns and potentialities with the development of negative refractive index materials.

See the following link for some good info on negative EM refraction index metamaterials (Padilla, Dimitri, Smith, 2006).:

http://people.ee.duke.edu/~drsmith/pubs_smith_group/padilla_materials_today_2006.pdf

Perhaps, the spacecraft shielding could be composed of a several kilometers to many kilometers thick frontal shield that would extend radially beyond the foot print of the crew cabin and hull of the spacecraft. This would enable extreme gamma factors for a long train-like spacecraft.

Alternatively, the shield might be composed of some yet to be devised super dense neutron crystal-like material(s) or perhaps some sort of quarkonium matter that is similar to or differing from theoretical stable strange matter. Such a material might make an excellent shield. A $10^{-15}$ meter thick sheet of neutron dense crystal materials would have a mass specific area of only 1 metric ton per square meter. A $10^{-12}$ meter thick sheet would have a mass of only 1,000 metric tons per square meter. Using even 1,000 square meters of such materials to shield the front of the spacecraft seems highly doable if such materials can be manufactured.

If a ten centimeter thick aluminum shield can attenuate the radiation by one percent, then a 10 meter thick shield can attenuate the radiation to a fraction of $(0.99)^{100} = 0.366$ of its impinging intensity; and a one kilometer thick aluminum shield can attenuate the radiation to a fraction of $[(0.99)^{10,000}] = 2.249 \times 10^{-44}$ of its impinging intensity, thereby making the radiation intensity problem a completely moot point.

The shield might alternatively take the form of a huge thin walled cylinder, perhaps constructed out of carbon nanotube materials, or boron nitride nanotube materials. The cylinder could contain an atmosphere that remains compressed as the spacecraft accelerates, much as the atmosphere on Earth, or even as the hot atmosphere of the planet Venus, remains compressed. The atmosphere, optionally in conjunction with a mini-magnetosphere could provide shielding for the craft from impinging hydrogen atoms, helium atoms, ions, and electrons.

Perhaps a large superconducting electromagnet or highly conducting conventional material-based electromagnet can be deployed to produce a significant magnetic field in order to divert the incident plasma. Forward aimed micro-lasers or nano-lasers could ionize the interstellar and intergalactic neutral atoms thereby rendering them divertible with respect to the spacecraft crew quarters and electronic control systems.

Thermal energy quasi-particle or phonon conductors have been proposed as theoretical mechanisms to greatly enhance heat conduction over traditional thermally conductive materials. Imagine if the energy impinging on the front of the spacecraft could be conducted very rapidly away from the front of the shield and exhausted out the back of the spacecraft such as by a photon rocket, electron rocket, ion rocket, or even a neutrino rocket whereupon the frontal shield would have a temperature maintained below the impinging radiation temperature (perhaps even below the ambient CMBR temperature). It is then plausible that the craft should be pulled forward by the radiation imbalance in an effectively perpetual manner so long as the process remains operative. Phonon conductors can in theory conduct heat from cold to hot material portions.

Perhaps certain neutronium crystalline materials might be stable from neutron decay outside of the environment of the extreme pressures and gravity fields of neutron stars.

Ideally, such neutron dense crystalline material, or stable strange material, somehow stable bottomonium, charmomium, or even Higgsinium could be fashioned into matter wave cloaks, or even matter wave and electromagnetic wave pull sails or pulling surfaces which can operate on extremely blue-shifted incident matter waves and EM waves. Note that matter wave negative refraction index meta-materials do not yet exist and are only a theoretical construct at present.

One potential way to deal with hard cosmic rays relative to a ship and her shield using ordinary periodic table atomic and molecular matter-based materials is to include a layered shield that is thick enough to cause all incoming particles to undergo collisions with the shield's atoms. The massive jets of decay products would then be diverted by electric and/or magnetic fields set up in vacancies between the shield's layers. Neutral particles that were produced, but which passed through an initial layer would be subject to transformation into decay jets that would include charged species which could then be diverted by other layers. The process could repeat itself until all of the energy was absorbed before it could irradiate the crew members or sensitive electronic equipment. Nanotech self-assembly repair mechanisms or other microbots could continually refashion the shielding components including field generation apparatus to compensate for radiation damage.

A good bulk shield material might be pure diamond. Improved forms of diamond that have higher heat conductivity and better refractory properties then the best natural diamonds have been produced within a laboratory setting. Such improved diamond consists of altered ratios of carbon isotopes.

We should not give up hope in the potential joys of extreme gamma factor interstellar manned space travel along with all of the mystique that such has to offer and as may have been overlooked or disregarded in traditional Special Relativity.

Relativistic velocity dust and gas shields for spacecraft might take the form of a cloud of smoke particles distributed in front of the spacecraft. The hardness, latent heat of vaporization, latent heat of fusion, latent heat of ionization, and density of the particles are important factors to consider in deploying a smoke shield. Particles of more than one elemental, isotopic, and molecular species can be included in the smoke shield. The mass, density, size distribution, and species of particles can be optimized for each range of gamma factors so as to most effectively block or divert the kinetic energy of interstellar dust grains. This is so because dust speck collisions producing differing explosive temperatures produce plasma and neutral particle jets with different characteristics and so the mass, density, and size spectrum of the smoke cloud particles can influence the attenuation and divergence of jets produced by the dust speck collisions.

At high gamma factors from about 100 to 300, the production of charmed, strange, and bottom, mesons and baryons becomes possible and any of the longer mean life time particles as such can travel macroscopic distances before colliding with other smoke particles' atoms and atomic nuclei. Some species of atomic nuclei are better at disrupting certain species of mesons and baryons, and heavy charged leptons, than others by transformative interactions.

Once we obtain a gamma factor above 100, the pressure of relativistic Doppler shifted cosmic microwave background radiation can easily dislodge the smoke cloud and so an electrodynamic mechanism may become necessary to retain the smoke cloud. The solution to this problem might be as simple as employing the Meissner effect to hold the smoke particles in place assuming a smoke cloud comprised of superconducting particles. Perhaps some sort of magnetic field generator can be used to control the distribution of the smoke cloud for cases where the smoke particles are magnetic, or magnetizable. The magnetic containment might take the form of a pulsed magnetic field that pulls the particles back into position when they start to drift apart, but not to the extent where the particles would all be pulled to one location within a magnetic field such as at one of its magnetic poles. The use of electric fields to hold the particles in stable configuration might be possible if the smoke particles could be electrically charged.

The Meissner effect could be easily employed to hold superconducting pebbles affixed for the case where the pebbles are large portions of superconducting materials. Such a Meissner shield could contain several layers of permanent magnets that are distributed in a plate-like manner ahead of the spacecraft. The superconducting pebbles would be distributed between the magnetic sheets or magnetic plates where the foremost magnetic sheet would hold a distribution of pebbles at the very front of the shielding mechanism in order to act as an initial collision buffer. The atomic, isotopic, and chemical composition of the pebbles could be judiciously chosen in order to disrupt the incoming dust particles.

Another option involves deploying a large balloon or series of balloons ahead of the spacecraft that are filled with gas so as to ablate or ionize the incoming dust particles. The balloons might employ nanotech self-assembly repair mechanisms or some other kind of self-healing fabric to repair the small holes produced by dust particles that punch through the balloons. Alternatively, macroscale microbots could continually scan the balloons for holes and make repairs with macroscale patches as necessary. The atomic and isotopic composition of the internal gas can be chosen so that the disruption of the kinetic energy jets produced by the colliding dust is maximized.

Note that the use of very thin balloons filled with very low pressure helium gas in space as nuclear warhead decoys has been suggested. If the balloons were thin enough, and contained a low enough pressure helium gas, the balloons might survive 100 meter proximity to the detonation of a one kiloton ABM nuclear warhead in space. According to the reasoning, the balloon's metallic fabric would be so thin that it would radiate away the heat generated by the blast so quickly that it would not get hot enough to melt or vaporize. The gamma rays and neutrons generated by the blast would deposit only a very small amount of its energy within the balloon membrane and its internal, rarified gas. The point is that such dust speck disrupting balloons might survive high but modest gamma factors and the resulting cosmic rays if a self-repair feature could be employed. Also note that a one kiloton nuclear device is generally considered to be a low yield tactical type of nuclear device. A human person or other similarly sized object such as any helium filled balloon caught out in the open and located within about 150 meters of such an exploding device would be immediately vaporized upon the devices detonation.

Alternatively, a series of membranous sheets or plates may be distributed ahead of the craft, where the first sheets in the series would act to ionize the dust specks. Judicious spacing between the sheets can be employed to permit divergence of the plasma and neutral particle jets produced by the collisions. If necessary, any of the previously described smoke distributions could be used to further disrupt the kinetic energy jets where the smoke particles would be instilled between the plates. The further option of including magnetic pebble distributions between the plates may also be of use here. Regardless, an electric and/or magnetic field distribution between the plates can be employed to cause plasma jets produced by particle collisions to experience greater diversion based on: 1) the invariant mass to charge ratio; 2) the momentum to charge ratio; or 3) the kinetic energy to charge ratio, spectral distribution of the plasma particles produced by dust impact collisions. The divergence effect would be analogous to that of a mass spectrometer.

Then there is the possibility of using lasing stations or microwave beams directed in front of the spacecraft in order to ionize or vaporize dust particles while they are still safely away from the spacecraft. For cases where the spacecraft is traveling in a very straight line, the beaming process is simple. For cases where the ship experiences high degrees of angular acceleration, the beam angle has to be adjusted so that the dust particles are vaporized or ionized while they are safely away from the ship. Charged particle beams may also be of help in this regard. For highly relativistic spacecraft, microwave and IR beams can be used because the frequency of the radiation relative to the dust specks located in the path of the spacecraft will be blue-shifted to from the soft U-V to hard U-V/soft x-ray frequencies.

Either way, I will happily settle for the 0.7 C of the ISV Venture Star bound for Pandora (Refer to the movie "Avatar", produced by James Cameron). They might even persuade me to undergo suspended animation, or hibernation, which are other ways of dealing with long voyage transit times. Personally, however, I'd rather be awake during the voyage.

Another possible remedy for the dust situation might involve deploying some sort of extreme magnetic field around the craft that would induce a dipole moment within the atoms of the dust particles and thus cause the dust to be pushed out of the way of the spacecraft. The magnetic field might be emitted from a long spindly electromagnet or perhaps from some form of extreme yet to be developed permanent magnet. The intensity of the magnetic field would be greatly amplified with respect to the dust particles due to a special relativistic effect, and so perhaps a 200 to 2,000 tesla field might work at extreme gamma factors.

Consider the scenario where a ship-based magnetic field could pull the particles around to the back end of the ship. Perhaps the particles could be made to collide with objects or a mass distribution that are at rest with respect to the ship's frame thereby producing large amounts of heat and BB thermal radiation which could be reflected and thus used as a photon rocket exhaust.

I do not know how to produce permanent magnets with field intensities so high. However, we can speculate that perhaps some form of yet to be developed materials would work such as exotic forms of somehow stabilized neutronium or quarkonium. Quarkonium might take the form of strange matter, charmed matter, or perhaps bottomonium. Another option might be Higgsinium, or monopolium.

Graphene may in theory be used to create 200 tesla electromagnets because of its extreme strength and conductivity. The most powerful continuously running electromagnets produce a maximum field intensity of from about 45 tesla to 48 tesla. These magnets are comprised of a liquid cooled hybrid construction that utilizes a combination of very good conventional conducting and superconducting materials within a system cables that are highly reinforced.

The energy stored within a magnetic field is proportional to the square of its intensity for a given differential volumetric element occupied by the field. The field of an electromagnet exerts a torque on the magnet's windings which is proportional to the square of the maximum magnetic field intensity. Thus, fields with the best cables we have currently cannot exceed about 50 Tesla. Perhaps liquid cooled carbon nanotube based cables or graphene based cables might be used to produce electromagnets that operate at 150 to 200 Tesla.

I throw these ideas out for anyone to bite on them and try to work with them. My gut feeling is that the dust problem can be remedied for extreme gamma factor craft. However, I would gladly be a crew member on a 0.1 C starship bound for Alpha Centauri if such a mission were to be launched this decade. I don't see this happening quite yet, though.

Regardless of whether electrodynamic plasma jet diffusive applications are included in the design of dust shields, the interspacing between layered dust and gas shields as forward deployed shields may optionally be filled at least in part by aerogel types of materials. Another optional diffusement mechanism can include membranous walled bubble types of foam. A good very low cos option would be the commonly used bubble wrap often used in mechanical insulation of parcels delivered by postal mail services or commercial parcel delivery services. Still yet another form of ultra-low mass diffusements may include aerated metallic materials such as foam core types of materials. A similar material was very recently reported to have a density of 0.9 $mg/cm^3$ and is an aerated form of a common nickel phosphorus alloy.

The above ultra-low mass diffusements may optionally be augmented by electric or magnetic field generators or emitters instilled within or in proximity to the bulk diffusement material(s). The electrodynamic fields can be used to filter and spread the plasma jets produced by impacting dust grains and highly energetic cosmic rays. Such filtering includes diffusement based on the electric charge of relativistic mass ratio, the electric charge to kinetic energy ratio, and the electric charge to momentum ratio, spectral distribution of the plasma debris created by the impacting dust particles and cosmic rays.

The next few pages include simple illustrations of various shielding mechanisms for relativistic spacecraft.

A

B

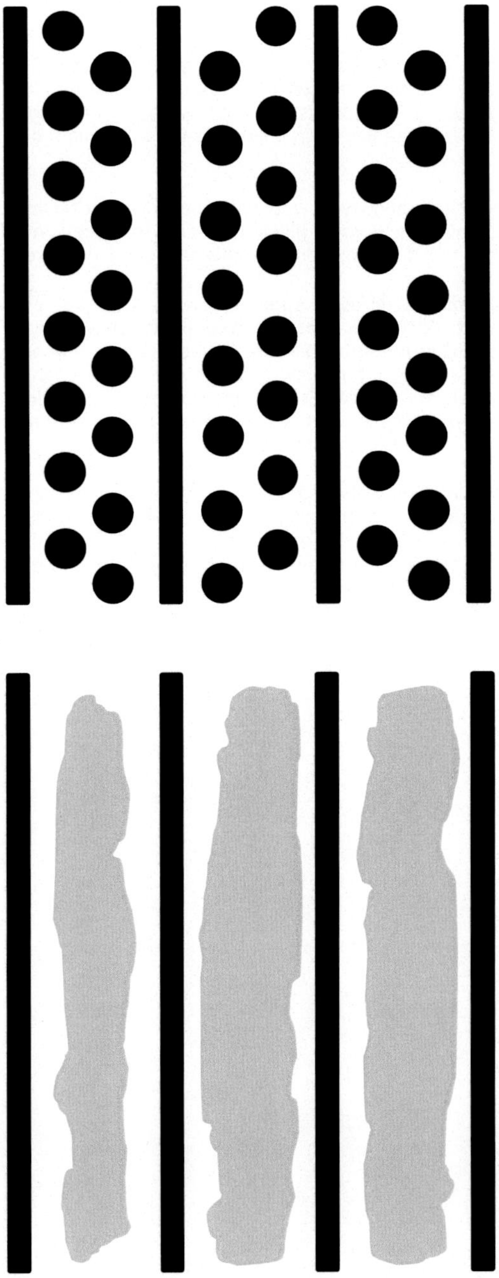

**FIGURES 23 A, B**: FIGURE A depicts a layered dust shield with a diffusive pebble based disrupter. FIGURE B depicts a layered smoke cloud based disrupter. In both figures, the vertical bars represent disrupter material(s) confinement walls.

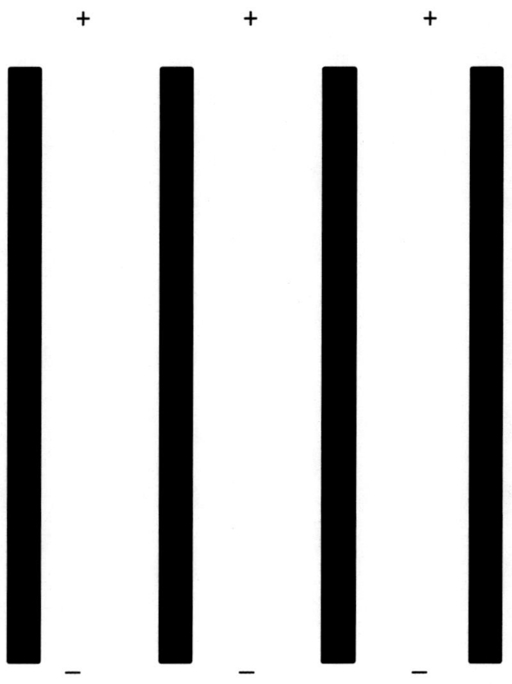


A

B

**FIGURES 24 A, B:** depict multi-layered shields. The + and – signs located outward and between the shield walls or the vertical bars, denote electrodynamic charges. The charges can be electric or magnetic, or serial combinations thereof, and of arbitrary serial patterns.

A

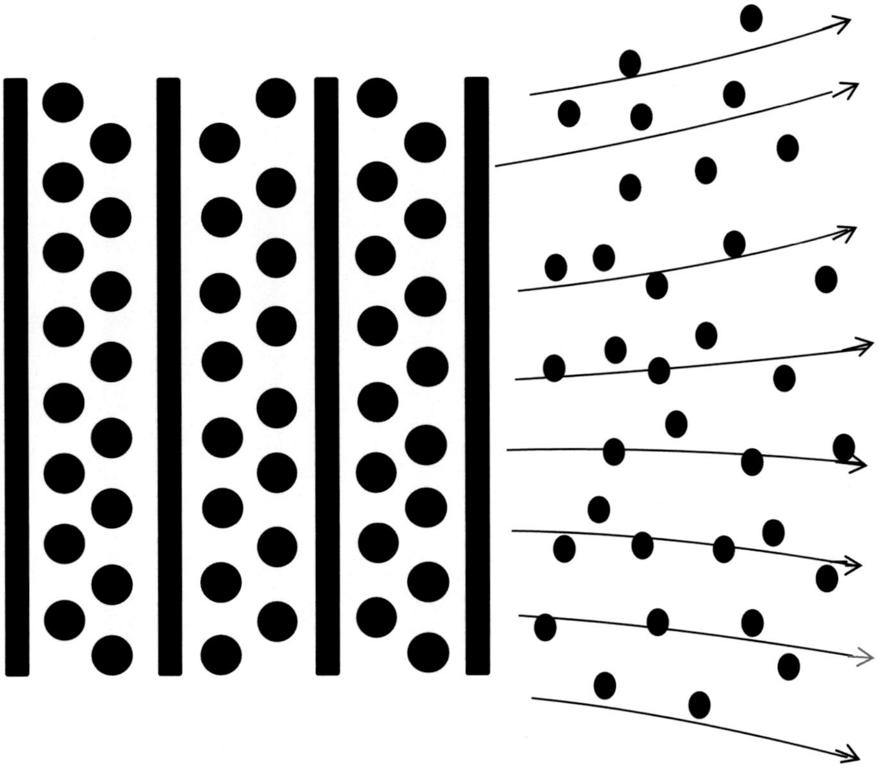

B

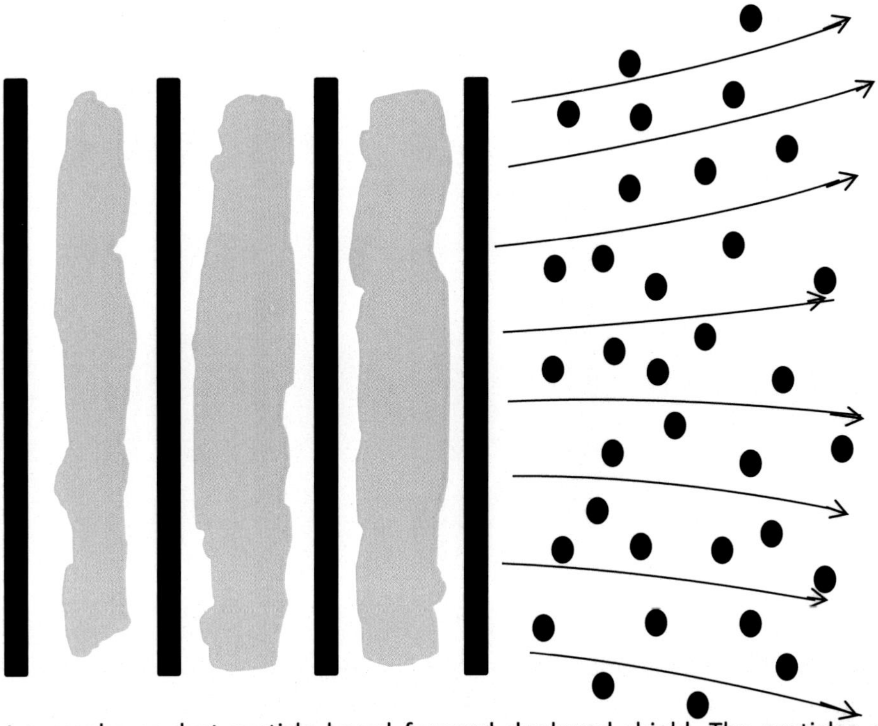

**FIGURES 25 A, B**: depict a smoke or dust particle based forward deployed shield. The particles are assumed to be comprised of superconducting materials and held in place by the magnetic field (curved arrows) through the Meissner Effect.

# Basic Shielding II: Ordinary Atoms-Based Space Shields And Drag Reduction Mechanisms For High Gamma Factors

Tantalum hafnium carbide, $Ta_4HfC_5$, has the highest melting point of all known compounds, 4,488 K (4,215 °C, 7,619 °F) and is almost as hard as diamond. The material would make an excellent blue-shifted; CMBR, star light, and quasar light, forward deployed shield for an interstellar ramjet.

Such a starship shield would plausibly retain its integrity for the case where the spacecraft would attain a gamma factor of 4,488 K/2.725 K = 1,647 or a velocity equal to 99.99999815 percent of the speed of light. Such velocities would facilitate manned missions to any location within the Milky Way Galaxy in under one contemporary healthy human productive life-time, ship's frame.

Provided that the shield could be actively cooled, the shield might endure higher gamma factors perhaps enabled by phonon diodic conducting materials. These are hypothetical materials that would conduct heat based acoustic or thermal vibrations very efficiently, perhaps even from cold to hot portions of a material in a non-zero net heat flow.

For a perfectly reflective surface that is normal to the impinging CMBR, the light pressure will be equal to $2\sigma T^4/C$. Here, $\sigma = 5.6704 \times 10^{-8}\ W\ m^{-2}\ K^{-4}$. For a one square kilometer shield that is normal to the spacecraft velocity vector, the force will be 153,600 Newtons or about 15 tons.

The following estimations are only first order approximations. So for rigorous mil-spec computations of heat loading and friction, relativistic reference frame rotation and distortion would be required as well as a precise analysis of the varying spectrum of the background non-CMBR photon sources such as star light, quasar light and the like. A rigorous accounting of such sources cannot be made at present because the frequency distribution of the various non-CMBR photon sources is spatial-temporal region specific. Also, the velocity of the starship with respect to a given spatial-temporal, non-CMBR, photon distribution would be required because of spacecraft velocity special relativistic aberration effects.

A square conical shaped forward deployed shield having an aspect ratio of 100 would reduce the drag by 254.65 fold or by a factor of $[\sin(90°/400)]^{-1}$. Such a shield would reduce the force acting against a square kilometer cross-section craft traveling at a gamma factor of 1,647 to only about 61.39 kilograms. The force would be reduced to only 6.139 kilograms for a 100,000 square meter cross-section craft. The force would be reduced to only 0.6139 kilograms for a 10,000 square meter cross-section craft. The force would be reduced to only 61.39 grams for a 1,000 square meter cross-section craft. The force would be reduced to only 6.139 grams for a 100 square meter craft. Finally, for a 10 square meter urban bus-like cross-section, the force would be reduced to only a paltry 0.6139 grams.

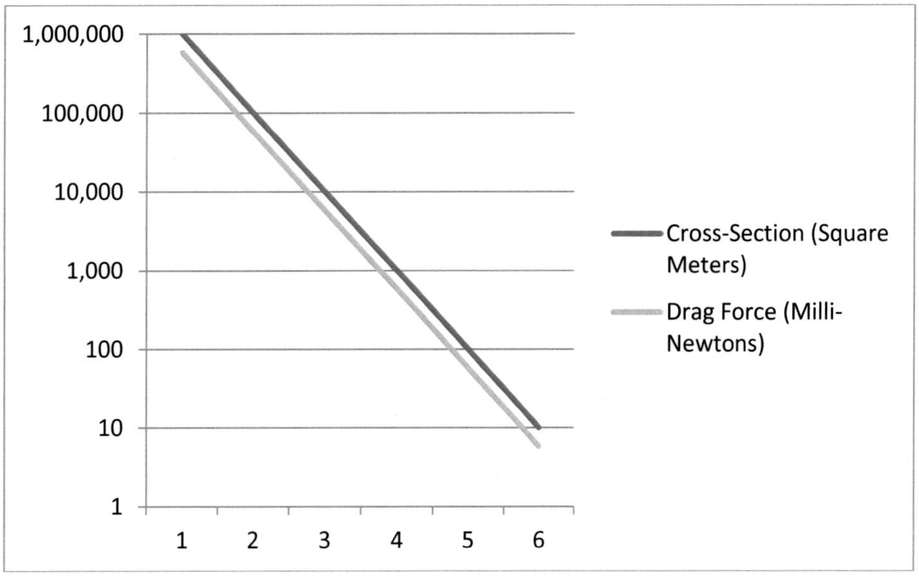


**GRAPH 1** is a graph of cross-sectional shield area and drag force for a conical shield having an aspect ratio of 100 for a spacecraft having a gamma factor of 1,647.

The drag can be good news for slowing the craft down. For example, a 1,000 square kilometer reflective drag chute deployed at γ = 1,647 would result in about 15,000 metric tons of drag force. Such a force would slow a space having an invariant mass of 15,000 metric tons traveling at a gamma factor of 1,647 with 1 G negative acceleration. Once the spacecraft gamma factor dropped to 164.7, a 10,000,000 square kilometer shield could be deployed for 1 G acceleration. Larger sails could be deployed for 1 G deceleration at lower gamma factors. Once the craft slowed to the point where CMBR sails would no longer be effective, then matter-antimatter rockets could perform final craft slow down.

The length of the cone would be 100 kilometers for the above very smooth space train fuselage type of craft having a 1,000,000 square meter cross-section. The length would be 31.6228 kilometers for a 100,000 square meter cross-section. The length of the cone would be 10 kilometers for a 10,000 square meter cross-section. The length would be 3.16228 kilometers for a 1,000 square meter cross-section. The length would be one kilometer for a 100 square meter cross-section. Finally, for a 10 square meter urban bus-like cross-section, the length would be only 0.316228 kilometers.

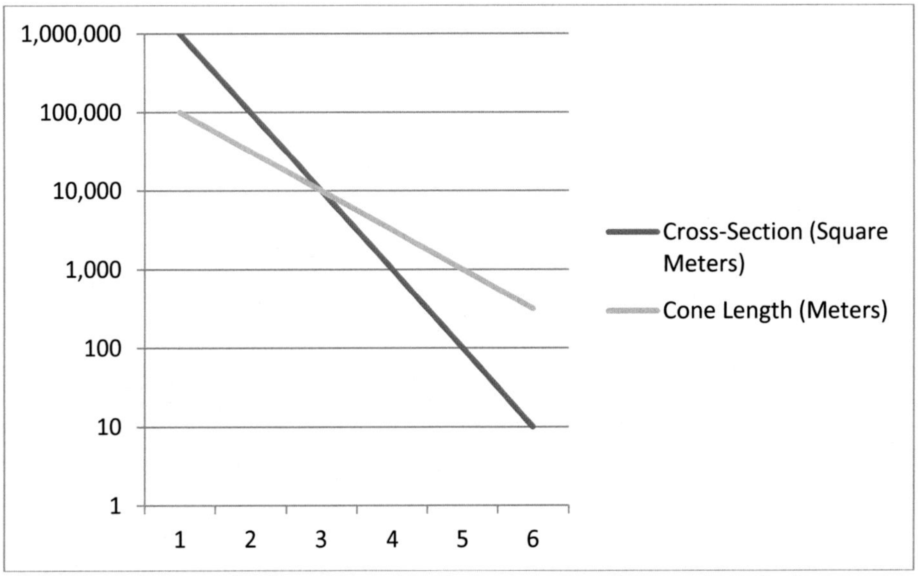


**GRAPH 2** is a graph of cross-sectional cone area and cone length for a cone having an aspect ratio of 100.

For the latter three examples, accelerations of roughly from one to 1,000 Earth G's should be possible given the bulk modulus of $Ta_4HfC_5$ at hundreds of time that of granite.

Hard rock permits Mount Everest to remain stable against compressive fatigue, and so it would seem that a material with well over 100 times the compressive strength of granite should enable conical structures having a length of a few hundred times Mount Everest's height to remain stable under 1 G loading. This would permit conical shield lengths of roughly (10 km)(300) = 3,000 km to withstand 1 G acceleration, 300 km to withstand 10 G acceleration, 30 km to withstand 100 G acceleration. Note that Mt Everest is the result of two colliding tectonic plates and the actual high of mountains is limited by the strength of a planet's gravitational field. For Earth, Mt Everest is nearly as tall as a mountain can grow.

A square conical shaped forward deployed shield having an aspect ratio of 10,000 would reduce the drag by 25,465 fold or by a factor of $[\text{Sin}\ (90^{\circ}/40{,}000)]^{-1}$. Such a shield would reduce the force acting against a square kilometer cross-section craft traveling at a gamma factor of 1,647 to only about 0.6139 kilograms. The force would be reduced to only 0.06139 kilograms for a 100,000 square meter cross-section craft. The force would be reduced to only 0.006139 kilograms for a 10,000 square meter cross-section craft. The force would be reduced to only 0.6139 grams for a 1,000 square meter cross-section craft. The force would be reduced to only 0.06139 grams for a 100 square meter craft. Finally, for a 10 square meter craft, the force would be reduced to only a paltry 0.006139 grams.

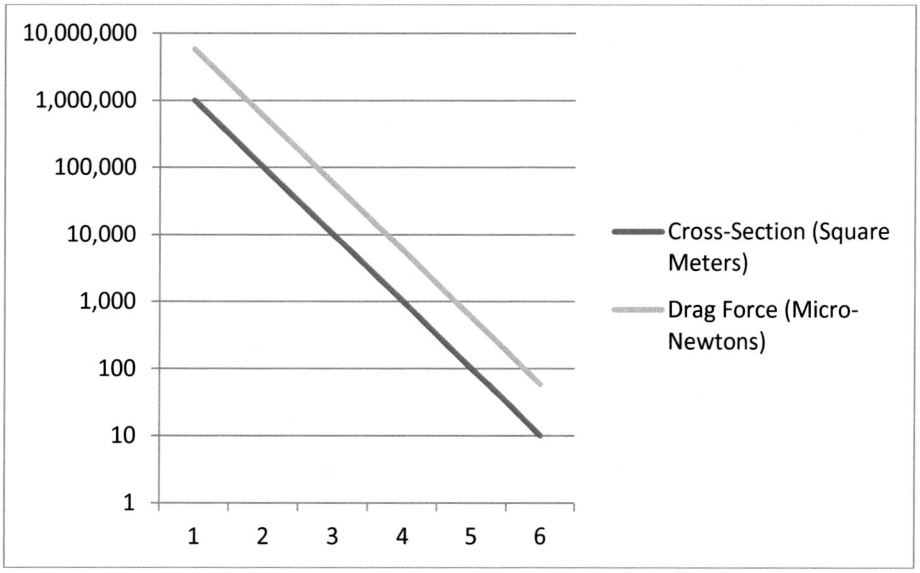


**GRAPH 3** is a graph of cross-sectional shield area and drag force for a conical shield having an aspect ratio of 10,000 for a spacecraft having a gamma factor of 1,647.

The drag can be good news for slowing the craft down. For example, a 1,000 square kilometer reflective drag chute deployed at γ = 1,647 would result in about 15,000 metric tons of drag force. Such a force would slow a space having an invariant mass of 15,000 metric tons traveling at a gamma factor of 1,647 with 1 G negative acceleration. Once the spacecraft gamma factor dropped to 164.7, a 10,000,000 square kilometer shield could be deployed for 1 G acceleration. Larger sails could be deployed for 1 G deceleration at lower gamma factors. Once the craft slowed to the point where CMBR sails would no longer be effective, then matter-antimatter rockets could perform final craft slow down.

The length of the cone would be 10,000 kilometers for the above very smooth space train fuselage type of craft having a 1,000,000 square meter cross-section. The length would be 3,162.28 kilometers for a 100,000 square meter cross-section. The length of the cone would be 1,000 kilometers for a 10,000 square meter cross-section. The length would be 316.228 kilometers for a 1,000 square meter cross-section. The length would be 100 kilometers for a 100 square meter cross-section. Finally, for a 10 square cross-section, the length would be only 31.6228 kilometers.

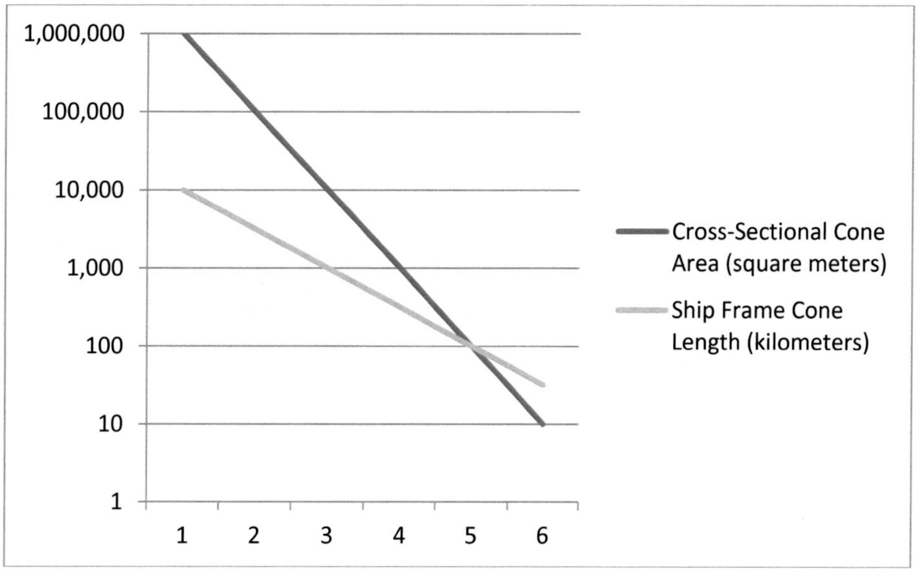


**GRAPH 4** is a graph of cross-sectional cone area and cone length for a cone having an aspect ratio of 10,000.

There is one huge caveat here and that is finding a way for the $Ta_4HfC_5$ shield to maintain its thermal-mechanical stability. A solution to this problem may include the instillation of a super high reflectivity coating on the cone, such that even as the cone is slightly ablated over time by the incident but extremely rarefied interstellar gas and dust, such ablation would act to polish the cone so that it can be all the more reflective, or at the very least, retain its original reflectivity within a safety margin.

Nanotechnology based self-assembly or micro-robotic based mechanisms might repair the shield either from the outside or perhaps from the inside, either within the conical inner vacancy or within the cone's material composition itself.

Can we develop ordinary periodic table element based materials that are even more refractory than $Ta_4HfC_5$ and which have a bulk modulus of perhaps 1 or 2 orders of magnitude greater than diamond, thus enabling even longer cones that operate under higher radiant flux loading? Diamond seems to be the hardest known material; however, calculations indicate that so-called carbon-nitride should be harder than diamond. How much harder remains to be seen as only microscopic portions of Beta carbon nitride have so far been produced. The latter samples are so small that a consistent measurement of its hardness has remained impossible to date.

Can we go harder than Beta carbon nitride? This is anyone's guess, but I will leave my positive feelings regarding this possibility to be experimentally refuted by future nanotechnologists, materials scientists, and physical chemists. After that, we will need stabilized white dwarf density matter, followed by neutronium which is made of pure atomic nucleus density neutronic matter, followed by a host of progressively denser quarkoniums starting with strange matter, then charmed matter, then bottom matter, perhaps followed by somehow producible top matter, then by Higgsinium assuming Higgsinos exist, and then monopolium. Monopolium is so dense in theory, that a cubic centimeter of the stuff would have 1/5 the mass of the Earth. Note that a one centimeter wide spherical lump of matter having the mass of the Earth would take the form of a black hole.

All of the above nuclear density and super-nuclear density forms of matter are still very theoretical (except for neutronium which neutron stars are made of), and even though I have hopes that they will be produced at some future time, I might have to wait millennia or perhaps even eons to witness such matter being produced in useful industrial bulk quantities. However, when we finally do produce such materials, we may be the envy of many of any starfaring extraterrestrial and ultraterrestrial races.

One mechanism for holding up an ordinary periodic table element based cone might include electrical charging of the cone so that it is extendable in a self-repulsive matter. An analogous magnetization of the cone might also work perhaps with the optional inclusion of the electrical charging mechanism.

Note that the heat of friction generated within the conical shield might be wicked away by diodic heat conducting materials, and converted to electrical energy via thermoelectric cells of unheard of efficiency, or perhaps stream driven turbo-electric systems. The turboelectric generators might optionally include multi-cycle stream systems with regenerative heating. Exotic photovoltaic systems might plausibly convert some of the light incident on the cone to electrical energy. As mentioned previously, very hot and large cooling radiators can be used for cooling fluids in order to enable spacecraft reactor power densities commensurate with 1 or more Gs of acceleration.

The virtues of using a square conical shield instead of a traditional circular cross-sectioned conical shield are such that the square cone has flat edges that are precisely parallel to the spacecraft's velocity vector. Thus, the effective impinging CMBR temperature with respect to these parallel sides permits no energy absorption from these surfaces thereby permitting heat generated within the obliquely angled surfaces to be exhausted by recessed propulsion mechanisms installed along the parallel sides or surfaces. However, the shields may also have a circular cross-section plan form, in which case the computed values given above would differ only slightly.

Now assume that the terminal gamma factor of the craft after final fuel burn-through is equal to 16,470. The temperature of the relativistic Doppler blue-shifted CMBR will be a whopping (2.725)(16,470) = 44,881 Kelvin. This is an increase in temperature of 10 fold over the 4,488 Kelvin which would melt the $Ta_4HfC_5$. We will clearly need to have a reflectivity for the rocket cone shield of at least 10,000 meaning that only 0.01 percent of the impinging light on the reflector is absorbed, all the rest skipping or reflecting harmlessly off of the cone.

Once again, for a perfectly reflective surface that is normal to the impinging CMBR, the light pressure will be $2\sigma T^4/C$. Here, $\sigma = 5.6704 \times 10^{-8}\ W\ m^{-2}\ K^{-4}$. The force will be 1,533,800,000 Newtons or about 150,000 tons for a one square kilometer shield that is normal to the spacecraft velocity vector. The force will be 153,380,000 Newtons or about 15,000 tons for a 0.1 square kilometer shield that is normal to the spacecraft velocity vector. The force will be 15,338,000 Newtons or about 1,500 tons for a 0.01 square kilometer shield that is normal to the spacecraft velocity vector. The force will be 1,533,800 Newtons or about 150 tons for a 0.001 square kilometer shield that is normal to the spacecraft velocity vector. The force will be 153,380 Newtons or about 15 tons for a 0.0001 square kilometer shield that is normal to the spacecraft velocity vector. Finally, for an area that is 0.00001 square kilometer, the drag will be a mere 1.5 metric tons.

Once again, the drag is good news for deceleration. For example, a 1,000 square kilometer reflective drag chute deployed at a gamma factor of 16,470 would produce 150,000,000 metric tons of drag force. The drag would slow a spacecraft having an invariant mass of 150,000,000 metric tons traveling at a gamma factor of 16,470 with 1 G negative acceleration. Once the spacecraft gamma factor dropped to 1,647, a 10,000,000 square kilometer chute could be deployed for 1 G negative acceleration. Larger chutes could be deployed for 1 G deceleration at lower gamma factors. Once the craft slowed to the point where CMBR chutes would no longer be effective, matter-antimatter rockets could bring the craft to rest. For a 150,000 metric ton spacecraft, a 1 square kilometer sail would induce 1 G of negative acceleration at a gamma factor of 16,470.

Including a square conical shape in the forward deployed shield having an aspect ratio of 100 would reduce the drag by 254.65 fold or by a factor of [Sin (90°/400)]$^{-1}$. Such a shield would reduce the force acting against a square kilometer cross-section craft traveling at a gamma factor of 16,470 to only about 613,900 kilograms force. The force would be reduced to only 61,390 kilograms metric tons force for a 100,000 square meter cross-section craft. The force would be reduced to only 6,139 kilograms force for a 10,000 square meter cross-section craft. The force would be reduced to only 613.9 kilograms force for a 1,000 square meter cross-section craft. The force would be reduced to only 61.39 kilograms force for a 100 square meter craft. Finally, for a 10 square meter area, the force would be reduced to a paltry 6.139 kilograms force.

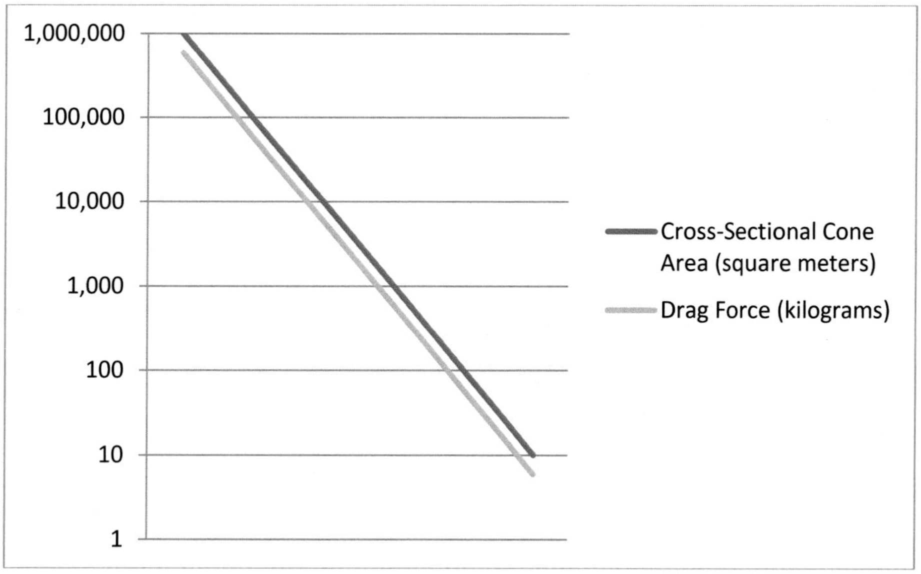


**GRAPH 5** is a graph of cross-sectional shield area and drag force for a conical shield having an aspect ratio of 100 for a spacecraft having a gamma factor of 16,470.

Once again, extreme shield aspect ratios such as for the first two examples in the previous paragraph start to become hard to bear for a 10 G to 100 G acceleration. Therefore, we would need to use shielding comprised of materials with a higher bulk modulus than pure diamond. One such extreme material might be Beta carbon nitride, especially perfect atomic lattice crystal forms of Beta carbon nitride. Such a material might be fabricated by nano-technology self-assembly or other nano-assembly processes. Perhaps a bulk modulus and hardness 10 times greater than that of diamond might be obtained thus permitting such spacecraft to be manufactured out of ordinary period table elements.

Once again, a diodic phonon conductor material could be used to wick away the heat generated by the absorbed CMBR thus reducing the shield temperature below the melting point of $Ta_4HfC_5$.

Note that any heat of friction generated within the conical shield that is wicked away by diodic heat conducting materials, might be converted to electrical energy by thermoelectric cells of unheard of efficiency, or perhaps by extreme power output, steam driven, turbo-electric systems.

The turboelectric generators might optionally include multi-cycle stream systems and regenerative heating. Exotic photovoltaic systems might convert some of the light incident on the cone to electrical energy.

Alternatively, multi-cycle liquid metal cooling might be used to cool the shield. Some such thermal working fluids may be liquid sodium or other high heat conductivity molten metals. The liquid sodium could then be cooled by a counter-flow water based steam mechanism, followed by a methanol cycle, and finally terminating with a nitrogen, $CO_2$, helium, or other cryogenic heat sink. The heat would have to go somewhere and perhaps would be used to drive turbo-electric generators, thermo-electric generators, or other mechanisms, which would provide electrical power for electrodynamic propulsion systems. Such propulsion systems may include photon, electron, proton, ion, pion, muon, and the like rockets. Thus, the auxiliary power sources could complement any primary matter-antimatter rocket propulsion system.

With appropriately elongated forward deployed shields, the impinging CMBR, star light, and quasar light would mostly skip or glance off the highly reflective conical shield. However, for cases of extreme heat, perhaps a carbon like residue could be continuously bled to the exterior of the shield. The heat of ionization and phase change resulting in the transition of the carbon from a liquid to a plasma state would cool the shield by ionic plasma cooling or by the plasma analogue of evaporative cooling.

Including a square conical shape in the forward deployed shield having an aspect ratio of 10,000 would reduce the drag by 25,465 fold or by a factor of [Sin (90°/40,000)]$^{-1}$. Such a shield would reduce the force acting against a square kilometer cross-section craft traveling at a gamma factor of 16,470 to only about 6,139 kilograms force. The force would be reduced to only 613.9 kilograms metric tons force for a 100,000 square meter cross-section craft. The force would be reduced to only 61.39 kilograms force for a 10,000 square meter cross-section craft. The force would be reduced to only 6.139 kilograms force for a 1,000 square meter cross-section craft. The force would be reduced to only 0.6139 kilograms force for a 100 square meter craft. And finally, for a 10 square meter area, the force would be reduced to a paltry 0.06139 kilograms force.

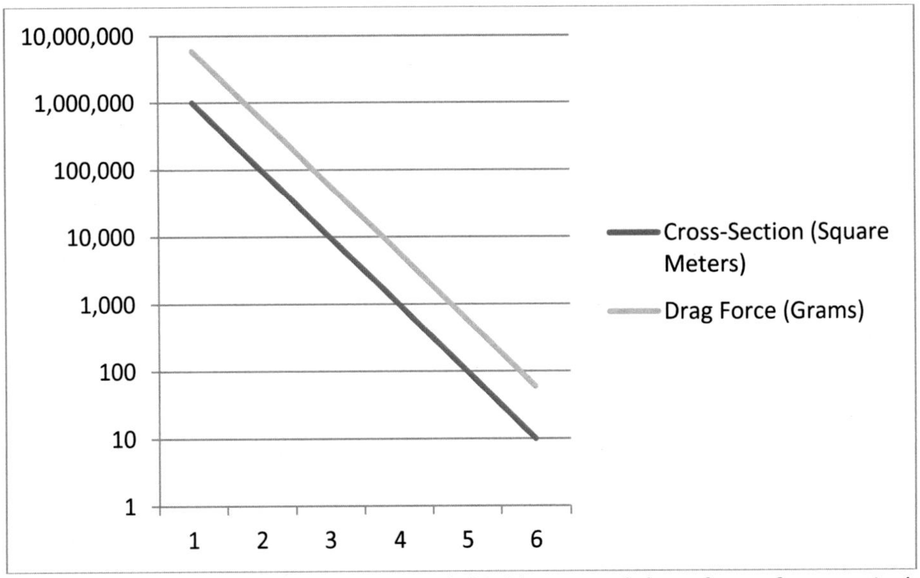


**GRAPH 6** is a graph of cross-sectional shield area and drag force for a conical shield having an aspect ratio of 10,000 for a spacecraft having a gamma factor of 16,470.

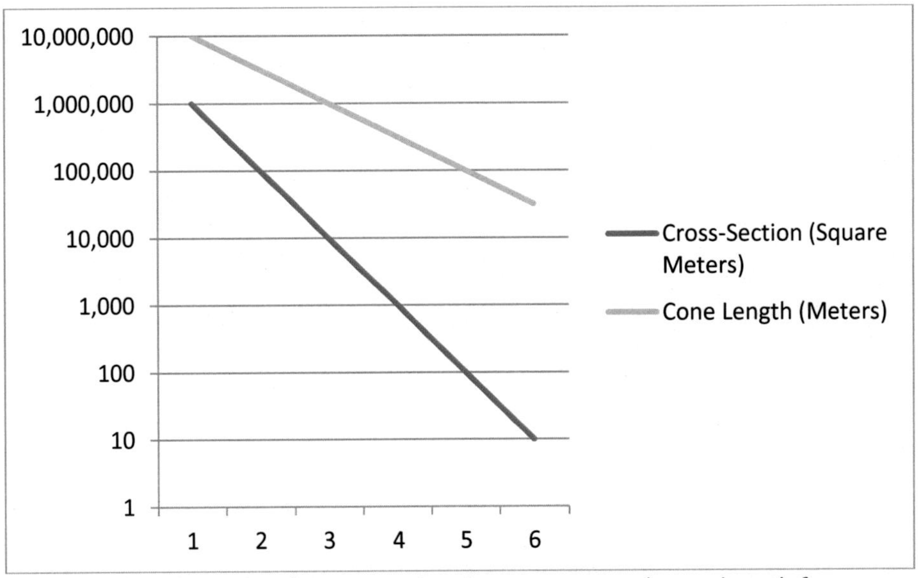


**GRAPH 7** is a graph of cross-sectional cone area and cone length for a cone having an aspect ratio of 10,000.

Including a square conical shape in the forward deployed shield having an aspect ratio of 1,000,000 would reduce the drag by 2,546,500 fold or by a factor of [Sin (90°/4,000,000)]$^{-1}$. Such a shield would reduce the force acting against a square kilometer cross-section craft traveling at a gamma factor of 16,470 to only about 61.39 kilograms force. The force would be reduced to only 6.139 kilograms metric tons force for a 100,000 square meter cross-section craft. The force would be reduced to only 0.6139 kilograms force for a 10,000 square meter cross-section craft. The force would be reduced to only 0.06139 kilograms force for a 1,000 square meter cross-section craft. The force would be reduced to only 0.00589 kilograms force for a 100 square meter craft. Finally, for a 10 square meter area, the force would be reduced to a paltry 0.006139 kilograms force.

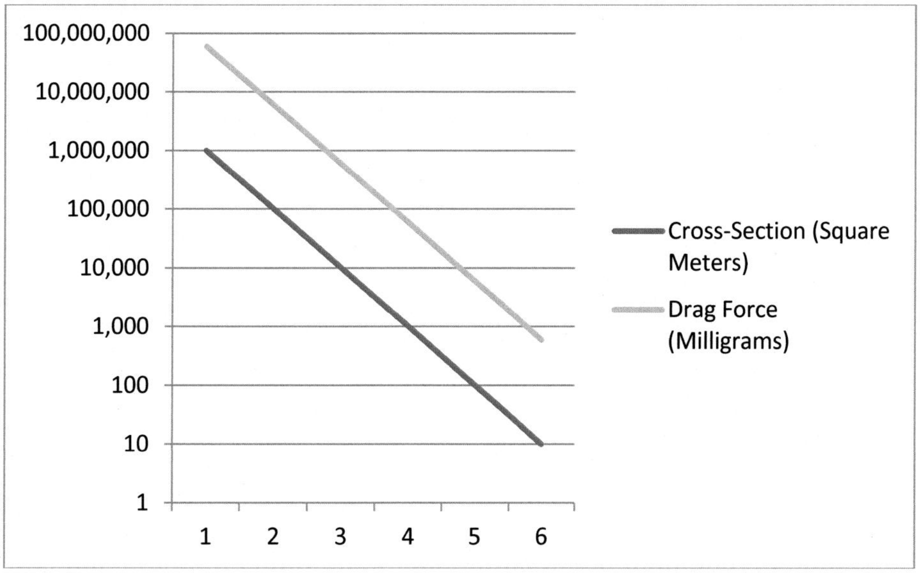


**GRAPH 8** is a graph of cross-sectional shield area and drag force for a conical shield having an aspect ratio of 1,000,000 for a spacecraft having a gamma factor of 16,470.

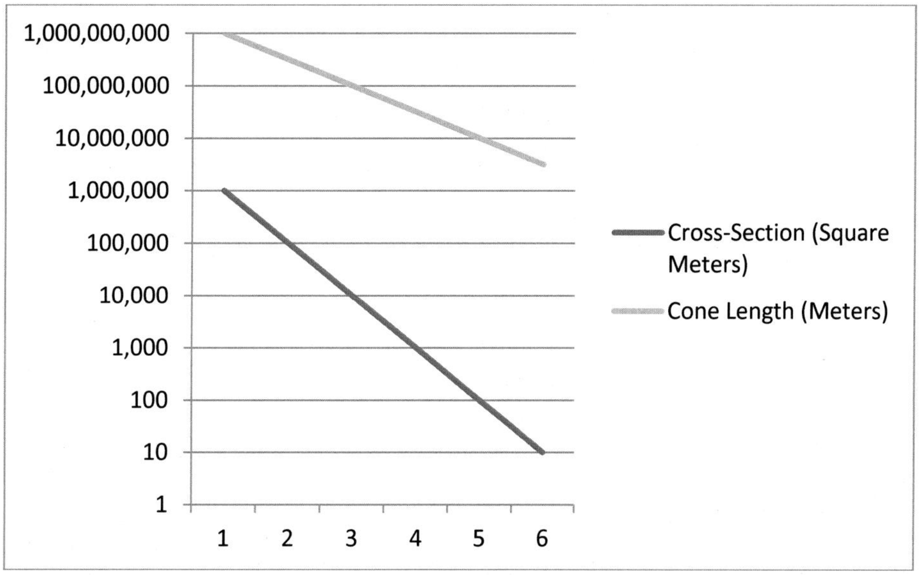


**GRAPH 9** is a graph of cross-sectional cone area and cone length for a cone having an aspect ratio of 1,000,000.

We might even increase the gamma factor to 52,083 to attain a velocity equal to about 0.9999999998 C or 99.99999998 percent of the speed of light in a vacuum. However, the temperature of the Doppler blue-shifted CMBR will be a whopping (2.725)(52,083) = 141,925 Kelvin. This is an increase in temperature of $10^{1.5}$ fold over the 4,488 Kelvin that would melt the $Ta_4HfC_5$.

A flat bow spacecraft would be impractical due to extreme astrodynamic drag forces. However, if accelerations were kept to no more than one G, but preferably to only a modest fraction of a G, forward deployed conical spacecraft shields made of plausible, ordinary periodic table atoms, based materials might do the trick. The gamma factor of the craft is so high that merely relaxing its propulsion systems or retracting its conical shield would result in rapid deceleration of a craft having a mass on the order of 100,000 metric tons. When the craft slowed to gamma factors in the 10,000 to 100 range, progressively larger drag chutes could be deployed. To slow to a stationary state, the craft might deploy reverse matter-antimatter rocket thrust, or reverse interstellar ramjet thrust when the light pressure impinging on the sail is reduced to trivial levels.

A square conical shaped forward deployed shield having an at rest aspect ratio of 100 would reduce the drag by 254.65 fold or by a factor of $[\mathrm{Sin}\ (90°/400)]^{-1}$. Such a shield would reduce the force acting on a 1.0 square kilometer cross-section craft traveling at a gamma factor of 52,083 to only about 61,390,000 kilograms force. The force would be reduced to only 6,139,000 kilograms force for a 0.1 square kilometer cross-section craft. The force would be reduced to only 613,900 kilograms force for a 0.01 square kilometer cross-section craft. The force would be reduced to only 61,390 kilograms force for a 0.001 square kilometer cross-section craft. The force would be reduced to only 6,139 kilograms force for a 0.0001 square kilometer cross-section craft. The force would be reduced to only 613.9 kilograms force for a 0.00001 square kilometer cross-section craft.

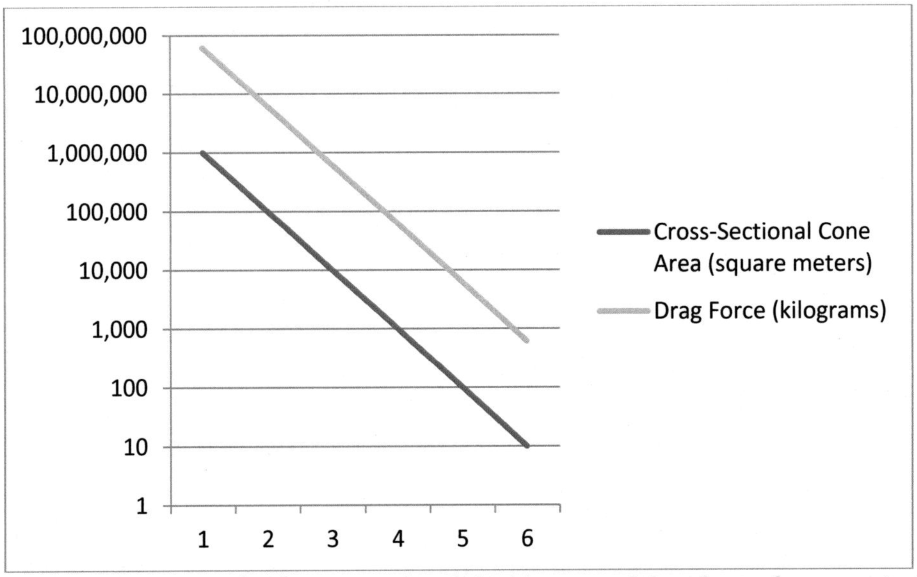


**GRAPH 10** is a graph of cross-sectional shield area and drag force for a conical shield having an at rest aspect ratio of 100 for a spacecraft having a gamma factor of 52,083.

The length of the square cone would be 100 kilometers for the above very smooth space train fuselage type of craft having a 1,000,000 square meter cross-section. The length would be 31.6228 kilometers for a 100,000 square meter cross-section. The length of the cone would be 10 kilometers for the above very smooth space train fuselage type of craft having a 10,000 square meter cross-section. The length would be 3.16228 kilometers for a 1,000 square meter cross-section. The length would be one kilometer for a 100 square meter cross-section. And finally, for a 10 square meter cross-section, the length would be only 0.316228 kilometers.

A square conical shaped forward deployed shield having an at rest aspect ratio of 10,000 would reduce the drag by 25,465 fold or by a factor of $[\mathrm{Sin}\ (90°/40{,}000)]^{-1}$. Such a shield would reduce the force acting on a 1.0 square kilometer cross-section craft traveling at a gamma factor of 52,083 to only about 613,960 kilograms force. The force would be reduced to only 61,396 kilograms force for a 0.1 square kilometer cross-section craft. The force would be reduced to only 6,139.60 kilograms force for a 0.01 square kilometer cross-section craft. The force would be reduced to only 613.96 kilograms force for a 0.001 square kilometer cross-section craft. The force would be reduced to only 61.396 kilograms force for a 0.0001 square kilometer cross-section craft. The force would be reduced to only 6.1396 kilograms force for a 0.00001 square kilometer cross-section craft.

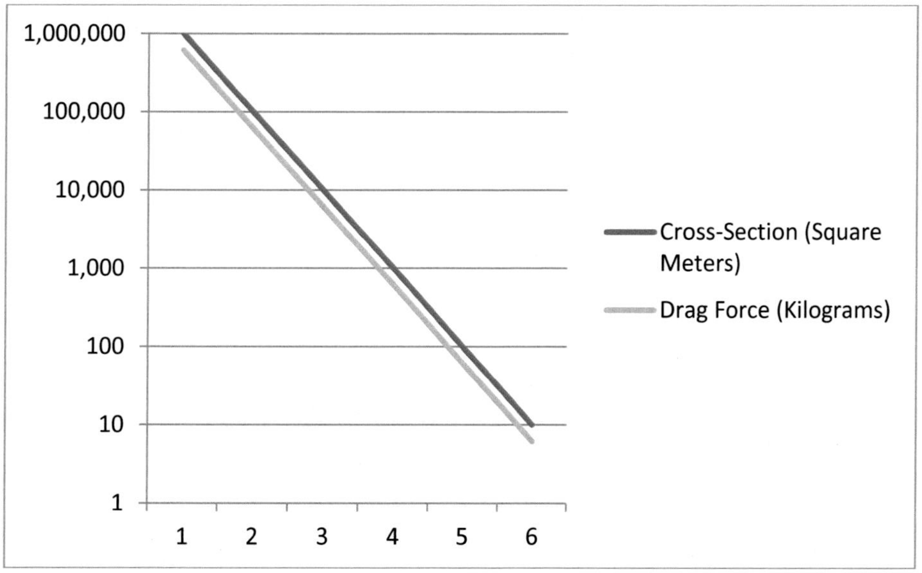


**GRAPH 11** is a graph of cross-sectional shield area and drag force for a conical shield having an at rest aspect ratio of 10,000 for a spacecraft having a gamma factor of 52,083.

A square conical shaped forward deployed shield having an at rest aspect ratio of 1,000,000 would reduce the drag by 2,546,500 fold or by a factor of $[\mathrm{Sin}\,(90^{\circ}/4{,}000{,}000)]^{-1}$. Such a shield would reduce the force acting on a 1.0 square kilometer cross-section craft traveling at a gamma factor of 52,083 to only about 6,139 kilograms force. The force would be reduced to only 613.9 kilograms force for a 0.1 square kilometer cross-section craft. The force would be reduced to only 61.39 kilograms force for a 0.01 square kilometer cross-section craft. The force would be reduced to only 6.139 kilograms force for a 0.001 square kilometer cross-section craft. The force would be reduced to only 0.6139 kilograms force for a 0.0001 square kilometer cross-section craft. The force would be reduced to only 0.06139 kilograms force for a 0.00001 square kilometer cross-section craft.

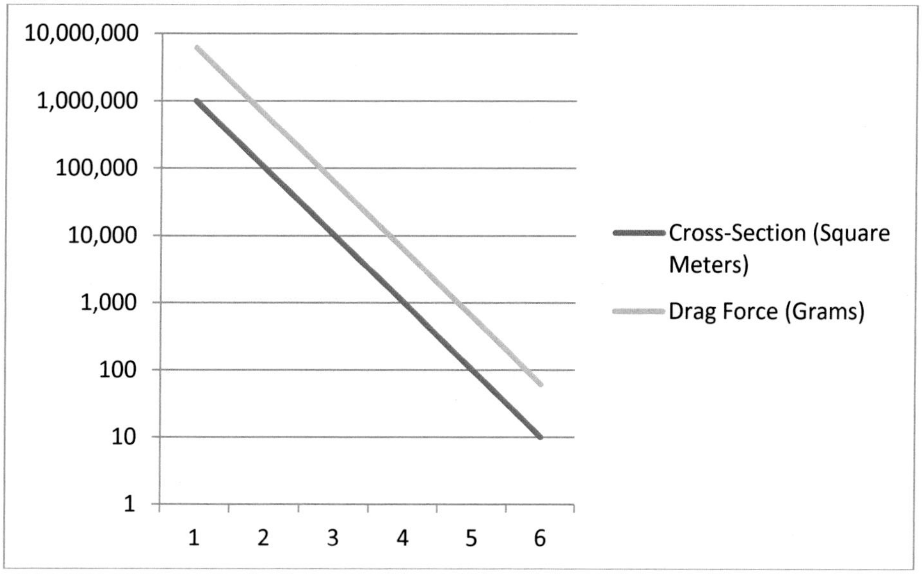


**GRAPH 12** is a graph of cross-sectional shield area and drag force for a conical shield having an at rest aspect ratio of 1,000,000 for a spacecraft having a gamma factor of 52,083.

A spacecraft having a gamma factor of 52,083 would certainly need a way to shield against interstellar and intergalactic ions, atoms, molecules, and dust specks with widths as high as micrometers. Perhaps forward aimed nano-scale lasers would suffice. The relativistic Doppler frequency shift of visible light lasers would result in hard x-ray beams with respect to the interstellar background massive medium. Even sub-millimeter infrared sources and short wave-length microwave beams could be effective at eliminating the cosmic ray and dust threat. Once ionized, these background massive species could be diverted electro-dynamically and perhaps even used as a reaction mass in electrodynamic-hydrodynamic-plasma-drive propulsion systems that could augment the matter-antimatter rockets.

A large spacecraft able to attain a gamma factor of 52,083 would easily enable human crewed missions that hop from Galaxy to Galaxy with the possibility of colonizing a large portion of the current cosmic light cone of 13.75 billion light years extent over the next 10 billion years or so.

The good news is that we might have all of the energy we need right here in our solar system to produce the large matter-antimatter mass ratios required for the above speculative scenarios!

Now consider the possibility of attaining a gamma factor of 500,000. The temperature of the Doppler blue-shifted CMBR will be a whopping = [(2.725) (500,000)] = 1,362,500 Kelvin. This is an increase in temperature of 303.59 fold over the 4,488 Kelvin that would melt the $Ta_4HfC_5$. An effective shield reflectivity of $303.59^4$ = 8,494,000,000 might be possible provided the reflective material absorbs at most one part in 8,494,000,000 of the incident radiation. By reflectivity, I mean the ratio of the reflected energy to the absorbed energy. This would be very hard to accomplish with ordinary period table elements. However, perhaps novel nanotechnology based assembly methods can produce exotic types of nanomaterials that are somehow exceptionally reflective to the wave front of the impinging electromagnetic radiation.

The reflectivity of the shield might be reducible to a mere 849,400 provided that a phononic diodic mechanism could wick away the thermal heat flux generated within the shield's outer surface 10,000 times faster than the heat is deposited. Still no small feat!

A flat bow spacecraft would be impractical due to the resulting extreme astrodynamic drag forces. However, if accelerations were kept to no more than one G, but preferably to only a modest fraction of a G, forward deployed conical spacecraft shields made of plausible ordinary periodic table atoms based materials might do the trick.

Once again, merely relaxing the craft's propulsion systems would result in rapid craft deceleration.

For a flat one square kilometer shield that is normal to the spacecraft velocity vector, the drag force will be 1,302,772,194,000,000 Newtons or about 133,000,000,000 tons. The force will be 13,300,000,000 tons for a flat 0.1 square kilometer shield that is normal to the spacecraft velocity vector. The force will be 1,330,000,000 tons for a flat 0.01 square kilometer shield that is normal to the spacecraft velocity vector. The force will be 133,000,000 tons for a flat 0.001 square kilometer shield that is normal to the spacecraft velocity vector. The force will be 13,300,000 tons for a flat 0.0001 square kilometer shield that is normal to the spacecraft velocity vector. Finally, for a flat 0.00001 square kilometer shield, the drag will be a mere 1,330,000 metric tons.

Including a square conical shape to the forward deployed shield having a rest aspect ratio of 100 should reduce the drag by 254.65 fold or by factor of $[\text{Sin}(90^\circ/400)]^{-1}$. A conical configuration would enhance the reflectivity of the shield due to grazing incidence effects.

Such a shield would reduce the force acting against a 1.0 square kilometer cross-section craft traveling at a gamma factor of 500,000 to only about 522,000,000 metric tons. The force acting against a 0.1 square kilometer cross-section craft traveling at a gamma factor of 500,000 would be reduced to only about 52,230,000 metric tons. The force acting against a 0.01 square kilometer cross-section craft traveling at a gamma factor of 500,000 would be reduced only about 5,223,000 metric tons. The force would be reduced to only 522,300 metric tons for a 0.001 square kilometer cross-section craft. The force would be reduced to only 52,230 metric tons for a 100 square meter cross-section craft. Finally, for a 10 square meters, the force would be reduced to only 5,223 metric tons.

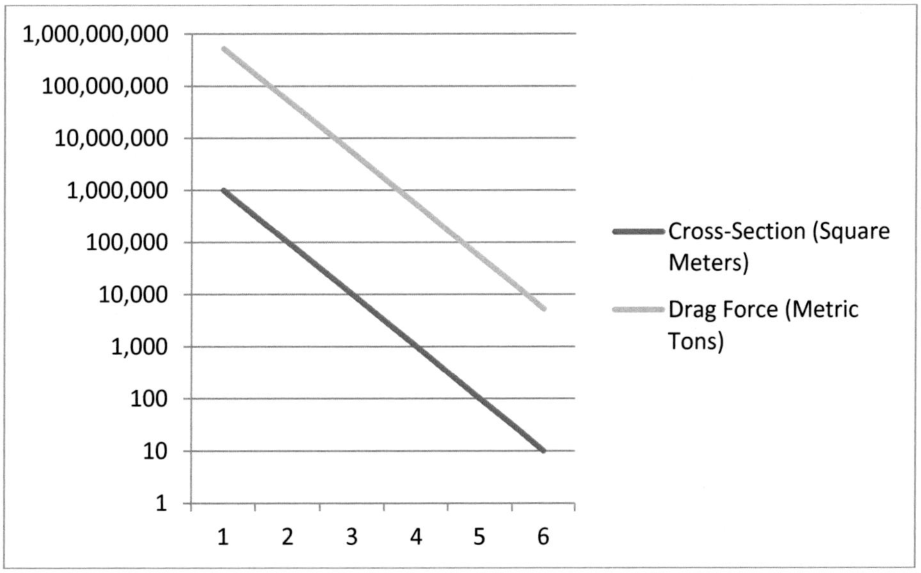


**GRAPH 13** is a graph of cross-sectional shield area and drag force for a conical shield having an at rest aspect ratio of 100 for a spacecraft having a gamma factor of 500,000.

The length of the square cone would be 100 kilometers for the above very smooth space train fuselage type of craft having a 1,000,000 square meter cross-section. The length would be 31.6228 kilometers For a 100,000 square meter cross-section. The length of the cone would be 10 kilometers for the above very smooth space train fuselage type of craft having a 10,000 square meter cross-section. The length would be 3.16228 kilometers for a 1,000 square meter cross-section. The length would be one kilometer for a 100 square meter cross-section,. And finally, for a 10 square meter urban bus-like cross-section, the length would be only 0.316228 kilometers.

Including a conical shape to the forward deployed shield having a rest aspect ratio of 10,000 should reduce the drag by 25,465 fold or by factor of $[\text{Sin } (90^{\circ}/40,000)]^{-1}$. A conical configuration would enhance the reflectivity of the shield due to grazing incidence effects.

Such a shield would reduce the force acting against a 1.0 square kilometer cross-section craft traveling at a gamma factor of 500,000 to only about 5,220,000 metric tons. The force acting against a 0.1 square kilometer cross-section craft traveling at a gamma factor of 500,000 would be reduced to only about 522,000 metric tons. The force acting against a 0.01 square kilometer cross-section craft traveling at a gamma factor of 500,000 would be reduced only about 52,200 metric tons. The force would be reduced to only 5,220 metric tons for a 0.001 square kilometer cross-section craft. The force would be reduced to only 522 metric tons for a 100 square meter cross-section craft. Finally, for a 10 square meters, the force would be reduced to only 52.2 metric tons.

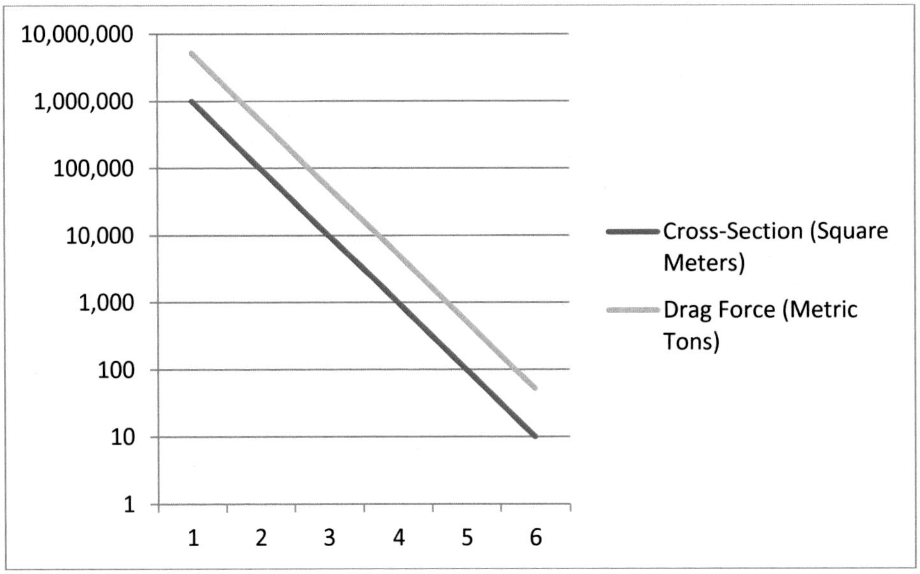


**GRAPH 14** is a graph of cross-sectional shield area and drag force for a conical shield having an at rest aspect ratio of 10,000 for a spacecraft having a gamma factor of 500,000.

Including a square conical shape to the forward deployed shield having a rest aspect ratio of 1,000,000 should reduce the drag by 2,546,500 fold or by a factor of $[\text{Sin}\ (90°/4{,}000{,}000)]^{-1}$. A conical configuration would enhance the reflectivity of the shield due to grazing incidence effects.

Such a shield would reduce the force acting against a 1.0 square kilometer cross-section craft traveling at a gamma factor of 500,000 to only about 52,200 metric tons. The force acting against a 0.1 square kilometer cross-section craft traveling at a gamma factor of 500,000 would be reduced to only about 5,220 metric tons. The force acting against a 0.01 square kilometer cross-section craft traveling at a gamma factor of 500,000 would be reduced only about 522 metric tons. The force would be reduced to only 52.2 metric tons for a 0.001 square kilometer cross-section craft. The force would be reduced to only 5.22 metric tons for a 100 square meter cross-section craft. Finally, for a 10 square meters, the force would be reduced to only 0.522 metric tons.

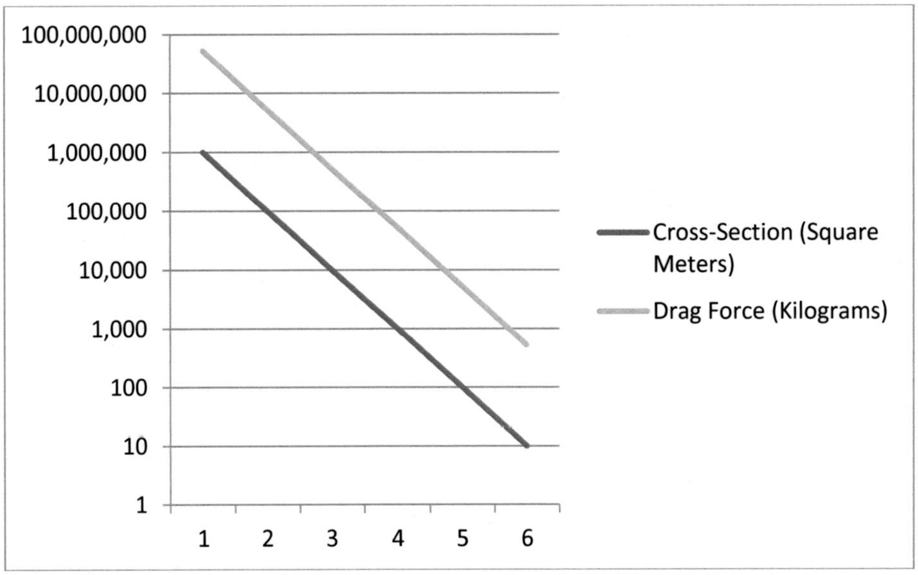


**GRAPH 15** is a graph of cross-sectional shield area and drag force for a conical shield having an at rest aspect ratio of 1,000,000 for a spacecraft having a gamma factor of 500,000.

A spacecraft having a gamma factor of 500,000 or a velocity of 99.9999999998 percent of the speed of light would certainly need a way to shield against interstellar and intergalactic ions, atoms, molecules, and dust specks with widths on the order of a micrometer or more. Perhaps forward aimed nano-scale lasers would suffice. The relativistic Doppler frequency shift of visible light lasers would result in gamma ray beams.

The heat wicked away from the shield's outer surface can be used to power EDHPD drive units placed at judicious locations along the shield. Ion, electron, pion, muon, or photon rockets that are protected within recessed pockets can be distributed along the length of the shield and used as a heat sink for the phononic diodes. Such rockets can be powered by turbo-electric, thermo-electric, or other novel electrical energy generation apparatus that utilizes thermal inputs.

Now, we will need to review some simple concepts before we estimate the maximum possible gamma factor for such a craft where the craft would be constructed of ordinary periodic table elements and isotopes.

To begin, note that the energy of a photon is equal to E = hc/λ. Thus, for a visible violet light wave such as a 400 nanometer photon, the photon energy will be (6.626 x $10^{-34}$ Js)(300,000,000 m/s)/(4 x $10^{-7}$m) = (4.9695 x $10^{-19}$ Joule)(1 eV)/(1.602 x$10^{-19}$Joule) = 3.102 eV.

The mass of the lightest fundamental fermion heavier than an electron is a muon. The muon is about 209 times the mass of an electron and thus has an invariant mass of 209 (0.511 MeV)/$C^2$ = 106.8 MeV/$C^2$.

The ratio of the energy equivalence of a muon with respect to the maximum photon energy of the greater portion of star light will be assumed equal to (106,800,000 eV)/(3.102 eV) = 34,429,000.

We will assume that the great majority of photons impinging on a spacecraft traveling at a gamma factor of 34,429,000 from interstellar and intergalactic sources including stars, quasars, and CMBR will have an energy less than 106.8 MeV with respect to the spacecraft.

The reader is likely about to question the purpose for the above brief seemingly superfluous digression. The reason for such is that the minimum energy for destructive interaction of the above back-ground photons with electronic components of the atoms that comprise the spacecraft shield is likely greater than 106.8 MeV thus corresponding to a plausible spacecraft gamma factor of at least 34,429,000.

Now you might ask, "How is this worthy of my consideration?" Well, imagine if the electrons as well as the nucleons within the atoms comprising the conical shield could be quantum-mechanically entangled with a power sink such as an energy collector that would power an ion, electron, pion, muon, or photon rocket, or perhaps a magneto-hydrodynamic-plasma-drive, an electro-hydrodynamic-plasma-drive, or an electro-magneto-hydrodynamic-plasma drive. The absolute limit for the gamma factor of the starship would seem to be a value at least as great as that associated with the minimum energy threshold cut-off value for the impinging back ground radiations for which impingement would result in destruction of the electronic and nucleonic components of the atoms comprising the shield. Thus, we might reasonably obtain gamma factors as great as 34,429,000 for matter-antimatter rocket powered spacecraft.

A mechanism to efficiently teleport the impinging photon energy absorbed by the electrons and nuclei of the atoms comprising the shield would be required. The process of transport would need to handle extreme power loads, even for a spacecraft conical shield that had a rest aspect ratio of 100. In addition, the materials of choice for construction of the shield would need to be very reflective to 106.8 MeV photons in the case where the photons would merely graze the reflector.

Provided that such shields could be perfectly polished, perhaps the incident gamma ray field would simply skip off the shield while depositing very, very, small portions of its energy into the shield itself. Such greatly reduced power input would go a long way to alleviating the need for quantum energy teleportation.

Note once again, merely relaxing the craft's propulsion systems would enable the craft to slow down.

Think of it! A gamma factor of almost 35 million might be possible for spacecraft having an ordinary periodic table element composition. There is a huge amount of engineering and material science as well as teleportation thermodynamics to be worked out in order for the above system to be made workable. However, the reality that we can intelligibly consider the plausibility of such systems with an awareness of the caveats gives me hope that they just might be fielded at some future time.

The spacecraft could travel several billion light-years in one contemporary human lifetime, ship's frame. Such distances are considered cosmic in scale.

At a gamma factor of 34,429,000, the CMBR will appear as (2.725)(34,429,000) K black body radiation or 93,819,000 Kelvin radiation. Clearly, extreme aspect ratio conical shields or extremely high power quantum energy teleportation schemes would be required.

Masahiro Hotta at Tohoku University in Japan has mathematically discovered a teleportation process which requires a classical feedback loop that carries a decoder signal to unlock the teleported energy. Thus, for the above conjectured spacecraft shield energy teleportation method to be possible for the extreme gamma factors considered, the rapid conveyance of the energy absorbed by the shield away from the shield to energy or heat sinks would be required, and the decoder signal may need to be transmitted over relatively short distances in order to close the teleportation loop in a timely manner. The heat sinks might take the form of photon, electron, proton, ion, muon, or charged pion rockets that are distributed in a high area density arrangement along the cone. Such mechanisms would need to be recessed within the cone in a similar manner that the depressed ends of the holes in a hand held cheese grater are recessed. Close proximity of these heat sinks with the differential area, conical element energy sources, would reduce the instantaneous rate of heat buildup.

A really big caveat is being able to maintain an energy balance. Should the teleportation mechanism fail for a given differential area element of the shield, a large energy explosion would result thereby destroying the shield. The shield element could be ionized in a time scale measured in nanoseconds thus instantly resulting in mission disaster. A continuous supply train of entangled states might be produced in a continuous piping like manner thus alleviating the need for what may otherwise require somehow instantaneous or at least highly superluminal energy teleportation. To facilitate the conveyance of the entangled states, perhaps an entangle states transference or cloning mechanism could be employed.

Another big caveat is that the process of energy teleportation to the heat sink can be accomplished in a manner that bi-passes undue astrodynamic drag.

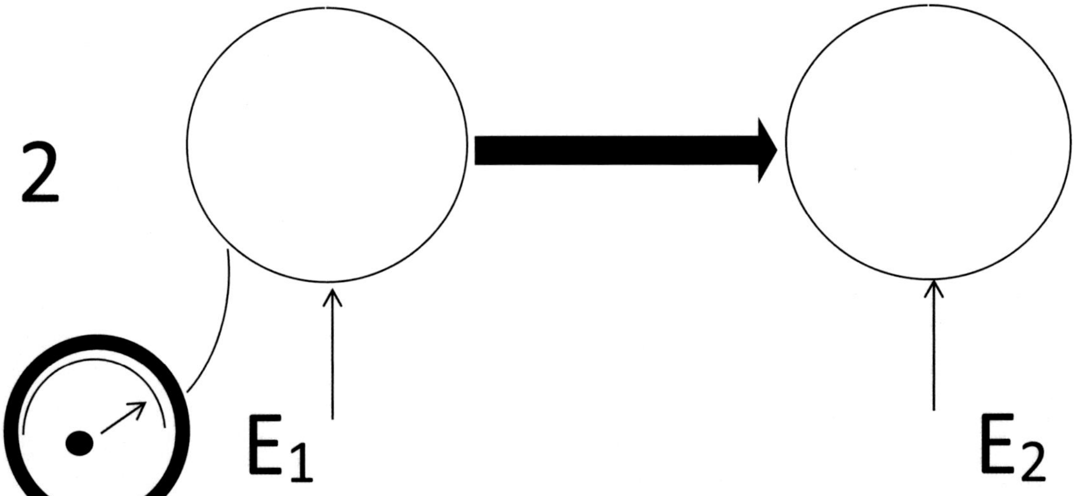


**FIGURE 26** is a very simple depiction of the energy teleportation process. The analogue meter indicates signal measurement. The circles represent the two locations over which the energy is to be teleported. The wavy arrow indicates the classical channel over which a required decoder signal is sent and $E_1$ and $E_2$ are the transmitted and received energy states.

**FIGURE 27 A, B** on next two pages illustrate conical spacecraft shields. Figure A depicts a conical shield for an extremely relativistic spacecraft. The recessed regions on either side of the upper portion of the diagram are heat sinks. The heat sinks can include ion, electron, proton, muon, tauon, charged baryon, and charged meson thrusters in both particle and antiparticle forms. Negatively and positively charged particles can be quickly intermixed in order to prevent particle fall back when the thrusting mechanism is electrostatically gridded. FIGURE B depicts a generic electrodynamic-hydrodynamic-plasma-drive conical shield . The plus signs indicate positively charged plasma derived from the interstellar or intergalactic medium. The obliquely oriented arrows of increasing length from top to bottom indicate the increasing velocity of the charged plasma with respect to the ship frame as the plasma is accelerated down the conical shield. The driving potential can in theory be embodied in; 1) electric fields, 2) magnetic fields, 3) co-located electric and magnetic fields, or 4) electromagnetic fields.

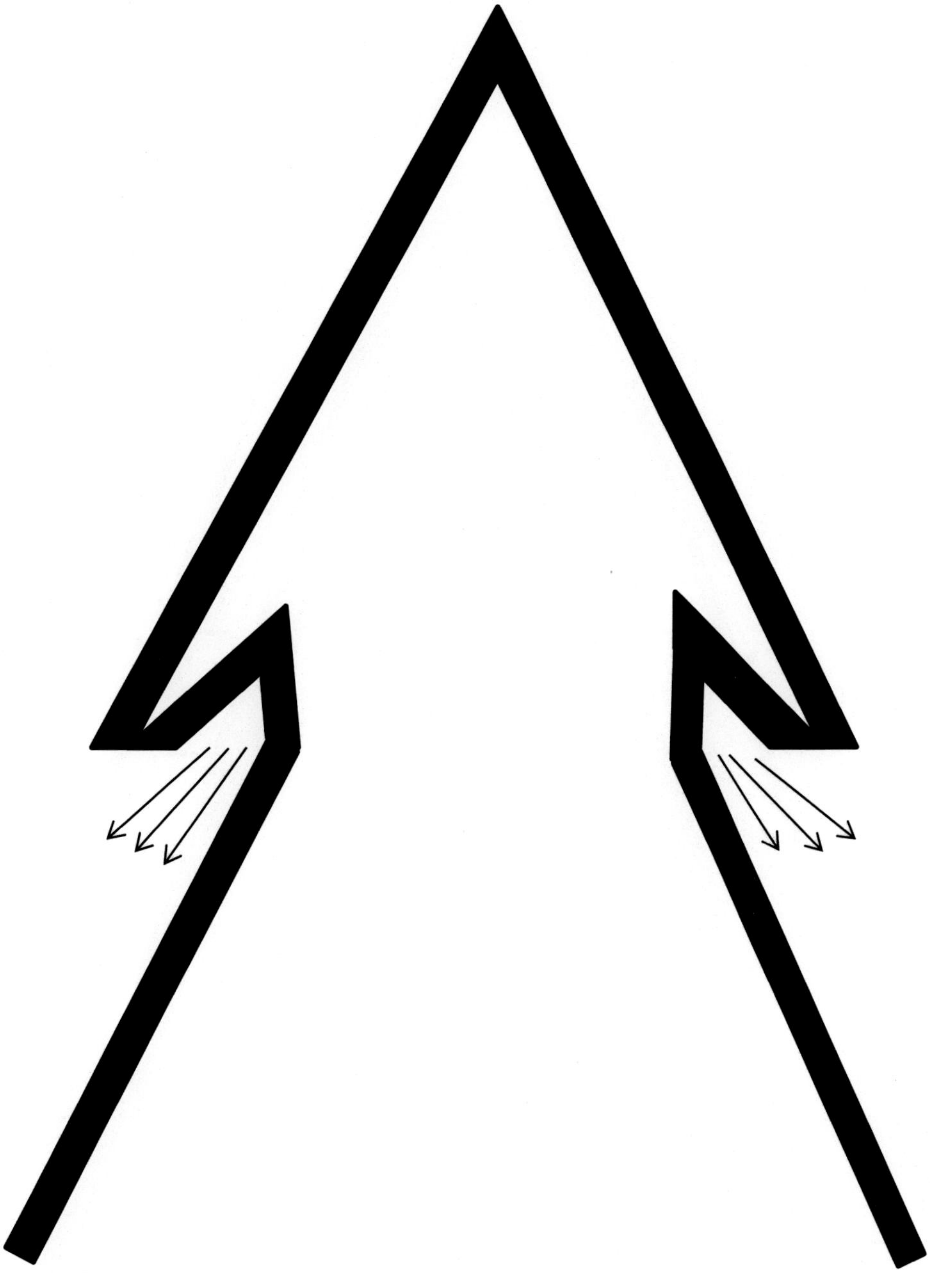

A

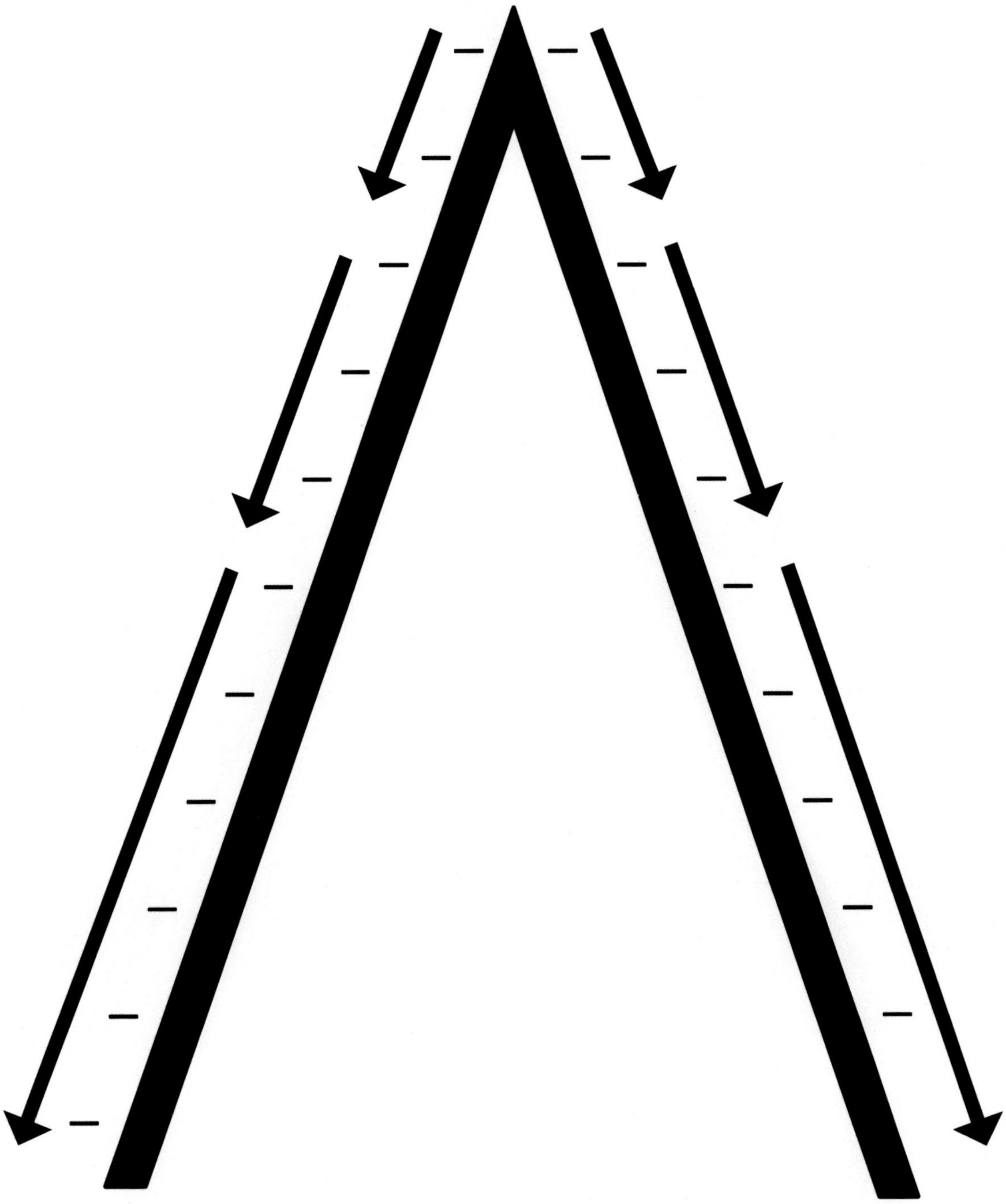

B

Once again, electrodynamic breaking could be used to slow the craft such as: 1) a conducting or superconducting coil based linear induction reverse magnetic sail; 2) a magnetic plasma bottle sail; 3) an electro-hydrodynamic-plasma-drive; 4) a magneto-hydrodynamic-plasma-drive; and 5) a reverse thrust interstellar ramjet mechanism.

Materials science and solid state physics in general will likely be a key consideration in the development of Humanity's first starships and will likely continue to be a key concern as more extreme spacecraft are designed and flown.

More extreme aspect ratio conical shields are conceivable, but the rest length of such shields would increase to many light-years. As a result, space trains with a small cross section, and huge masses, may be our best option for extreme gamma factor starships made of ordinary periodic table elements. If we can develop atomic nucleus density materials and denser materials, then the problem of providing adequate thermal shielding will be easier to solve.

Note that the conical shields anticipated herein would experience a Lorentz Length Contraction by a factor of gamma. However, the length of the shields in all cases would be the same with respect to the spacecraft. Therefore, we need not lengthen the shields compensate for relativistic length contraction of the spacecraft shields at high gamma factors, unless we desired to drastically reduce astro-dynamic CMBR and star and quasar light at extreme gamma factors.

Now suppose we had a spacecraft that accelerated to a gamma factor of $10^{12}$ and that we had a shield somehow made of high end performance quarkonium(s) or perhaps Higgsinium(s).

For such a spacecraft, the CMBR temperature impinging on a conical shield would have an effective temperature of 2.725 trillion K. Once again, the light pressure will be $2\sigma T^4/C$ for a perfectly reflective surface that is normal to the impinging CMBR. For a flat bowed spacecraft having a one square kilometer shield traveling at a gamma of one trillion, the drag imposed by CMBR would be (1,000,000 $m^2$)(2)[5.6704 x $10^{-8}$ W $m^{-2}$ $K^{-4}$] {[2.725 x $10^{12}$ K]$^4$}/(300,000,000 m/s) = 2.0844 x $10^{40}$ Newtons. Now, suppose that we shielded the spacecraft with a $10^{20}$ kilometer long shield. The drag would be reduced to 8.1854 x $10^{19}$ Newtons or by a factor of {sin {(90)/[4 x $10^{20}$]}}$^{-1}$. This force is a whopping 8.34395 x $10^{15}$ metric tons. The shield would be 10 million light-years long. We might build such space trains, even without mass energy wave cloaking, and may accelerate such trains by a uniform electro-dynamic-hydrodynamic-plasma-drive mechanism.

Assume that a spacecraft could obtain an additive gamma factor gain of 10 billion for every $R_o$ distance of travel through space where $R_o$ is the current radius of the observable universe. We will assume that the spacecraft accelerates at a constant 0.667 G = 6.545 m/s$^2$, and that it takes about $T_o$ = 13.75 billion years background to travel a distance of $R_o$.

Now for present purposes,

$$\gamma = [ad/C^2] + 1 = \{\{[6.545 \text{ m/s}^2]\{[1.375 \times 10^{26} \text{ m}]\}/\{[3 \times 10^8 \text{ m/s}]^2\}\} - 1 = 10^{10}.$$

The CMBR is currently black body in form with an effective temperature of 2.725 K. Assume a constant rate of space-time expansion. In 13.75 billion years from now, the CMBR background temperature will be reduced by a factor of 2 to a value of (1/2)(2.725) K = 1.3625 K. In 10 $T_o$ into the future, the effective temperature will be equal to 0.2725 K. In 1,000 $T_o$, the effective temperature will be 0.002725 K. In 1,000,000,000 $T_o$, the effective CMBR temperature will be only 0.000000002725 K.

In general, after the craft has traveled [$10^N$]{[13.75 x $10^9$]} billion light-years, the background temperature of the CMBR will have dropped by N orders of magnitude. For example, for N = 1, the universe will be 137.5 billion years old and the CMBR will be 0.2725 K. For N = 2, the universe will be 1.375 trillion years old and the CMBR will be 0.02725 K. For N = 3, the universe will be 13.75 trillion years old and the CMBR will be 0.002725 K. In other words, for a universal age of (M)($T_o$), the CMBR will have cooled to a temperature of (2.725)/M. At present, M is equal to one.

Now assume that a spacecraft has reached a gamma factor of 10 billion during the current era. The onslaught of black body CMBR will be a stupendous 2,725,000,000 Kelvin. Thus, very reflective, refractory, and streamlined forward spacecraft surfaces made of: 1) stabilized neutronium; 2) quarkoniums such those made from; up, down, charmed, strange, bottom, and/or top matter; 3), Higgsinium(s); or 4) monopolium(s) would be of great help here.

The really good news is that for each order of magnitude increase in the spacecraft gamma factor, the CMBR will cool by an order of magnitude. Ironically, the above acceleration profile would permit the spacecraft to remain forwardly bathed in a CMBR, star-light, Quasar light, and the like background radiation having has a temperature with respect to the spacecraft of only 2,725,000,000 Kelvin. The CMBR, starlight, and Quasar light backgrounds, will in theory remain no hotter than 2,725,000,000 Kelvin with respect to the spacecraft should the spacecraft accelerate at about 1 G to an ensemble, e1, gamma factor,. The CMBR, starlight, and Quasar light backgrounds, will in theory remain no hotter than about 2,725,000,000 Kelvin with respect to the spacecraft should the spacecraft accelerate at about 1 G to an infinity scrapper, i1, gamma factor. Taken to more extreme limits, the CMBR, starlight, and Quasar light backgrounds, will in theory remain no hotter than 2,725,000,000 Kelvin should the spacecraft accelerate at about 1 G to an Aleph 0, gamma factor. Taken to even more extreme limits, the CMBR, starlight, and Quasar light backgrounds, will in theory remain no hotter than 2,725,000,000 Kelvin with respect to the spacecraft should the spacecraft accelerate at about 1 G to an Aleph 1, gamma factor. Taken to yet still more extreme limits, the CMBR, starlight, and Quasar light backgrounds, will in theory remain no hotter than 2,725,000,000 Kelvin with respect to the spacecraft should the spacecraft accelerate at about 1 G to an Aleph 2, gamma factor. And thus, the progression might continue forever to ever higher transfinite Aleph number ordinal terms.

Now, I am not suggesting that we will ever obtain infinite gamma factors by inertial acceleration. However, such infinite values are an ideal that we can approach in the ever open ended problem of determining how to obtain ever higher gamma factors.

Now, assume again that a spacecraft has reached a gamma factor of 1,647 during the current era. The onslaught of black body CMBR will be a roasty, toasty, but modest 2,725 Kelvin. Thus, very reflective, refractory, and streamlined forward spacecraft surfaces made of the best current state of the art refractory materials would work here.

Now tantalum hafnium carbide, $Ta_4HfC_5$, has the highest melting point of all known compounds, 4,488 K (4215 °C, 7619 °F), and is almost as hard as diamond. Such a material would make an excellent blue-shifted; CMBR, star light, and Quasar light, forward deployed shield for a relativistic rocket.

Such a shield would plausibly retain its integrity for a spacecraft that would attain a gamma factor of 4,488 /2.725 K = 1,647 or a velocity equal to 99.99999815 percent of the speed of light. Such velocities would facilitate manned missions to any location within the observable universe in giant space arks or world zonds.

Provided that the shield could be actively cooled, the shield might endure higher gamma factors sooner perhaps enabled by phonon diodic conducting materials which are hypothetical materials that would conduct heat based acoustic or thermal vibrations very efficiently, perhaps even permitting net positive heat transfer from cold to hot portions of a material.

Once again, for a perfectly reflective surface that is normal to the impinging CMBR, the light pressure will be $2\sigma T^4/C$. Here, $\sigma = 5.6704 \times 10^{-8}$ W $m^{-2}$ $K^{-4}$. The force will be 153,600 Newtons or about 15 tons for a one square kilometer shield that is normal to the spacecraft velocity vector.

Including a conical shape to the forward deployed shield having an aspect ratio of 100 should reduce the drag by 254.65 fold or by a factor of $[\mathrm{Sin}\ (90°/400)]^{-1}$. Such a shield would reduce the force acting against a square kilometer cross-section craft traveling at a gamma factor of 1,647 to only about 613.9 kilograms. The force would be reduced to only 61.39 kilograms for a 10,000 square meter cross-section craft. The force would be reduced to only 613.9 grams for a 1,000 square meter cross-section craft. The force would be reduced to only 61.39 grams for a 100 square meter craft. The force would be reduced to only a paltry 6.139 grams for a 10 square meter craft.

Thus, the length of the cone would be 10 kilometers for a very smooth space train fuselage type of craft having a 10,000 square meter cross-section. The length would be 3.16228 kilometers for a 1,000 square meter cross-section,. The length would be 1 kilometers For a 100 square meter cross-section. And finally, the length would be only 0.316228 kilometers for a 10 square meter urban bus like cross-section.

For the latter three examples, accelerations of roughly from one to 10 Earth G's should be possible given that the bulk modulus of $Ta_4HfC_5$ is hundreds of time higher than granite.

Hard rock permits Mount Everest to remain stable against compressive fatigue, and so it would seem that a material with well over 100 times the compressive strength of granite should enable conical structures of a length of a few hundred times Mount Everest's height to remain stable under 1 G loading. This would permit conical shield lengths of roughly; (10 km)(300) = 3,000 km to withstand 1 G acceleration, 300 km to withstand 10 G acceleration, 30 km to withstand 100 G acceleration.

There is one huge caveat here and that is finding a way for the $Ta_4HfC_5$ shield to maintain its thermal-mechanical stability. The answer to this problem is the instillation of a super high reflectivity coating on the cone, such that even as the cone is slightly ablated over time by the incident but extremely rarefied interstellar gas and dust, such ablation would only act to polish the cone so that it can be all the more reflective, or at the very least, retain its original reflectivity within a safety design margin.

Nanotechnology based self-assembly or micro-robotic based mechanisms may repair the shield either from the outside or from the inside, either within the conical inner vacancy or, within the cone's material composition itself.

Now, consider a spacecraft that accelerates at only $1.07738 \times 10^{-6} m/s^2 \approx 0.000000011$ G in a perpetual manner after having been fairly quickly accelerated to a gamma factor of:

$$\gamma = [ad/C^2] + 1 = \{[1.07738 \times 10^{-6} m/s^2][1.375 \times 10^{26} m]/\{[3 \times 10^8 m/s]^2\}\} + 1 = 1,647.$$

After the craft has traveled $[10^N]\{[13.75 \times 10^9]\}$ billion light-years, the background temperature of the CMBR will have dropped by N orders of magnitude. For example, for N = 1, the universe will be 137.5 billion years old and the CMBR will be 0.2725 K. For N = 2, the universe will be 1.375 trillion years old and the CMBR will be 0.02725 K. For N = 3, the universe will be 13.75 trillion years old and the CMBR will be 0.002725 K. In other words, for a universal age of $(M)(T_o)$, the CMBR will have cooled to a temperature of (2.725)/M. At present, M is equal to one.

The point is that there will theoretically always be opportunities to achieve ever higher gamma factors, assuming continuous universal space time expansion and rarefication of the interstellar and intergalactic medium. Our intrepid cosmic descendants in the trillions plus years to follow will likely have evolved to a radically different physiology but will also likely have evolved in technology so as to perpetually survive and thrive in an ever cooling universe. Perhaps a symmetry breaking event that would separate the electric and magnetic forces from the current electromagnetic unified field will occur, thus providing a fresh new supply of entropy and mass energy to support our future descendants or whatever species may follow should we become extinct. However, with regard to human progress, I am an eternal optimist.

As a final note, recall the assumption of conical shields. The cross-sectional profiles of the shields can be judiciously selected from a set of otherwise arbitrary shapes. Some applications may best be served by pyramidal or tetrahedral shaped shields. For some systems, wedge shaped shields may be preferred. Moreover, the cross-sectional shape can very along the length of the shield. It is even possible to include a variable cross-sectional shape adjustment mechanism along with the previously mentioned optional means for shield elongation in the ship frame. Such additional morphological degrees of freedom can enable more appropriate interaction of the shield with the interstellar and intergalactic medium including but not limited to natural variations in: 1) plasma distribution by density, charge, and species; 2) neutral gas distribution by density and species; 3) ambient stellar and quasar light distributions according to energy spectrum and power flux density; and 4) and interstellar and intergalactic magnetic field intensity and vector field orientation. Such considerations can be important for drag reduction, and for certain conditions, intentional increases in drag such as for re-routing and deceleration. Such natural variations can degrade or enhance spacecraft performances including but not limited to propulsion system efficiency.

Although we have focused on gas compression guns, magnetic mass drivers are also workable. In the end, it is a matter of which system works best.

For example, pulsed magnetic fields can lose energy due to differential volumetric field start-up and field decay as a given section of a solenoidal magnetic field generator is activated.

Precisely controlling powerful magnetic field barrels can be problematic and induce harmonic oscillations in the gun and adversely result in magnetic field based stress on the overall barrel based on interaction with the solar, lunar, Earth-based, and/or other magnetic fields.

An electron gas discharge results in rapid charge neutralization or at least the electron gas gun configuration can be designed to neutralize the plasma charge with protons or other positive ions.

Additionally, explosive plasma charge can be tailored as can the shape of the projectile so that some of the plasma slides between the round and the inside of the barrel to prevent barrel contact with the inner wall of the barrel.

Still further, whether the gas gun or magnetic mass driver is chosen to propel the round, the gun can be comprised on a linear series of aligned and disconnected segments. Such a design for gas guns allows more easily buffering between the pellet and tbe inner wall of the barrel segments. As the round passes through a given segment, plasma may be discharges from the inner surface of the segments to keep the round from contacting the inner segment walls.

Further still, electron gas guns may recycle a portion of their electrons. For example, electrons in the plasma that imbed into the barrel walls may be applied for the next shot for each projectile firing. The thermal energy of the plasma that is deposited in the barrel walls may also be at least partially recycled..

The electron gas gun as a binary propulsion effect. The first effect is electron gas thermal pressure acting to push the round forward. The second effect is the coulombic propulsion of the round as a result of field effect repulsion.

Note that hybrid guns having magnetic field effect mass-driving and plasma discharge methods are also of worthy consideration. So, there may in principle be three significantly large driving force components to accelerate the rounds down the barrel.

# Enablement By Extreme Materials Made From Ordinary Elements

A brief description of extreme materials composed of ordinary periodic table elements is given below. These materials although generally currently prohibitively expensive might be mass producible by nano-technology self-assembly or perhaps be widespread mining operations throughout the solar system.

#### A) Carbon Nanotubes

In 2000, a multiwall nanotube was observed to have a tensile strength of 63 gigapascals. This translates into a cable with a cross-sectional area of one square millimeter being able to hold 6,422 kg or 6.422 metric tons. A cable with a cross-section of one square centimeter could support 642 metric tons. A cable with a cross-section of one square inch could support 4,143 metric tons.

The table below was adapted from the same Wikipedia page and summarizes the relative mechanical strength properties of CNTs

Comparison of Mechanical Properties

Material Young's modulus (TPa) Tensile strength (GPa) Elongation at break (%)

| | | | |
|---|---|---|---|
| SWNT | ~1 (from 1 to 5) | 13–53E | 16 |
| Armchair SWNT | 0.94T | 126.2T | 23.1 |
| Zigzag SWNT | 0.94T | 94.5T | 15.6–17.5 |
| Chiral SWNT | 0.92 | | |
| MWNT | 0.27E–0.8E–0.95E | 11E–63E–150E | |
| Stainless steel | 0.186E–0.214E | 0.38E–1.55E | 15–50 |
| Kevlar–29&149 | 0.06E–0.18E | 3.6E–3.8E | ~2 |

$^{E}$ = Experimental Observation; $^{T}$ = Theoretical Prediction
Chart adapted from Wikipedia at Courtesy of Wikipedia at: http://en.wikipedia.org/wiki/Carbon_nanotube

Metallic CNTs can carry an electric current density of [4 x ($10^9$)] amperes/($cm^2$). This is more than 1,000 times greater than metals such as copper (Wikipedia, 2011).

CNTs have theoretical temperature stability for temperatures as great as 2800° C in a vacuum and about 750° C in air.

This material may be of use in forward deployed starship shields as well as in metalized thin membrane like fabrics or grids for application in light sails. When charged, such sails can be used for Project Medusa style propulsion systems.

**B) Graphene**

Graphene consists of one atom-thick sheets of carbon atoms arranged in a repeating hexagonal pattern.

The electrical resistivity of graphene has been measured at ($10^{-6}$)[(Ohms)($cm^2$)] and thus has a conductivity that is higher than that of silver.

Graphene has been shown to have a bulk strength of 130 gigapascals or about (1,300,000)(14.7) pounds per square inch or 19.11 million PSI. This is a whopping 9,555 tons per square inch.

This material may be of use in forward deployed starship shields as well as in metalized thin membrane like fabrics or grids for application in light sails. When charged, such sails can be used for Project Medusa style propulsion systems.

**C) Graphene-Oxide Paper**

Graphene-oxide paper has been shown to have a tensile strength of 32 gigapascals or about (320,000)(14.7) pounds per square inch or 4.704 million PSI (Wikipedia, 2011). This material may also be of use in forward deployed starship shields as well as in metalized thin membrane like fabrics or grids for application in light sails. When charged, such sails can be used for Project Medusa style propulsion systems.

**D) Beta carbon nitride**

Beta carbon nitride (β-C3N4) is predicted to be harder than diamond and is crystalline in composition and formed by extremely short and strong chemical bonds between carbon and nitrogen atoms. Because the bonds are so strong, the possibility of producing ultra-strong threads composed of Beta carbon nitride becomes irresistible to researchers. So far, only nano-sized crystals of Beta carbon nitride have been produced (Wikipedia, 2010). This material may also be of use in forward deployed starship shields as well as in metalized thin membrane like fabrics or grids for application in light sails. When charged, such sails can be used for Project Medusa style propulsion systems.

**E) Heterodiamond**

Another material with extremely strong atomic bonds is heterodiamond. This material is composed of carbon, boron, and nitrogen atoms. The bulk modulus of heterodiamond is 282 Gpa and is surpassed only by that of diamond. Heterodiamond is extremely hard and possesses an excellent heat resistance. Heterodiamond may find application in starship forward shielding and also Project Orion style pusher plates. When fashioned into metalized thin membrane like fabrics or grids, the material can find application in light sails. When charged, such sails can be used for Project Medusa style propulsion systems.

**F) Diamond** in naturally occurring forms is the hardest known natural substance. Provided bulk portions of diamond can be fabricated, such a material would be useful for forward deployed starship shields and Project Orion pusher plates. Natural diamond is an excellent heat conductor and so it could be used to conduct heat from the outer surface of a Project Orion style pusher plate.

**G) Aggregated Diamond Nanorods**, or **ADNRs**, are a nanocrystalline form of diamond, also referred to as "nanodiamond" or hyperdiamond. In 2003, nanodiamond was produced by compression of graphite in 2003 and was found to be much harder than bulk diamond. Later produced by compression of fullerene, it was confirmed as the hardest and least compressible known material and had an isothermal bulk modulus of 491 gigapascals (GPa). Conventional diamond has a modulus of 442–446 Gpa.

Photovolataic, photo-electric, and/or thermo-electric mechanisms may be used to collect Doppler blue-shifted CMBR and star-light to provide power for electrodynamic-hydrodynamic-plasma-drive systems, and/or chargon rockets or photon rockets. The chargons may be collected by the electrostatic hairbrush mechanism and/or be produced within ship-board accelerators, fusion reactors, and/or matter-antimatter reactors.

Since one hair-brush can be disposed behind another, the opportunity for multi-stage electrodynamic-hydrodynamic-plasma-drive mechanisms is available for suitably designed space craft.

The hairbrush bristles can include any suitably refractory and radiation resistant materials and would likely require construction from somehow stabilized neutroniums, quarkoniums, higgsiniums, mono-higgsiniums, and the like nuclear or super-nuclear density materials. Quarkoniums may include; un-differentiated up-downium, strange matter, charmonium, bottomonium, and toponium, provided some method(s) can be developed to stably bind top quarks with itself or other quark flavors.

Such a propulsion system, although still a science fiction dream, is none-the-less, lexicographically plausible, at least to a degree commensurate with the rationalization of the above proposition. The caveats are numerous and include the following as non-limiting examples.

First, a method of immediately converting intake chargon invariant mass and kinetic energy to propulsive energy in a gainful manner is needed for the respective sub-class of propulsion methods. Second, the ability to sort, parse, or otherwise direct the intake chargons in suitable proximities to the enhanced electric field flux regions without undue astrodynamic drag is required. Third is the required capability to suitably interact with the background chargon intake stream(s) without undue astrodynamic drag in cases where the intake stream(s) is captured in the spacecraft reference frame and used for fusion reactions and/or matter-antimatter reactions. Fourth is the required capability to maintain a suitable electric charge on the bristles or spikes amidst the intake of background chargons, especially in cases where the chargon kinetic energy and invariant mass is absorbed by the bristles. Fifth, the materials of construction as alluded to previously must be suitably refractory and radiation resistant. Sixth, a means is required for maintaining the species of the materials of composition of the bristles amidst the onslaught of extremely hard photonic and massive species based radiation intake.

The above prose is just a brief summary of my ideas regarding the subject propulsion methods. The ideas more or less popped into my mind all at once as was out purchasing fireworks at a local stand for July 4 celebrations. Of course, the caveats need to be met, but if it can be, then we have a really cool, or should I say, hot propulsion system. Speaking of hot, it is about 100 plus degree Fahrenheit outside as I compose this post. We need to find alternate homes for humanity in the event that Earth becomes uninhabitable.

Now assume that there is only one very large bristle within the hair brush and that the electric flux at the tip of the bristle is concentrated in the manner described above. Assume that the bristle is conical for astrodynamic drag reduction.

All positively charged species in front of the space craft and within the light cone of the space craft have an electrical field relative to the space craft's electrically charged conical member expressible as:

$$E = [1/(4\pi\varepsilon_0)][\gamma_0 q\mathbf{R}]/\{\{[\gamma_0^2 R^2 \cos^2\theta] + [R^2 \sin^2\theta]\}^{3/2}\} = \{[1/(4\pi\varepsilon_0)]\{q[1-(v_0^2/c^2)]]/\{\{1 - [(v_0^2/c^2)\sin^2\theta]\}^{3/2}\}\ [\dot{\mathbf{R}}/R^2]$$

where $\gamma_0$ is the Lorentz factor of the space craft, q is the charge of a given incident particle, **R** is the effective radial coordinate vector of the forwardly located charged particle with respect to the charged member, R is the effective distance of the forwardly incident charged particle from the charged member, θ is the angle between the space craft and cone's velocity vector with respect to the incident forwardly displaced charged particle, $v_0$ is the velocity of the space craft and cone relative to the background, and $\dot{\mathbf{R}}$ is the unit vector pointing in the direction from the axis of the cone to the forwardly displaced charged particle.

All positively charged species in front of the space craft and within the light cone of the space craft have a magnetic field relative to the space craft's electrically charged member that is expressible as:

$B = \{[\mu_0/(4\pi)]\{[qv[1-(v_0^2/c^2)]]\sin\theta\}/\{\{1 - [(v_0^2/c^2)\sin^2\theta]\}^{3/2}\}\ [\acute{\boldsymbol{\phi}}/R^2]$ where $\acute{\boldsymbol{\phi}}$ aims counterclockwise as the space craft crew faces the incoming charge.

where $\gamma_0$ is the Lorentz factor of the space craft, q is the charge of a given incident particle, **R** is the effective radial coordinate vector of the forwardly located charged particle with respect to the charged member, R is the effective distance of the forwardly incident charged particle from the charged member, θ is the angle between the space craft and cone's velocity vector with respect to the incident forwardly displaced charged particle, $v_0$ is the velocity of the space craft and cone relative to the background, and $\dot{\mathbf{R}}$ is the unit vector pointing in the direction from the axis of the cone to the forwardly displaced charged particle.

Now $F = \mathbf{F_1} + \mathbf{F_2} + \ldots = Q\mathbf{E}$.

The electric force exerted on a given differential surface area charge element of the charged cone, Q, by a single intake particle of positive charge, q, not taking into account the pointed shape effect concentration of the electric field flux at the tip of the bristle, is equal to is equal to:

$F_{elect} = Q\,\{[1/(4\pi\varepsilon_0)][\gamma q\,\mathbf{R}]/\{\{[\gamma^2 R^2 \cos^2\theta] + [R^2 \sin^2\theta]\}^{3/2}\}\}$.

Here, q is in integer multiples of positronic charge in electric charge units expressed in units of Coulomb.

The incremental step-wise series of forces exerted on the charged particle as it is pulled into the charged conical surface by a given differential surface area charge element of the charged cone, Q, not taking into account the pointed shape effect concentration of the electric field flux at the tip of the bristle is:

$\Sigma F_h\,(h=1, h=n_{cone}) = Q\,\{\Sigma\,(h=1, h=n_{cone})\,\{[1/(4\pi\varepsilon_0)][\gamma_h q\mathbf{R}_h]/\{\{[\gamma_h^2 R_h^2 \cos^2\theta_h] + [R_h^2 \sin^2\theta_h]\}^{3/2}\}\}\}$.

Consider a stream of k positively charged intake particles impinging on the cone where the particles are distributed in the background frame over a unit spatial interval in the background frame. The electric force exerted by the particles on the differential unit of conical surface charge not taking into account the pointed shape effect concentration of the electric field flux at the tip of the bristle is equal to:

$F_{elect} = Q\,\{\Sigma(k=1, k=o)\,\{\Sigma\,(h=1, h=n_{cone})\,\{[1/(4\pi\varepsilon_0)][\gamma_{h,k}\,(sq_{h,k})\,\mathbf{R_{h,k}}]/\{\{[\gamma_{h,k}{}^2 R_{h,k}{}^2 \cos^2\theta_{h,k}] + [R_{h,k}{}^2 \sin^2\theta_{h,k}]\}^{3/2}\}\}\}\}(\gamma)(v/C)$,

where s is a positive integer representing the number of net electronic charge units.

The total electric force exerted on the charged cone by the stream of k charged particles where the particles are distributed in the background frame over a unit spatial interval in the background frame not taking into account the pointed shape effect concentration of the electric field flux at the tip of the bristle is equal to:

$F_{elect} = \{\Sigma\,(u=1, u=n_{cone})\,[Q_{u,cone}(x,y,x,t)]\{\Sigma(k=1, k=m)\,\{\Sigma\,(h=1, h=n)\,\{[1/(4\pi\varepsilon_0)][\gamma_{h,k}\,(sq_{h,k})\,\mathbf{R}_{h,k}]/\{\{[\gamma_{h,k}{}^2 R_{h,k}{}^2 \cos^2\theta_{h,k}] + [R_{h,k}{}^2 \sin^2\theta_{h,k}]\}^{3/2}\}\}\}\}\}(\gamma)(v/C)$

where the total number of differential area surface charges on the charged member is equal to $n_{cone}$.

The propulsive force due to the beamed exhaust produced by the energy conversion of the charged mass absorbed into the conical surface is most simply expressed as:

$<[(S)(A)]>/C = <\{d\,\{\Sigma\,(u=1, u=n_{cone1})\,[M_{uc}C^2]\,(\gamma)(v/C)\}/dt\}>/C$,

where $<[(S)(A)]>$ is the ship-time-averaged light speed exhaust thrust based power, $n_{cone1}$ is the number of charged intake particles per unit background spatial interval in the background reference frame, $M_{uc}$ is the slowed mass of the uth intake charged particle in the ship's reference frame or the invariant mass of the particle applied in the space-craft's reference frame for particles being absorbed into the conical surface, and t is the background frame time. Here, S is the electromagnetic radiation flux density and A the photonic exhaust source cross-sectional area.

The electro-dynamic-hydrodynamic-plasma-drive feature produces a driving thrust force equal to:

$F_{edhpd} = \{\{\{d[(M_{0j})(\Delta v_{j\text{-}rel\text{-}edhpd})(\Delta\gamma_{j\text{-}rel\text{-}edhpd})\,(\gamma_j)(v_j/C)]/dt\}\,[\cos\beta]\}\{(V_{ce}/\,V_j) = [(z_{present}/\,z_j)^3] = [(a_{present}/\,a_j)^3]\}\}$

$= \{\{\{d[(M_{0j})(\Delta v_{j\text{-}rel\text{-}edhpd})\{\{1/\{[1 - [(\Delta v_{j\text{-}rel\text{-}edhpd}/C)^2]]^{1/2}\}\}\,\{\{1/\{[1 - [(v_j/C)^2]]^{1/2}\}\}(v_j/C)]/dt\}\,[\cos\beta]\}\,\{(V_{ce}/\,V_j) = [(z_{present}/\,z_j)^3] = [(a_{present}/\,a_j)^3]\}\}$

where $\Delta v_{j\text{-}rel\text{-}edhpd}$ is the change in relative velocity of the charged interstellar and intergalactic massive species with respect to the cone, $\Delta\gamma_{j\text{-}rel\text{-}edhpd}$ is the change in relative Lorentz factor of the charged interstellar and intergalactic massive species with respect to the cone, and β is the angular divergence of the walls of the cone relative to the conical axis of rotation. For example, a value of β = 30 degrees would imply that a slice of the cone along its greatest width would form an equilateral triangle. A value of β almost equal to 90 degrees would imply that the cone is almost a flat sheet having surface normals pointing parallel and anti-parallel to the space craft velocity vector. A value of β equal zero degrees would imply that the cone is a cylinder of constant radius oriented length-wise in a parallel direction with respect to the space craft velocity vector.

The factor, $\{(V_{ce}/ V_j) = [(z_{present}/ z_j)^3] = [(a_{present}/ a_j)^3]\}$, is a trinary equivalence operator that allows accounting for universal space-time expansion based rarefication of the intergalactic medium. The variables $V_{ce}$, $V_j$, $z_{present}$, $z_j$, $a_{present}$, and $a_j$ are the present era volume of the universe, the volume of the future universe during the jth time step, the CMBR red-shift in the current era universe, the CMBR red-shift in the jth time step, the present age of the universe, and the age of the universe in the jth time step.

Consider again the electric hair brush bristle drive but this time where the intake chargons are atleast partially processed for exothermic fuel yields.

The the net drive force per intake particle is equal to that produced by harnessing the exothermic consumption of the intake chargon plus the electric force exerted on a given differential surface area charge element of the charged cone, Q, by a single intake particle of positive charge, q, taking into account the pointed shape effect induced concentration of the electric field flux at the tip of the bristle.

The associated drive force is abstractly equal to is equal to:

$F = \{Q\,[f(\text{shape}, \gamma, \text{dist})]\ \{[1/(4\pi\varepsilon_0)][\gamma q\,\mathbf{R}]/\{\{[\gamma^2 R^2 \cos^2\theta] + [R^2 \sin^2\theta]\}^{3/2}\}\}\}$ + (Particle reaction drive force) + (Drag Force).

where [f(shape, γ, dist)] is a field modifying function of shape factor, gamma, and volumetric charge distribution within the bristle.

Here, q is in integer multiples of positronic charge in electric charge units expressed in units of Coulomb.

The incremental step-wise series of forces exerted on the charged particle as it is pulled into the charged conical surface by a given differential surface area charge element of the charged cone, Q, taking into account the pointed shape effect concentration of the electric field flux at the tip of the bristle is:

$\Sigma F_h\,(h=1, h=n_{cone}) = Q\ \{\Sigma\,(h=1, h=n_{cone})\{\,[1/(4\pi\varepsilon_0)][\gamma_h q\mathbf{R}_h]/\{\{[\gamma_h^2 R_h^2 \cos^2\theta_h] + [R_h^2 \sin^2\theta_h]\}^{3/2}\}\}\}\ [f(\text{shape}, \gamma, \text{dist})]$.

So, the associated drive force is:

$\Sigma F_h\,(h=1, h=n_{cone})$ + (Particle reaction drive force) + (Drag Force) = $Q\ \{\Sigma\,(h=1, h=n_{cone})\{\,[1/(4\pi\varepsilon_0)][\gamma_h q\mathbf{R}_h]/\{\{[\gamma_h^2 R_h^2 \cos^2\theta_h] + [R_h^2 \sin^2\theta_h]\}^{3/2}\}\}\}\ [f(\text{shape}, \gamma, \text{dist})]$ + (Particle reaction drive force) + (Drag Force).

Consider a stream of k positively charged intake particles impinging on the cone where the particles are distributed in the background frame over a unit spatial interval in the background frame. The electric force exerted by the particles on the differential unit of conical surface charge taking into account the pointed shape effect concentration of the electric field flux at the tip of the bristle is equal to:

$F_{elect} = \{Q\ \{\Sigma(k=1, k=o)\ \{\Sigma\,(h=1, h=n_{cone})\ \{[1/(4\pi\varepsilon_0)][\gamma_{h,k}\,(sq_{h,k})\,\mathbf{R}_{h,k}]/\{\{[\gamma_{h,k}{}^2 R_{h,k}^2 \cos^2\theta_{h,k}] + [R_{h,k}{}^2 \sin^2\theta_{h,k}]\}^{3/2}\}\}\}\}(\gamma)(v/C)\}$ $[f(\text{shape}, \gamma, \text{dist})]$,

where s is a positive integer representing the number of net electronic charge units.

So, the associated drive force is:

$F_{elect}$ + (Particles reaction drive forces) + (Drag Forces)= $\{Q\ \{\Sigma(k=1, k=o)\ \{\Sigma\,(h=1, h=n_{cone})\ \{[1/(4\pi\varepsilon_0)][\gamma_{h,k}\,(sq_{h,k})\,\mathbf{R}_{h,k}]/\{\{[\gamma_{h,k}{}^2 R_{h,k}^2 \cos^2\theta_{h,k}] + [R_{h,k}{}^2 \sin^2\theta_{h,k}]\}^{3/2}\}\}\}\}(\gamma)(v/C)\}\ [f(\text{shape}, \gamma, \text{dist})]$ + (Particles reaction drive force) + (Drag Forces).

The total electric force exerted on the charged cone by the stream of k particles where the particles are distributed in the background frame over a unit spatial interval in the background frame taking into account the pointed shape effect concentration of the electric field flux at the tip of the bristle is equal to:

$F_{elect} = \{\{\Sigma\,(u=1, u=n_{cone})\ [Q_{u,cone}(x,y,x,t)]\{\Sigma(k=1, k=m)\ \{\Sigma\,(h=1, h=n)\ \{[1/(4\pi\varepsilon_0)][\gamma_{h,k}\,(sq_{h,k})\,\mathbf{R}_{h,k}]/\{\{[\gamma_{h,k}{}^2 R_{h,k}{}^2 \cos^2\theta_{h,k}] + [R_{h,k}{}^2 \sin^2\theta_{h,k}]\}^{3/2}\}\}\}\}\}(\gamma)(v/C)\ [f(\text{shape}, \gamma, \text{dist})]\}$

where the total number of differential area surface charges on the central charged member is equal to $n_{cone}$.

So, the associated drive force is:

$$F_{elect} + \text{(Particles reaction drive force)} + \text{(Drag Forces)}. = \{\{\Sigma\ (u = 1, u = n_{cone})\ [Q_{u,cone}(x,y,x,t)]\{\Sigma(k = 1, k = m)\ \{\Sigma\ (h = 1, h = n)\ \{[1/(4\pi\varepsilon_0)][\gamma_{h,k}\ (sq_{h,k})\ \mathbf{R}_{h,k}]/\{\{[\gamma_{h,k}{}^2 R_{h,k}{}^2 \cos^2\theta_{h,k}] + [R_{h,k}{}^2 \sin^2\theta_{h,k}]\}^{3/2}\}\}\}\}(\gamma)(v/C)\ [f(\text{shape}, \gamma, \text{dist})]\}\} + \text{(Particles reaction drive force)} + \text{(Drag Forces)}.$$

The total step-wise spatially integrated force or energy gain by the space craft based on the electric charge attraction mechanism is therefore equal to:

$$\int[F_{elect} + \text{(Particles reaction drive force)} + \text{(Drag Forces)}.]\bullet dx = \{\int (x_{i1}, x_{i2})\ \{\{\Sigma\ (j = 1, j = w)\ \{\Sigma\ (u = 1, u = n_{cone})\ [Q_{u,cone}(x,y,x,t)]\{\Sigma(k = 1, k = m)\ \{\Sigma\ (h = 1, h = n)\ \{[1/(4\pi\varepsilon_0)][\gamma_{h,k}\ (sq_{h,k})\ \mathbf{R}_{h,k}]/\{\{[\gamma_{h,k}{}^2 R_{h,k}{}^2 \cos^2\theta_{h,k}] + [R_{h,k}{}^2 \sin^2\theta_{h,k}]\}^{3/2}\}\}\}\}(\gamma_j)(v_j/C)\}\ [f(\text{shape}, \gamma, \text{dist})]\}\} + \text{(Particles reaction drive force)} + \text{(Drag Forces)}\}. \bullet dx_j\}$$

Here: $\gamma_j$, $v_j$, $\text{shape}_j$, and $\text{dist}_j$ are the space craft translational Lorentz factor, space craft translational velocity, bristle electric field shape factor, and bristle electrostatic charge distribution; over the jth background spatial interval traveled.

The component of energy gained by the space craft due to the beamed exhaust produced by the energy conversion of the charged mass absorbed into the conical surface is most simply expressed as;

$$E_{gaintrocket}\ \{\int (x_{i1}, x_{i2})\ \{\Sigma\ (j = 1, j = w)\ [\langle[(S)(A)]_j\rangle/C]\} \bullet dx\} = \{\int (x_{i1}, x_{i2})\ \{\Sigma\ (j = 1, j = w)\ \{\langle\{d\ \{\Sigma\ (u = 1, u = n_{cone1})\ [M_{uc}C^2]\}/dt\}\rangle/C\}\ (\gamma_j)(v_j/C)\}\bullet dx_j\}$$

The electro-dynamic-hydrodynamic-plasma-drive feature produces an energy gain component of:

$$E_{gaintedhpd}\ \{\int (x_{i1}, x_{i2})\ \{\Sigma\ (j = 1, j = w)\ \{\{d[(M_{0j})(\Delta v_{j\text{-rel-edhpd}})(\Delta\gamma_{j\text{-rel-edhpd}})]/dt\}\ [\cos\beta]\}\ (\gamma_j)(v_j/C)\} \bullet dx_j\}$$

Taking into account all three propulsive mechanisms and universal expansion, as well as drag energy, the total energy gain by the space craft becomes:

$$E_{gaintotal} + \text{(Drag Energies)} = \{\{\{\int (x_{i1}, x_{i2})\ \{\Sigma\ (j = 1, j = w)\{\Sigma\ (u = 1, u = n_{cone})\ [Q_{u,cone}(x,y,x,t)]\{\Sigma(k = 1, k = m)\ \{\Sigma\ (h = 1, h = n)\ \{[1/(4\pi\varepsilon_0)][\gamma_{h,k}\ (sq_{h,k})\ \mathbf{R}_{h,k}]/\{\{[\gamma_{h,k}{}^2 R_{h,k}{}^2 \cos^2\theta_{h,k}] + [R_{h,k}{}^2 \sin^2\theta_{h,k}]\}^{3/2}\}\}\}\}(\gamma_j)(v_j/C)\}\ [f(\text{shape}, \gamma, \text{dist})]\}\} \bullet dx_j\}$$

$$+ \{\int (x_{i1}, x_{i2})\ \{\Sigma\ (j = 1, j = w)\ \{\langle\{d\ \{\Sigma\ (u = 1, u = n_{cone1})\ [M_{uc}C^2]\}/dt\}\rangle/C\}\ (\gamma_j)(v_j/C)\}\bullet dx_j\}$$

$$+ \{\int (x_{i1}, x_{i2})\ \{\Sigma\ (j = 1, j = w)\ \{\{d[(M_{0j})(\Delta v_{j\text{-rel-edhpd}})(\Delta\gamma_{j\text{-rel-edhpd}})]/dt\}\ [\cos\beta]\}\ (\gamma_j)(v_j/C)\} \bullet dx_j\}\}\ \{(V_{ce}/V_j) = [(z_{present}/z_j)^3] = [(a_{present}/a_j)^3]\}\} + \text{(Drag Energies)}$$

The Lorentz factor achieved by the space craft is equal to:

$$\text{Gamma} = \{\{M_0 C^2 + \{\{\{\int (x_{i1}, x_{i2})\ \{\Sigma\ (j = 1, j = w)\{\Sigma\ (u = 1, u = n_{cone})\ [Q_{u,cone}(x,y,x,t)]\{\Sigma(k = 1, k = m)\ \{\Sigma\ (h = 1, h = n)\ \{[1/(4\pi\varepsilon_0)][\gamma_{h,k}\ (sq_{h,k})\ \mathbf{R}_{h,k}]/\{\{[\gamma_{h,k}{}^2 R_{h,k}{}^2 \cos^2\theta_{h,k}] + [R_{h,k}{}^2 \sin^2\theta_{h,k}]\}^{3/2}\}\}\}\}(\gamma_j)(v_j/C)\}\ [f(\text{shape}, \gamma, \text{dist})]\}\} \bullet dx_j\}$$

$$+ \{\int (x_{i1}, x_{i2})\ \{\Sigma\ (j = 1, j = w)\{\langle\{d\ \{\Sigma\ (u = 1, u = n_{cone1})\ [M_{uc}C^2]\}/dt\}\rangle/C\}(\gamma_j)(v_j/C)\}\bullet dx_j\}$$

$$+ \{\int (x_{i1}, x_{i2})\ \{\Sigma\ (j = 1, j = w)\ \{\{d[(M_{0j})(\Delta v_{j\text{-rel-edhpd}})(\Delta\gamma_{j\text{-rel-edhpd}})]/dt\}\ [\cos\beta]\}\ (\gamma_j)(v_j/C)\} \bullet dx_j\}\}\ \{(V_{ce}/V_j) = [(z_{present}/z_j)^3] = [(a_{present}/a_j)^3]\}\}\} + \text{(Drag Energies)}\}\ /[M_0 C^2]$$

The accrued velocity is provided by:

$$v = c\{\{-\{1/\{\ \{\{M_0 C^2 + \{\{\{\int (x_{i1}, x_{i2})\ \{\Sigma\ (j = 1, j = w)\ \{\Sigma\ (u = 1, u = n_{cone})\ [Q_{u,cone}(x,y,x,t)]\{\Sigma(k = 1, k = m)\ \{\Sigma\ (h = 1, h = n)\ \{[1/(4\pi\varepsilon_0)][\gamma_{h,k}\ (sq_{h,k})\ \mathbf{R}_{h,k}]/\{\{[\gamma_{h,k}{}^2 R_{h,k}{}^2 \cos^2\theta_{h,k}] + [R_{h,k}{}^2 \sin^2\theta_{h,k}]\}^{3/2}\}\}\}\}\ (\gamma_j)(v_j/C)\}\ [f(\text{shape}, \gamma, \text{dist})]\}\}\bullet dx_j\}$$

$$+ \{\int (x_{i1}, x_{i2})\ \{\Sigma\ (j = 1, j = w)\ \{\langle\{d\ \{\Sigma\ (u = 1, u = n_{cone1})\ [M_{uc}C^2]\}/dt\}\rangle/C\}(\gamma_j)(v_j/C)\}\bullet dx_j\}$$

$$+ \{\int (x_{i1}, x_{i2})\ \{\Sigma\ (j = 1, j = w)\ \{\{d[(M_{0j})(\Delta v_{j\text{-rel-edhpd}})(\Delta\gamma_{j\text{-rel-edhpd}})]/dt\}\ [\cos\beta]\}\ (\gamma_j)(v_j/C)\} \bullet dx_j\}\}\ \{(V_{ce}/V_j) = [(z_{present}/z_j)^3] = [(a_{present}/a_j)^3]\}\}\} + \text{(Drag Energies)}\}/[M_0 C^2]\ \}^2\}\} + 1\}^{1/2}\}.$$

We now consider the case where there are, b, bristles in a hairbrush for which the bristles are distributed in a single planar pattern of arbitrary shape and where the array has arbitrary planar dimensions. We will assume that each bristle is electrically charged.

The acquired space craft kinetic energy gain taking into account universal space-time expansion becomes:

$E_{gaintotal}$ + (Drag Energies )= $\{\{\{\int (x_{i1}, x_{i2}) \{\Sigma (j = 1, j = w) \{\Sigma (b = 1, b = z) \{\Sigma (u = 1, u = n_{cone}) [Q_{u,b,cone}(x,y,x,t)]\{\Sigma(k = 1, k = m) \{\Sigma (h =1, h = n) \{[1/(4\pi\varepsilon_0)][\gamma_{h,k,b} (sq_{h,k,b}) \mathbf{R}_{h,k}]/\{\{[\gamma_{h,k,b}{}^2 R_{h,k,b}{}^2 \cos^2 \theta_{h,k,b}] + [R_{h,k,b}{}^2 \sin^2 \theta_{h,k,b}]\}^{3/2}\}\}\}\}(\gamma_j)(v_j/C)\} [f(shape_j, \gamma_j, dist_j)]\} \bullet dx_j\}$

$+ \{\int (x_{i1}, x_{i2}) \{ \Sigma (j = 1, j = w) \{<\{d \{\Sigma (u = 1, u = n_{cone1}) [M_{uc}C^2] \}/dt\}>/C\}(\gamma_j)(v_j/C)\}\bullet dx_j\}$

$+ \{ \int (x_{i1}, x_{i2}) \{\Sigma (j = 1, j = w) \{\{d[(M_{0j})(\Delta v_{j\text{-rel-edhpd}})(\Delta\gamma_{j\text{-rel-edhpd}})]/dt\} [\cos \beta]\} (\gamma_j)(v_j/C)\} \bullet dx_j\}\} \{(V_{ce}/ V_j) = [(z_{present}/ z_j)^3] = [(a_{present}/ a_j)^3]\}\}$ + (Drag Energies)

The Lorentz factor achieved by the space craft is equal to:

Gamma = $\{\{M_0 C^2 + \{\{\{\int (x_{i1}, x_{i2}) \{\Sigma (j = 1, j = w) \{\Sigma (b = 1, b = z) \{\Sigma (u = 1, u = n_{cone}) [Q_{u,b,cone}(x,y,x,t)]\{\Sigma(k = 1, k = m) \{\Sigma (h =1, h = n) \{[1/(4\pi\varepsilon_0)][\gamma_{h,k,b} (sq_{h,k,b}) \mathbf{R}_{h,k}]/\{\{[\gamma_{h,k,b}{}^2 R_{h,k,b}{}^2 \cos^2 \theta_{h,k,b}] + [R_{h,k,b}{}^2 \sin^2 \theta_{h,k,b}]\}^{3/2}\}\}\}\}(\gamma_j)(v_j/C)\} [f(shape_j, \gamma_j, dist_j)]\} \bullet dx_j\}$

$+ \{\int (x_{i1}, x_{i2}) \{ \Sigma (j = 1, j = w) \{<\{d \{\Sigma (u = 1, u = n_{cone1}) [M_{uc}C^2] \}/dt\}>/C\}(\gamma_j)(v_j/C)\}\bullet dx_j\}$

$+ \{ \int (x_{i1}, x_{i2}) \{\Sigma (j = 1, j = w) \{\{d[(M_{0j})(\Delta v_{j\text{-rel-edhpd}})(\Delta\gamma_{j\text{-rel-edhpd}})]/dt\} [\cos \beta]\} (\gamma_j)(v_j/C)\} \bullet dx_j\}\} \{(V_{ce}/ V_j) = [(z_{present}/ z_j)^3] = [(a_{present}/ a_j)^3]\}\}\}$ + (Drag Enerngies)$\}/[M_0 C^2]$

The accrued velocity is equal to:

$v = c\{\{-\{1/\{ \{ \{\{M_0 C^2 + \{\{\{\int (x_{i1}, x_{i2}) \{\Sigma (j = 1, j = w) \{\Sigma (b = 1, b = z) \{\Sigma (u = 1, u = n_{cone}) [Q_{u,b,cone}(x,y,x,t)]\{\Sigma(k = 1, k = m) \{\Sigma (h =1, h = n) \{[1/(4\pi\varepsilon_0)][\gamma_{h,k,b} (sq_{h,k,b}) \mathbf{R}_{h,k}]/\{\{[\gamma_{h,k,b}{}^2 R_{h,k,b}{}^2 \cos^2 \theta_{h,k,b}] + [R_{h,k,b}{}^2 \sin^2 \theta_{h,k,b}]\}^{3/2}\}\}\}\}(\gamma_j)(v_j/C)\} [f(shape_j, \gamma_j, dist_j)]\} \bullet dx_j\}$

$+ \{\int (x_{i1}, x_{i2}) \{ \Sigma (j = 1, j = w) \{<\{d \{\Sigma (u = 1, u = n_{cone1}) [M_{uc}C^2] \}/dt\}>/C\}(\gamma_j)(v_j/C)\}\bullet dx_j\}\}$

$+ \{ \int (x_{i1}, x_{i2}) \{\Sigma (j = 1, j = w) \{\{d[(M_{0j})(\Delta v_{j\text{-rel-edhpd}})(\Delta\gamma_{j\text{-rel-edhpd}})]/dt\} [\cos \beta]\} (\gamma_j)(v_j/C)\} \bullet dx_j\}\}\} \{(V_{ce}/ V_j) = [(z_{present}/ z_j)^3] = [(a_{present}/ a_j)^3]\}\}\}$ + (Drag Energies)$\} /[M_0 C^2] \}^2\}\} + 1\}^{1/2}\}$.

The above formulas are only numerical approximations, but can provide good results when used with modern computers, programs, and spreadsheets.

In the case where a detonated nuclear bomb produces additional charged plasma for which the electrons are diverted to positively charged hair-brush arrays while the positive ions are diverted to a negatively charged hair bush arrays, and where a fraction of the blast ground state neutron flux, (Gound state $N_{flux,}$), and a fraction of the blast excited state neutron flux, (Excited state $N_{flux,}$), and a fraction of photon flux produced in the blast, $Ph_{flux}$, is converted to spacecraft thrust directly, and/or indirectely via electrical propulsion systems, the spacecraft thrust force becomes:

$F_{elect}$ = $\{[(electrons):(Q_{natural})]$ $\{[1/(4\pi\varepsilon_0)][\gamma_{average,electrons,natural,ship\text{-}frame} \mathbf{R}]/\{\{[\gamma_{average,electrons,natural,ship\text{-}frame}{}^2 R^2 \cos^2 \theta] + [R^2 \sin^2 \theta]\}^{3/2}\}\}\}$ $+\{[(electrons):(Q_{bomb})]$ $\{[1/(4\pi\varepsilon_0)][\gamma_{average,bomb,ship\text{-}frame} \mathbf{R}]/\{\{[\gamma_{average,bomb,ship\text{-}frame}{}^2 R^2 \cos^2 \theta] + [R^2 \sin^2 \theta]\}^{3/2}\}\}\}$ + {[Σ(n = 1; n = number of positive background ion species):(nth ion species)]: $(Q_{natural})]$ $\{[1/(4\pi\varepsilon_0)][\gamma_{average,nth\text{-}ion\text{-}species,natural,ship\text{-}frame} \mathbf{R}]/\{\{[\gamma_{average,nth\text{-}ion\text{-}species,natural,ship\text{-}frame}{}^2 R^2 \cos^2 \theta] + [R^2 \sin^2 \theta]\}^{3/2}\}\}\}$ + {[Σ(m = 1; m = number of positive bomb ion species):(mth ion species)]: $(Q_{bomb})]$ $\{[1/(4\pi\varepsilon_0)][\gamma_{average,mth\text{-}ion\text{-}species,bomb,ship\text{-}frame} \mathbf{R}]/\{\{[\gamma_{average,mth\text{-}ion\text{-}species,bomb,ship\text{-}frame}{}^2 R^2 \cos^2 \theta] + [R^2 \sin^2 \theta]\}^{3/2}\}\}\}$ + {[Σ(n = 1; n = number of negative background ion species):(nth ion species)]: $(Q_{natural})]$ $\{[1/(4\pi\varepsilon_0)][\gamma_{average,nth\text{-}ion\text{-}species,natural,ship\text{-}frame} \mathbf{R}]/\{\{[\gamma_{average,nth\text{-}ion\text{-}species,bomb,ship\text{-}frame}{}^2 R^2 \cos^2 \theta] + [R^2 \sin^2 \theta]\}^{3/2}\}\}\}$ + {[Σ(m = 1; m = number of negative bomb ion species):(mth ion species)]: $(Q_{bomb})]$ $\{[1/(4\pi\varepsilon_0)][\gamma_{average,mth\text{-}ion\text{-}species,bomb,ship\text{-}frame} \mathbf{R}]/\{\{[\gamma_{average,mth\text{-}ion\text{-}species,bomb,ship\text{-}frame}{}^2 R^2 \cos^2 \theta] + [R^2 \sin^2 \theta]\}^{3/2}\}\}\}$ + {d{[(Gound state $N_{flux}$)($v_{average,ship\text{-}frame}$)($\gamma_{average,ship\text{-}frame}$)] for ground state neutrons}/$dt_{ship}$} + {d {Σ(z = 1; z = number of excited neutron states):[(zth excited state $N_{flux,}$)($v_{average,ship\text{-}frame}$)($\gamma_{average,ship\text{-}frame}$)] for excited state neutrons}/dt} + [<$S_{background,photons}$>/c] + [[<$S_{bomb,photons}$>/c] ($Ph_{flux}$)]

For inclusuin if astrodynamuc drag, we consider all above mechanism have non-zero drag components but where the spacecraft is positively accelerated, the propulsion thrust is greater than the drag.

Accordingly, we produce the following formula for net thrust.

$F_{elect}$ = {{[(electrons):($Q_{natural}$)] {[1/(4π$\varepsilon_0$)][$\gamma_{\text{average,electrons,natural,ship-frame}}$ **R**]/{{[$\gamma_{\text{average,electrons,natural,ship-frame}}{}^2$ $R^2 \cos^2\theta$] + [$R^2 \sin^2\theta$]}$^{3/2}$}}} +{[(electrons):($Q_{bomb}$)] {[1/(4π$\varepsilon_0$)][$\gamma_{\text{average,bomb,ship-frame}}$ **R**]/{{[$\gamma_{\text{average,bomb,ship-frame}}{}^2$ $R^2 \cos^2\theta$] + [$R^2 \sin^2\theta$]}$^{3/2}$}}} + {[Σ(n = 1; n = number of positive background ion species):(nth ion species)]: ($Q_{natural}$)] {[1/(4π$\varepsilon_0$)][$\gamma_{\text{average,nth-ion-species,natural,ship-frame}}$ **R**]/{{[$\gamma_{\text{average,nth-ion-species,natural,ship-frame}}{}^2$ $R^2 \cos^2\theta$] + [$R^2 \sin^2\theta$]}$^{3/2}$}}} + {[Σ(m = 1; m = number of positive bomb ion species):(mth ion species)]: ($Q_{bomb}$)] {[1/(4π$\varepsilon_0$)][$\gamma_{\text{average,mth-ion-species,bomb,ship-frame}}$ **R**]/{{[$\gamma_{\text{average,mth-ion-species,bomb,ship-frame}}{}^2$ $R^2 \cos^2\theta$] + [$R^2 \sin^2\theta$]}$^{3/2}$}}} + {[Σ(n = 1; n = number of negative background ion species):(nth ion species)]: ($Q_{natural}$)] {[1/(4π$\varepsilon_0$)][$\gamma_{\text{average,nth-ion-species,natural,ship-frame}}$ **R**]/{{[$\gamma_{\text{average,nth-ion-species,bomb,ship-frame}}{}^2$ $R^2 \cos^2\theta$] + [$R^2 \sin^2\theta$]}$^{3/2}$}}} + {[Σ(m = 1; m = number of negative bomb ion species):(mth ion species)]: ($Q_{bomb}$)] {[1/(4π$\varepsilon_0$)][$\gamma_{\text{average,mth-ion-species,bomb,ship-frame}}$ **R**]/{{[$\gamma_{\text{average,mth-ion-species,bomb,ship-frame}}{}^2$ $R^2 \cos^2\theta$] + [$R^2 \sin^2\theta$]}$^{3/2}$}}} + {d{[(Gound state $N_{flux}$)($v_{average,ship\text{-}frame}$)($\gamma_{average,ship\text{-}frame}$)] for ground state neutrons}/$dt_{ship}$} + {d {Σ(z = 1; z = number of excited neutron states):[(zth excited state $N_{flux,}$)($v_{average,ship\text{-}frame}$)($\gamma_{average,ship\text{-}frame}$)] for excited state neutrons}/dt} + [<$S_{background,photons}$>/c] + [[<$S_{bomb,photons}$>/c] ($Ph_{flux}$)]}

+ {Drag based on the following propulsive force inducing mechanisms {{[(electrons):($Q_{natural}$)] {[1/(4π$\varepsilon_0$)][$\gamma_{\text{average,electrons,natural,ship-frame}}$ **R**]/{{[$\gamma_{\text{average,electrons,natural,ship-frame}}{}^2$ $R^2 \cos^2\theta$] + [$R^2 \sin^2\theta$]}$^{3/2}$}}} +{[(electrons):($Q_{bomb}$)] {[1/(4π$\varepsilon_0$)][$\gamma_{\text{average,bomb,ship-frame}}$ **R**]/{{[$\gamma_{\text{average,bomb,ship-frame}}{}^2$ $R^2 \cos^2\theta$] + [$R^2 \sin^2\theta$]}$^{3/2}$}}} + {[Σ(n = 1; n = number of positive background ion species):(nth ion species)]: ($Q_{natural}$)] {[1/(4π$\varepsilon_0$)][$\gamma_{\text{average,nth-ion-species,natural,ship-frame}}$ **R**]/{{[$\gamma_{\text{average,nth-ion-species,natural,ship-frame}}{}^2$ $R^2 \cos^2\theta$] + [$R^2 \sin^2\theta$]}$^{3/2}$}}} + {[Σ(m = 1; m = number of positive bomb ion species):(mth ion species)]: ($Q_{bomb}$)] {[1/(4π$\varepsilon_0$)][$\gamma_{\text{average,mth-ion-species,bomb,ship-frame}}$ **R**]/{{[$\gamma_{\text{average,mth-ion-species,bomb,ship-frame}}{}^2$ $R^2 \cos^2\theta$] + [$R^2 \sin^2\theta$]}$^{3/2}$}}} + {[Σ(n = 1; n = number of negative background ion species):(nth ion species)]: ($Q_{natural}$)] {[1/(4π$\varepsilon_0$)][$\gamma_{\text{average,nth-ion-species,natural,ship-frame}}$ **R**]/{{[$\gamma_{\text{average,nth-ion-species,bomb,ship-frame}}{}^2$ $R^2 \cos^2\theta$] + [$R^2 \sin^2\theta$]}$^{3/2}$}}} + {[Σ(m = 1; m = number of negative bomb ion species):(mth ion species)]: ($Q_{bomb}$)] {[1/(4π$\varepsilon_0$)][$\gamma_{\text{average,mth-ion-species,bomb,ship-frame}}$ **R**]/{{[$\gamma_{\text{average,mth-ion-species,bomb,ship-frame}}{}^2$ $R^2 \cos^2\theta$] + [$R^2 \sin^2\theta$]}$^{3/2}$}}} + {d{[(Gound state $N_{flux}$)($v_{average,ship\text{-}frame}$)($\gamma_{average,ship\text{-}frame}$)] for ground state neutrons}/$dt_{ship}$} + {d {Σ(z = 1; z = number of excited neutron states):[(zth excited state $N_{flux,}$)($v_{average,ship\text{-}frame}$)($\gamma_{average,ship\text{-}frame}$)] for excited state neutrons}/dt} + [<$S_{background,photons}$>/c] + [[<$S_{bomb,photons}$>/c] ($Ph_{flux}$)]}}

Now, we can consider scenanrios for which the bombs explode alongside a non-nuclear pellet. Accordingly, the non-nuclear pellet would be ionized by the atomic blast of the nuclear pellets to this liberate more electrical charge for use as a reaction mass. We thus obtain the following formula for net propulsion power.

$F_{elect}$ = {{[(electrons):($Q_{natural}$)] {[1/(4π$\varepsilon_0$)][$\gamma_{\text{average,electrons,natural,ship-frame}}$ **R**]/{{[$\gamma_{\text{average,electrons,natural,ship-frame}}{}^2$ $R^2 \cos^2\theta$] + [$R^2 \sin^2\theta$]}$^{3/2}$}}} +{[(electrons):($Q_{bomb}$)] {[1/(4π$\varepsilon_0$)][$\gamma_{\text{average,bomb,ship-frame}}$ **R**]/{{[$\gamma_{\text{average,bomb,ship-frame}}{}^2$ $R^2 \cos^2\theta$] + [$R^2 \sin^2\theta$]}$^{3/2}$}}} +{[(electrons):($Q_{\text{non-nuclear pellet}}$)] {[1/(4π$\varepsilon_0$)][$\gamma_{\text{average,non-nuclear pellet,ship-frame}}$ **R**]/{{[$\gamma_{\text{average,non-nuclear pellet,ship-frame}}{}^2$ $R^2 \cos^2\theta$] + [$R^2 \sin^2\theta$]}$^{3/2}$}}} + {[Σ(n = 1; n = number of positive background ion species):(nth ion species)]: ($Q_{natural}$)] {[1/(4π$\varepsilon_0$)][$\gamma_{\text{average,nth-ion-species,natural,ship-frame}}$ **R**]/{{[$\gamma_{\text{average,nth-ion-species,natural,ship-frame}}{}^2$ $R^2 \cos^2\theta$] + [$R^2 \sin^2\theta$]}$^{3/2}$}}} + {[Σ(m = 1; m = number of positive bomb ion species):(mth ion species)]: ($Q_{bomb}$)] {[1/(4π$\varepsilon_0$)][$\gamma_{\text{average,mth-ion-species,bomb,ship-frame}}$ **R**]/{{[$\gamma_{\text{average,mth-ion-species,bomb,ship-frame}}{}^2$ $R^2 \cos^2\theta$] + [$R^2 \sin^2\theta$]}$^{3/2}$}}} + {[Σ(g = 1; g = number of positive bomb heated non-nuclear pellet sourced ion species):(gth ion species)]: ($Q_{\text{non-nuclear pellet}}$)] {[1/(4π$\varepsilon_0$)][$\gamma_{\text{average,gth-ion-species,non-nuclear pellet,,ship-frame}}$ **R**]/{{[$\gamma_{\text{average,gth-ion-species,non-nuclear pellet,ship-frame}}{}^2$ $R^2 \cos^2\theta$] + [$R^2 \sin^2\theta$]}$^{3/2}$}}} +{[Σ(e = 1; e = number of negative non-nuclear pellet heated non-nuclear pellet sourced ion species):(eth ion species)]: ($Q_{\text{non-nuclear pellet}}$)] {[1/(4π$\varepsilon_0$)][$\gamma_{\text{average,gth-ion-species,non-nuclear pellet,,ship-frame}}$ **R**]/{{[$\gamma_{\text{average,gth-ion-species,non-nuclear pellet,ship-frame}}{}^2$ $R^2 \cos^2\theta$] + [$R^2 \sin^2\theta$]}$^{3/2}$}}} + {[Σ(n = 1; n = number of negative background ion species):(nth ion species)]: ($Q_{natural}$)] {[1/(4π$\varepsilon_0$)][$\gamma_{\text{average,nth-ion-species,natural,ship-frame}}$ **R**]/{{[$\gamma_{\text{average,nth-ion-species,bomb,ship-frame}}{}^2$ $R^2 \cos^2\theta$] + [$R^2 \sin^2\theta$]}$^{3/2}$}}} + {[Σ(m = 1; m = number of negative bomb ion species):(mth ion species)]: ($Q_{bomb}$)] {[1/(4π$\varepsilon_0$)][$\gamma_{\text{average,mth-ion-species,bomb,ship-frame}}$ **R**]/{{[$\gamma_{\text{average,mth-ion-species,bomb,ship-frame}}{}^2$ $R^2 \cos^2\theta$] + [$R^2 \sin^2\theta$]}$^{3/2}$}}} + {d{[(Gound state $N_{flux}$)($v_{average,ship\text{-}frame}$)($\gamma_{average,ship\text{-}frame}$)] for ground state neutrons}/$dt_{ship}$} + {d {Σ(z = 1; z = number of excited neutron states):[(zth excited state $N_{flux,}$)($v_{average,ship\text{-}frame}$)($\gamma_{average,ship\text{-}frame}$)] for excited state neutrons}/dt} + [<$S_{background,photons}$>/c] + [[<$S_{bomb,photons}$>/c] ($Ph_{flux}$)] + [[<$S_{\text{non-nuclear pellet,photons}}$>/c] ($Ph_{flux}$)]}

+ {Drag based on the following propulsive force inducing mechanisms {{[(electrons):($Q_{natural}$)] {[1/(4π$\varepsilon_0$)][$\gamma_{\text{average,electrons,natural,ship-frame}}$ **R**]/{{[$\gamma_{\text{average,electrons,natural,ship-frame}}{}^2$ $R^2 \cos^2\theta$] + [$R^2 \sin^2\theta$]}$^{3/2}$}}} +{[(electrons):($Q_{bomb}$)] {[1/(4π$\varepsilon_0$)][$\gamma_{\text{average,bomb,ship-frame}}$ **R**]/{{[$\gamma_{\text{average,bomb,ship-frame}}{}^2$ $R^2 \cos^2\theta$] + [$R^2 \sin^2\theta$]}$^{3/2}$}}} +{[(electrons):($Q_{\text{non-nuclear pellet}}$)] {[1/(4π$\varepsilon_0$)][$\gamma_{\text{average,non-nuclear pellet,ship-frame}}$ **R**]/{{[$\gamma_{\text{average,non-nuclear pellet,ship-frame}}{}^2$ $R^2 \cos^2\theta$] + [$R^2 \sin^2\theta$]}$^{3/2}$}}} + {[Σ(n = 1; n = number of positive background ion species):(nth ion species)]: ($Q_{natural}$)] {[1/(4π$\varepsilon_0$)][$\gamma_{\text{average,nth-ion-species,natural,ship-frame}}$ **R**]/{{[$\gamma_{\text{average,nth-ion-species,natural,ship-frame}}{}^2$ $R^2 \cos^2\theta$] + [$R^2 \sin^2\theta$]}$^{3/2}$}}} + {[Σ(m = 1; m = number of positive bomb ion species):(mth ion species)]: ($Q_{bomb}$)] {[1/(4π$\varepsilon_0$)][$\gamma_{\text{average,mth-ion-species,bomb,ship-frame}}$ **R**]/{{[$\gamma_{\text{average,mth-ion-species,bomb,ship-frame}}{}^2$ $R^2 \cos^2\theta$] + [$R^2 \sin^2\theta$]}$^{3/2}$}}} + {[Σ(g = 1; g = number of positive bomb heated non-nuclear pellet sourced ion species):(gth ion species)]: ($Q_{\text{non-nuclear pellet}}$)] {[1/(4π$\varepsilon_0$)][$\gamma_{\text{average,gth-ion-species,non-nuclear pellet,,ship-frame}}$ **R**]/{{[$\gamma_{\text{average,gth-ion-species,non-nuclear pellet,ship-frame}}{}^2$ $R^2 \cos^2\theta$] + [$R^2 \sin^2\theta$]}$^{3/2}$}}} +{[Σ(e = 1; e = number of negative non-nuclear pellet heated non-nuclear pellet sourced ion species):(eth ion species)]: ($Q_{\text{non-nuclear pellet}}$)] {[1/(4π$\varepsilon_0$)][$\gamma_{\text{average,gth-ion-species,non-nuclear pellet,,ship-frame}}$ **R**]/{{[$\gamma_{\text{average,gth-ion-species,non-nuclear pellet,ship-frame}}{}^2$ $R^2 \cos^2\theta$] + [$R^2 \sin^2\theta$]}$^{3/2}$}}} + {[Σ(n = 1; n = number of negative background ion species):(nth ion species)]: ($Q_{natural}$)] {[1/(4π$\varepsilon_0$)][$\gamma_{\text{average,nth-ion-species,natural,ship-frame}}$ **R**]/{{[$\gamma_{\text{average,nth-ion-species,bomb,ship-frame}}{}^2$ $R^2 \cos^2\theta$] +

$[R^2 \sin^2 \theta]\}^{3/2}\}\}\}$ + {[Σ(m = 1; m = number of negative bomb ion species):(mth ion species)]: ($Q_{bomb}$)] {[1/($4\pi\varepsilon_0$)][$\gamma_{average,mth\text{-}ion\text{-}species,bomb,ship\text{-}frame}$ R]/{{[$\gamma_{average,mth\text{-}ion\text{-}species,bomb,ship\text{-}frame}{}^{2}\ R^2 \cos^2 \theta$] + [$R^2 \sin^2 \theta$]}$^{3/2}$}}} + {d{[(Gound state $N_{flux}$)($v_{average,ship\text{-}frame}$)($\gamma_{average,ship\text{-}frame}$)] for ground state neutrons}/$dt_{ship}$} + {d {Σ(z = 1; z = number of excited neutron states):[(zth excited state $N_{flux,}$)($v_{average,ship\text{-}frame}$)($\gamma_{average,ship\text{-}frame}$)] for excited state neutrons}/dt} + [$<S_{background,photons}>$/c] + [[$<S_{bomb,photons}>$/c] ($Ph_{flux}$)] + [[$<S_{non\text{-}nuclear\ pellet,photons}>$/c] ($Ph_{flux}$)]}}

**RELATIVISTIC PINWHEEL DYNAMO EFFECT AUGMENTED BY ELECTRODYNAMIC HAIR BRUSH.**

The non-relativistic Lorentz Force can be defined as:

$$F = q\,(E + v \times B)$$

Where q is the charge being acted on, E is the electric field, v is the velocity of the loop through the magnetic field, and B is the magnetic field and v x B is the cross-product of v and B.

For a non-relativistic dynamo, the electromotive force on a wire loop is:

$\varepsilon = (1/q) \int F \bullet dL = \int (E + v \times B) \bullet dL$ where $F \bullet dL$ is the scalar product or dot product of F and dL.

dL here is a differential or infinitesimal arc length along the loop.

Here, the line integral is evaluated along the length of the loop or the path of the loop.

For N loops of the same shape and size which is effectively N turns of a single loop, the electromotive force on a wire loop is:

$\varepsilon = N\,[(1/q) \int F \bullet dL] = N\,[\int (E + v \times B) \bullet dL]$ where $F \bullet dL$ is the scalar product or dot product of F and dL.

For N loops for which the size and shape of the loops can vary, the following approximation describes the voltage within the coil:

$$\varepsilon = \Sigma\,(n = 1, n = N_{coil})\ [(1/q) \int F \bullet dL]_n = \Sigma\,(n = 1, n = N_{coil})\ [\int (E + v \times B) \bullet dL]_n$$

Now, for the jth time step in the loop frame or corresponding travel interval, the relativistic Lorentz force may be expressed as:

$$F = q\ \{\{Ex_j + \{\{\{1 - [(v_j/C)^2]\}^{-1/2}\}[E_{y,j} - (v_j B_{z,j})]\} + \{\{\{1 - [(v_j/C)^2]\}^{-1/2}\}[E_{z,j} + (v_j B_{y,j})]\}\} + \{v \times \{\{Bx_j + \{\{1 - [(v_j/C)^2]\}^{-1/2}\{B_{y,j} + \{[[v_j/[C^2]]E_{z,j}\}\}\} + \{\{1 - [(v_j/C)^2]\}^{-1/2}\{B_{z,j} - \{[[v_j/[C^2]]E_{y,j}\}\}\}\}\}\}\}$$

For a relativistic dynamo, the electromotive force on a wire loop in the loop frame is:

$$\varepsilon = (1/q) \int \bullet dL = \int \{\{Ex_j + \{\{\{1 - [(v_j/C)^2]\}^{-1/2}\}[E_{y,j} - (v_j B_{z,j})]\} + \{\{\{1 - [(v_j/C)^2]\}^{-1/2}\}[E_{z,j} + (v_j B_{y,j})]\}\} + \{v \times \{\{Bx_j + \{\{1 - [(v_j/C)^2]\}^{-1/2}\{B_{y,j} + \{[[v_j/[C^2]]E_{z,j}\}\}\} + \{\{1 - [(v_j/C)^2]\}^{-1/2}\{B_{z,j} - \{[[v_j/[C^2]]E_{y,j}\}\}\}\}\}\}\} \bullet dL$$

For N loops of the same shape and size which is effectively N turns of a single loop, the electromotive force on the coil in the coil frame is:

$\varepsilon = N\,[(1/q) \int F \bullet dL] = N\ \{\int\{\{Ex_j + \{\{\{1 - [(v_j/C)^2]\}^{-1/2}\}[E_{y,j} - (v_j B_{z,j})]\} + \{\{\{1 - [(v_j/C)^2]\}^{-1/2}\}[E_{z,j} + (v_j B_{y,j})]\}\} + \{v \times \{\{Bx_j + \{\{1 - [(v_j/C)^2]\}^{-1/2}\{B_{y,j} + \{[[v_j/[C^2]]E_{z,j}\}\}\} + \{\{1 - [(v_j/C)^2]\}^{-1/2}\{B_{z,j} - \{[[v_j/[C^2]]E_{y,j}\}\}\}\}\}\}\} \bullet dL\}$ where $F \bullet dL$ is the scalar product or dot product of F and dL.

For N loops for which the size and shape of the loops can vary, the following approximation describes the voltage within the coil in the coil frame:

$$\varepsilon = \Sigma\ (n = 1, n = N_{coil})\ [(1/q) \int F \bullet dL]_n = \Sigma\ (n = 1, n = N_{coil})\ \{\int\{\{Ex_j + \{\{\{1 - [(v_j /C)^2]\}^{-1/2}\}[E_{y,j} - (v_jB_{z,j})]\}\} + \{\{\{1 - [(v_j /C)^2]\}^{-1/2}\}[E_{z,j} + (v_j B_{y,j})]\}\} + \{v \times \{\{Bx_j + \{\{1 - [(v_j/C)^2]\}^{-1/2} \{B_{y,j} + \{[[v_j/[C^2]]E_{z,j}\}\}\} + \{\{1 - [(v_j/C)^2]\}^{-1/2} \{B_{z,j} - \{[[v_j/[C^2]]E_{y,j}\}\}\}\}\}\} \bullet dL\}_n.$$

Taking into account the composite velocity of the rotor blades and coils contained within the rotor blades, we obtain an approximate but more detailed description of the electromotive force generated within the coils.

Accordingly; the EMF for N uniformly sized and shaped coils the coil in the coil frame becomes:

$$\varepsilon = N\ [(1/q) \int F \bullet dL] = N\ \{\int\{\{Ex_j + \{\{\{1 - [((V_{t,j} +_{\text{(relativistic composition)}} V_{r,j})\ /C)^2]\}^{-1/2}\}[E_{y,j} - (v_jB_{z,j})]\} +_{\text{(relativistic composition)}} \{\{\{1 - [((V_{t,j} +_{\text{(relativistic composition)}} V_{r,j})_j /C)^2]\}^{-1/2}\}[E_{z,j} + ((V_{t,j} +_{\text{(relativistic composition)}} V_{r,j})B_{y,j})]\}\} + \{(V_{t,j} +_{\text{(relativistic composition)}} V_{r,j}) \times \{\{Bx_j + \{\{1 - [((V_{t,j} +_{\text{(relativistic composition)}} V_{r,j})/C)^2]\}^{-1/2} \{B_{y,j} + \{[[(V_{t,j} +_{\text{(relativistic composition)}} V_{r,j})/[C^2]]E_{z,j}\}\}\} + \{\{1 - [((V_{t,j} +_{\text{(relativistic composition)}} V_{r,j})/C)^2]\}^{-1/2} \{B_{z,j} - \{[[(V_{t,j} +_{\text{(relativistic composition)}} V_{r,j})/[C^2]]E_{y,j}\}\}\}\}\}\} \bullet dL\};$$

where F • dL is the scalar product or dot product of F and dL.

For N loops for which the size and shape of the loops can vary, the following approximation describes the voltage within the coil in the coil in the coil frame:

$$\varepsilon = \Sigma\ (n = 1, n = N_{coil})\ [(1/q) \int F \bullet dL]_n = \Sigma\ (n = 1, n = N_{coil})\ \{\int\{\{Ex_j + \{\{\{1 - [((V_{t,j} +_{\text{(relativistic composition)}} V_{r,j})\ /C)^2]\}^{-1/2}\}[E_{y,j} - (v_jB_{z,j})]\} + \{\{\{1 - [((V_{t,j} +_{\text{(relativistic composition)}} V_{r,j})_j /C)^2]\}^{-1/2}\}[E_{z,j} + ((V_{t,j} +_{\text{(relativistic composition)}} V_{r,j})B_{y,j})]\}\} + \{(V_{t,j} +_{\text{(relativistic composition)}} V_{r,j}) \times \{\{Bx_j + \{\{1 - [((V_{t,j} +_{\text{(relativistic composition)}} V_{r,j})/C)^2]\}^{-1/2} \{B_{y,j} + \{[[(V_{t,j} +_{\text{(relativistic composition)}} V_{r,j})/[C^2]]E_{z,j}\}\}\} + \{\{1 - [((V_{t,j} +_{\text{(relativistic composition)}} V_{r,j})/C)^2]\}^{-1/2} \{B_{z,j} - \{[[(V_{t,j} +_{\text{(relativistic composition)}} V_{r,j})/[C^2]]E_{y,j}\}\}\}\}\}\} \bullet dL\}_n;$$

where F • dL is the scalar product or dot product of F and dL.

Formulating explicitly for the rotor velocity components in the spacecraft reference frame, we obtain;

Accordingly; the EMF for N uniformly sized and shaped coils the coil in the coil frame becomes:

$$\varepsilon = N\ [(1/q) \int F \bullet dL] = N\ \{\int\{\{Ex_j + \{\{\{1 - [(((\{V_{t,j} + V_{rpar,j} + \{\{\{1 - \{[IV_{t,j}\ I\ ^2]/[C\ ^2]\}\}\ ^{1/2}\}(V_{rperp,j})\}\}/\{1 + \{[V_{t,j} \text{ dot } V_{r,j}]/[C\ ^2]\}\}\})\ /C)^2]\}^{-1/2}\}[E_{y,j} - (v_jB_{z,j})]\} + \{\{\{1 - [(((\{V_{t,j} + V_{rpar,j} + \{\{\{1 - \{[IV_{t,j}\ I\ ^2]/[C\ ^2]\}\}\ ^{1/2}\}(V_{rperp,j})\}\}/\{1 + \{[V_{t,j} \text{ dot } V_{r,j}]/[C\ ^2]\}\}\})\ /C)^2]\}^{-1/2}\}[E_{z,j} + (((\{V_{t,j} + V_{rpar,j} + \{\{\{1 - \{[IV_{t,j}\ I\ ^2]/[C\ ^2]\}\}\ ^{1/2}\}(V_{rperp,j})\}\}/\{1 + \{[V_{t,j} \text{ dot } V_{r,j}]/[C\ ^2]\}\}\})B_{y,j})]\}\} + \{((\{V_{t,j} + V_{rpar,j} + \{\{\{1 - \{[IV_{t,j}\ I\ ^2]/[C\ ^2]\}\}\ ^{1/2}\}(V_{rperp,j})\}\}/\{1 + \{[V_{t,j} \text{ dot } V_{r,j}]/[C\ ^2]\}\}\}) \times \{\{Bx_j + \{\{1 - [(((\{V_{t,j} + V_{rpar,j} + \{\{\{1 - \{[IV_{t,j}I\ ^2]/[C\ ^2]\}\}\ ^{1/2}\}(V_{rperp,j})\}\}/\{1 + \{[V_{t,j} \text{ dot } V_{r,j}]/[C\ ^2]\}\}\})/C)^2]\}^{-1/2} \{B_{y,j} + \{[[((\{V_{t,j} + V_{rpar,j} + \{\{\{1 - \{[IV_{t,j}\ I\ ^2]/[C\ ^2]\}\}\ ^{1/2}\}(V_{rperp,j})\}\}/\{1 + \{[V_{t,j} \text{ dot } V_{r,j}]/[C\ ^2]\}\}\})/[C^2]]E_{z,j}\}\}\} + \{\{1 - [(((\{V_{t,j} + V_{rpar,j} + \{\{\{1 - \{[IV_{t,j}I\ ^2]/[C\ ^2]\}\}\ ^{1/2}\}(V_{rperp,j})\}\}/\{1 + \{[V_{t,j} \text{ dot } V_{r,j}]/[C\ ^2]\}\}\})/C)^2]\}^{-1/2} \{B_{z,j} - \{[[((\{V_{t,j} + V_{rpar,j} + \{\{\{1 - \{[IV_{t,j}I\ ^2]/[C\ ^2]\}\}\ ^{1/2}\}(V_{rperp,j})\}\}/\{1 + \{[V_{t,j} \text{ dot } V_{r,j}]/[C\ ^2]\}\}\})/[C^2]]E_{y,j}\}\}\}\}\}\} \bullet dL\};$$

For N loops for which the size and shape of the loops can vary, the following approximation describes the voltage within the coil in the coil in the coil frame:

$$\varepsilon = \Sigma\ (n = 1, n = N_{coil})\ [(1/q) \int F \bullet dL]_n = \Sigma\ (n = 1, n = N_{coil})\ \{\int\{\{Ex_j + \{\{\{1 - [(((\{V_{t,j} + V_{rpar,j} + \{\{\{1 - \{[IV_{t,j}\ I\ ^2]/[C\ ^2]\}\}\ ^{1/2}\}(V_{rperp,j})\}\}/\{1 + \{[V_{t,j} \text{ dot } V_{r,j}]/[C\ ^2]\}\}\})\ /C)^2]\}^{-1/2}\}[E_{y,j} - (v_jB_{z,j})]\} + \{\{\{1 - [(((\{V_{t,j} + V_{rpar,j} + \{\{\{1 - \{[IV_{t,j}\ I\ ^2]/[C\ ^2]\}\}\ ^{1/2}\}(V_{rperp,j})\}\}/\{1 + \{[V_{t,j} \text{ dot } V_{r,j}]/[C\ ^2]\}\}\})\ /C)^2]\}^{-1/2}\}[E_{z,j} + (((\{V_{t,j} + V_{rpar,j} + \{\{\{1 - \{[IV_{t,j}\ I\ ^2]/[C\ ^2]\}\}\ ^{1/2}\}(V_{rperp,j})\}\}/\{1 + \{[V_{t,j} \text{ dot } V_{r,j}]/[C\ ^2]\}\}\})B_{y,j})]\}\} + \{((\{V_{t,j} + V_{rpar,j} + \{\{\{1 - \{[IV_{t,j}\ I\ ^2]/[C\ ^2]\}\}\ ^{1/2}\}(V_{rperp,j})\}\}/\{1 + \{[V_{t,j} \text{ dot } V_{r,j}]/[C\ ^2]\}\}\}) \times \{\{Bx_j + \{\{1 - [(((\{V_{t,j} + V_{rpar,j} + \{\{\{1 - \{[IV_{t,j}I\ ^2]/[C\ ^2]\}\}\ ^{1/2}\}(V_{rperp,j})\}\}/\{1 + \{[V_{t,j} \text{ dot } V_{r,j}]/[C\ ^2]\}\}\})/C)^2]\}^{-1/2} \{B_{y,j} + \{[[((\{V_{t,j} + V_{rpar,j} + \{\{\{1 - \{[IV_{t,j}\ I\ ^2]/[C\ ^2]\}\}\ ^{1/2}\}(V_{rperp,j})\}\}/\{1 + \{[V_{t,j} \text{ dot } V_{r,j}]/[C\ ^2]\}\}\})/[C^2]]E_{z,j}\}\}\} + \{\{1 - [(((\{V_{t,j} + V_{rpar,j} + \{\{\{1 - \{[IV_{t,j}I\ ^2]/[C\ ^2]\}\}\ ^{1/2}\}(V_{rperp,j})\}\}/\{1 + \{[V_{t,j} \text{ dot } V_{r,j}]/[C\ ^2]\}\}\})/C)^2]\}^{-1/2} \{B_{z,j} - \{[[((\{V_{t,j} + V_{rpar,j} + \{\{\{1 - \{[IV_{t,j}I\ ^2]/[C\ ^2]\}\}\ ^{1/2}\}(V_{rperp,j})\}\}/\{1 + \{[V_{t,j} \text{ dot } V_{r,j}]/[C\ ^2]\}\}\})/[C^2]]E_{y,j}\}\}\}\}\}\} \bullet dL\}_n;$$

where F • dL is the scalar product or dot product of F and dL.

Now, electrical power though a conductive loop can be expressed as:

$P = QV/t = (I)(V);$

where P is the work done per unit of time.

Accordingly; the power for N uniformly sized and shaped coils the coil in the coil frame including the electric hair brush mechanism with detonated bombs becomes:

$P_{electric}$ = ($I_{electric}$){N [(1/q) ∫ F • dL]} = ($I_{electric}$){N{∫{{$Ex_j$ + {{{1 – [(({{$V_{t,j}$ + $V_{rpar,j}$ + {{{1 – {[I$V_{t,j}$ I $^{2}$]/[C $^{2}$]}} $^{1/2}$}($V_{rperp,j}$)}}}/{1 + {[$V_{t,j}$ dot $V_{r,j}$]/[C $^{2}$]}}}}) /C)$^{2}$]}$^{-1/2}$ }[$E_{y,j}$ - ($v_j B_{z,j}$)]} + {{{1 – [(({{$V_{t,j}$ + $V_{rpar,j}$ + {{{1 – {[I$V_{t,j}$ I $^{2}$]/[C $^{2}$]}} $^{1/2}$}($V_{rperp,j}$)}}}/{1 + {[$V_{t,j}$ dot $V_{r,j}$]/[C $^{2}$]}}}}) /C)$^{2}$]}$^{-1/2}$ }[$E_{z,j}$ + (({{$V_{t,j}$ + $V_{rpar,j}$ + {{{1 – {[I$V_{t,j}$ I $^{2}$]/[C $^{2}$]}} $^{1/2}$}($V_{rperp,j}$)}}}/{1 + {[$V_{t,j}$ dot $V_{r,j}$]/[C $^{2}$]}}}})$B_{y,j}$)]}} + {({{$V_{t,j}$ + $V_{rpar,j}$ + {{{1 – {[I$V_{t,j}$ I $^{2}$]/[C $^{2}$]}} $^{1/2}$}($V_{rperp,j}$)}}}/{1 + {[$V_{t,j}$ dot $V_{r,j}$]/[C $^{2}$]}}}}) x {{$Bx_j$ + {{1 – [(({{$V_{t,j}$ + $V_{rpar,j}$ + {{{1 – {[I$V_{t,j}$I $^{2}$]/[C $^{2}$]}} $^{1/2}$}($V_{rperp,j}$)}}}/{1 + {[$V_{t,j}$ dot $V_{r,j}$]/[C $^{2}$]}}}})/C)$^{2}$]}$^{-1/2}$ {$B_{y,j}$ + {[[({{$V_{t,j}$ + $V_{rpar,j}$ + {{{1 – {[I$V_{t,j}$ I $^{2}$]/[C $^{2}$]}} $^{1/2}$}($V_{rperp,j}$)}}}/{1 + {[$V_{t,j}$ dot $V_{r,j}$]/[C $^{2}$]}}}})/[C$^{2}$]]$E_{z,j}$}}} + {{1 – [(({{$V_{t,j}$ + $V_{rpar,j}$ + {{{1 – {[I$V_{t,j}$I $^{2}$]/[C $^{2}$]}} $^{1/2}$}($V_{rperp,j}$)}}}/{1 + {[$V_{t,j}$ dot $V_{r,j}$]/[C $^{2}$]}}}})/C)$^{2}$]}$^{-1/2}$ {$B_{z,j}$ - {[[({{$V_{t,j}$ + $V_{rpar,j}$ + {{{1 – {[I$V_{t,j}$I $^{2}$]/[C $^{2}$]}} $^{1/2}$}($V_{rperp,j}$)}}}/{1 + {[$V_{t,j}$ dot $V_{r,j}$]/[C $^{2}$]}}}})/[C$^{2}$]]$E_{y,j}$}}}}}}} • dL}}

+{{d {∫ {{{[(electrons):($Q_{natural}$)] {[1/(4π$\varepsilon_0$)][$\gamma_{average,electrons,natural,ship-frame}$ **R**]/{{[$\gamma_{average,electrons,natural,ship-frame}$ $^{2}$ $R^2$ cos$^2$ θ] + [$R^2$ sin$^2$ θ]}$^{3/2}$}}} +{[(electrons):($Q_{bomb}$)] {[1/(4π$\varepsilon_0$)][$\gamma_{average,bomb,ship-frame}$ **R**]/{{[$\gamma_{average,bomb,ship-frame}$$^{2}$ $R^2$ cos$^2$ θ] + [$R^2$ sin$^2$ θ]}$^{3/2}$}}} + {[Σ(n = 1; n = number of positive background ion species):(nth ion species)]: ($Q_{natural}$)] {[1/(4π$\varepsilon_0$)][$\gamma_{average,nth-ion-species,natural,ship-frame}$ **R**]/{{[$\gamma_{average,nth-ion-species,natural,ship-frame}$ $^{2}$ $R^2$ cos$^2$ θ] + [$R^2$ sin$^2$ θ]}$^{3/2}$}}} + {[Σ(m = 1; m = number of positive bomb ion species):(mth ion species)]: ($Q_{bomb}$)] {[1/(4π$\varepsilon_0$)][$\gamma_{average,mth-ion-species,bomb,ship-frame}$ **R**]/{{[$\gamma_{average,mth-ion-species,bomb,ship-frame}$ $^{2}$ $R^2$ cos$^2$ θ] + [$R^2$ sin$^2$ θ]}$^{3/2}$}}} + {[Σ(n = 1; n = number of negative background ion species):(nth ion species)]: ($Q_{natural}$)] {[1/(4π$\varepsilon_0$)][$\gamma_{average,nth-ion-species,natural,ship-frame}$ **R**]/{{[$\gamma_{average,nth-ion-species,bomb,ship-frame}$ $^{2}$ $R^2$ cos$^2$ θ] + [$R^2$ sin$^2$ θ]}$^{3/2}$}}} + {[Σ(m = 1; m = number of negative bomb ion species):(mth ion species)]: ($Q_{bomb}$)] {[1/(4π$\varepsilon_0$)][$\gamma_{average,mth-ion-species,bomb,ship-frame}$ **R**]/{{[$\gamma_{average,mth-ion-species,bomb,ship-frame}$ $^{2}$ $R^2$ cos$^2$ θ] + [$R^2$ sin$^2$ θ]}$^{3/2}$}}} + {d{[(Gound state $N_{flux}$)($v_{average,ship-frame}$)($\gamma_{average,ship-frame}$)] for ground state neutrons}/$dt_{ship}$} + {d {Σ(z = 1; z = number of excited neutron states):[(zth excited state $N_{flux,}$)($v_{average,ship-frame}$)($\gamma_{average,ship-frame}$)] for excited state neutrons}/dt} + [<$S_{background,photons}$>/c] + [[<$S_{bomb,photons}$>/c] ($Ph_{flux}$)]}

+ {Drag based on the following propulsive force inducing mechanisms {{[(electrons):($Q_{natural}$)] {[1/(4π$\varepsilon_0$)][$\gamma_{average,electrons,natural,ship-frame}$ **R**]/{{[$\gamma_{average,electrons,natural,ship-frame}$ $^{2}$ $R^2$ cos$^2$ θ] + [$R^2$ sin$^2$ θ]}$^{3/2}$}}} +{[(electrons):($Q_{bomb}$)] {[1/(4π$\varepsilon_0$)][$\gamma_{average,bomb,ship-frame}$ **R**]/{{[$\gamma_{average,bomb,ship-frame}$$^{2}$ $R^2$ cos$^2$ θ] + [$R^2$ sin$^2$ θ]}$^{3/2}$}}} + {[Σ(n = 1; n = number of positive background ion species):(nth ion species)]: ($Q_{natural}$)] {[1/(4π$\varepsilon_0$)][$\gamma_{average,nth-ion-species,natural,ship-frame}$ **R**]/{{[$\gamma_{average,nth-ion-species,natural,ship-frame}$ $^{2}$ $R^2$ cos$^2$ θ] + [$R^2$ sin$^2$ θ]}$^{3/2}$}}} + {[Σ(m = 1; m = number of positive bomb ion species):(mth ion species)]: ($Q_{bomb}$)] {[1/(4π$\varepsilon_0$)][$\gamma_{average,mth-ion-species,bomb,ship-frame}$ **R**]/{{[$\gamma_{average,mth-ion-species,bomb,ship-frame}$ $^{2}$ $R^2$ cos$^2$ θ] + [$R^2$ sin$^2$ θ]}$^{3/2}$}}} + {[Σ(n = 1; n = number of negative background ion species):(nth ion species)]: ($Q_{natural}$)] {[1/(4π$\varepsilon_0$)][$\gamma_{average,nth-ion-species,natural,ship-frame}$ **R**]/{{[$\gamma_{average,nth-ion-species,bomb,ship-frame}$ $^{2}$ $R^2$ cos$^2$ θ] + [$R^2$ sin$^2$ θ]}$^{3/2}$}}} + {[Σ(m = 1; m = number of negative bomb ion species):(mth ion species)]: ($Q_{bomb}$)] {[1/(4π$\varepsilon_0$)][$\gamma_{average,mth-ion-species,bomb,ship-frame}$ **R**]/{{[$\gamma_{average,mth-ion-species,bomb,ship-frame}$ $^{2}$ $R^2$ cos$^2$ θ] + [$R^2$ sin$^2$ θ]}$^{3/2}$}}} + {d{[(Gound state $N_{flux}$)($v_{average,ship-frame}$)($\gamma_{average,ship-frame}$)] for ground state neutrons}/$dt_{ship}$} + {d {Σ(z = 1; z = number of excited neutron states):[(zth excited state $N_{flux,}$)($v_{average,ship-frame}$)($\gamma_{average,ship-frame}$)] for excited state neutrons}/dt} + [<$S_{background,photons}$>/c] + [[<$S_{bomb,photons}$>/c] ($Ph_{flux}$)]}}} • dr}/dt}(Bristle-frame translational gamma)}

Here, dt is bristle-time.

Here, gamma in the second and third portion of the above formula is with respect to the incident chargons relative to the hair brush bristles. In cases where the bristles are mounted on the rotors, the interpretation of gamma is adjusted accordingly.

For cases where the bombs ionize a greater inert mass pellet, we obtain the following results:

$P_{electric}$ = ($I_{electric}$){N [(1/q) ∫ F • dL]} = ($I_{electric}$){N{∫{{$Ex_j$ + {{{1 – [(({{$V_{t,j}$ + $V_{rpar,j}$ + {{{1 – {[I$V_{t,j}$ I $^{2}$]/[C $^{2}$]}} $^{1/2}$}($V_{rperp,j}$)}}}/{1 + {[$V_{t,j}$ dot $V_{r,j}$]/[C $^{2}$]}}}}) /C)$^{2}$]}$^{-1/2}$ }[$E_{y,j}$ - ($v_j B_{z,j}$)]} + {{{1 – [(({{$V_{t,j}$ + $V_{rpar,j}$ + {{{1 – {[I$V_{t,j}$ I $^{2}$]/[C $^{2}$]}} $^{1/2}$}($V_{rperp,j}$)}}}/{1 + {[$V_{t,j}$ dot $V_{r,j}$]/[C $^{2}$]}}}}) /C)$^{2}$]}$^{-1/2}$ }[$E_{z,j}$ + (({{$V_{t,j}$ + $V_{rpar,j}$ + {{{1 – {[I$V_{t,j}$ I $^{2}$]/[C $^{2}$]}} $^{1/2}$}($V_{rperp,j}$)}}}/{1 + {[$V_{t,j}$ dot $V_{r,j}$]/[C $^{2}$]}}}})$B_{y,j}$)]}} + {({{$V_{t,j}$ + $V_{rpar,j}$ + {{{1 – {[I$V_{t,j}$ I $^{2}$]/[C $^{2}$]}} $^{1/2}$}($V_{rperp,j}$)}}}/{1 + {[$V_{t,j}$ dot $V_{r,j}$]/[C $^{2}$]}}}}) x {{$Bx_j$ + {{1 – [(({{$V_{t,j}$ + $V_{rpar,j}$ + {{{1 – {[I$V_{t,j}$I $^{2}$]/[C $^{2}$]}} $^{1/2}$}($V_{rperp,j}$)}}}/{1 + {[$V_{t,j}$ dot $V_{r,j}$]/[C $^{2}$]}}}})/C)$^{2}$]}$^{-1/2}$ {$B_{y,j}$ + {[[({{$V_{t,j}$ + $V_{rpar,j}$ + {{{1 – {[I$V_{t,j}$ I $^{2}$]/[C $^{2}$]}} $^{1/2}$}($V_{rperp,j}$)}}}/{1 + {[$V_{t,j}$ dot $V_{r,j}$]/[C $^{2}$]}}}})/[C$^{2}$]]$E_{z,j}$}}} + {{1 – [(({{$V_{t,j}$ + $V_{rpar,j}$ + {{{1 – {[I$V_{t,j}$I $^{2}$]/[C $^{2}$]}} $^{1/2}$}($V_{rperp,j}$)}}}/{1 + {[$V_{t,j}$ dot $V_{r,j}$]/[C $^{2}$]}}}})/C)$^{2}$]}$^{-1/2}$ {$B_{z,j}$ - {[[({{$V_{t,j}$ + $V_{rpar,j}$ + {{{1 – {[I$V_{t,j}$I $^{2}$]/[C $^{2}$]}} $^{1/2}$}($V_{rperp,j}$)}}}/{1 + {[$V_{t,j}$ dot $V_{r,j}$]/[C $^{2}$]}}}})/[C$^{2}$]]$E_{y,j}$}}}}}}} • dL}}

{{d{∫{{{{[(electrons):($Q_{natural}$)] {[1/(4π$\varepsilon_0$)][$\gamma_{average,electrons,natural,ship-frame}$ **R**]/{{[$\gamma_{average,electrons,natural,ship-frame}$ $^{2}$ $R^2$ cos$^2$ θ] + [$R^2$ sin$^2$ θ]}$^{3/2}$}}} +{[(electrons):($Q_{bomb}$)] {[1/(4π$\varepsilon_0$)][$\gamma_{average,bomb,ship-frame}$ **R**]/{{[$\gamma_{average,bomb,ship-frame}$$^{2}$ $R^2$ cos$^2$ θ] + [$R^2$ sin$^2$ θ]}$^{3/2}$}}} +{[(electrons):($Q_{non-nuclear\ pellet}$)] {[1/(4π$\varepsilon_0$)][$\gamma_{average,non-nuclear\ pellet,ship-frame}$ **R**]/{{[$\gamma_{average,non-nuclear\ pellet,ship-frame}$$^{2}$ $R^2$ cos$^2$ θ] + [$R^2$ sin$^2$ θ]}$^{3/2}$}}} + {[Σ(n = 1; n = number of positive background ion species):(nth ion species)]: ($Q_{natural}$)] {[1/(4π$\varepsilon_0$)][$\gamma_{average,nth-ion-species,natural,ship-frame}$ **R**]/{{[γ

$_{average,nth-ion-species,natural,ship-frame}$ $^2$ $R^2 \cos^2\theta$] + [$R^2 \sin^2\theta$]}$^{3/2}$}}} + {[Σ(m = 1; m = number of positive bomb ion species):(mth ion species)]: ($Q_{bomb}$)] {[1/(4π$\varepsilon_0$)][$\gamma_{average,mth-ion-species,bomb,ship-frame}$ **R**]/{{[$\gamma_{average,mth-ion-species,bomb,ship-frame}$ $^2$ $R^2 \cos^2\theta$] + [$R^2 \sin^2\theta$]}$^{3/2}$}}} + {[Σ(g = 1; g = number of positive bomb heated non-nuclear pellet sourced ion species):(gth ion species)]: ($Q_{non-nuclear\ pellet}$)] {[1/(4π$\varepsilon_0$)][$\gamma_{average,gth-ion-species,non-nuclear\ pellet,,ship-frame}$ **R**]/{{[$\gamma_{average,gth-ion-species,non-nuclear\ pellet,ship-frame}$ $^2$ $R^2 \cos^2\theta$] + [$R^2 \sin^2\theta$]}$^{3/2}$}}} +{[Σ(e = 1; e = number of negative non-nuclear pellet heated non-nuclear pellet sourced ion species):(eth ion species)]: ($Q_{non-nuclear\ pellet}$)] {[1/(4π$\varepsilon_0$)][$\gamma_{average,gth-ion-species,non-nuclear\ pellet,,ship-frame}$ **R**]/{{[$\gamma_{average,gth-ion-species,non-nuclear\ pellet,ship-frame}$ $^2$ $R^2 \cos^2\theta$] + [$R^2 \sin^2\theta$]}$^{3/2}$}}} + {[Σ(n = 1; n = number of negative background ion species):(nth ion species)]: ($Q_{natural}$)] {[1/(4π$\varepsilon_0$)][$\gamma_{average,nth-ion-species,natural,ship-frame}$ **R**]/{{[$\gamma_{average,nth-ion-species,bomb,ship-frame}$ $^2$ $R^2 \cos^2\theta$] + [$R^2 \sin^2\theta$]}$^{3/2}$}}} + {[Σ(m = 1; m = number of negative bomb ion species):(mth ion species)]: ($Q_{bomb}$)] {[1/(4π$\varepsilon_0$)][$\gamma_{average,mth-ion-species,bomb,ship-frame}$ **R**]/{{[$\gamma_{average,mth-ion-species,bomb,ship-frame}$ $^2$ $R^2 \cos^2\theta$] + [$R^2 \sin^2\theta$]}$^{3/2}$}}} + {d{[(Gound state $N_{flux}$)($v_{average,ship-frame}$)($\gamma_{average,ship-frame}$)] for ground state neutrons}/dt$_{ship}$} + {d {Σ(z = 1; z = number of excited neutron states):[(zth excited state $N_{flux,}$)($v_{average,ship-frame}$)($\gamma_{average,ship-frame}$)] for excited state neutrons}/dt} + [<$S_{background,photons}$>/c] + [[<$S_{bomb,photons}$>/c] ($Ph_{flux}$)] + [[<$S_{non-nuclear\ pellet,photons}$>/c] ($Ph_{flux}$)]}

+ {Drag based on the following propulsive force inducing mechanisms {{[(electrons):($Q_{natural}$)] {[1/(4π$\varepsilon_0$)][$\gamma_{average,electrons,natural,ship-frame}$ **R**]/{{[$\gamma_{average,electrons,natural,ship-frame}$ $^2$ $R^2 \cos^2\theta$] + [$R^2 \sin^2\theta$]}$^{3/2}$}}} +{[(electrons):($Q_{bomb}$)] {[1/(4π$\varepsilon_0$)][$\gamma_{average,bomb,ship-frame}$ **R**]/{{[$\gamma_{average,bomb,ship-frame}$$^2$ $R^2 \cos^2\theta$] + [$R^2 \sin^2\theta$]}$^{3/2}$}}} +{[(electrons):($Q_{non-nuclear\ pellet}$)] {[1/(4π$\varepsilon_0$)][$\gamma_{average,non-nuclear\ pellet,ship-frame}$ **R**]/{{[$\gamma_{average,non-nuclear\ pellet,ship-frame}$$^2$ $R^2 \cos^2\theta$] + [$R^2 \sin^2\theta$]}$^{3/2}$}}} + {[Σ(n = 1; n = number of positive background ion species):(nth ion species)]: ($Q_{natural}$)] {[1/(4π$\varepsilon_0$)][$\gamma_{average,nth-ion-species,natural,ship-frame}$ **R**]/{{[$\gamma_{average,nth-ion-species,natural,ship-frame}$ $^2$ $R^2 \cos^2\theta$] + [$R^2 \sin^2\theta$]}$^{3/2}$}}} + {[Σ(m = 1; m = number of positive bomb ion species):(mth ion species)]: ($Q_{bomb}$)] {[1/(4π$\varepsilon_0$)][$\gamma_{average,mth-ion-species,bomb,ship-frame}$ **R**]/{{[$\gamma_{average,mth-ion-species,bomb,ship-frame}$ $^2$ $R^2 \cos^2\theta$] + [$R^2 \sin^2\theta$]}$^{3/2}$}}} + {[Σ(g = 1; g = number of positive bomb heated non-nuclear pellet sourced ion species):(gth ion species)]: ($Q_{non-nuclear\ pellet}$)] {[1/(4π$\varepsilon_0$)][$\gamma_{average,gth-ion-species,non-nuclear\ pellet,,ship-frame}$ **R**]/{{[$\gamma_{average,gth-ion-species,non-nuclear\ pellet,ship-frame}$ $^2$ $R^2 \cos^2\theta$] + [$R^2 \sin^2\theta$]}$^{3/2}$}}} +{[Σ(e = 1; e = number of negative non-nuclear pellet heated non-nuclear pellet sourced ion species):(eth ion species)]: ($Q_{non-nuclear\ pellet}$)] {[1/(4π$\varepsilon_0$)][$\gamma_{average,gth-ion-species,non-nuclear\ pellet,,ship-frame}$ **R**]/{{[$\gamma_{average,gth-ion-species,non-nuclear\ pellet,ship-frame}$ $^2$ $R^2 \cos^2\theta$] + [$R^2 \sin^2\theta$]}$^{3/2}$}}} + {[Σ(n = 1; n = number of negative background ion species):(nth ion species)]: ($Q_{natural}$)] {[1/(4π$\varepsilon_0$)][$\gamma_{average,nth-ion-species,natural,ship-frame}$ **R**]/{{[$\gamma_{average,nth-ion-species,bomb,ship-frame}$ $^2$ $R^2 \cos^2\theta$] + [$R^2 \sin^2\theta$]}$^{3/2}$}}} + {[Σ(m = 1; m = number of negative bomb ion species):(mth ion species)]: ($Q_{bomb}$)] {[1/(4π$\varepsilon_0$)][$\gamma_{average,mth-ion-species,bomb,ship-frame}$ **R**]/{{[$\gamma_{average,mth-ion-species,bomb,ship-frame}$ $^2$ $R^2 \cos^2\theta$] + [$R^2 \sin^2\theta$]}$^{3/2}$}}} + {d{[(Gound state $N_{flux}$)($v_{average,ship-frame}$)($\gamma_{average,ship-frame}$)] for ground state neutrons}/dt$_{ship}$} + {d {Σ(z = 1; z = number of excited neutron states):[(zth excited state $N_{flux,}$)($v_{average,ship-frame}$)($\gamma_{average,ship-frame}$)] for excited state neutrons}/dt} + [<$S_{background,photons}$>/c] + [[<$S_{bomb,photons}$>/c] ($Ph_{flux}$)] + [[<$S_{non-nuclear\ pellet,photons}$>/c] ($Ph_{flux}$)]}}• dr}/dt} (Bristle-frame translational gamma)}

Here, dt is bristle-time and where F • dL is the scalar product or dot product of F and dL.

For N loops for which the size and shape of the loops can vary, the following approximation describes the power generated by the coil in the coil in the coil frame:

$P_{electric,j}$ = ($I_{electric,j}$){Σ(n = 1, n = $N_{coil}$)[(1/q) ∫ $F_j$ • dL]$_n$} = ($I_{electric,j}$){Σ(n = 1, n = $N_{coil}$){∫{{$Ex_j$ + {{{1 – [(({{$V_{t,j}$ + $V_{rpar,j}$ + {{{1 – {[I$V_{t,j}$ I $^2$]/[C $^2$]}} $^{1/2}$}($V_{rperp,j}$)}}/{1 + {[$V_{t,j}$ dot $V_{r,j}$]/[C $^2$]}}}) /C)$^2$]}$^{-1/2}$}[$E_{y,j}$ - ($v_j B_{z,j}$)]} + {{{1 – [(({{$V_{t,j}$ + $V_{rpar,j}$ + {{{1 – {[I$V_{t,j}$ I $^2$]/[C $^2$]}} $^{1/2}$}($V_{rperp,j}$)}}/{1 + {[$V_{t,j}$ dot $V_{r,j}$]/[C $^2$]}}})$_j$ /C)$^2$]}$^{-1/2}$}[$E_{z,j}$ + (({{$V_{t,j}$ + $V_{rpar,j}$ + {{{1 – {[I$V_{t,j}$ I $^2$]/[C $^2$]}} $^{1/2}$}($V_{rperp,j}$)}}/{1 + {[$V_{t,j}$ dot $V_{r,j}$]/[C $^2$]}}})$B_{y,j}$)]}} + {({{$V_{t,j}$ + $V_{rpar,j}$ + {{{1 – {[I$V_{t,j}$ I $^2$]/[C $^2$]}} $^{1/2}$}($V_{rperp,j}$)}}/{1 + {[$V_{t,j}$ dot $V_{r,j}$]/[C $^2$]}}}) x {{$Bx_j$ + {{1 – [(({{$V_{t,j}$ + $V_{rpar,j}$ + {{{1 – {[I$V_{t,j}$ I $^2$]/[C $^2$]}} $^{1/2}$}($V_{rperp,j}$)}}/{1 + {[$V_{t,j}$ dot $V_{r,j}$]/[C $^2$]}}})/C)$^2$]}$^{-1/2}$ {$B_{y,j}$ + {[[(({{$V_{t,j}$ + $V_{rpar,j}$ + {{{1 – {[I$V_{t,j}$ I $^2$]/[C $^2$]}} $^{1/2}$}($V_{rperp,j}$)}}/{1 + {[$V_{t,j}$ dot $V_{r,j}$]/[C $^2$]}}})/[C$^2$]]$E_{z,j}$}}} + {{1 – [(({{$V_{t,j}$ + $V_{rpar,j}$ + {{{1 – {[I$V_{t,j}$ I $^2$]/[C $^2$]}} $^{1/2}$}($V_{rperp,j}$)}}/{1 + {[$V_{t,j}$ dot $V_{r,j}$]/[C $^2$]}}})/C)$^2$]}$^{-1/2}$ {$B_{z,j}$ - {[[(({{$V_{t,j}$ + $V_{rpar,j}$ + {{{1 – {[I$V_{t,j}$ I $^2$]/[C $^2$]}} $^{1/2}$}($V_{rperp,j}$)}}/{1 + {[$V_{t,j}$ dot $V_{r,j}$]/[C $^2$]}}})/[C$^2$]]$E_{y,j}$}}}}}}$_n$ • dL}}

+{{d {∫ {{{[(electrons):($Q_{natural}$)] {[1/(4π$\varepsilon_0$)][$\gamma_{average,electrons,natural,ship-frame}$ **R**]/{{[$\gamma_{average,electrons,natural,ship-frame}$ $^2$ $R^2 \cos^2\theta$] + [$R^2 \sin^2\theta$]}$^{3/2}$}}} +{[(electrons):($Q_{bomb}$)] {[1/(4π$\varepsilon_0$)][$\gamma_{average,bomb,ship-frame}$ **R**]/{{[$\gamma_{average,bomb,ship-frame}$$^2$ $R^2 \cos^2\theta$] + [$R^2 \sin^2\theta$]}$^{3/2}$}}} + {[Σ(n = 1; n = number of positive background ion species):(nth ion species)]: ($Q_{natural}$)] {[1/(4π$\varepsilon_0$)][$\gamma_{average,nth-ion-species,natural,ship-frame}$ **R**]/{{[$\gamma_{average,nth-ion-species,natural,ship-frame}$ $^2$ $R^2 \cos^2\theta$] + [$R^2 \sin^2\theta$]}$^{3/2}$}}} + {[Σ(m = 1; m = number of positive bomb ion species):(mth ion species)]: ($Q_{bomb}$)] {[1/(4π$\varepsilon_0$)][$\gamma_{average,mth-ion-species,bomb,ship-frame}$ **R**]/{{[$\gamma_{average,mth-ion-species,bomb,ship-frame}$ $^2$ $R^2 \cos^2\theta$] + [$R^2 \sin^2\theta$]}$^{3/2}$}}} + {[Σ(n = 1; n = number of negative background ion species):(nth ion species)]: ($Q_{natural}$)] {[1/(4π$\varepsilon_0$)][$\gamma_{average,nth-ion-species,natural,ship-frame}$ **R**]/{{[$\gamma_{average,nth-ion-species,bomb,ship-frame}$ $^2$ $R^2 \cos^2\theta$] + [$R^2 \sin^2\theta$]}$^{3/2}$}}} + {[Σ(m = 1; m = number of negative bomb ion species):(mth ion species)]: ($Q_{bomb}$)] {[1/(4π$\varepsilon_0$)][$\gamma_{average,mth-ion-species,bomb,ship-frame}$ **R**]/{{[$\gamma_{average,mth-ion-species,bomb,ship-frame}$ $^2$ $R^2 \cos^2\theta$] + [$R^2 \sin^2\theta$]}$^{3/2}$}}} + {d{[(Gound state $N_{flux}$)($v_{average,ship-frame}$)($\gamma_{average,ship-frame}$)] for ground state neutrons}/dt$_{ship}$} + {d {Σ(z = 1; z = number of excited neutron states):[(zth excited state $N_{flux,}$)($v_{average,ship-frame}$)($\gamma_{average,ship-frame}$)] for excited state neutrons}/dt} + [<$S_{background,photons}$>/c] + [[<$S_{bomb,photons}$>/c] ($Ph_{flux}$)]}

+ {Drag based on the following propulsive force inducing mechanisms {{[(electrons):($Q_{natural}$)] {[1/(4π$\varepsilon_0$)][$\gamma_{average,electrons,natural,ship-frame}$ **R**]/{{[$\gamma_{average,electrons,natural,ship-frame}$ $^2$ $R^2 \cos^2\theta$] + [$R^2 \sin^2\theta$]}$^{3/2}$}}} +{[(electrons):($Q_{bomb}$)] {[1/(4π$\varepsilon_0$)][$\gamma_{average,bomb,ship-frame}$ **R**]/{{[$\gamma$

$_{average,bomb,ship\text{-}frame}{}^{2}$ $R^2 \cos^2 θ$] + [$R^2 \sin^2 θ$]}$^{3/2}$}}} + {[Σ(n = 1; n = number of positive background ion species):(nth ion species)]: ($Q_{natural}$)] {[1/(4πε$_0$)][$γ_{average,nth\text{-}ion\text{-}species,natural,ship\text{-}frame}$ **R**]/{{[$γ_{average,nth\text{-}ion\text{-}species,natural,ship\text{-}frame}{}^{2}$ $R^2 \cos^2 θ$] + [$R^2 \sin^2 θ$]}$^{3/2}$}}} + {[Σ(m = 1; m = number of positive bomb ion species):(mth ion species)]: ($Q_{bomb}$)] {[1/(4πε$_0$)][$γ_{average,mth\text{-}ion\text{-}species,bomb,ship\text{-}frame}$ **R**]/{{[$γ_{average,mth\text{-}ion\text{-}species,bomb,ship\text{-}frame}{}^{2}$ $R^2 \cos^2 θ$] + [$R^2 \sin^2 θ$]}$^{3/2}$}}} + {[Σ(n = 1; n = number of negative background ion species):(nth ion species)]: ($Q_{natural}$)] {[1/(4πε$_0$)][$γ_{average,nth\text{-}ion\text{-}species,natural,ship\text{-}frame}$ **R**]/{{[$γ_{average,nth\text{-}ion\text{-}species,bomb,ship\text{-}frame}{}^{2}$ $R^2 \cos^2 θ$] + [$R^2 \sin^2 θ$]}$^{3/2}$}}} + {[Σ(m = 1; m = number of negative bomb ion species):(mth ion species)]: ($Q_{bomb}$)] {[1/(4πε$_0$)][$γ_{average,mth\text{-}ion\text{-}species,bomb,ship\text{-}frame}$ **R**]/{{[$γ_{average,mth\text{-}ion\text{-}species,bomb,ship\text{-}frame}{}^{2}$ $R^2 \cos^2 θ$] + [$R^2 \sin^2 θ$]}$^{3/2}$}}} + {d{[(Gound state $N_{flux}$)($v_{average,ship\text{-}frame}$)($γ_{average,ship\text{-}frame}$)] for ground state neutrons}/$dt_{ship}$} + {d {Σ(z = 1; z = number of excited neutron states):[(zth excited state $N_{flux}$,)($v_{average,ship\text{-}frame}$)($γ_{average,ship\text{-}frame}$)] for excited state neutrons}/dt} + [<$S_{background,photons}$>/c] + [[<$S_{bomb,photons}$>/c] ($Ph_{flux}$)]}}} • dr}/dt}(Bristle-frame translational gamma)}

where F • dL is the scalar product or dot product of F and dL and where dt in the second and third portions of the above formula is the bristle time .

For N loops for which the size and shape of the loops can vary and where the bombs ionize inert pellets, the following approximation describes the power generated by the coil in the coil in the coil frame:

$P_{electric,j}$ = ($I_{electric,j}$){Σ(n = 1, n = $N_{coil}$)[(1/q) ∫ $F_j$ • dL]$_n$} = ($I_{electric,j}$){Σ(n = 1, n = $N_{coil}$)[∫{{$Ex_j$ + {{{1 – [(({{$V_{t,j}$ + $V_{rpar,j}$ + {{{1 – {[I$V_{t,j}$ I $^2$]/[C $^2$]}} $^{1/2}$}($V_{rperp,j}$)}}}/{1 + {[$V_{t,j}$ dot $V_{r,j}$]/[C $^2$]}}}}) /C)$^2$]}$^{-1/2}$ }[$E_{y,j}$ - ($v_j B_{z,j}$)]} + {{{1 – [(({{$V_{t,j}$ + $V_{rpar,j}$ + {{{1 – {[I$V_{t,j}$ I $^2$]/[C $^2$]}} $^{1/2}$}($V_{rperp,j}$)}}}/{1 + {[$V_{t,j}$ dot $V_{r,j}$]/[C $^2$]}}}})$_j$ /C)$^2$]}$^{-1/2}$ }[$E_{z,j}$ + (({{$V_{t,j}$ + $V_{rpar,j}$ + {{{1 – {[I$V_{t,j}$ I $^2$]/[C $^2$]}} $^{1/2}$}($V_{rperp,j}$)}}}/{1 + {[$V_{t,j}$ dot $V_{r,j}$]/[C $^2$]}}}}$B_{y,j}$)]}} + {(({{$V_{t,j}$ + $V_{rpar,j}$ + {{{1 – {[I$V_{t,j}$ I $^2$]/[C $^2$]}} $^{1/2}$}($V_{rperp,j}$)}}}/{1 + {[$V_{t,j}$ dot $V_{r,j}$]/[C $^2$]}}}}) x {{$Bx_j$ + {{1 – [(({{$V_{t,j}$ + $V_{rpar,j}$ + {{{1 – {[I$V_{t,j}$ I $^2$]/[C $^2$]}} $^{1/2}$}($V_{rperp,j}$)}}}/{1 + {[$V_{t,j}$ dot $V_{r,j}$]/[C $^2$]}}}})/C)$^2$]}$^{-1/2}$ {$B_{y,j}$ + {[[(({{$V_{t,j}$ + $V_{rpar,j}$ + {{{1 – {[I$V_{t,j}$ I $^2$]/[C $^2$]}} $^{1/2}$}($V_{rperp,j}$)}}}/{1 + {[$V_{t,j}$ dot $V_{r,j}$]/[C $^2$]}}}})/[C$^2$]]$E_{z,j}$}}} + {{1 – [(({{$V_{t,j}$ + $V_{rpar,j}$ + {{{1 – {[I$V_{t,j}$ I $^2$]/[C $^2$]}} $^{1/2}$}($V_{rperp,j}$)}}}/{1 + {[$V_{t,j}$ dot $V_{r,j}$]/[C $^2$]}}}})/C)$^2$]}$^{-1/2}$ {$B_{z,j}$ - {[[(({{$V_{t,j}$ + $V_{rpar,j}$ + {{{1 – {[I$V_{t,j}$ I $^2$]/[C $^2$]}} $^{1/2}$}($V_{rperp,j}$)}}}/{1 + {[$V_{t,j}$ dot $V_{r,j}$]/[C $^2$]}}}})/[C$^2$]]$E_{y,j}$}}}}}}}$_n$ • dL}}

{{d{∫{{{{[(electrons):($Q_{natural}$)] {[1/(4πε$_0$)][$γ_{average,electrons,natural,ship\text{-}frame}$ **R**]/{{[$γ_{average,electrons,natural,ship\text{-}frame}{}^{2}$ $R^2 \cos^2 θ$] + [$R^2 \sin^2 θ$]}$^{3/2}$}}} +{[(electrons):($Q_{bomb}$)] {[1/(4πε$_0$)][$γ_{average,bomb,ship\text{-}frame}$ **R**]/{{[$γ_{average,bomb,ship\text{-}frame}{}^{2}$ $R^2 \cos^2 θ$] + [$R^2 \sin^2 θ$]}$^{3/2}$}}} +{[(electrons):($Q_{non\text{-}nuclear\ pellet}$)] {[1/(4πε$_0$)][$γ_{average,non\text{-}nuclear\ pellet,ship\text{-}frame}$ **R**]/{{[$γ_{average,non\text{-}nuclear\ pellet,ship\text{-}frame}{}^{2}$ $R^2 \cos^2 θ$] + [$R^2 \sin^2 θ$]}$^{3/2}$}}} + {[Σ(n = 1; n = number of positive background ion species):(nth ion species)]: ($Q_{natural}$)] {[1/(4πε$_0$)][$γ_{average,nth\text{-}ion\text{-}species,natural,ship\text{-}frame}$ **R**]/{{[$γ_{average,nth\text{-}ion\text{-}species,natural,ship\text{-}frame}{}^{2}$ $R^2 \cos^2 θ$] + [$R^2 \sin^2 θ$]}$^{3/2}$}}} + {[Σ(m = 1; m = number of positive bomb ion species):(mth ion species)]: ($Q_{bomb}$)] {[1/(4πε$_0$)][$γ_{average,mth\text{-}ion\text{-}species,bomb,ship\text{-}frame}$ **R**]/{{[$γ_{average,mth\text{-}ion\text{-}species,bomb,ship\text{-}frame}{}^{2}$ $R^2 \cos^2 θ$] + [$R^2 \sin^2 θ$]}$^{3/2}$}}} + {[Σ(g = 1; g = number of positive bomb heated non-nuclear pellet sourced ion species):(gth ion species)]: ($Q_{non\text{-}nuclear\ pellet}$)] {[1/(4πε$_0$)][$γ_{average,gth\text{-}ion\text{-}species,non\text{-}nuclear\ pellet,,ship\text{-}frame}$ **R**]/{{[$γ_{average,gth\text{-}ion\text{-}species,non\text{-}nuclear\ pellet,ship\text{-}frame}{}^{2}$ $R^2 \cos^2 θ$] + [$R^2 \sin^2 θ$]}$^{3/2}$}}} +{[Σ(e = 1; e = number of negative non-nuclear pellet heated non-nuclear pellet sourced ion species):(eth ion species)]: ($Q_{non\text{-}nuclear\ pellet}$)] {[1/(4πε$_0$)][$γ_{average,gth\text{-}ion\text{-}species,non\text{-}nuclear\ pellet,,ship\text{-}frame}$ **R**]/{{[$γ_{average,gth\text{-}ion\text{-}species,non\text{-}nuclear\ pellet,ship\text{-}frame}{}^{2}$ $R^2 \cos^2 θ$] + [$R^2 \sin^2 θ$]}$^{3/2}$}}} + {[Σ(n = 1; n = number of negative background ion species):(nth ion species)]: ($Q_{natural}$)] {[1/(4πε$_0$)][$γ_{average,nth\text{-}ion\text{-}species,natural,ship\text{-}frame}$ **R**]/{{[$γ_{average,nth\text{-}ion\text{-}species,bomb,ship\text{-}frame}{}^{2}$ $R^2 \cos^2 θ$] + [$R^2 \sin^2 θ$]}$^{3/2}$}}} + {[Σ(m = 1; m = number of negative bomb ion species):(mth ion species)]: ($Q_{bomb}$)] {[1/(4πε$_0$)][$γ_{average,mth\text{-}ion\text{-}species,bomb,ship\text{-}frame}$ **R**]/{{[$γ_{average,mth\text{-}ion\text{-}species,bomb,ship\text{-}frame}{}^{2}$ $R^2 \cos^2 θ$] + [$R^2 \sin^2 θ$]}$^{3/2}$}}} + {d{[(Gound state $N_{flux}$)($v_{average,ship\text{-}frame}$)($γ_{average,ship\text{-}frame}$)] for ground state neutrons}/$dt_{ship}$} + {d {Σ(z = 1; z = number of excited neutron states):[(zth excited state $N_{flux}$,)($v_{average,ship\text{-}frame}$)($γ_{average,ship\text{-}frame}$)] for excited state neutrons}/dt} + [<$S_{background,photons}$>/c] + [[<$S_{bomb,photons}$>/c] ($Ph_{flux}$)] + [[<$S_{non\text{-}nuclear\ pellet,photons}$>/c] ($Ph_{flux}$)]}

+ {Drag based on the following propulsive force inducing mechanisms {{[(electrons):($Q_{natural}$)] {[1/(4πε$_0$)][$γ_{average,electrons,natural,ship\text{-}frame}$ **R**]/{{[$γ_{average,electrons,natural,ship\text{-}frame}{}^{2}$ $R^2 \cos^2 θ$] + [$R^2 \sin^2 θ$]}$^{3/2}$}}} +{[(electrons):($Q_{bomb}$)] {[1/(4πε$_0$)][$γ_{average,bomb,ship\text{-}frame}$ **R**]/{{[$γ_{average,bomb,ship\text{-}frame}{}^{2}$ $R^2 \cos^2 θ$] + [$R^2 \sin^2 θ$]}$^{3/2}$}}} +{[(electrons):($Q_{non\text{-}nuclear\ pellet}$)] {[1/(4πε$_0$)][$γ_{average,non\text{-}nuclear\ pellet,ship\text{-}frame}$ **R**]/{{[$γ_{average,non\text{-}nuclear\ pellet,ship\text{-}frame}{}^{2}$ $R^2 \cos^2 θ$] + [$R^2 \sin^2 θ$]}$^{3/2}$}}} + {[Σ(n = 1; n = number of positive background ion species):(nth ion species)]: ($Q_{natural}$)] {[1/(4πε$_0$)][$γ_{average,nth\text{-}ion\text{-}species,natural,ship\text{-}frame}$ **R**]/{{[$γ_{average,nth\text{-}ion\text{-}species,natural,ship\text{-}frame}{}^{2}$ $R^2 \cos^2 θ$] + [$R^2 \sin^2 θ$]}$^{3/2}$}}} + {[Σ(m = 1; m = number of positive bomb ion species):(mth ion species)]: ($Q_{bomb}$)] {[1/(4πε$_0$)][$γ_{average,mth\text{-}ion\text{-}species,bomb,ship\text{-}frame}$ **R**]/{{[$γ_{average,mth\text{-}ion\text{-}species,bomb,ship\text{-}frame}{}^{2}$ $R^2 \cos^2 θ$] + [$R^2 \sin^2 θ$]}$^{3/2}$}}} + {[Σ(g = 1; g = number of positive bomb heated non-nuclear pellet sourced ion species):(gth ion species)]: ($Q_{non\text{-}nuclear\ pellet}$)] {[1/(4πε$_0$)][$γ_{average,gth\text{-}ion\text{-}species,non\text{-}nuclear\ pellet,,ship\text{-}frame}$ **R**]/{{[$γ_{average,gth\text{-}ion\text{-}species,non\text{-}nuclear\ pellet,ship\text{-}frame}{}^{2}$ $R^2 \cos^2 θ$] + [$R^2 \sin^2 θ$]}$^{3/2}$}}} +{[Σ(e = 1; e = number of negative non-nuclear pellet heated non-nuclear pellet sourced ion species):(eth ion species)]: ($Q_{non\text{-}nuclear\ pellet}$)] {[1/(4πε$_0$)][$γ_{average,gth\text{-}ion\text{-}species,non\text{-}nuclear\ pellet,,ship\text{-}frame}$ **R**]/{{[$γ_{average,gth\text{-}ion\text{-}species,non\text{-}nuclear\ pellet,ship\text{-}frame}{}^{2}$ $R^2 \cos^2 θ$] + [$R^2 \sin^2 θ$]}$^{3/2}$}}} + {[Σ(n = 1; n = number of negative background ion species):(nth ion species)]: ($Q_{natural}$)] {[1/(4πε$_0$)][$γ_{average,nth\text{-}ion\text{-}species,natural,ship\text{-}frame}$ **R**]/{{[$γ_{average,nth\text{-}ion\text{-}species,bomb,ship\text{-}frame}{}^{2}$ $R^2 \cos^2 θ$] + [$R^2 \sin^2 θ$]}$^{3/2}$}}} + {[Σ(m = 1; m = number of negative bomb ion species):(mth ion species)]: ($Q_{bomb}$)] {[1/(4πε$_0$)][$γ_{average,mth\text{-}ion\text{-}species,bomb,ship\text{-}frame}$ **R**]/{{[$γ_{average,mth\text{-}ion\text{-}species,bomb,ship\text{-}frame}{}^{2}$ $R^2 \cos^2 θ$] + [$R^2 \sin^2 θ$]}$^{3/2}$}}} + {d{[(Gound state $N_{flux}$)($v_{average,ship\text{-}frame}$)($γ_{average,ship\text{-}frame}$)] for ground state neutrons}/$dt_{ship}$} + {d {Σ(z = 1; z = number of excited neutron states):[(zth excited state

$N_{flux,}$)($v_{average,ship-frame}$)($\gamma_{average,ship-frame}$)] for excited state neutrons}/dt} + [$<S_{background,photons}>$/c] + [[$<S_{bomb,photons}>$/c] ($Ph_{flux}$)] + [[$<S_{non-nuclear\ pellet,photons}>$/c] ($Ph_{flux}$)]]}}• dr}/dt} (Bristle-frame translational gamma)}

Here, dt is bristle-time and where F • dL is the scalar product or dot product of F and dL.

The electrical energy generated over the jth time step without the bristle mechanism is thus equal to:

$E_{coil}$ = [d $P_{electric,j}$/$dt_{coils}$] {1/{{1 - [[($V_t$ $+_{(relativistic\ composition)}$ $V_r$)/C]$^2$]}$^{1/2}$}}

= {d{($I_{electric,j}$)}{Σ(n = 1, n = $N_{coil}$)[(1/q) ∫ $F_j$ • dL]$_n$}}} = {d{($I_{electric,j}$)}{Σ(n = 1, n = $N_{coil}$){∫{{$Ex_j$ + {{{1 – [((({{$V_{t,j}$ + $V_{rpar,j}$ + {{{1 – {[I$V_{t,j}$ I $^2$]/[C $^2$]}} $^{1/2}$}($V_{rperp,j}$)}}}/{1 + {[$V_{t,j}$ dot $V_{r,j}$]/[C $^2$]}}}}) /C)$^2$]}$^{-1/2}$ }[$E_{y,j}$ - ($v_jB_{z,j}$)]} + {{{1 – [((({{$V_{t,j}$ + $V_{rpar,j}$ + {{{1 – {[I$V_{t,j}$ I $^2$]/[C $^2$]}} $^{1/2}$}($V_{rperp,j}$)}}}/{1 + {[$V_{t,j}$ dot $V_{r,j}$]/[C $^2$]}}}})$_j$ /C)$^2$]}$^{-1/2}$ }[$E_{z,j}$ + ((({$V_{t,j}$ + $V_{rpar,j}$ + {{{1 – {[I$V_{t,j}$ I $^2$]/[C $^2$]}} $^{1/2}$}($V_{rperp,j}$)}}}/{1 + {[$V_{t,j}$ dot $V_{r,j}$]/[C $^2$]}}}})$B_{y,j}$)]}} + {((({$V_{t,j}$ + $V_{rpar,j}$ + {{{1 – {[I$V_{t,j}$ I $^2$]/[C $^2$]}} $^{1/2}$}($V_{rperp,j}$)}}}/{1 + {[$V_{t,j}$ dot $V_{r,j}$]/[C $^2$]}}}}) x {{$Bx_j$ + {{1 – [((({{$V_{t,j}$ + $V_{rpar,j}$ + {{{1 – {[I$V_{t,j}$ I $^2$]/[C $^2$]}} $^{1/2}$}($V_{rperp,j}$)}}}/{1 + {[$V_{t,j}$ dot $V_{r,j}$]/[C $^2$]}}}})/C)$^2$]}$^{-1/2}$ {$B_{y,j}$ + {[[((({$V_{t,j}$ + $V_{rpar,j}$ + {{{1 – {[I$V_{t,j}$ I $^2$]/[C $^2$]}} $^{1/2}$}($V_{rperp,j}$)}}}/{1 + {[$V_{t,j}$ dot $V_{r,j}$]/[C $^2$]}}}})/[C$^2$]]$E_{z,j}$}}} + {{1 – [((({{$V_{t,j}$ + $V_{rpar,j}$ + {{{1 – {[I$V_{t,j}$ I $^2$]/[C $^2$]}} $^{1/2}$}($V_{rperp,j}$)}}}/{1 + {[$V_{t,j}$ dot $V_{r,j}$]/[C $^2$]}}}})/C)$^2$]}$^{-1/2}$ {$B_{z,j}$ - {[[((({$V_{t,j}$ + $V_{rpar,j}$ + {{{1 – {[I$V_{t,j}$ I $^2$]/[C $^2$]}} $^{1/2}$}($V_{rperp,j}$)}}}/{1 + {[$V_{t,j}$ dot $V_{r,j}$]/[C $^2$]}}}})/[C$^2$]]$E_{y,j}$}}}}}}}$_n$ • dL}}}/$dt_{coil(s)}$} {1/{{1 - [[($V_{t,j}$ $+_{(relativistic\ composition)}$ $V_{r,j}$)/C]$^2$]}$^{1/2}$}}

= {d{($I_{electric,j}$)}{Σ(n = 1, n = $N_{coil}$)[(1/q) ∫ $F_j$ • dL]$_n$}}} = {d{($I_{electric,j}$)}{Σ(n = 1, n = $N_{coil}$){∫{{$Ex_j$ + {{{1 – [((({{$V_{t,j}$ + $V_{rpar,j}$ + {{{1 – {[I$V_{t,j}$ I $^2$]/[C $^2$]}} $^{1/2}$}($V_{rperp,j}$)}}}/{1 + {[$V_{t,j}$ dot $V_{r,j}$]/[C $^2$]}}}}) /C)$^2$]}$^{-1/2}$ }[$E_{y,j}$ - ($v_jB_{z,j}$)]} + {{{1 – [((({{$V_{t,j}$ + $V_{rpar,j}$ + {{{1 – {[I$V_{t,j}$ I $^2$]/[C $^2$]}} $^{1/2}$}($V_{rperp,j}$)}}}/{1 + {[$V_{t,j}$ dot $V_{r,j}$]/[C $^2$]}}}})$_j$ /C)$^2$]}$^{-1/2}$ }[$E_{z,j}$ + ((({$V_{t,j}$ + $V_{rpar,j}$ + {{{1 – {[I$V_{t,j}$ I $^2$]/[C $^2$]}} $^{1/2}$}($V_{rperp,j}$)}}}/{1 + {[$V_{t,j}$ dot $V_{r,j}$]/[C $^2$]}}}})$B_{y,j}$)]}} + {((({$V_{t,j}$ + $V_{rpar,j}$ + {{{1 – {[I$V_{t,j}$ I $^2$]/[C $^2$]}} $^{1/2}$}($V_{rperp,j}$)}}}/{1 + {[$V_{t,j}$ dot $V_{r,j}$]/[C $^2$]}}}}) x {{$Bx_j$ + {{1 – [((({{$V_{t,j}$ + $V_{rpar,j}$ + {{{1 – {[I$V_{t,j}$ I $^2$]/[C $^2$]}} $^{1/2}$}($V_{rperp,j}$)}}}/{1 + {[$V_{t,j}$ dot $V_{r,j}$]/[C $^2$]}}}})/C)$^2$]}$^{-1/2}$ {$B_{y,j}$ + {[[((({$V_{t,j}$ + $V_{rpar,j}$ + {{{1 – {[I$V_{t,j}$ I $^2$]/[C $^2$]}} $^{1/2}$}($V_{rperp,j}$)}}}/{1 + {[$V_{t,j}$ dot $V_{r,j}$]/[C $^2$]}}}})/[C$^2$]]$E_{z,j}$}}} + {{1 – [((({{$V_{t,j}$ + $V_{rpar,j}$ + {{{1 – {[I$V_{t,j}$ I $^2$]/[C $^2$]}} $^{1/2}$}($V_{rperp,j}$)}}}/{1 + {[$V_{t,j}$ dot $V_{r,j}$]/[C $^2$]}}}})/C)$^2$]}$^{-1/2}$ {$B_{z,j}$ - {[[((({$V_{t,j}$ + $V_{rpar,j}$ + {{{1 – {[I$V_{t,j}$ I $^2$]/[C $^2$]}} $^{1/2}$}($V_{rperp,j}$)}}}/{1 + {[$V_{t,j}$ dot $V_{r,j}$]/[C $^2$]}}}})/[C$^2$]]$E_{y,j}$}}}}}}}$_n$ • dL}}}/$dt_{coil(s)}$} {1/{{1 - [[($V_{t,j}$ $+_{(relativistic\ composition)}$ $V_{r,j}$)/C]$^2$]}$^{1/2}$}{{1/{1 – {{{$V_{t,j}$ + $V_{rpar,j}$ + {{{1 – {[I$V_{t,j}$I $^2$]/[C $^2$]}} $^{1/2}$}($V_{rperp,j}$)}}/{1 + {[[($V_{t,j}$)($V_{r,j}$) cosine (Θ)]]/[C $^2$]}}}}/C} $^2$}}} ½}.

Now, the rotor blade power intake and generation is affected by blade and coil pitch angle, surface area, coil radial extension width and length profiles, coil conductivity, and other factors. Thus, we will modify the latter formula without the bristle feature to yield;

$E_{coil}$ = [d $P_{electric,j}$/$dt_{coils}$] {1/{{1 - [[($V_t$ $+_{(relativistic\ composition)}$ $V_r$)/C]$^2$]}$^{1/2}$}}{f[(Pitch angle),(Surface area),(Coil radial extension width and length profile),(Coil conductivity),[Σ Other factors]]}

= {d{($I_{electric,j}$)}{Σ(n = 1, n = $N_{coil}$)[(1/q) ∫ $F_j$ • dL]$_n$}}}{f[(Pitch angle),(Surface area),(Coil radial extension width and length profile),(Coil conductivity),[Σ Other factors]]} = {{d{($I_{electric,j}$)}{Σ(n = 1, n = $N_{coil}$){∫{{$Ex_j$ + {{{1 – [((({{$V_{t,j}$ + $V_{rpar,j}$ + {{{1 – {[I$V_{t,j}$ I $^2$]/[C $^2$]}} $^{1/2}$}($V_{rperp,j}$)}}}/{1 + {[$V_{t,j}$ dot $V_{r,j}$]/[C $^2$]}}}}) /C)$^2$]}$^{-1/2}$ }[$E_{y,j}$ - ($v_jB_{z,j}$)]} + {{{1 – [((({{$V_{t,j}$ + $V_{rpar,j}$ + {{{1 – {[I$V_{t,j}$ I $^2$]/[C $^2$]}} $^{1/2}$}($V_{rperp,j}$)}}}/{1 + {[$V_{t,j}$ dot $V_{r,j}$]/[C $^2$]}}}})$_j$ /C)$^2$]}$^{-1/2}$ }[$E_{z,j}$ + ((({$V_{t,j}$ + $V_{rpar,j}$ + {{{1 – {[I$V_{t,j}$ I $^2$]/[C $^2$]}} $^{1/2}$}($V_{rperp,j}$)}}}/{1 + {[$V_{t,j}$ dot $V_{r,j}$]/[C $^2$]}}}})$B_{y,j}$)]}} + {((({$V_{t,j}$ + $V_{rpar,j}$ + {{{1 – {[I$V_{t,j}$ I $^2$]/[C $^2$]}} $^{1/2}$}($V_{rperp,j}$)}}}/{1 + {[$V_{t,j}$ dot $V_{r,j}$]/[C $^2$]}}}}) x {{$Bx_j$ + {{1 – [((({{$V_{t,j}$ + $V_{rpar,j}$ + {{{1 – {[I$V_{t,j}$ I $^2$]/[C $^2$]}} $^{1/2}$}($V_{rperp,j}$)}}}/{1 + {[$V_{t,j}$ dot $V_{r,j}$]/[C $^2$]}}}})/C)$^2$]}$^{-1/2}$ {$B_{y,j}$ + {[[((({$V_{t,j}$ + $V_{rpar,j}$ + {{{1 – {[I$V_{t,j}$ I $^2$]/[C $^2$]}} $^{1/2}$}($V_{rperp,j}$)}}}/{1 + {[$V_{t,j}$ dot $V_{r,j}$]/[C $^2$]}}}})/[C$^2$]]$E_{z,j}$}}} + {{1 – [((({{$V_{t,j}$ + $V_{rpar,j}$ + {{{1 – {[I$V_{t,j}$ I $^2$]/[C $^2$]}} $^{1/2}$}($V_{rperp,j}$)}}}/{1 + {[$V_{t,j}$ dot $V_{r,j}$]/[C $^2$]}}}})/C)$^2$]}$^{-1/2}$ {$B_{z,j}$ - {[[((({$V_{t,j}$ + $V_{rpar,j}$ + {{{1 – {[I$V_{t,j}$ I $^2$]/[C $^2$]}} $^{1/2}$}($V_{rperp,j}$)}}}/{1 + {[$V_{t,j}$ dot $V_{r,j}$]/[C $^2$]}}}})/[C$^2$]]$E_{y,j}$}}}}}}}$_n$ • dL}}}/$dt_{coil(s)}$} {1/{{1 - [[($V_{t,j}$ $+_{(relativistic\ composition)}$ $V_{r,j}$)/C]$^2$]}$^{1/2}$}}{f[(Pitch angle),(Surface area),(Coil radial extension width and length profile),(Coil conductivity),[Σ Other factors]]}}

= {{d{($I_{electric,j}$)}{Σ(n = 1, n = $N_{coil}$)[(1/q) ∫ $F_j$ • dL]$_n$}}} = {d{($I_{electric,j}$)}{Σ(n = 1, n = $N_{coil}$){∫{{$Ex_j$ + {{{1 – [((({{$V_{t,j}$ + $V_{rpar,j}$ + {{{1 – {[I$V_{t,j}$ I $^2$]/[C $^2$]}} $^{1/2}$}($V_{rperp,j}$)}}}/{1 + {[$V_{t,j}$ dot $V_{r,j}$]/[C $^2$]}}}}) /C)$^2$]}$^{-1/2}$ }[$E_{y,j}$ - ($v_jB_{z,j}$)]} + {{{1 – [((({{$V_{t,j}$ + $V_{rpar,j}$ + {{{1 – {[I$V_{t,j}$ I $^2$]/[C $^2$]}} $^{1/2}$}($V_{rperp,j}$)}}}/{1 + {[$V_{t,j}$ dot $V_{r,j}$]/[C $^2$]}}}})$_j$ /C)$^2$]}$^{-1/2}$ }[$E_{z,j}$ + ((({$V_{t,j}$ + $V_{rpar,j}$ + {{{1 – {[I$V_{t,j}$ I $^2$]/[C $^2$]}} $^{1/2}$}($V_{rperp,j}$)}}}/{1 + {[$V_{t,j}$ dot $V_{r,j}$]/[C $^2$]}}}})$B_{y,j}$)]}} + {((({$V_{t,j}$ + $V_{rpar,j}$ + {{{1 – {[I$V_{t,j}$ I $^2$]/[C $^2$]}} $^{1/2}$}($V_{rperp,j}$)}}}/{1 + {[$V_{t,j}$ dot $V_{r,j}$]/[C $^2$]}}}}) x {{$Bx_j$ + {{1 – [((({{$V_{t,j}$ + $V_{rpar,j}$ + {{{1 – {[I$V_{t,j}$ I $^2$]/[C $^2$]}} $^{1/2}$}($V_{rperp,j}$)}}}/{1 + {[$V_{t,j}$ dot $V_{r,j}$]/[C $^2$]}}}})/C)$^2$]}$^{-1/2}$ {$B_{y,j}$ + {[[((({$V_{t,j}$ + $V_{rpar,j}$ + {{{1 – {[I$V_{t,j}$ I $^2$]/[C $^2$]}} $^{1/2}$}($V_{rperp,j}$)}}}/{1 + {[$V_{t,j}$ dot $V_{r,j}$]/[C $^2$]}}}})/[C$^2$]]$E_{z,j}$}}} + {{1 – [((({{$V_{t,j}$ + $V_{rpar,j}$ + {{{1 – {[I$V_{t,j}$ I $^2$]/[C $^2$]}} $^{1/2}$}($V_{rperp,j}$)}}}/{1 + {[$V_{t,j}$ dot $V_{r,j}$]/[C $^2$]}}}})/C)$^2$]}$^{-1/2}$ {$B_{z,j}$ - {[[((({$V_{t,j}$ + $V_{rpar,j}$ + {{{1 – {[I$V_{t,j}$ I $^2$]/[C $^2$]}} $^{1/2}$}($V_{rperp,j}$)}}}/{1 + {[$V_{t,j}$ dot $V_{r,j}$]/[C $^2$]}}}})/[C$^2$]]$E_{y,j}$}}}}}}}$_n$ • dL}}}/$dt_{coil(s)}$} {1/{{1 - [[($V_{t,j}$ $+_{(relativistic\ composition)}$ $V_{r,j}$)/C]$^2$]}$^{1/2}$}{{1/{1 – {{{$V_{t,j}$ + $V_{rpar,j}$ + {{{1 – {[I$V_{t,j}$I $^2$]/[C $^2$]}} $^{1/2}$}($V_{rperp,j}$)}}/{1 + {[[($V_{t,j}$)($V_{r,j}$) cosine (Θ)]]/[C $^2$]}}}}/C} $^2$}}} ½}{f[(Pitch angle),(Surface area),(Coil radial extension width and length profile),(Coil conductivity),[Σ Other factors]]}}.

Now F = $\mathbf{F_1} + \mathbf{F_2} + ... = Q\mathbf{E}$.

Therefore, the electric force exerted on a given differential surface area charge element of the central axial charge member, Q, by a single intake particle of positive charge, q, is initially equal to:

$F = Q \{[1/(4\pi\varepsilon_0)][\gamma q \mathbf{R}]/\{\{[\gamma^2 R^2 \cos^2\theta] + [R^2 \sin^2\theta]\}^{3/2}\}\}$.

Here, q is in integer multiples of positronic or protonic electric charge units expressed in units of Coulomb.

The incremental step-wise series of forces exerted on the charged particle as it is pulled into the charged axial surface by a given differential surface area charge element of the charged axial surface, Q is:

$\Sigma F_h\,(h = 1, h = n_{axial}) = Q\,\{\Sigma\,(h = 1, h = n_{axial})\,\{[1/(4\pi\varepsilon_0)][\gamma_h q \mathbf{R}_h]/\{\{[\gamma_h^2 R_h^2 \cos^2\theta_h] + [R_h^2 \sin^2\theta_h]\}^{3/2}\}\}\}$.

For a stream of m positively charged intake particles entering the cone where the particles are distributed in the background frame over a unit spatial interval in the background frame; the electric force exerted by the intake particles on the differential axial charge element is equal to:

For a stream of k positively charged intake particles entering the cone where the particles are distributed in the background frame over a unit spatial interval in the background frame; the electric force exerted by the particles on the differential axial charge element is equal to:

$F_{elect} = Q\,\{\Sigma(k = 1, k = o)\,\{\Sigma\,(h = 1, h = n_{axial})\,\{[1/(4\pi\varepsilon_0)][\gamma_{h,k}\,(sq_{h,k})\,\mathbf{R_{h,k}}]/\{\{[\gamma_{h,k}{}^2 R_{h,k}{}^2 \cos^2\theta_{h,k}] + [R_{h,k}{}^2 \sin^2\theta_{h,k}]\}^{3/2}\}\}\}\}(\gamma)(v/C)$,

where s is a positive integer representing the number of net electronic charge units.

The total electric force exerted on the axial charged member by the stream of particles where the particles are distributed in the background frame over a unit spatial interval in the background frame is equal to:

The total electric force exerted on the axial charged member by the stream of k charged particles where the particles are distributed in the background frame over a unit spatial interval in the background frame is equal to.

$F_{elect} = \{\Sigma\,(u = 1, u = n_{axial})\,[Q_{u,axial}(x,y,z,t)]\{\Sigma(k = 1, k = m)\,\{\Sigma\,(h = 1, h = n)\,\{[1/(4\pi\varepsilon_0)][\gamma_{h,k}\,(sq_{h,k})\,\mathbf{R}_{h,k}]/\{\{[\gamma_{h,k}{}^2 R_{h,k}{}^2 \cos^2\theta_{h,k}] + [R_{h,k}{}^2 \sin^2\theta_{h,k}]\}^{3/2}\}\}\}\}\}(\gamma)(v/C)$

The electric force exerted on a given differential surface area charge element of the charged cone, Q, by a single intake particle of negative charge, q, is equal to:

$F = Q \{[1/(4\pi\varepsilon_0)][\gamma q \mathbf{R}]/\{\{[\gamma^2 R^2 \cos^2\theta] + [R^2 \sin^2\theta]\}^{3/2}\}\}$.

Here, q is in integer multiples of electronic charge in electric charge units expressed in units of Coulomb.

The incremental step-wise series of forces exerted on the charged particle as it is pulled into the charged conical surface by a given differential surface area charge element of the charged conical surface, Q is:

$\Sigma F_h\,(h = 1, h = n_{cone}) = Q\,\{\Sigma\,(h = 1, h = n_{cone})\,\{[1/(4\pi\varepsilon_0)][\gamma_h q \mathbf{R}_h]/\{\{[\gamma_h^2 R_h^2 \cos^2\theta_h] + [R_h^2 \sin^2\theta_h]\}^{3/2}\}\}\}$.

For a stream of k positively charged intake particles entering the cone where the particles are distributed in the background frame over a unit spatial interval in the background frame; the electric force exerted by the particles on the differential conical charge element is equal to:

$F_{elect} = Q\,\{\Sigma(k = 1, k = o)\,\{\Sigma\,(h = 1, h = n_{cone})\,\{[1/(4\pi\varepsilon_0)][\gamma_{h,k}\,(sq_{h,k})\,\mathbf{R_{h,k}}]/\{\{[\gamma_{h,k}{}^2 R_{h,k}{}^2 \cos^2\theta_{h,k}] + [R_{h,k}{}^2 \sin^2\theta_{h,k}]\}^{3/2}\}\}\}\}(\gamma)(v/C)$

where s is a positive integer representing the number of net electronic charge units.

The total electric force exerted on the charged conical member by the stream of k charged particles where the particles are distributed in the background frame over a unit spatial interval in the background frame is equal to:

$F_{elect} = \{\Sigma\,(u = 1, u = n_{cone})\,[Q_{u,cone}(x,y,z,t)]\{\Sigma(k = 1, k = m)\,\{\Sigma\,(h = 1, h = n)\,\{[1/(4\pi\varepsilon_0)][\gamma_{h,k}\,(sq_{h,k})\,\mathbf{R}_{h,k}]/\{\{[\gamma_{h,k}{}^2 R_{h,k}{}^2 \cos^2\theta_{h,k}] + [R_{h,k}{}^2 \sin^2\theta_{h,k}]\}^{3/2}\}\}\}\}\}(\gamma)(v/C)$

The following factors or terms are non-dimensional scalar quantities.

When the bristles' electric flux density enhancements are taken into account, each of the above equations for overall axial member pulling effects and/or overall conical pulling effects on the intake chargons is modified by the factoral suffix;

{Σ (i = 0, i = n) {Σ (r = 1, r = N) [f(bristle shape$_{r,i}$, intake chargon $\gamma_{r,i}$, bristle electric-field density dist$_{r,i}$, total bristle electric field energy$_{r,i}$)]/N}}.

For the axial bristles, we obtain:

{Σ (i = 0, i = n) {Σ (r = 1, r = N) [f(bristle shape$_{r,i}$, intake chargon $\gamma_{r,i}$, bristle electric-field density dist$_{r,i}$, total bristle electric field energy$_{r,i}$)]/$N_{axial}$}}.

For the cone bristles, we obtain:

{Σ (i = 0, i = n) {Σ (r = 1, r = N) [f(bristle shape$_{r,i}$, intake chargon $\gamma_{r,i}$, bristle electric-field density dist$_{r,i}$, total bristle electric field energy$_{r,i}$)]/$N_{conical}$}}.

Here, $N_{axial}$ and $N_{conical}$ are the numbers of axial and conical bristles. Thus, the above sigma functions are averages which take into account the collective electric flux concentration effects of the entire respective sets of bristles.

The factors, {Σ (i = 0, i = n) {Σ (r = 1, r = N) [f(magnetic$_{r,i}$)]/$N_{axial}$}}, and {Σ (i = 0, i = n) {Σ (r = 1, r = N) [f(magnetic$_{r,i}$)]/$N_{conical}$}} take into account the average collective effects of magnetic field elements associated with the focusing of the intake chargons onto the bristle tips.

Taken together, the two factors describing the bristle effects and operations are for axial bristles:

{{Σ (i = 0, i = n) {Σ (r = 1, r = N) [f(magnetic$_{r,i}$)]/$N_{axial}$}}{Σ (i = 0, i = n) {Σ (r = 1, r = N) [f(bristle shape$_{r,i}$, intake chargon $\gamma_{r,i}$, bristle electric-field density dist$_{r,i}$, total bristle electric field energy$_{r,i}$)]/$N_{axial}$}}}

and for the cone bristles,

{{Σ (i = 0, i = n) {Σ (r = 1, r = N) [f(bristle shape$_{r,i}$, intake chargon $\gamma_{r,i}$, bristle electric-field density dist$_{r,i}$, total bristle electric field energy$_{r,i}$)]/$N_{conical}$}}{Σ (i = 0, i = n) {Σ (r = 1, r = N) [f(magnetic$_{r,i}$)]/$N_{conical}$}}}.

The sub-functions bristle shape$_{r,i}$, intake chargon $\gamma_{r,i}$, bristle electric-field density dist$_{r,i}$, total bristle electric field energy$_{r,I}$ account for each bristle: the bristle shape, the Lorentz factor of the intake chargons as modified by the enhanced electric flux densities near the tips of the bristles, the bristle electric field density distribution, and total bristle electric field based energy.

For cases where no bristles are included in the conical design, i.e., the bristle mechanism is neglected, the above modification factors for the bristles may be set to one thus indicating a non-operative function. Setting the bristle terms equal to zero would be in error because then the energy gain expressions which include the bristle factors would become equal to zero, thus wrongly implying that the associated terms for chargon drive yield zero drive energy input for chargon drive modes.

All of the above mentioned intake chargon drives as also plausibly include mirror matter model analogues. We use the prefix operator [SM, MMM] in the formulations below for gamma and velocity to indicate both Standard Model and Mirror Matter Model charge mechanisms.

The charged intake matter that enters the cone is assumed to be completely absorbed into either the cone surface or the central axial member and converted to energy under the condition where astrodynamic drag energy is completely absorbed and recycled so that the net astrodynamic drag imposed by such intake matter is completely neutralized. The energy of conversion of the intake charged mass is assumed to be converted to a virtually 100 percent efficient light speed exhaust stream such as a photonic exhaust, neutrino exhaust, and/or gravitational radiation based exhaust that is close to perfectly collimated.

Therefore, the force produced by the beamed exhaust produced by the energy conversion of the charged mass absorbed into the axial member is equal to;

<S>/C = <{[SM, MMM]: d {Σ (u = 1, u = $n_{axial}$) [$M_{ua}C^2$] (γ)(v/C)}/dt}>/C,

where <S> is the ship time averaged light speed exhaust thrust based power, $n_{axial}$ is the number of charged intake particles per unit background spatial interval in the background reference frame, $M_{ua}$ is the slowed mass of the uth intake particle in the ship's reference frame or the invariant mass of the particle applied in the space-craft's reference frame for particles being absorbed into the axial member, and t is the background frame time .

The force produced by the beamed exhaust produced by the energy conversion of the charged mass absorbed into the conical surface is equal to;

$<S>/C = <\{[SM, MMM]: d\{\Sigma (u = 1, u = n_{cone}) [M_{uc}C^2] (\gamma)(v/C)\}/dt\}>/C,$

where <S> is the ship time averaged light speed exhaust thrust based power, $n_{conel}$ is the number of charged intake particles per unit background spatial interval in the background reference frame, $M_{uc}$ is the slowed mass of the uth intake particle in the ship's reference frame or the invariant mass of the particle applied in the space-craft's reference frame for particles being absorbed into the conical surface, and t is the background frame time .

The above mentioned axial and conical intake chargon conversion to light speed exhaust mechanisms also plausibly include mirror matter model analogues. We use the prefix operator [SM, MMM] in the formulations below for gamma and velocity to indicate both Standard Model and Mirror Matter Model charge mechanisms.

The total power with respect to the sail's reference frame for a given sail Lorentz factor taking into account the bomb based bristle mechanism is therefore:

$P = \{\{\int(y_1,y_2) \{\int(x_1,x_2) \{\int (0, \pi) \{\{\{(T_{cmbr}) /\{\gamma [1 + [(v/C) \cos \theta]]\}\}^4\} \sigma e\} d\theta\}dx\}dy\}(2)\}$ + {[Standard Model and Mirror Matter Model]$d\{\int < \{\Sigma (u = 1, u = n_{axial}) (2)[Q_{u,axial}(x,y,z,t)]\{\Sigma(k = 1, k = m) \{\Sigma (h =1, h = n) \{[1/(4\pi\varepsilon_0)][\gamma_{h,k} (sq_{h,k}) \mathbf{R}_{h,k}]/\{\{[\gamma_{h,k}{}^2 R_{h,k}{}^2 \cos^2 \theta_{h,k}] + [R_{h,k}{}^2 \sin^2 \theta_{h,k}]\}^{3/2}\}\}\}\}(\gamma)(v/C)> \cdot dx\}/dt\}$ + {[Standard Model and Mirror Matter Model]$d\{\int < \{\Sigma (u = 1, u = n_{cone}) (2)[Q_{u,cone}(x,y,z,t)]\{\Sigma(k = 1, k = m) \{\Sigma (h =1, h = n) \{[1/(4\pi\varepsilon_0)][\gamma_{h,k} (sq_{h,k}) \mathbf{R}_{h,k}]/\{\{[\gamma_{h,k}{}^2 R_{h,k}{}^2 \cos^2 \theta_{h,k}] + [R_{h,k}{}^2 \sin^2 \theta_{h,k}]\}^{3/2}\}\}\}\}(\gamma)(v/C)> \cdot dx\}/dt\}$ + $\{[SM, MMM]: d \{\Sigma (u = 1, u = n_{axial}) [M_{ua}C^2] (\gamma)(v/C)\}/dt\}$ + $\{[SM, MMM]: d \{\Sigma (u = 1, u = n_{cone}) [M_{uc}C^2] (\gamma)(v/C)\}/dt_{ship}\}$

+{{d {∫ {{{[(electrons):($Q_{natural}$)] $\{[1/(4\pi\varepsilon_0)][\gamma_{average,electrons,natural,ship\text{-}frame}\ \mathbf{R}]/\{\{[\gamma_{average,electrons,natural,ship\text{-}frame}{}^2 R^2 \cos^2 \theta] + [R^2 \sin^2 \theta]\}^{3/2}\}\}$ +{[(electrons):($Q_{bomb}$)] $\{[1/(4\pi\varepsilon_0)][\gamma_{average,bomb,ship\text{-}frame}\ \mathbf{R}]/\{\{[\gamma_{average,bomb,ship\text{-}frame}{}^2 R^2 \cos^2 \theta] + [R^2 \sin^2 \theta]\}^{3/2}\}\}$ + {[Σ(n = 1; n = number of positive background ion species):(nth ion species)]: ($Q_{natural}$)] $\{[1/(4\pi\varepsilon_0)][\gamma_{average,nth\text{-}ion\text{-}species,natural,ship\text{-}frame}\ \mathbf{R}]/\{\{[\gamma_{average,nth\text{-}ion\text{-}species,natural,ship\text{-}frame}{}^2 R^2 \cos^2 \theta] + [R^2 \sin^2 \theta]\}^{3/2}\}\}$ + {[Σ(m = 1; m = number of positive bomb ion species):(mth ion species)]: ($Q_{bomb}$)] $\{[1/(4\pi\varepsilon_0)][\gamma_{average,mth\text{-}ion\text{-}species,bomb,ship\text{-}frame}\ \mathbf{R}]/\{\{[\gamma_{average,mth\text{-}ion\text{-}species,bomb,ship\text{-}frame}{}^2 R^2 \cos^2 \theta] + [R^2 \sin^2 \theta]\}^{3/2}\}\}$ + {[Σ(n = 1; n = number of negative background ion species):(nth ion species)]: ($Q_{natural}$)] $\{[1/(4\pi\varepsilon_0)][\gamma_{average,nth\text{-}ion\text{-}species,natural,ship\text{-}frame}\ \mathbf{R}]/\{\{[\gamma_{average,nth\text{-}ion\text{-}species,bomb,ship\text{-}frame}{}^2 R^2 \cos^2 \theta] + [R^2 \sin^2 \theta]\}^{3/2}\}\}$ + {[Σ(m = 1; m = number of negative bomb ion species):(mth ion species)]: ($Q_{bomb}$)] $\{[1/(4\pi\varepsilon_0)][\gamma_{average,mth\text{-}ion\text{-}species,bomb,ship\text{-}frame}\ \mathbf{R}]/\{\{[\gamma_{average,mth\text{-}ion\text{-}species,bomb,ship\text{-}frame}{}^2 R^2 \cos^2 \theta] + [R^2 \sin^2 \theta]\}^{3/2}\}\}$ + {d{[(Gound state $N_{flux}$)($v_{average,ship\text{-}frame}$)($\gamma_{average,ship\text{-}frame}$)] for ground state neutrons}/$dt_{ship}$} + {d {Σ(z = 1; z = number of excited neutron states):[(zth excited state $N_{flux}$,)($v_{average,ship\text{-}frame}$)($\gamma_{average,ship\text{-}frame}$)] for excited state neutrons}/dt} + [<$S_{background,photons}$>/c] + [[<$S_{bomb,photons}$>/c] ($Ph_{flux}$)]}

+ {Drag based on the following propulsive force inducing mechanisms {{[(electrons):($Q_{natural}$)] $\{[1/(4\pi\varepsilon_0)][\gamma_{average,electrons,natural,ship\text{-}frame}\ \mathbf{R}]/\{\{[\gamma_{average,electrons,natural,ship\text{-}frame}{}^2 R^2 \cos^2 \theta] + [R^2 \sin^2 \theta]\}^{3/2}\}\}$ +{[(electrons):($Q_{bomb}$)] $\{[1/(4\pi\varepsilon_0)][\gamma_{average,bomb,ship\text{-}frame}\ \mathbf{R}]/\{\{[\gamma_{average,bomb,ship\text{-}frame}{}^2 R^2 \cos^2 \theta] + [R^2 \sin^2 \theta]\}^{3/2}\}\}$ + {[Σ(n = 1; n = number of positive background ion species):(nth ion species)]: ($Q_{natural}$)] $\{[1/(4\pi\varepsilon_0)][\gamma_{average,nth\text{-}ion\text{-}species,natural,ship\text{-}frame}\ \mathbf{R}]/\{\{[\gamma_{average,nth\text{-}ion\text{-}species,natural,ship\text{-}frame}{}^2 R^2 \cos^2 \theta] + [R^2 \sin^2 \theta]\}^{3/2}\}\}$ + {[Σ(m = 1; m = number of positive bomb ion species):(mth ion species)]: ($Q_{bomb}$)] $\{[1/(4\pi\varepsilon_0)][\gamma_{average,mth\text{-}ion\text{-}species,bomb,ship\text{-}frame}\ \mathbf{R}]/\{\{[\gamma_{average,mth\text{-}ion\text{-}species,bomb,ship\text{-}frame}{}^2 R^2 \cos^2 \theta] + [R^2 \sin^2 \theta]\}^{3/2}\}\}$ + {[Σ(n = 1; n = number of negative background ion species):(nth ion species)]: ($Q_{natural}$)] $\{[1/(4\pi\varepsilon_0)][\gamma_{average,nth\text{-}ion\text{-}species,natural,ship\text{-}frame}\ \mathbf{R}]/\{\{[\gamma_{average,nth\text{-}ion\text{-}species,bomb,ship\text{-}frame}{}^2 R^2 \cos^2 \theta] + [R^2 \sin^2 \theta]\}^{3/2}\}\}$ + {[Σ(m = 1; m = number of negative bomb ion species):(mth ion species)]: ($Q_{bomb}$)] $\{[1/(4\pi\varepsilon_0)][\gamma_{average,mth\text{-}ion\text{-}species,bomb,ship\text{-}frame}\ \mathbf{R}]/\{\{[\gamma_{average,mth\text{-}ion\text{-}species,bomb,ship\text{-}frame}{}^2 R^2 \cos^2 \theta] + [R^2 \sin^2 \theta]\}^{3/2}\}\}$ + {d{[(Gound state $N_{flux}$)($v_{average,ship\text{-}frame}$)($\gamma_{average,ship\text{-}frame}$)] for ground state neutrons}/$dt_{ship}$} + {d {Σ(z = 1; z = number of excited neutron states):[(zth excited state $N_{flux}$,)($v_{average,ship\text{-}frame}$)($\gamma_{average,ship\text{-}frame}$)] for excited state neutrons}/dt} + [<$S_{background,photons}$>/c] + [[<$S_{bomb,photons}$>/c] ($Ph_{flux}$)]}}} ● dr}/dt}(Bristle-frame translational gamma)}

$= \{\{\int(y_1,y_2) \{\int(x_1,x_2) \{\int (0, \pi) \{\{\{(T_{cmbr}) /\{\{1/\{[1 + [(v/C)^2 ]]^{1/2}\}\} [1 + [(v/C) \cos \theta]]\}\}^4\} \sigma e\} d\theta\}dx\}dy\}(2)\}$ + {[Standard Model and Mirror Matter Model] $d\{\int < \{\Sigma (u = 1, u = n_{axial}) (2)[Q_{u,axial}(x,y,z,t)]\{\Sigma(k = 1, k = m) \{\Sigma (h =1, h = n) \{[1/(4\pi\varepsilon_0)]\{ 1/\{1 - [(v_{h,k}/C)^2 ]\}^{1/2}\} (sq_{h,k}) \mathbf{R}_{h,k}\}/\{\{\{ \{1/\{1 - [(v_{h,k}/C)^2 ]\}^{1/2}\}^2\} R_{h,k}{}^2 \cos^2 \theta_{h,k}\} + [R_{h,k}{}^2 \sin^2 \theta_{h,k}]\}^{3/2}\}\}\}\}\{1/\{1 -[(v/C)^2 ]\}^{1/2}\} (v/C)> \cdot dx\} /dt\}$ + {[Standard Model and Mirror Matter Model] $d\{\int < \{\Sigma (u = 1, u = n_{cone}) (2)[Q_{u,cone}(x,y,z,t)]\{\Sigma(k = 1, k = m) \{\Sigma (h =1, h = n) \{[1/(4\pi\varepsilon_0)]\{ 1/\{1 - [(v_{h,k}/C)^2 ]\}^{1/2}\} (sq_{h,k}) \mathbf{R}_{h,k}\}/\{\{\{\{1/\{1 - [(v_{h,k}/C)^2 ]\}^{1/2}\}^2\} R_{h,k}{}^2 \cos^2 \theta_{h,k}\} + [R_{h,k}{}^2 \sin^2 \theta_{h,k}]\}^{3/2}\}\}\}\}\{1/\{1 - [(v/C)^2 ]\}^{1/2}\} (v/C)> \cdot dx\}/dt\}$

+ {[SM, MMM]: d {Σ (u = 1, u = $n_{axial}$) [$M_{ua}C^2$] {1/{1 - [$(v/C)^2$ ]} $^{1/2}$} (v/C)}/dt} + {[SM, MMM]: d {Σ (u = 1, u = $n_{cone}$) [$M_{uc}C^2$] {1/{1 - [$(v/C)^2$ ]} $^{1/2}$} (v/C)}/$dt_{ship}$}

+{{d {∫ {{{[(electrons):($Q_{natural}$)] {[1/(4π$\varepsilon_0$)][$\gamma_{average,electrons,natural,ship\text{-}frame}$ **R**]/{{[$\gamma_{average,electrons,natural,ship\text{-}frame}$ $^2$ $R^2$ $\cos^2$ θ] + [$R^2$ $\sin^2$ θ]}$^{3/2}$}}} +{[(electrons):($Q_{bomb}$)] {[1/(4π$\varepsilon_0$)][$\gamma_{average,bomb,ship\text{-}frame}$ **R**]/{{[$\gamma_{average,bomb,ship\text{-}frame}$$^2$ $R^2$ $\cos^2$ θ] + [$R^2$ $\sin^2$ θ]}$^{3/2}$}}} + {[Σ(n = 1; n = number of positive background ion species):(nth ion species)]: ($Q_{natural}$)] {[1/(4π$\varepsilon_0$)][$\gamma_{average,nth\text{-}ion\text{-}species,natural,ship\text{-}frame}$ **R**]/{{[$\gamma_{average,nth\text{-}ion\text{-}species,natural,ship\text{-}frame}$ $^2$ $R^2$ $\cos^2$ θ] + [$R^2$ $\sin^2$ θ]}$^{3/2}$}}} + {[Σ(m = 1; m = number of positive bomb ion species):(mth ion species)]: ($Q_{bomb}$)] {[1/(4π$\varepsilon_0$)][$\gamma_{average,mth\text{-}ion\text{-}species,bomb,ship\text{-}frame}$ **R**]/{{[$\gamma_{average,mth\text{-}ion\text{-}species,bomb,ship\text{-}frame}$ $^2$ $R^2$ $\cos^2$ θ] + [$R^2$ $\sin^2$ θ]}$^{3/2}$}}} + {[Σ(n = 1; n = number of negative background ion species):(nth ion species)]: ($Q_{natural}$)] {[1/(4π$\varepsilon_0$)][$\gamma_{average,nth\text{-}ion\text{-}species,natural,ship\text{-}frame}$ **R**]/{{[$\gamma_{average,nth\text{-}ion\text{-}species,bomb,ship\text{-}frame}$ $^2$ $R^2$ $\cos^2$ θ] + [$R^2$ $\sin^2$ θ]}$^{3/2}$}}} + {[Σ(m = 1; m = number of negative bomb ion species):(mth ion species)]: ($Q_{bomb}$)] {[1/(4π$\varepsilon_0$)][$\gamma_{average,mth\text{-}ion\text{-}species,bomb,ship\text{-}frame}$ **R**]/{{[$\gamma_{average,mth\text{-}ion\text{-}species,bomb,ship\text{-}frame}$ $^2$ $R^2$ $\cos^2$ θ] + [$R^2$ $\sin^2$ θ]}$^{3/2}$}}} + {d{[(Gound state $N_{flux}$)($v_{average,ship\text{-}frame}$)($\gamma_{average,ship\text{-}frame}$)] for ground state neutrons}/$dt_{ship}$} + {d {Σ(z = 1; z = number of excited neutron states):[(zth excited state $N_{flux,}$)($v_{average,ship\text{-}frame}$)($\gamma_{average,ship\text{-}frame}$)] for excited state neutrons}/dt} + [$<S_{background,photons}>$/c] + [[$<S_{bomb,photons}>$/c] ($Ph_{flux}$)]}

+ {Drag based on the following propulsive force inducing mechanisms {{[(electrons):($Q_{natural}$)] {[1/(4π$\varepsilon_0$)][$\gamma_{average,electrons,natural,ship\text{-}frame}$ **R**]/{{[$\gamma_{average,electrons,natural,ship\text{-}frame}$ $^2$ $R^2$ $\cos^2$ θ] + [$R^2$ $\sin^2$ θ]}$^{3/2}$}}} +{[(electrons):($Q_{bomb}$)] {[1/(4π$\varepsilon_0$)][$\gamma_{average,bomb,ship\text{-}frame}$ **R**]/{{[$\gamma_{average,bomb,ship\text{-}frame}$$^2$ $R^2$ $\cos^2$ θ] + [$R^2$ $\sin^2$ θ]}$^{3/2}$}}} + {[Σ(n = 1; n = number of positive background ion species):(nth ion species)]: ($Q_{natural}$)] {[1/(4π$\varepsilon_0$)][$\gamma_{average,nth\text{-}ion\text{-}species,natural,ship\text{-}frame}$ **R**]/{{[$\gamma_{average,nth\text{-}ion\text{-}species,natural,ship\text{-}frame}$ $^2$ $R^2$ $\cos^2$ θ] + [$R^2$ $\sin^2$ θ]}$^{3/2}$}}} + {[Σ(m = 1; m = number of positive bomb ion species):(mth ion species)]: ($Q_{bomb}$)] {[1/(4π$\varepsilon_0$)][$\gamma_{average,mth\text{-}ion\text{-}species,bomb,ship\text{-}frame}$ **R**]/{{[$\gamma_{average,mth\text{-}ion\text{-}species,bomb,ship\text{-}frame}$ $^2$ $R^2$ $\cos^2$ θ] + [$R^2$ $\sin^2$ θ]}$^{3/2}$}}} + {[Σ(n = 1; n = number of negative background ion species):(nth ion species)]: ($Q_{natural}$)] {[1/(4π$\varepsilon_0$)][$\gamma_{average,nth\text{-}ion\text{-}species,natural,ship\text{-}frame}$ **R**]/{{[$\gamma_{average,nth\text{-}ion\text{-}species,bomb,ship\text{-}frame}$ $^2$ $R^2$ $\cos^2$ θ] + [$R^2$ $\sin^2$ θ]}$^{3/2}$}}} + {[Σ(m = 1; m = number of negative bomb ion species):(mth ion species)]: ($Q_{bomb}$)] {[1/(4π$\varepsilon_0$)][$\gamma_{average,mth\text{-}ion\text{-}species,bomb,ship\text{-}frame}$ **R**]/{{[$\gamma_{average,mth\text{-}ion\text{-}species,bomb,ship\text{-}frame}$ $^2$ $R^2$ $\cos^2$ θ] + [$R^2$ $\sin^2$ θ]}$^{3/2}$}}} + {d{[(Gound state $N_{flux}$)($v_{average,ship\text{-}frame}$)($\gamma_{average,ship\text{-}frame}$)] for ground state neutrons}/$dt_{ship}$} + {d {Σ(z = 1; z = number of excited neutron states):[(zth excited state $N_{flux,}$)($v_{average,ship\text{-}frame}$)($\gamma_{average,ship\text{-}frame}$)] for excited state neutrons}/dt} + [$<S_{background,photons}>$/c] + [[$<S_{bomb,photons}>$/c] ($Ph_{flux}$)]}}} • dr}/dt}(Bristle-frame translational gamma)}

Here, dt is bristle-time.

The total power with respect to the sail's reference frame for a given sail Lorentz factor taking into account the bomb based bristle mechanism for which inert pellets are ionized is therefore:

P = {{∫($y_1$,$y_2$) {∫($x_1$,$x_2$) {∫ (0, π) {{{($T_{cmbr}$) /{γ [1 + [(v/C) cos θ]]}}$^4$} σ e} dθ}dx}dy}(2)} + {[Standard Model and Mirror Matter Model]d{∫ < {Σ (u = 1, u = $n_{axial}$) (2)[$Q_{u,axial}$(x,y,z,t)]{Σ(k = 1, k = m) {Σ (h =1, h = n) {[1/(4π$\varepsilon_0$)][$\gamma_{h,k}$ ($sq_{h,k}$) $\mathbf{R}_{h,k}$]/{{[$\gamma_{h,k}$ $^2$ $R_{h,k}$ $^2$ $\cos^2$ $\theta_{h,k}$] + [$R_{h,k}$ $^2$ $\sin^2$ $\theta_{h,k}$]}$^{3/2}$}}}}}(γ)(v/C)> · dx}/dt} + {[Standard Model and Mirror Matter Model]d{∫ < {Σ (u = 1, u = $n_{cone}$) (2)[$Q_{u,cone}$(x,y,z,t)]{Σ(k = 1, k = m) {Σ (h =1, h = n) {[1/(4π$\varepsilon_0$)][$\gamma_{h,k}$ ($sq_{h,k}$) $\mathbf{R}_{h,k}$]/{{[$\gamma_{h,k}$ $^2$ $R_{h,k}$ $^2$ $\cos^2$ $\theta_{h,k}$] + [$R_{h,k}$ $^2$ $\sin^2$ $\theta_{h,k}$]}$^{3/2}$}}}}}(γ)(v/C)> · dx}/dt} + {[SM, MMM]: d {Σ (u = 1, u = $n_{axial}$) [$M_{ua}C^2$] (γ)(v/C)}/dt} + {[SM, MMM]: d {Σ (u = 1, u = $n_{cone}$) [$M_{uc}C^2$] (γ)(v/C)}/$dt_{ship}$}

+ {{d{∫{{{[(electrons):($Q_{natural}$)] {[1/(4π$\varepsilon_0$)][$\gamma_{average,electrons,natural,ship\text{-}frame}$ **R**]/{{[$\gamma_{average,electrons,natural,ship\text{-}frame}$ $^2$ $R^2$ $\cos^2$ θ] + [$R^2$ $\sin^2$ θ]}$^{3/2}$}}} +{[(electrons):($Q_{bomb}$)] {[1/(4π$\varepsilon_0$)][$\gamma_{average,bomb,ship\text{-}frame}$ **R**]/{{[$\gamma_{average,bomb,ship\text{-}frame}$$^2$ $R^2$ $\cos^2$ θ] + [$R^2$ $\sin^2$ θ]}$^{3/2}$}}} +{[(electrons):($Q_{non\text{-}nuclear\ pellet}$)] {[1/(4π$\varepsilon_0$)][$\gamma_{average,non\text{-}nuclear\ pellet,ship\text{-}frame}$ **R**]/{{[$\gamma_{average,non\text{-}nuclear\ pellet,ship\text{-}frame}$$^2$ $R^2$ $\cos^2$ θ] + [$R^2$ $\sin^2$ θ]}$^{3/2}$}}} + {[Σ(n = 1; n = number of positive background ion species):(nth ion species)]: ($Q_{natural}$)] {[1/(4π$\varepsilon_0$)][$\gamma_{average,nth\text{-}ion\text{-}species,natural,ship\text{-}frame}$ **R**]/{{[$\gamma_{average,nth\text{-}ion\text{-}species,natural,ship\text{-}frame}$ $^2$ $R^2$ $\cos^2$ θ] + [$R^2$ $\sin^2$ θ]}$^{3/2}$}}} + {[Σ(m = 1; m = number of positive bomb ion species):(mth ion species)]: ($Q_{bomb}$)] {[1/(4π$\varepsilon_0$)][$\gamma_{average,mth\text{-}ion\text{-}species,bomb,ship\text{-}frame}$ **R**]/{{[$\gamma_{average,mth\text{-}ion\text{-}species,bomb,ship\text{-}frame}$ $^2$ $R^2$ $\cos^2$ θ] + [$R^2$ $\sin^2$ θ]}$^{3/2}$}}} + {[Σ(g = 1; g = number of positive bomb heated non-nuclear pellet sourced ion species):(gth ion species)]: ($Q_{non\text{-}nuclear\ pellet}$)] {[1/(4π$\varepsilon_0$)][$\gamma_{average,gth\text{-}ion\text{-}species,non\text{-}nuclear\ pellet,,ship\text{-}frame}$ **R**]/{{[$\gamma_{average,gth\text{-}ion\text{-}species,non\text{-}nuclear\ pellet,ship\text{-}frame}$ $^2$ $R^2$ $\cos^2$ θ] + [$R^2$ $\sin^2$ θ]}$^{3/2}$}}} +{[Σ(e = 1; e = number of negative non-nuclear pellet heated non-nuclear pellet sourced ion species):(eth ion species)]: ($Q_{non\text{-}nuclear\ pellet}$)] {[1/(4π$\varepsilon_0$)][$\gamma_{average,gth\text{-}ion\text{-}species,non\text{-}nuclear\ pellet,,ship\text{-}frame}$ **R**]/{{[$\gamma_{average,gth\text{-}ion\text{-}species,non\text{-}nuclear\ pellet,ship\text{-}frame}$ $^2$ $R^2$ $\cos^2$ θ] + [$R^2$ $\sin^2$ θ]}$^{3/2}$}}} + {[Σ(n = 1; n = number of negative background ion species):(nth ion species)]: ($Q_{natural}$)] {[1/(4π$\varepsilon_0$)][$\gamma_{average,nth\text{-}ion\text{-}species,natural,ship\text{-}frame}$ **R**]/{{[$\gamma_{average,nth\text{-}ion\text{-}species,bomb,ship\text{-}frame}$ $^2$ $R^2$ $\cos^2$ θ] + [$R^2$ $\sin^2$ θ]}$^{3/2}$}}} + {[Σ(m = 1; m = number of negative bomb ion species):(mth ion species)]: ($Q_{bomb}$)] {[1/(4π$\varepsilon_0$)][$\gamma_{average,mth\text{-}ion\text{-}species,bomb,ship\text{-}frame}$ **R**]/{{[$\gamma_{average,mth\text{-}ion\text{-}species,bomb,ship\text{-}frame}$ $^2$ $R^2$ $\cos^2$ θ] + [$R^2$ $\sin^2$ θ]}$^{3/2}$}}} + {d{[(Gound state $N_{flux}$)($v_{average,ship\text{-}frame}$)($\gamma_{average,ship\text{-}frame}$)] for ground state neutrons}/$dt_{ship}$} + {d {Σ(z = 1; z = number of excited neutron states):[(zth excited state $N_{flux,}$)($v_{average,ship\text{-}frame}$)($\gamma_{average,ship\text{-}frame}$)] for excited state neutrons}/dt} + [$<S_{background,photons}>$/c] + [[$<S_{bomb,photons}>$/c] ($Ph_{flux}$)] + [[$<S_{non\text{-}nuclear\ pellet,photons}>$/c] ($Ph_{flux}$)]}

+ {Drag based on the following propulsive force inducing mechanisms {{[(electrons):($Q_{natural}$)] {[1/(4π$ε_0$)][$γ_{average,electrons,natural,ship-frame}$ **R**]/{{[$γ_{average,electrons,natural,ship-frame}{}^2$ $R^2 \cos^2 θ$] + [$R^2 \sin^2 θ$]}$^{3/2}$}}} +{[(electrons):($Q_{bomb}$)] {[1/(4π$ε_0$)][$γ_{average,bomb,ship-frame}$ **R**]/{{[$γ_{average,bomb,ship-frame}{}^2$ $R^2 \cos^2 θ$] + [$R^2 \sin^2 θ$]}$^{3/2}$}}} +{[(electrons):($Q_{non-nuclear\ pellet}$)] {[1/(4π$ε_0$)][$γ_{average,non-nuclear\ pellet,ship-frame}$ **R**]/{{[$γ_{average,non-nuclear\ pellet,ship-frame}{}^2$ $R^2 \cos^2 θ$] + [$R^2 \sin^2 θ$]}$^{3/2}$}}} + {[Σ(n = 1; n = number of positive background ion species):(nth ion species)]: ($Q_{natural}$)] {[1/(4π$ε_0$)][$γ_{average,nth-ion-species,natural,ship-frame}$ **R**]/{{[$γ_{average,nth-ion-species,natural,ship-frame}{}^2$ $R^2 \cos^2 θ$] + [$R^2 \sin^2 θ$]}$^{3/2}$}}} + {[Σ(m = 1; m = number of positive bomb ion species):(mth ion species)]: ($Q_{bomb}$)] {[1/(4π$ε_0$)][$γ_{average,mth-ion-species,bomb,ship-frame}$ **R**]/{{[$γ_{average,mth-ion-species,bomb,ship-frame}{}^2$ $R^2 \cos^2 θ$] + [$R^2 \sin^2 θ$]}$^{3/2}$}}} + {[Σ(g = 1; g = number of positive bomb heated non-nuclear pellet sourced ion species):(gth ion species)]: ($Q_{non-nuclear\ pellet}$)] {[1/(4π$ε_0$)][$γ_{average,gth-ion-species,non-nuclear\ pellet,,ship-frame}$ **R**]/{{[$γ_{average,gth-ion-species,non-nuclear\ pellet,ship-frame}{}^2$ $R^2 \cos^2 θ$] + [$R^2 \sin^2 θ$]}$^{3/2}$}}} +{[Σ(e = 1; e = number of negative non-nuclear pellet heated non-nuclear pellet sourced ion species):(eth ion species)]: ($Q_{non-nuclear\ pellet}$)] {[1/(4π$ε_0$)][$γ_{average,gth-ion-species,non-nuclear\ pellet,,ship-frame}$ **R**]/{{[$γ_{average,gth-ion-species,non-nuclear\ pellet,ship-frame}{}^2$ $R^2 \cos^2 θ$] + [$R^2 \sin^2 θ$]}$^{3/2}$}}} + {[Σ(n = 1; n = number of negative background ion species):(nth ion species)]: ($Q_{natural}$)] {[1/(4π$ε_0$)][$γ_{average,nth-ion-species,natural,ship-frame}$ **R**]/{{[$γ_{average,nth-ion-species,bomb,ship-frame}{}^2$ $R^2 \cos^2 θ$] + [$R^2 \sin^2 θ$]}$^{3/2}$}}} + {[Σ(m = 1; m = number of negative bomb ion species):(mth ion species)]: ($Q_{bomb}$)] {[1/(4π$ε_0$)][$γ_{average,mth-ion-species,bomb,ship-frame}$ **R**]/{{[$γ_{average,mth-ion-species,bomb,ship-frame}{}^2$ $R^2 \cos^2 θ$] + [$R^2 \sin^2 θ$]}$^{3/2}$}}} + {d{[(Gound state $N_{flux}$)($v_{average,ship-frame}$)($γ_{average,ship-frame}$)] for ground state neutrons}/$dt_{ship}$} + {d {Σ(z = 1; z = number of excited neutron states):[(zth excited state $N_{flux,}$)($v_{average,ship-frame}$)($γ_{average,ship-frame}$)] for excited state neutrons}/dt} + [<$S_{background,photons}$>/c] + [[<$S_{bomb,photons}$>/c] ($Ph_{flux}$)] + [[<$S_{non-nuclear\ pellet,photons}$>/c] ($Ph_{flux}$)]}}}● dr}/dt} (Bristle-frame translational gamma)}

= {{∫($y_1$,$y_2$) {∫($x_1$,$x_2$) {∫ (0, π) {{{($T_{cmbr}$) /{{1/{[1 + [(v/C)$^2$ ]] $^{1/2}$}} [1 + [(v/C) cos θ]]}}$^4$} σ e} dθ}dx}dy}(2)} + {[Standard Model and Mirror Matter Model] d{∫ < {Σ (u = 1, u = $n_{axial}$) (2)[$Q_{u,axial}$(x,y,z,t)]{Σ(k = 1, k = m) {Σ (h =1, h = n) {[1/(4π$ε_0$)]{ {1/{1 - [($v_{h,k}$/C)$^2$ ]} $^{1/2}$} (sq $_{h,k}$) **R** $_{h,k}$}/{{{ {1/{1 - [($v_{h,k}$/C)$^2$ ]} $^{1/2}$}$^2$} $R_{h,k}{}^2 \cos^2 θ_{h,k}$} + [$R_{h,k}{}^2 \sin^2 θ_{h,k}$]}$^{3/2}$}}}}}{1/{1 -[(v/C)$^2$ ]} $^{1/2}$} (v/C)> · dx} /dt} + {[Standard Model and Mirror Matter Model] d{∫ < {Σ (u = 1, u = $n_{cone}$) (2)[$Q_{u,cone}$(x,y,z,t)]{Σ(k = 1, k = m) {Σ (h =1, h = n) {[1/(4π$ε_0$)]{ {1/{1 - [($v_{h,k}$/C)$^2$ ]} $^{1/2}$} (sq $_{h,k}$) **R** $_{h,k}$}/{{{{1/{1 - [($v_{h,k}$/C)$^2$ ]} $^{1/2}$}$^2$} $R_{h,k}{}^2 \cos^2 θ_{h,k}$} + [$R_{h,k}{}^2 \sin^2 θ_{h,k}$]}$^{3/2}$}}}}}{1/{1 - [(v/C)$^2$ ]} $^{1/2}$} (v/C)> · dx}/dt} + {[SM, MMM]: d {Σ (u = 1, u = $n_{axial}$) [$M_{ua}C^2$] {1/{1 - [(v/C)$^2$ ]} $^{1/2}$} (v/C)}/dt} + {[SM, MMM]: d {Σ (u = 1, u = $n_{cone}$) [$M_{uc}C^2$] {1/{1 - [(v/C)$^2$ ]} $^{1/2}$} (v/C)}/$dt_{ship}$}

+ {{d{∫{{{{[(electrons):($Q_{natural}$)] {[1/(4π$ε_0$)][$γ_{average,electrons,natural,ship-frame}$ **R**]/{{[$γ_{average,electrons,natural,ship-frame}{}^2$ $R^2 \cos^2 θ$] + [$R^2 \sin^2 θ$]}$^{3/2}$}}} +{[(electrons):($Q_{bomb}$)] {[1/(4π$ε_0$)][$γ_{average,bomb,ship-frame}$ **R**]/{{[$γ_{average,bomb,ship-frame}{}^2$ $R^2 \cos^2 θ$] + [$R^2 \sin^2 θ$]}$^{3/2}$}}} +{[(electrons):($Q_{non-nuclear\ pellet}$)] {[1/(4π$ε_0$)][$γ_{average,non-nuclear\ pellet,ship-frame}$ **R**]/{{[$γ_{average,non-nuclear\ pellet,ship-frame}{}^2$ $R^2 \cos^2 θ$] + [$R^2 \sin^2 θ$]}$^{3/2}$}}} + {[Σ(n = 1; n = number of positive background ion species):(nth ion species)]: ($Q_{natural}$)] {[1/(4π$ε_0$)][$γ_{average,nth-ion-species,natural,ship-frame}$ **R**]/{{[$γ_{average,nth-ion-species,natural,ship-frame}{}^2$ $R^2 \cos^2 θ$] + [$R^2 \sin^2 θ$]}$^{3/2}$}}} + {[Σ(m = 1; m = number of positive bomb ion species):(mth ion species)]: ($Q_{bomb}$)] {[1/(4π$ε_0$)][$γ_{average,mth-ion-species,bomb,ship-frame}$ **R**]/{{[$γ_{average,mth-ion-species,bomb,ship-frame}{}^2$ $R^2 \cos^2 θ$] + [$R^2 \sin^2 θ$]}$^{3/2}$}}} + {[Σ(g = 1; g = number of positive bomb heated non-nuclear pellet sourced ion species):(gth ion species)]: ($Q_{non-nuclear\ pellet}$)] {[1/(4π$ε_0$)][$γ_{average,gth-ion-species,non-nuclear\ pellet,,ship-frame}$ **R**]/{{[$γ_{average,gth-ion-species,non-nuclear\ pellet,ship-frame}{}^2$ $R^2 \cos^2 θ$] + [$R^2 \sin^2 θ$]}$^{3/2}$}}} +{[Σ(e = 1; e = number of negative non-nuclear pellet heated non-nuclear pellet sourced ion species):(eth ion species)]: ($Q_{non-nuclear\ pellet}$)] {[1/(4π$ε_0$)][$γ_{average,gth-ion-species,non-nuclear\ pellet,,ship-frame}$ **R**]/{{[$γ_{average,gth-ion-species,non-nuclear\ pellet,ship-frame}{}^2$ $R^2 \cos^2 θ$] + [$R^2 \sin^2 θ$]}$^{3/2}$}}} + {[Σ(n = 1; n = number of negative background ion species):(nth ion species)]: ($Q_{natural}$)] {[1/(4π$ε_0$)][$γ_{average,nth-ion-species,natural,ship-frame}$ **R**]/{{[$γ_{average,nth-ion-species,bomb,ship-frame}{}^2$ $R^2 \cos^2 θ$] + [$R^2 \sin^2 θ$]}$^{3/2}$}}} + {[Σ(m = 1; m = number of negative bomb ion species):(mth ion species)]: ($Q_{bomb}$)] {[1/(4π$ε_0$)][$γ_{average,mth-ion-species,bomb,ship-frame}$ **R**]/{{[$γ_{average,mth-ion-species,bomb,ship-frame}{}^2$ $R^2 \cos^2 θ$] + [$R^2 \sin^2 θ$]}$^{3/2}$}}} + {d{[(Gound state $N_{flux}$)($v_{average,ship-frame}$)($γ_{average,ship-frame}$)] for ground state neutrons}/$dt_{ship}$} + {d {Σ(z = 1; z = number of excited neutron states):[(zth excited state $N_{flux,}$)($v_{average,ship-frame}$)($γ_{average,ship-frame}$)] for excited state neutrons}/dt} + [<$S_{background,photons}$>/c] + [[<$S_{bomb,photons}$>/c] ($Ph_{flux}$)] + [[<$S_{non-nuclear\ pellet,photons}$>/c] ($Ph_{flux}$)]}

+ {Drag based on the following propulsive force inducing mechanisms {{[(electrons):($Q_{natural}$)] {[1/(4π$ε_0$)][$γ_{average,electrons,natural,ship-frame}$ **R**]/{{[$γ_{average,electrons,natural,ship-frame}{}^2$ $R^2 \cos^2 θ$] + [$R^2 \sin^2 θ$]}$^{3/2}$}}} +{[(electrons):($Q_{bomb}$)] {[1/(4π$ε_0$)][$γ_{average,bomb,ship-frame}$ **R**]/{{[$γ_{average,bomb,ship-frame}{}^2$ $R^2 \cos^2 θ$] + [$R^2 \sin^2 θ$]}$^{3/2}$}}} +{[(electrons):($Q_{non-nuclear\ pellet}$)] {[1/(4π$ε_0$)][$γ_{average,non-nuclear\ pellet,ship-frame}$ **R**]/{{[$γ_{average,non-nuclear\ pellet,ship-frame}{}^2$ $R^2 \cos^2 θ$] + [$R^2 \sin^2 θ$]}$^{3/2}$}}} + {[Σ(n = 1; n = number of positive background ion species):(nth ion species)]: ($Q_{natural}$)] {[1/(4π$ε_0$)][$γ_{average,nth-ion-species,natural,ship-frame}$ **R**]/{{[$γ_{average,nth-ion-species,natural,ship-frame}{}^2$ $R^2 \cos^2 θ$] + [$R^2 \sin^2 θ$]}$^{3/2}$}}} + {[Σ(m = 1; m = number of positive bomb ion species):(mth ion species)]: ($Q_{bomb}$)] {[1/(4π$ε_0$)][$γ_{average,mth-ion-species,bomb,ship-frame}$ **R**]/{{[$γ_{average,mth-ion-species,bomb,ship-frame}{}^2$ $R^2 \cos^2 θ$] + [$R^2 \sin^2 θ$]}$^{3/2}$}}} + {[Σ(g = 1; g = number of positive bomb heated non-nuclear pellet sourced ion species):(gth ion species)]: ($Q_{non-nuclear\ pellet}$)] {[1/(4π$ε_0$)][$γ_{average,gth-ion-species,non-nuclear\ pellet,,ship-frame}$ **R**]/{{[$γ_{average,gth-ion-species,non-nuclear\ pellet,ship-frame}{}^2$ $R^2 \cos^2 θ$] + [$R^2 \sin^2 θ$]}$^{3/2}$}}} +{[Σ(e = 1; e = number of negative non-nuclear pellet heated non-nuclear pellet sourced ion species):(eth ion species)]: ($Q_{non-nuclear\ pellet}$)] {[1/(4π$ε_0$)][$γ_{average,gth-ion-species,non-nuclear\ pellet,,ship-frame}$ **R**]/{{[$γ_{average,gth-ion-species,non-nuclear\ pellet,ship-frame}{}^2$ $R^2 \cos^2 θ$] + [$R^2 \sin^2 θ$]}$^{3/2}$}}} + {[Σ(n = 1; n = number of negative background ion species):(nth ion species)]: ($Q_{natural}$)] {[1/(4π$ε_0$)][$γ_{average,nth-ion-species,natural,ship-frame}$ **R**]/{{[$γ_{average,nth-ion-species,bomb,ship-frame}{}^2$ $R^2 \cos^2 θ$] + [$R^2 \sin^2 θ$]}$^{3/2}$}}} + {[Σ(m = 1; m = number of negative bomb ion species):(mth ion species)]: ($Q_{bomb}$)] {[1/(4π$ε_0$)][$γ_{average,mth-ion-species,bomb,ship-frame}$ **R**]/{{[$γ_{average,mth-ion-species,bomb,ship-frame}{}^2$ $R^2 \cos^2 θ$] + [$R^2 \sin^2 θ$]}$^{3/2}$}}} + {d{[(Gound state $N_{flux}$)($v_{average,ship-$

$_{frame})(\gamma_{average,ship\text{-}frame})]$ for ground state neutrons}/$dt_{ship}$} + {d {Σ(z = 1; z = number of excited neutron states):[(zth excited state $N_{flux},)(v_{average,ship\text{-}frame})(\gamma_{average,ship\text{-}frame})]$ for excited state neutrons}/dt} + $[<S_{background,photons}>/c]$ + $[[<S_{bomb,photons}>/c]\ (Ph_{flux})]$ + $[[<S_{non\text{-}nuclear\ pellet,photons}>/c]\ (Ph_{flux})]]\}\}\bullet$ dr}/dt} (Bristle-frame translational gamma)}

Here, dt is bristle-time.

It is plausible that the bristle function may permit hidden super-relativistic-quantum-electrodynamics-based propulsion for which various clever tricks can be employed to exceed Planck Power propulsion limits. To the extent propulsion power can exceed the Planck Power level without the production of a black hole state, the spacecraft acceleration in its own reference frame becomes unlimited.

One way to avoid black hole formation from continuous super-Planck-Power drive systems would be to engage small sized power units at levels greater than the Planck Power. However, the power for activated propulsion units would need to be turned off before the time integrated propulsion power would cause a black hole to form. However, this is easier to do than you might think even for constant super-Planck-Power propulsion. How so? Simply by designing into a propulsion slab or block array or matrix of super-Planck-Power units for which suitably separated units would switch on and off.

For example, if two units can produce power levels of 10 Planck Power Units where the units are square-like of one square meter but separated by more than about 10 meters for on-center distances, and the units are sswitched on for less than the time it takes light to travel one tenth of the smallest dimensional extension of a unit, then a black hole state will not form and the propulsion power level of 10 Planck Power Units can be perpetually maintained.

As another example, if two units can produce power levels of 1,000 Planck Power Units where the units are square-like of one square meter but separated by more than about 1,000 meters for on-center distances, and the units are sswitched on for less than the time it takes light to travel 0.001 of the smallest dimensional extension of a unit, then a black hole state will not form and the propulsion power level of 1,000 Planck Power Units can be perpetually maintained.

As another example, if two units can produce power levels of 1,000,000 Planck Power Units where the units are square-like of one square meter but separated by more than about 1,000,000 meters for on-center distances, and the units are sswitched on for less than the time it takes light to travel 0.000001 of the smallest dimensional extension of a unit, then a black hole state will not form and the propulsion power level of 1,000,000 Planck Power Units can be perpetually maintained.

As another example, if two units can produce power levels of 1,000,000,000 Planck Power Units where the units are square-like of one square meter but separated by more than about 1,000,000,000 meters for on-center distances, and the units are sswitched on for less than the time it takes light to travel 0.000000001 of the smallest dimensional extension of a unit, then a black hole state will not form and the propulsion power level of 1,000,000,000 Planck Power Units can be perpetually maintained.

As another example, if two units can produce power levels of 10 EXP 12 Planck Power Units where the units are square-like of one square meter but separated by more than about 10 EXP 12 meters for on-center distances, and the units are sswitched on for less than the time it takes light to travel 10 EXP - 12 of the smallest dimensional extension of a unit, then a black hole state will not form and the propulsion power level of 10 EXP 12 Planck Power Units can be perpetually maintained.

As another example, if two units can produce power levels of 10 EXP 15 Planck Power Units where the units are square-like of one square meter but separated by more than about 10 EXP 15 meters for on-center distances, and the units are sswitched on for less than the time it takes light to travel 10 EXP - 15 of the smallest dimensional extension of a unit, then a black hole state will not form and the propulsion power level of 10 EXP 15 Planck Power Units can be perpetually maintained.

As another example, if two units can produce power levels of 10 EXP 18 Planck Power Units where the units are square-like of one square meter but separated by more than about 10 EXP 18 meters for on-center distances, and the units are sswitched on for less than the time it takes light to travel 10 EXP - 18 of the smallest dimensional extension of a unit, then a black hole state will not form and the propulsion power level of 10 EXP 18 Planck Power Units can be perpetually maintained.

As another example, if two units can produce power levels of 10 EXP 21 Planck Power Units where the units are square-like of one square meter but separated by more than about 10 EXP 21 meters for on-center distances, and the units are sswitched on for less than the time it takes light to travel 10 EXP - 21 of the smallest dimensional extension of a unit, then a black hole state will not form and the propulsion power level of 10 EXP 21 Planck Power Units can be perpetually maintained.

As another example, if two units can produce power levels of 10 EXP 24 Planck Power Units where the units are square-like of one square meter but separated by more than about 10 EXP 24 meters for on-center distances, and the units are sswitched on for less than the time it takes light to travel 10 EXP - 24 of the smallest dimensional extension of a unit, then a black hole state will not form and the propulsion power level of 10 EXP 24 Planck Power Units can be perpetually maintained.

As another example, if two units can produce power levels of 10 EXP 27 Planck Power Units where the units are square-like of one square meter but separated by more than about 10 EXP 27 meters for on-center distances, and the units are sswitched on for less than the time it takes light to travel 10 EXP - 27 of the smallest dimensional extension of a unit, then a black hole state will not form and the propulsion power level of 10 EXP 27 Planck Power Units can be perpetually maintained.

As another example, if two units can produce power levels of 10 EXP 30 Planck Power Units where the units are square-like of one square meter but separated by more than about 10 EXP 30 meters for on-center distances, and the units are sswitched on for less than the time it takes light to travel 10 EXP - 30 of the smallest dimensional extension of a unit, then a black hole state will not form and the propulsion power level of 10 EXP 30 Planck Power Units can be perpetually maintained.

As another example, if two units can produce power levels of 10 EXP 33 Planck Power Units where the units are square-like of one square meter but separated by more than about 10 EXP 33 meters for on-center distances, and the units are sswitched on for less than the time it takes light to travel 10 EXP - 33 of the smallest dimensional extension of a unit, then a black hole state will not form and the propulsion power level of 10 EXP 33 Planck Power Units can be perpetually maintained.

We are limited to analogues for the above one meter square driving plates for which light would take time to travel $1.616255(18)\times10^{-35}$ meter. The later spatial interval is the Planck Length = $\{[h/(2\pi)]\ G/(c^3)\}^{1/2}$ and is theoretically the shortest possible distance. The time to travel this distance is the Planck Time = $5.391247(60)\times10^{-44}$ second = $\{[h/(2\pi)]\ G/(c^5)\}^{1/2}$. This corresponds to a on center separation distance of $[1/\ [1.616255(18)\times10^{-35}]]$ meters = $6.1871418 \times 10^{34}$ meters and $6.1871418 \times 10^{34}$ Planck Power Units. The Planck Power Unit = $3.628 \times 10^{52}$ watts = $c^5/G$.

Now, a fascinating electrically charged rotor system would consist of a rotor or pin-wheel of plasma rotating at near the speed of light. The plasma may be electrons, protons, ions, and the like and may be magnetically affixed to the spacecraft. Alternatively, the plasma may be electrically affixed to the spacecraft. The near light-speed rotors of plasma would greatly increased the relativistic Lorentz turing force compared to a non-rotating solid sail of the same mass, eventhough the mass of the plasma rotor could be ultra-relativistic in the spacecraft reference frame. The radius of the plasma rotor may be as great as 10,000 km for greatly reduced synchron radiation and associated losses.

Another mechanism would utilize explosions off-set to the star-bound side of the craft. Accordingly, the craft would be turned by explosions set off properly behind the craft and the thrust would be captured by one or more specially dedicated sails. Note that relativistic aberration would require that the explosions be adjusted in location with respect to the craft.

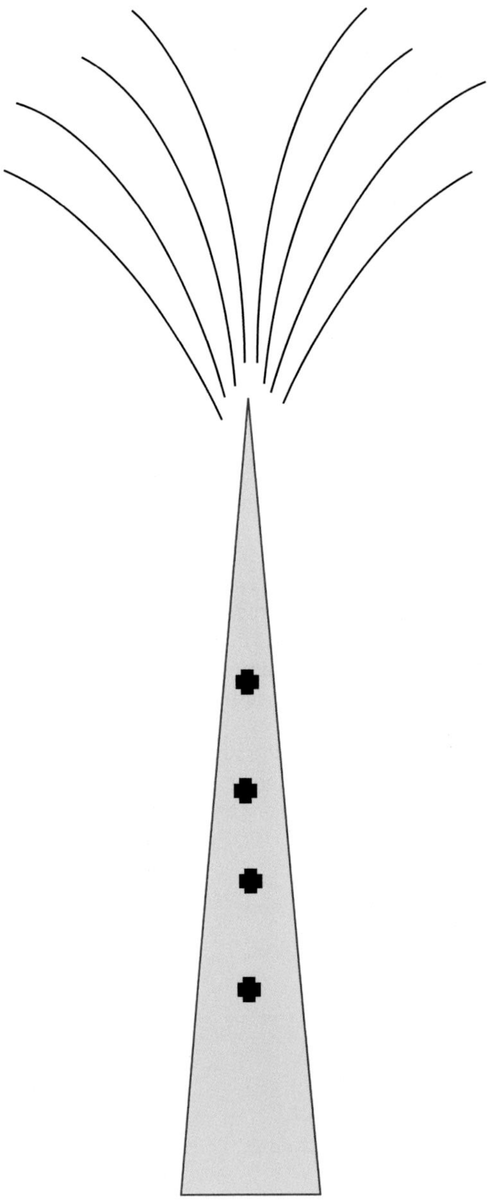

**FIGURE 28** illustrates the amplification of the electric flux density near the tip of an electrostatically charged bristle as a subcomponent of the electrostatic hairbrush propulsion mechanism. The smoothly curved black lines in the upper portion of the diagram depict the concentrated electric flux. The heavy black plus signs within the yellow triangle represent the high degree of positive charge instilled within the bristle. The bristle may optionally be negatively charged. Only one bristle is shown.

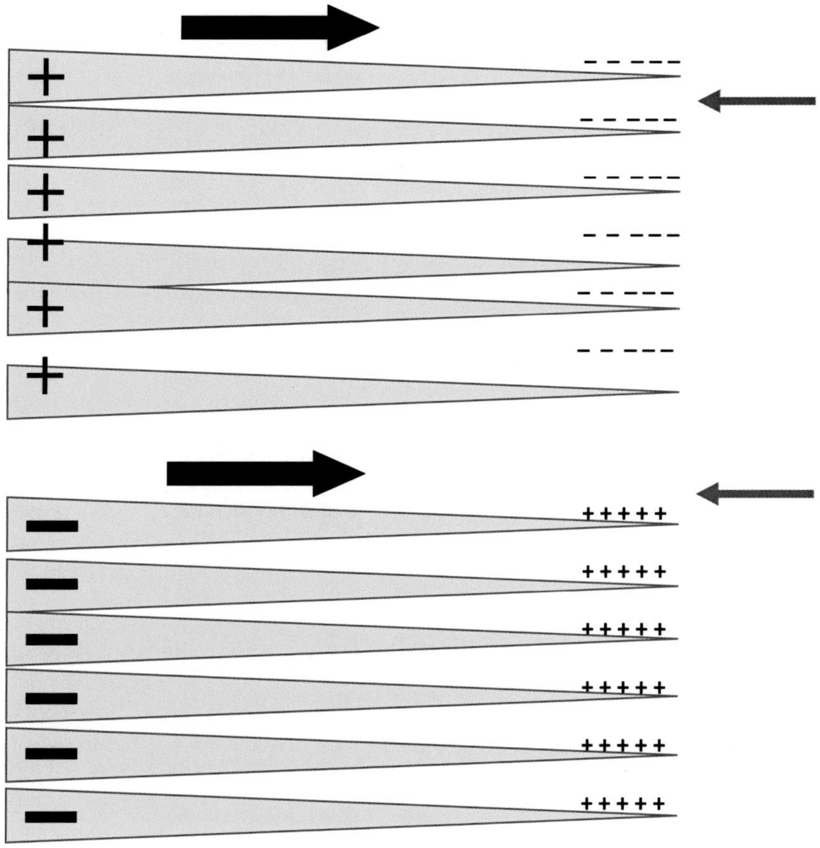

**FIGURE 29** depicts a linear disposed multi-bristle electrostatic hairbrush propulsion mechanism. The large plus and minus signs within the elongated yellow triangles (bristles) depict the sign of the bristle charge. The small plus and minus signs on the right side of the page indicate the attracted incident background chargons.

Note that such fringe based electric flux compressions are usually considered a second order effect and thus may validly but loosely be described as micro-classical-electrodynamic effects. The usefulness of the conceptualized propulsion systems described in this chapter requires that the proposed methods do indeed have benefits over at least some systems that lack the fringe effect. However, if precise control of chargon intake sufficient to direct the charged species into the enhanced flux concentrations can be maintained and the bristle charge and form can be suitably maintained amidst any intake therein of incident charge, it is safe to say that such systems have merit in some cases. The magnitude and specific functional form of the fringe effects of the bristles are somewhat arbitrary and can vary as a function of bristle point shape, sharpness, and other bristle configutational aspects.

Evidently, one or more primary bristles may incorporate secondary branches in the form of smaller bristles, which can in turn incorporate trinary branches in the form of smaller bristles, and so on. This way, chargons not incident on the very tip, or nearly so, of one or more primary bristles can be permitted elongated pathways for electric field based acceleration and thus feel the full bulk effect of the primary bristles' electrical field assisted acceleration. Ideally, an infinitely repeating fractal pattern would be most useful here, however, the quantized nature of electric charge, and its embodying chargons, and also, most likely space and time would pose a upper finite limit to the number of branching levels.

Chaotic phenomenon might be harnessed to effectively filter and draw the incident chargons into the enhanced flux regions at and near the tips of the bristles. Perhaps so-called strange attractors may be of use to bias the inflow patterns of incident chargons for cases where the incident chargon flux pattern is not completely random. Thus, for even mildly ordered influx patterns, perhaps the chargons can be funneled into the bristle tips in a controlled manner thereby promoting feasibility of the hair brush method. Forward insitu sensing of the incoming chargon distribution by species, velocity, and density may be of use in adjusting the electrodynamic scoop mechanisms so as to enable a practical strange attractor type of mechanism to operate even in cases where the chargon pre-incident flux patterns as such are nearly completely ramdomized.

Perhaps the best use of the proposed fuel pellet runways is for boosting spacecraft to sufficient velocities such that the craft can then meaningfully collect naturally occurring background fuel and energy for increased spacecraft Lorentz factors.

An interesting scenario would involve strange attractors and chaotic processes to enable self-assembled explosive fuel pellet streams.

Accordingly, molecular and/or macroscopic machinery would assemble hydrogen and helium fuel pellets. The molecular and macroscopic machinery can be self-replicating.

The linear extent or curvilinear extent of the pellet runways may grow at near the speed of light commensurate with relativistic propulsion machinery depositing seeds of assembly machinery.

In theory, the pellet runways may span more than one cosmic light-cone radius and into adjacent light-cone scale portions of our universe. This would require compatible profiles of universal expansion.

Herein we also consider electrodynamic-hydrodynamic-plasma-drive mechanisms of various configurations.

Accordingly, we consider systems that affix a forward plasms bow-shock located in from of the bow of a relativistic spacecraft.

Additionally, we consider a forwardly exhausted rock plume that is directed into the bow-shock.

The rocket plume is conditioned such that it will be back reflected by the bow shock and further accelerated by LINAC style mass-drivers then further accelerated by magnetic fields produced by a linear magnetic induction coil and ejected sternward as a plasma.

The magnetic induction coil as super-conducting may have a pulsed or otherwise moderated magnetic field to produce thermal heating of the plasma.

Other configurations of the craft may include sternward plasma compression and pressurization which then causes a powerful additional thrust component. The sternward plasma may optionally be directed onto an astrospike to apply and unbalanced squeezing force to provide thrust.

The plasma forwardly incident upon the bow-shock may optionally be from high temperature chemical combustion, nuclear thermal rockets, chemical combustion augmented by nuclear thermal heating, nuclear fission reactor electric propulsion, PV, TE, or turbo-electric powered mechanisms that power forwardly incident ion or electron streams, matter-antimatter reactor powered turbo-electric generators that power forwardly incident ion or electron streams, nuclear fusion reactor powered turbo-electric generators that power forwardly incident ion or electron streams, and other mechanisms.

Electrical propulsion systems may include:

1) Electro-hydrodynamic-plasma-drives
2) Magneto-hydrodynamic-plasma-drives
3) Electro-magneto-hydrodynamic-plasma-drives
4) Electromagnetic-hydrodynamic-plasma-drives
5) Direct magnetic field effect propulsion
6) Ion rockets
7) Electron rockets
8) Magnetic background plasma sails
9) Magnetic-background plasma bottle sail
10) Combinations of the above nine systems and others.

The fuel required to power the above systems may be at least in part as a non-limiting option collected from a pellet run-way deployed in from of the spacecraft or behind in such a way that the fuel pellets catch up with the spacecraft and are collected by the spacecraft.

Fuel pellets may include chemical fuels, nuclear fission fuels, nuclear fusion fuels, matter-antimatter fuels, or other fuels as may become known in the future.

In principle, the bow shock is limited mathematically to energy densities that would cause the formation of a black hole. However, if maximally charged and maximally rotating black holes have proverbial "hair", the additional electrodynamic energy can be installed within a black-hole based bow shock.

Note that currently accepted black hole physics holds that black holes are completely and only defined by their mass, spin, and electrical charge. Thus, is it commonly stated that black holes have no hair. A recent computationally intensive black hole simulation suggest that black holes may have occasionally occurring hair wisps. Thus, these black holes, which must be maximally rotating and charged may provide additional energy density than would otherwise be possible. However, nature may or may not comply with these numerical simulations.

A fascinating aspect of energy densities greater than those within a black hole event horizon may imply the possibility for super-luminal travel. The arguments here would be highly technical but the general idea here is to attain energy densities greater than that of black holes relative to event horizon radial coordinates.

Note that under normal circumstances, the craft would need fabrication of neutron element of neutronium, quarkonium, solid higgsinium, a combination of higgsinos and magnetic monopoles or other exotic materials for employing a black hole density electrodynamic bow shock, that is, unless the bow shock was far enough in ahead of the craft, or suitably small.

It may be possible that wormholes can be fabricated of real positive mass and positive energy materials instead of requiring negative mass and negative energy inner linings to hold the wormholes open.

The super-relativistic energy densities of black holes with hair may imply wormhole like behavior in some aspects of black hole thermodynamics and couplings of the black hole style bow shock to the spacecraft.

**NEUTRONIUM**

Now the force of attraction between nucleons is about 10,000 newtons. Therefore, the tensile strength of neutronium would be approximately (10,000 newtons)[$(10^{15})^2$] = $10^{34}$ Newtons per square meter.

Nature already has precedence for neutronium in the context of neutron stars. However, the neutronium in such stars is continuously being regenerated under the enormous gravitationally self-induced pressures within neutron stars.

Large assemblages of neutrons and protons have been proven to exist in the form of the atomic nuclei of heavy periodic table elements.

It is not inconceivable that a primarily neutronic material that is doped with exotic mesons, exotic baryons, electrons, muons, tauons, positrons, anti-muons, anti-tauons, still undiscovered higgsinos, gluinos, photinos, winos, and the like non-Standard-Model fermions could be stabilized against decay, even under low pressured conditions.

Exotic mesons and exotic baryons may be comprised of strange, charmed, bottom, top, anti-strange, anti-charmed, anti-bottom, and anti-top quarks (provided top quarks can be stably bound) but only in cases where particles and antiparticles do not come in contact with each other. Alternatively, the exotic mesons and baryons may be comprised of up and/or down quarks and one or more of the following: strange, charmed, bottom, top, anti-strange, anti-charmed, anti-bottom, and anti-top quarks but only in cases where particles and antiparticles do not come in contact with each other.

Composite stabilizer particles may conceivably be composed of any combination of charged leptons, exotic mesons, exotic baryons, Minimally Supersymmetric Standard Model fermions, and/or Extended Supersymmetric Models fermions in crystalline, quasi-crystalline, and/or amorphous material patterns.

Another stabilizing mechanism may include judicious neutronic crystalline patterns for the bulk neutronium with or without the additional doping mechanism.

**QUARKONIUM**

The yield strength of protons and neutrons is about 100 times greater than the force of attraction amongst these nucleons within the composition of the atomic nucleus. The latter force of attraction is about 10,000 newtons. Therefore, the yield strength of protons and neutrons is about 1,000,000 newtons. A differential cross-sectional area of a column of up-down quarkonium where the column is 1,000 $^{1/2}$ neutrons by 1,000 $^{1/2}$ neutrons wide is about (3)[1,000 $^{1/2}$] up and/or down quarks wide by (3)[1,000 $^{1/2}$] up and/or down quarks thick. The tensile strength of typical up-down quarkonium will thereby be about {{[3]} $^{2}$}(100) times greater than that of neutronium or on the order of 1,000 times greater than that of neutronium. There are likely a large number of possible up-down quarkonium types considering the range of plausible crystalline patterns, quarkonium excited or isomer states and the like. Thus, we may have some freedom in the designed strength of specific quarkoniums by up to 3 or perhaps even 4 orders of magnitude. Note the analogous cases of comparison for the element carbon which includes soft pencil graphite or activated charcoal or the much stronger forms of carbonaceous supermaterials. Here, the yield strength of the materials spans a range of several orders of magnitude.

Quarkoniums in the form of quark nuggets left over as relics from the early stages of the Big Bang are plausible and have been the subject of serious theoretical studies. The quark nuggets would theoretically be on the order of 1 to 2 meters in size and have a mass of roughly that of Earth's Moon.

Some theories of stellar evolution involving intermediate stages between neutron stars and black holes have posited the existence of quark stars. Either way, many theoretical considerations regarding the interior of neutron stars suggest the existence of a quarkonium core within at least some neutron stars.

We can intelligibly speculate that bulk bottom-quarks-based matter or bottomonium has a similar maximum density as strange matter or charmed matter and that it also has similar tensile strength. This may likely be an underestimate of bottomonium density and tensile strength since the mass of the bottom quark is significantly greater than that of either the charmed quark or the strange quark. However, we will proceed with the assumption on the basis that it provides a lower plausible bonding limit for maximum sail gamma factors.

Single particle bottomonium has been observed in a laboratory setting as a multi-quark composition of bottom quarks and lighter quarks. The observed particles are generally comprised of two or three quarks and are highly unstable. Essentially, these particles generally decay so rapidly that its existence is inferred indirectly by the detection of predicted decay products and related sequential transformations.

**Charmed-Strangeium.**

Now the invariant masses of the Charmed and Strange quarks are 1,270 MeV/(C $^{2}$) and 101 MeV/(C $^{2}$) respectively while the invariant masses of the up and down quarks are 1.7-3.3 MeV/(C $^{2}$) and 4.1-5.8 MeV/(C $^{2}$), respectively. So, it is plausible to assume as a first order estimate that the compressive strength of Charmed-Strangeium may be as high as {[[[(2)(1,270)] + 101]/3]$^{3}$}/{[[[(2)(1.7 to 3.3)] + (4.1 to 5.8)]/3] $^{3}$} times that of the proton or 682,246,693/70.6157 to 682,246,693/15.625 or 9,661,402 to 43,663,788 times as strong as the proton. Here, we assume an electrically neutral quarkonium.

Such a quarkonium will have a compressive strength of [9.661402 x (10 $^{42}$)] newtons per square meter to [4.3663788 x (10 $^{43}$)[ to newtons per square meter . This is about equal to 100,000 to 440,000 times the strength needed to hold open a wormhole for cases where the quarkonium wall would be thick enough so that the quarkonium compressive strength is the driving support mechanism.

Now, we assume that the quark on quark bounding strength of Charmed-Strangeium is the same as that within the proton. Thus, the plausible compressive strength of Charmed-Strangeium is {{[[(2)(1,720)] + 101]/3} $^{2}$}}/{[[[(2)(1.7 to 3.3)] + (4.1 to 5.8)]/3] $^{2}$} times greater than that of the proton or 1,393,186 /[6.25 to 17.0844] times stronger or in the low end of the range 81,547 times to in the high-end of range, 222,909 times stronger than the proton. This compressive strength is equal to [8.1447 x (10 $^{40}$)] newtons per square meter to [2.22090 x (10 $^{41}$)] newtons per square meter. The is in the range of about 800 to 2,200 times the strength of material needed to hold open our conjectural 2 meter wide wormhole.

Note, that by compressive strength, we assume a force per unit of area or pressure above and in the cases presented below.

**Charmed-Bottomonium.**

Now, the invariant masses of the Charmed and Bottom quarks are 1,270 MeV/($C^2$) and 4.19 GeV(MS) or 4.67 GeV(1S), respectively while, once again, the invariant masses of the up and down quarks are 1.7-3.3 MeV/($C^2$) and 4.1-5.8 MeV/($C^2$), respectively. So, it is plausible to assume as a first order estimate that the compressive strength of Charmed-Bottomonium may be as high as {[[[(2)(1,270)] + [4,190 (MS) or 4,670 (1S)]]/3] $^3$}/{[[[(2)(1.7 to 3.3)] + (4.1 to 5.8)]/3] $^3$} times that of the proton or [1.12897 x ($10^{10}$)]/70.6157 to [1.38817 x ($10^{10}$)]/15.625 or 159,875,211 to 888,427,522 times as strong as the proton. Here, we assume an electrically neutral quarkonium.

Such quarkonium will have a compressive strength ranging from [1.59875 x ($10^{44}$)] to newtons per square meter to [8.88427522 x ($10^{44}$)] to newtons per square meter. This is about equal to 1,600,000 to, 8,900,000 times the strength needed to hold open a wormhole for cases where the quarkonium wall would be thick enough so that the quarkonium compressive strength is the driving support mechanism.

Now, we assume that the quark on quark bounding strength of Charmed-Bottomonium is the same as that within the proton. Thus, the plausible compressive strength of Charmed-Botomomium is {{[[[(2)(1,720)] + [4,190 (MS) or 4,670 (1S)]]/3} $^2$}}/{[[[(2)(1.7 to 3.3)] + (4.1 to 5.8)]/3] $^2$} times greater than that of the proton or 378,621 times stronger in the low end of the range to in the high-end of the range,1,169,281 times stronger than the proton. This compressive strength is equal to [3.786 x ($10^{41}$)] newtons per square meter to [1.169 x ($10^{42}$)] newtons per square meter. The is in the range of about 1,000 to 4,700 times the strength of material needed to hold open our conjectural 2 meter wide wormhole.

**Top-Bottomonium.**

Now, the invariant masses of the Bottom quark and Top quark are 1,270 MeV/($C^2$)] and 172 GeV/($C^2$), respectively while, still once again, the invariant masses of the up and down quarks are 1.7-3.3 MeV/($C^2$) and 4.1-5.8 MeV/($C^2$), respectively. So, it is plausible to assume as a first order estimate that the compressive strength of bulk Top-Bottomonium may be as high as {[[[(2)(172,000)] + [4,190 (MS) or 4,670 (1S)]]/3] $^3$}/{[[[(2)(1.7 to 3.3)] + (4.1 to 5.8)]/3] $^3$} times that of the proton or [1.12897 x ($10^{15}$)]/70.6157 to [1.38817 x ($10^{15}$)]/15.625 or [1.5987 x ($10^{13}$)] to [8.8843 x ($10^{13}$)] times as strong as the proton. Here, we assume an electrically neutral quarkonium.

Such quarkonium will have a compressive strength ranging from [1.5987 x ($10^{49}$)] to newtons per square meter to [8.8843 x ($10^{49}$)] to newtons per square meter . This is about equal to [1.5987 x ($10^{11}$)] to [8.8843 x ($10^{11}$)] times the strength needed to hold open a wormhole for cases where the quarkonium wall would be thick enough so that the quarkonium compressive strength is the driving support mechanism.

Now, we assume that the quark on quark bounding strength of Top-Bottomonium is the same as that within the proton. Thus, the plausible compressive strength of Top-Botomomium is {{[[[(2)(172,000)] + [4,190 (MS) or 4,670 (1S)]]/3} $^2$}}/{[[[(2)(1.7 to 3.3)] + (4.1 to 5.8)]/3] $^3$} times greater than that of the proton or 788,476,286 times stronger in the low end of the range to in the high-end of the range, 2,161,258,144 times stronger than the proton. This compressive strength is equal to 7.88 x $10^{44}$ newtons per square meter to 2.1617 x $10^{45}$ newtons per square meter. The is in the range of about 1,900,000 to 8,645,000 ttimes the strength of material needed to hold open our conjectural 2 meter wide wormhole.

**Pure Toponium.**

Now, the invariant masses of the Top quark and its antimatter counterpart is 172 GeV/($C^2$) while, still once again, the invariant masses of the up and down quarks are 1.7-3.3 MeV/($C^2$) and 4.1-5.8 MeV/($C^2$), respectively. So, it is plausible to assume as a first order estimate that the compressive strength of bulk Toponium may be as high as {[[[(2)(172,000)]]]/2] $^3$}/{[[[(2)(1.7 to 3.3)] + (4.1 to 5.8)]/3] $^3$} times that of the proton or [5.0884 x ($10^{15}$)]/70.6157 to [5.0884. x ($10^{15}$)]/15.625 or [7.2058 x ($10^{13}$)] to [3.2566 x ($10^{14}$)] times as strong as the proton. Here, we assume an electrically neutral quarkonium.

Such quarkonium will have a compressive strength ranging from [1.5987 x ($10^{49}$)] to newtons per square meter to [8.8843 x ($10^{49}$)] to newtons per square meter . This is about equal to [7.2058 x ($10^{11}$)] to [3.2566 x ($10^{12}$)] times the strength needed to hold open a wormhole for cases where the quarkonium wall would be thick enough so that the quarkonium compressive strength is the driving support mechanism.

Now, we assume that the quark on quark bounding strength of Top-Bottomonium is the same as that within the proton. Thus, the plausible compressive strength of Top-Botomomium is {{[[[(2)(172,000)] ]]/2} $^2$}}/{[[[(2)(1.7 to 3.3)] +

(4.1 to 5.8)]/3] $^{2}$} times greater than that of the proton or 1,731,633,715 times stronger in the low end of the range to in the high-end of the range, 4,733,440,000 times stronger than the proton. This compressive strength is equal to 1.731 x $10^{45}$ newtons per square meter to 4.733 x $10^{45}$ newtons per square meter. The is in the range of about 17,300,000 to 47,000,000 times the strength of material needed to hold open our conjectural 2 meter wide wormhole. Thus, pure toponium, assuming that stable forms of it could be manufactured, is plausibly the strongest Standard Model material.

**MONO-HIGGSINIUM**

The mass of the monopole has been theoretically estimated to be between 600 GeV/[C $^{2}$) and as high as $10^{17}$ GeV/(C $^{2}$). The Planck Mass is equal to:

$$Mp = \{[h/2\pi]C/G\}^{1/2} \approx 1.2209 \times 10^{19} \text{ GeV/(C}^{2}).$$

Thus, the width of a 600 GeV/C $^{2}$) monopole would be about [0.000511 GeV/(C $^{2}$)]/[600 GeV/C $^{2}$)] angstroms = 0.000000851 angstroms.

The mass of a light higgsino would be about equal to 100 GeV/(C $^{2}$)in a lower mass range and thus would have a width about equal to [0.000511 GeV/(C $^{2}$)]/[100 GeV/(C $^{2}$)] = .00000511 angstroms.

Thus, a unit cell of solid crystalline materials formed of equal parts by particle number of low end mass range monopolium, low end mass range higgsinos, and low end mass range anti-higgsinos would have a width of about [0.000000851 + 0.00000511] anstroms = 0.000005961 angstroms.

A plausible general planar crystalline form for the mono-higgsinium is as follows:

H M H M- H M H M-

M- H- M H- M- H- M H-

H M H M- H M H M-

M- H- M H- M- H- M H-

H M H M- H M H M-

M- H- M H- M- H- M H-

The bond strength between the higgsinos and antihiggsinos should be about equal to:

$$F = \{1/[(4)(pi)(epsilon\ naught)]\}[(q_1)(q_2)]/[r^{2}]$$

$$= \{[8.987551787 \times 10^{9}]\ N\ (m^{2})\ (Coulomb^{-2}\}\{[-1.602176565(35)\times(10^{-19})]\ Coulombs\}\ \{[1.602176565(35)\times(10^{-19})]\ Coulombs\}/\{[5.961 \times (10^{-16})m]^{2}\} = 649.278\ Newtons.$$

The bond strength between the monopoles and anti-monopoles should be roughly the same. Therefore, the tensile strength of the material made with the lightest proposed higgsinos and anti-higgsinos and the lightest monopoles and anti-monopoles should be {649.278/[0.5961 $^{2}$]} newtons per square femtometer or 1,827 newtons per square femtometer. This binding force and tensile strength is about equal to that of pure neutronium but perhaps much less than that of quarkonium. However, the laboratory frame energy required to transmute either a low-mass-range monopole and a low mass range higgsino is likely to be much larger than the invariant masses of either of these two particles and associated anti-particles. Thus, a conical shield and/or sail having an aspect ratio far larger than that of the proposed neutronium and quarkonium shields may enable gamma factors at least as high as that for the above quarkoniums if not much higher. Then again, perhaps the gamma factors for the low-mass-range monohiggsiniums would be smaller than that for quarkoniums.

The density of the low-mass-range mono-higgsinium would be about:

$$Rho = \{[(10^{11}) + [(6)(10^{11})]]\ [1.602\times(10^{-19})]\ J\}/\{[5.961 \times (10^{-16})m]^{3}\}$$

$$= [5.882 \times (10^{18})]\ \text{metric tons per cubic meter.}$$

Now consider a higgsino having a mass near in the high end for the range of predicted mass for the particle and a low end of range mass for the magnetic monopole.

The mass of the heavy higgsino would be about equal to 1,000 GeV/(C $^2$) and thus would have a width about equal to [0.000511 GeV/(C $^2$)]/[1,000 GeV/(C $^2$)] =0.000000511 angstroms.

Thus, a unit cell of solid crystalline materials formed of equal parts by particle number of low-end mass monopoles, high-end mass higgsinos, and high-end mass antihiggsinos, would have a plausible width of about [0.000000851 + 0.000000511] angstroms = 0.000001362 angstroms.

The bond strength between the higgsinos and antihiggsinos within monohiggsinium should be about equal to:

F = {1/[(4)(pi)($\varepsilon_0$)]}[($q_1$)($q_2$)]/[r $^2$]

= {[8.987551787 x 10 $^9$] N (m $^2$) (Coulomb $^{-2}$)}{[-1.602176565(35)×(10 $^{-19}$)] Coulombs} {[1.602176565(35) × (10 $^{-19}$)] Coulombs}/{[1.362 x (10 $^{-16}$)m] $^2$} = 12,437 Newtons.

The bond strength between the monopoles and anti-monopoles should be roughly the same. Therefore, the tensile strength of the material made with the high mass; higgsinos and anti-hggsinos, and the lightest monopoles and antimonopoles, should be {12,436/[0.1362 $^2$]} Newtons per square femtometer or 670,424 Newtons per square femtometer. This binding force and tensile strength is about equal to the yield strength of the proton and neutron and is thus roughly equal to 100 times the yield strength of neutronium. However, the laboratory frame energy required to transmute either a low mass range monopole and a high mass range higgsino is likely to be much larger than the invariant masses of either of these two particles and associated anti-particles.

The density of this mono-higgsinium would be about:

Rho = {[(10 $^{12}$) + [(6)(10 $^{11}$)]] [1.602×(10 $^{-19}$)] J}/{[1.362 x (10 $^{-16}$)m] $^3$}

= [1.1272 x )10 $^{21}$)] metric tons per cubic meter.

The specific crystalline pattern depicted above would be ion-like in form in a manner analogous to bi-atomic salts such as ordinary table salt or Sodium Chloride.

Nature seems to have provided humanity and humanoid ETI the ability to produce permanent wormhole infrastructures that are maintainable by neutronium, quarkonium, and mono-higgsinium linings. The bulk modulus of these materials seems roughly near the limit required to hold wormholes open with real positive mass.

At the very least, wormholes having an inner diameter equal to the width of a playground sliding board may be possible. Thus, folks could travel by "sliding down" such wormholes.

The above conceptual positive mass linings occurred to me late yesterday evening and I have waiting to find the time to bring the ideas to the attention of my readership.

**X-Y MATERIAL**

The X and Y bosons are predicted by the Georgi-Glashow model, a grand unified theory. This model proposes another force. The X and Y bosons would be analogous to the W and Z bosons of the weak force.

An X boson would decay into two up quarks or an anti-down quark and a positron.

A Y boson would decay into a positron and an anti-up-quark, a down quark, and an up quark, or an anti-down-quark and an anti-electron-neutrino.

The mass of the X and Y bosons would be extreme at $10^{15}$ GeV/$c^2$ = {[1.602 x $10^{-19}$]($10^{24}$)Joules}/{{[3 x $10^8$]m/s}$^2$} = 1.78 x $10^{-12}$ kg.

The width of a somehow stabilized valence arrangement of one X boson per three Y bosons would plausibly be about (0.000511)/($10^{15}$) times that of a typical neutral periodic table atom or 5.11 x $10^{-19}$ times that of a typical atom. Thus, the diameter of such a valence arrangement would about 5.11 x $10^{-29}$ meters.

We can imagine a string comprised of (0.0001){[5.11 x $10^{-29}$]$^{-1}$ } = 1.9569 x $10^{26}$ X-Y boson atoms which would be about 0.0001 meter long and have a mass of 1.3933 x $10^{15}$ kilograms and which would be 10 X-Y bosonic atoms wide and 10 X-Y bosonic atoms thick. Now a black hole having a mass of 1.3933 x $10^{15}$ kilograms will have a radius of about 2.37468 x $10^{-13}$ meters. We could therefore plausibly compact the 0.0001 meter long string into a conical configuration having a lateral height about equal to $10^{-11}$ meters and a basic width of $10^{-15}$ meters and comprised of a cross-woven grid of X-Y boson threads separated by about 2 x $10^{-22}$ meters. The lateral area of the cone will be 1.5708 x $10^{-26}$ square meters. The reasoning behind this assumption is that the 0.0001 meter thread could be partitioned into $10^{-7}$ sub-threads where each sub-thread has a cross-sectional area of 10 X-Y bosonic atoms by 10 X-Y bosonic atoms and is $10^{-11}$ meters long. A $10^{-18}$ meter by $10^{-18}$ meter section of the fabric will have a mass of [1.3933 x $10^{15}$]($10^{-14}$)(10,000) kg = 139,330 kg. This is the equivalent of 139,330 kg per $10^{-36}$ square meters. Thus, corrected mass of the cone will be (139,330 kg) [1.5708 x $10^{-26}$]/($10^{-36}$) = 2.1885 x $10^{15}$ kg.

The bond strength between the X and Y bosons should be roughly equal to:

$$F = \{1/[4\pi\varepsilon_0]\}[q_1 q_2]/[r^2]$$
$$= \{[8.987551787 \times 10^9] \text{ N m}^2 \text{ C}^{-2}\}[-1.602176565(35)\times10^{-19} \text{ Coulombs}]$$
$$[1.602176565(35)\times10^{-19} \text{ Coulombs}]/\{[5.11 \times 10^{-29}\text{m}]^2\} = 8.83528 \times 10^{28} \text{ Newtons.}$$

Therefore, the yield strength of each thread should be about ($10^2$)[8.83528 x $10^{28}$] Newtons = 8.83528 x $10^{30}$ Newtons.

Such an X-Y material cone could be used as the leading tip of an astrodynamic space train like starship and could, therefore, bear the forward brunt of the cosmic ray, dust particle, gas and plasma particle, and photonic onslaught from the vacuum of space. The X-Y material could be the proverbial tip of the spear for truly extreme gamma factor starships.

One example of a plausible crystalline material formed from X and a Y particle has the following pattern.

| | | | | | | | |
|---|---|---|---|---|---|---|---|
| X $Y_3$ | Y | X $Y_3$ | Y | X $Y_3$ | Y | X $Y_3$ | Y |
| Y | X $Y_3$ | Y | X $Y_3$ | Y | X $Y_3$ | Y | X $Y_3$ |
| X $Y_3$ | Y | X $Y_3$ | Y | X $Y_3$ | Y | X $Y_3$ | Y |
| Y | X $Y_3$ | Y | X $Y_3$ | Y | X $Y_3$ | Y | X $Y_3$ |
| X $Y_3$ | Y | X $Y_3$ | Y | X $Y_3$ | Y | X $Y_3$ | Y |
| Y | X $Y_3$ | Y | X $Y_3$ | Y | X $Y_3$ | Y | X $Y_3$ |

Another example of a plausible crystalline material formed from X and Y particles has the following pattern.

| | | | | | | | |
|---|---|---|---|---|---|---|---|
| -X-$Y_3$ | -Y | -X-$Y_3$ | -Y | -X-$Y_3$ | -Y | -X-$Y_3$ | -Y |
| -Y | -X-$Y_3$ | -Y | -X-$Y_3$ | -Y | -X-$Y_3$ | -Y | -X-$Y_3$ |
| -X-$Y_3$ | -Y | -X-$Y_3$ | -Y | -X-$Y_3$ | -Y | -X-$Y_3$ | -Y |
| -Y | -X-$Y_3$ | -Y | -X-$Y_3$ | -Y | -X-$Y_3$ | -Y | -X-$Y_3$ |
| -X-$Y_3$ | -Y | -X-$Y_3$ | -Y | -X-$Y_3$ | -Y | -X-$Y_3$ | -Y |
| -Y | -X-$Y_3$ | -Y | -X-$Y_3$ | -Y | -X-$Y_3$ | -Y | -X-$Y_3$ |

Again, antimatter versions of any of the above conjectured materials are possible and can be used as rocket fuel when combined with carried aboard Fermi-Dirac annihilation partner materials and/or such materials extracted from the interstellar or intergalactic medium. Such materials would likely require artificial manufacture and instillation with in the interstellar and/or intergalactic medium.

## MONOPOLIUM

The mass of the still theoretical magnetic monopole is at most about $10^{17}$ GeV/c$^2$ = {[1.602 x $10^{-19}$]($10^{26}$)Joules}/{{[3 x $10^8$]m/s}$^2$} = 1.78 x $10^{-10}$ kg.. The width of a stable valence relationship between a South monopole and a North monopole would seem to be about (0.000511)/($10^{17}$) times that of a typical neutral period table atom or 5.11 x $10^{-21}$ times that of a typical period table atom. Thus, the diameter of such a valence arrangement would be 5.11 x $10^{-31}$ meters which is approaching that of the Planck Length at:

$$l_p = \{[h(2\pi)]G/[C^3]\}^{1/2} = 1.616199 \times 10^{-35} \text{ meters.}$$

## X AND/OR Y MONOPOLIUM

A unit cell of solid crystalline materials formed of equal parts by particle of high-end mass range monopoles ($10^{17}$ GeV/c$^2$ per particle), and $10^{15}$ GeV/c$^2$ X particles would have a width of about [5.11 x $10^{-29}$] + [5.11 x $10^{-31}$] anstroms = 5.1611 x $10^{-29}$ meters.

An example of a planar crystalline pattern for the monohiggsinium is:

X M X M- X M X M-

M- X- M X- M- X- M X-

X M X M- X M X M-

M- X- M X- M- X- M X-

X M X M- X M X M-

M- X- M X- M- X- M X-

Another example of a planar crystalline form for the mono-higgsinium is as follows:

Y M Y M- Y M Y M-

M- Y- M Y- M- Y- M Y-

Y M Y M- Y M Y M-

M- Y- M Y- M- Y- M Y-

Y M Y M- Y M Y M-

M- Y- M Y- M- Y- M Y-

Another material plausibly formed from high-end of mass range monopoles and X and Y particles having a similar density and mechanical strength has the following crystalline pattern.

| | | | | | | | |
|---|---|---|---|---|---|---|---|
| $X Y_3$ | M | $X Y_3$ | M- | $X Y_3$ | M | $X Y_3$ | M- |
| M- | Y | M | Y | M- | Y | M | Y |
| $X Y_3$ | M | $X Y_3$ | M- | $X Y_3$ | M | $X Y_3$ | M- |
| M- | Y | M | Y | M- | Y | M | Y |
| $X Y_3$ | M | $X Y_3$ | M- | $X Y_3$ | M | $X Y_3$ | M- |
| M- | Y | M | Y | M- | Y | M | Y |

Still another material plausibly formed from high-end of mass range monopoles and X and Y particles having a similar density and mechanical strength has the following crystalline pattern.

| | | | | | | | |
|---|---|---|---|---|---|---|---|
| -X-$Y_3$ | M | $XY_3$ | M- | -X-$Y_3$ | M | $XY_3$ | M- |
| M- | -Y | M | Y | M- | -Y | M | Y |
| -X-$Y_3$ | M | $XY_3$ | M- | -X-$Y_3$ | M | $XY_3$ | M- |
| M- | -Y | M | Y | M- | -Y | M | Y |
| -X-$Y_3$ | M | $XY_3$ | M- | -X-$Y_3$ | M | $XY_3$ | M- |
| M- | -Y | M | Y | M- | -Y | M | -Y |

Still yet another material plausibly formed from high-end of mass range monopoles and X and Y particles having a similar density and mechanical strength has the following crystalline pattern.

| | | | | | | | |
|---|---|---|---|---|---|---|---|
| Y | M | -X | M- | Y | M | -X | M- |
| M- | Y- | M | X | M- | Y- | M | X |
| Y | M | -X | M- | Y | M | -X | M- |
| M- | Y- | M | X | M- | Y- | M | X |
| Y | M | -X | M- | Y | M | -X | M- |
| M- | Y- | M | X- | M- | Y- | M | X- |

Still yet another material plausibly formed from high-end of mass range monopoles and X and Y particles having a similar density and mechanical strength has the following crystalline pattern.

| | | | | | | | |
|---|---|---|---|---|---|---|---|
| $XY_3$ | M | Y | M- | $XY_3$ | M | Y | M- |
| M- | $XY_3$ | M | Y | M- | $XY_3$ | M | Y |
| $XY_3$ | M | Y | M- | $XY_3$ | M | Y | M- |
| M- | $XY_3$ | M | Y | M- | $XY_3$ | M | Y |
| $XY_3$ | M | Y | M- | $XY_3$ | M | Y | M- |
| M- | $XY_3$ | M | Y | M- | $XY_3$ | M | Y |

Still yet another material plausibly formed from high-end of mass range monopoles and X and Y particles having a similar density and mechanical strength has the following crystalline pattern.

| | | | | | | | |
|---|---|---|---|---|---|---|---|
| $-X-Y_3$ | M | -Y | M- | $-X-Y_3$ | M | -Y | M- |
| M- | $-X-Y_3$ | M | -Y | M- | $-X-Y_3$ | M | -Y |
| $-X-Y_3$ | M | -Y | M- | $-X-Y_3$ | M | -Y | M- |
| M- | $-X-Y_3$ | M | -Y | M- | $-X-Y_3$ | M | -Y |
| $-X-Y_3$ | M | -Y | M- | $-X-Y_3$ | M | -Y | M- |
| M- | $-X-Y_3$ | M | -Y | M- | $-X-Y_3$ | M | -Y |

The specific crystalline patterns depicted above would resemble ionic-bonds in a manner analogous to bi-atomic salts such as ordinary table salt or Sodium Chloride.

Note that the negative signs included in the above crystalline depictions indicate antimatter versions of the particle species as denoted by the upper-case letters.

Again, antimatter versions of any of the above conjectured materials are possible and can be used as rocket fuel when combined with carried aboard Fermi-Dirac annihilation partner materials and/or such materials extracted from the interstellar or intergalactic medium. Such materials would likely require artificial manufacture and instillation with in the interstellar and/or intergalactic medium. Such materials may include magnetic monopoles and antimonopoles, X, Y, anti-X, and/or anti-Y bosons.

Now, what might the reader ask is the point to this digression? The answer is elementary my dear Watson! A very narrow tipped cone on the bow of a starship traveling at a sufficient gamma factor might be used to spear one or more ultramicroscopic wormholes which may theoretically continuously form and then disappear on the scale of the Planck Length, the Planck Area, and the Planck Volume on time periods roughly equal to the Planck Time.

Accordingly, the spacecraft would spear open one of these wormholes before the wormhole could collapse. The spike-like cone needle would pry the wormhole open while the rest of a conical shield would proceed to plow into the expanding wormhole. Eventually, a 2 meter diameter or larger quarkonium portion of the shield would wedge into the wormhole to hold the wormhole open.

The quarkonium shield would include attometer self-assembly technology to configure itself into a uniformly stable cylinder to line the wormhole.

The X-Y material, Monopolium, and Mono-X-Y materials described above would lead the tip of the spear.

**FIGURE 30** is a cross-sectional diagram which shows forward thrusted rocket plasma plume (red)reacting off of and backwardly reflected by an extremely high electrodynamic energy density plasma bow shock (orange) set up by electrodynamic fields produced by the spacecraft as the craft travels through space at ultra-relativistic velocities. The bow shock may be optionally facilitated by a linear induction superconducting coil (light-blue) that produces extremely large internal electrical currents. The bow shock is affixed to the spacecraft by electrodynamic fields.

**FIGURE 31** is a cross-sectional diagram which shows forward thrusted rocket plasma plume (red)reacting off of and backwardly reflected by an extremely high electrodynamic energy density plasma bow shock (orange) set up by electrodynamic fields produced by the spacecraft as the craft travels through space at ultra-relativistic velocities. The bow shock may be optionally facilitated by a linear induction superconducting coil that produces extremely large internal electrical currents. The bow shock is affixed to the spacecraft by electrodynamic fields. The backwardly reflected exhaust wraps around the forwardly oriented cone in the stern of the craft where the plasma builds up pressure and is back-reflected to produce net positive thrust.

**FIGURE 32** on the previous page is a cross-sectional diagram which shows forward thrusted rocket plasma plume (red)reacting off of and backwardly reflected by an extremely high electrodynamic energy density plasma bow shock (orange) set up by electrodynamic fields produced by the spacecraft as the craft travels through space at ultra-relativistic velocities. The bow shock may be optionally facilitated by a linear induction superconducting coil that produces extremely large internal electrical currents. The bow shock is affixed to the spacecraft by electrodynamic fields. The backwardly reflected exhaust wraps around the forwardly oriented heart shaped astrospike in the stern of the craft where the plasma builds up pressure and squeezes the astrospike to produce net positive thrust.

# Building Bombs Enroute.

***In this chapter, we cover mechanisms for assembling nuclear devices enroute from materials brought along from the start of a mission and/or materials collected during a flight.***

Nuclear fissile materials and non-fissile bomb components may be brought along from the start of a mission and converted into nuclear propulsive devices. The same can be said for the production of fusion bombs such as thermonuclear bombs.

It may even be the case where materials such fissile feed-stock or fusion fuel can be collected from the ambient environment during a flight along with materials to be used in the non-fissile portions of the bombs. Ideally, the ambient materials intaked would be somehow funneled into a chamber for which the associated ions, neutral atoms, electrons, and dust particles would be fully decelerated for which the energy of friction and collision would be captured and immediately or nearly so converted to electrical power to energize secondary electrical propulsion systems. Such a spacecraft would maintain its velocity as virtually constant over instantaneous time steps because the collision and frictional energy would be converted to propulsion energy.

The above being stated, it then becomes plausible that the nuclear bombs fashioned from the naturally occurring or artificially disposed background massive species could then be used for propulsion thus resulting in a net velocity gain for the spacecraft. As such, the craft would operate more or less as a pulse drive interstellar ramjet.

In order to extract the required energy from the background intake species, one option is to ionize all neutral atoms and neutral molecules funneled into a linear induction generating chamber as well as the same for dust particles. Another option is to simply let the intake matter impact or be absorbed by a thermal mass for which the heat would then be used to drive mechanical electric generators by working steam effluents, photo-voltaic cells, and/or thermos-electric cells.

In reality, mechanical electrical generators and associated housings could be layered with Dewar style insulation and hierarchies of PV cells and/or thermoelectric cells. This way, there would be many layers of waste heat capture for which the generated electrical power could then be directed to energize one or more secondary electrical propulsion systems.

Such a system may plausibly experience more or less constant invariant mass with uniform or increasing ship-time dependent acceleration.

Now, as mentioned in the previous chapter, a phenomenon which herein is referred to as slap-back can occur for which impinging plasma rebounding of the spacecraft pusher-plate is slapped back into the pusher plate by subsequently impinging plasma. Slap-back is likely not significantly affected by impinging neutron or impinging photons because neutrons are not easily reflected being chargeless and photons are generally within the x-ray and gamma ray range and are thus absorbed by the pusher plate because of their penetrating power instead of being easily reflected. For chargon force exertion on the plate, we will include the factor {Chargon slap-back) which modifies in abstract notation the force exerted on the plate. It is conceivable that this factor could be less than, equal to, or greater than one depending on the chargon flux patterns at or near the pusher plate.

Provided a steady state spacecraft invariant mass could be attained, the following formula provides a detailed account of the energy accrued by the space craft over a spatial interval of r(x,y,z) for which tbe spacecraft has attained steady state invariant mass and where only identical bombs are used.

$$E_{kinetic,steady,state} = \{\{\int\langle\{d\ P_{combined,incident}/dt\}\rangle \bullet d\ r(x,y,z)\}_{steady,state}\} = \{\{\int\langle\{d\{\int[[(\alpha = 0, \alpha = [[f(\pi)]_{max,chargon}]]\{(2)\{(A_{capture,chargon})/[(4)(\pi)(r^2)]\}\ \{\sum(i = 1, i = N)\ \{\{\{((k_B)(T_{ith,species,average})(N_{ith,species})^2) - \{[(m_{0,ith,species})(c^2)]^2\}\}/(c^2)\}^{1/2}\}\}(\text{Chargon angular correction factor})\ \{\text{Chargon slap-back})\ (\cos\alpha_i)d\alpha\} + \{\int[(\alpha = 0, \alpha = [[f(\pi)]_{max,neutron}]]\{\{(A_{capture,neutron})/[(4)(\pi)(r^2)]\}\ \{\{\{\{\sum(j = 1, j = N_{max})\{\{(M_{0,neutron})(\gamma_j)(c^2)\} - [(M_{0,neutron})(c^2)]\}\}^{2}\} - \{[(M_{0,neutron})(c^2)]^2\}\}/(c^2)\}^{1/2}\}(\text{Neutron},j)\}\ (\text{Neutron angular correction factor})\ (\cos\alpha_i)d\alpha\} + \{\int[(\alpha = 0, \alpha = [[f(\pi)]_{max,photon,nuclear}]]\{\{(A_{capture,photon})/[(4)(\pi)(r^2)]\}\ \{\sum(jk = 1, k = N_{photon,nuclear})\ \{(F_k/c) = (h\nu_k/c) = (h/\lambda_k)\}(\text{Photon,nuclear},k)\}\}\ (\text{Nuclear photon angular correction factor})(\cos\alpha_i)d\alpha\} + \{\int[(\alpha = 0, \alpha = [[f(\pi)]_{max,photon,thermal}]]\{d\{(A_{capture,photon})\ \{\{[5.670373 \times 10^{-8}]\ (J\ m^{-2}\ s^{-1}\ K^{-4})\}[T_{plasma}^{4}]/[3 \times 10^{8}\ m/s]\}\}/dt\}(\text{Thermal photon angular correction factor})\ (\cos\alpha_i)d\alpha\}/dt\}\rangle \bullet d\ r(x,y,z)\}_{steady,state}\}$$

or

$E_{kinetic\ ,steady,state}$= {{∫<{d $P_{combined,incident}$/dt}>• d r(x,y,z)} $_{steady,state}$} = {{∫<{d{∫[(α = 0, α = [[f(π)] $_{max,chargon}$]]{(2){($A_{capture,chargon}$)/[(4)(π)($r^2$)]} {∑(i = 1, i = N) {∑(i = 1, i = N){[(($k_B$)($T_{ith,species,average}$)($N_{ith,species}$)$^2$)/($c^2$)] – [($m_{0,ith,species}$ $^2$)($c^2$)] }$^{1/2}$}}} (Chargon angular correction factor) {Chargon slap-back) (cos $\alpha_i$)dα} + {∫[(α = 0, α = [[f(π)] $_{max,neutron}$]]{{($A_{capture,neutron}$)/[(4)(π)($r^2$)]} {{{{{∑(j = 1, j = $N_{max}$) {{($M_{0,neutron}$) {{1 – [($v_j^2$)/($c^2$)]}$^{1/2}$} ($c^2$)} – [($M_{0,neutron}$)($c^2$)]}} $^2$}/($c^2$)} – [($M_{0,neutron}$ $^2$)($c^2$)] }$^{1/2}$}(Neutron,j)} (Neutron angular correction factor) (cos $\alpha_i$)dα} +{∫[(α = 0, α = [[f(π)] $_{max,photon,nuclear}$]]{{($A_{capture,photon}$)/[(4)(π)($r^2$)]} {∑(jk = 1, k = $N_{photon,nuclear}$) {($E_k$/c) = ($h\nu_k$/c) = ($h/\lambda_k$)}(Photon,nuclear,k)}} (Nuclear photon angular correction factor) (cos $\alpha_i$)dα} + {∫[(α = 0, α = [[f(π)] $_{max,photon,thermal}$]]{d{($A_{capture,photon}$) {{[5.670373 x $10^{-8}$] (J $m^{-2}$ $s^{-1}$ $K^{-4}$)}[$T_{plasma}^4$]/[3 x $10^8$ m/s]}}/dt}(Thermal photon angular correction factor) (cos $\alpha_i$)dα}/dt}>• d r(x,y,z)} $_{steady,state}$}.

The accrued gamma factor for the spacecraft while in mass ratio burn down mode is equal to:

$\gamma_{accrued}$ = {{ 1 - {{{C Tanh {[$I_{sp}$/C) ln ($M_0$/M1)}}/C}$^2$}}$^{-(1/2)}$} + {{$E_{kinetic,steady,state}$}/[($M_0$)[$C^2$]]}

= {{ 1 - {{{C Tanh {[$I_{sp}$/C) ln ($M_0$/M1)}}/C}$^2$}}$^{-(1/2)}$} + {{{∫{d $P_{combined,incident}$/dt} • d r(x,y,z)}$_{steady,state}$}/[($M_0$)[$C^2$]]}

= {{ 1 - {{{C Tanh {[$I_{sp}$/C) ln ($M_0$/M1)}}/C}$^2$}}$^{-(1/2)}$} + {{{∫<{d{∫[(α = 0, α = [[f(π)] $_{max,chargon}$]]{(2){($A_{capture,chargon}$)/[(4)(π)($r^2$)]} {∑(i = 1, i = N) {{{{($k_B$)($T_{ith,species,average}$)($N_{ith,species}$)$^2$)– {[($m_{0,ith,species}$)($c^2$)]$^2$}}/($c^2$)}$^{1/2}$}}}(Chargon angular correction factor) {Chargon slap-back) (cos $\alpha_i$)dα} + {∫[(α = 0, α = [[f(π)] $_{max,neutron}$]]{{($A_{capture,neutron}$)/[(4)(π)($r^2$)]} {{{{{∑(j = 1, j = $N_{max}$){{($M_{0,neutron}$)($\gamma_j$)($c^2$)} – [($M_{0,neutron}$)($c^2$)]}} $^2$}– {[($M_{0,neutron}$)($c^2$)]$^2$}}/($c^2$)}$^{1/2}$}(Neutron,j)} (Neutron angular correction factor) (cos $\alpha_i$)dα} + {∫[(α = 0, α = [[f(π)] $_{max,photon,nuclear}$]]{{($A_{capture,photon}$)/[(4)(π)($r^2$)]} {∑(jk = 1, k = $N_{photon,nuclear}$) {($E_k$/c) = ($h\nu_k$/c) = ($h/\lambda_k$)}(Photon,nuclear,k)}} (Nuclear photon angular correction factor)(cos $\alpha_i$)dα} + {∫[(α = 0, α = [[f(π)] $_{max,photon,thermal}$]]{d{($A_{capture,photon}$) {{[5.670373 x $10^{-8}$] (J $m^{-2}$ $s^{-1}$ $K^{-4}$)}[$T_{plasma}^4$]/[3 x $10^8$ m/s]}}/dt}(Thermal photon angular correction factor) (cos $\alpha_i$)dα}/dt}>• d r(x,y,z)} $_{steady,state}$}/[($M_0$)[$C^2$]]}

= {{ 1 - {{{C Tanh {[$I_{sp}$/C) ln ($M_0$/M1)}}/C}$^2$}}$^{-(1/2)}$} +{{{∫<{d{∫[(α = 0, α = [[f(π)] $_{max,chargon}$]]{ (2){($A_{capture,chargon}$)/[(4)(π)($r^2$)]} {∑(i = 1, i = N) {∑(i = 1, i = N){[(($k_B$)($T_{ith,species,average}$)($N_{ith,species}$)$^2$)/($c^2$)] – [($m_{0,ith,species}$ $^2$)($c^2$)] }$^{1/2}$}}} (Chargon angular correction factor) {Chargon slap-back) (cos $\alpha_i$)dα} + {∫[(α = 0, α = [[f(π)] $_{max,neutron}$]]{{($A_{capture,neutron}$)/[(4)(π)($r^2$)]} {{{{{∑(j = 1, j = $N_{max}$) {{($M_{0,neutron}$) {{1 – [($v_j^2$)/($c^2$)]}$^{1/2}$} ($c^2$)} – [($M_{0,neutron}$)($c^2$)]}} $^2$}/($c^2$)} – [($M_{0,neutron}$ $^2$)($c^2$)] }$^{1/2}$}(Neutron,j)} (Neutron angular correction factor) (cos $\alpha_i$)dα} +{∫[(α = 0, α = [[f(π)] $_{max,photon,nuclear}$]]{{($A_{capture,photon}$)/[(4)(π)($r^2$)]} {∑(jk = 1, k = $N_{photon,nuclear}$) {($E_k$/c) = ($h\nu_k$/c) = ($h/\lambda_k$)}(Photon,nuclear,k)}} (Nuclear photon angular correction factor) (cos $\alpha_i$)dα} + {∫[(α = 0, α = [[f(π)] $_{max,photon,thermal}$]]{d{($A_{capture,photon}$) {{[5.670373 x $10^{-8}$] (J $m^{-2}$ $s^{-1}$ $K^{-4}$)}[$T_{plasma}^4$]/[3 x $10^8$ m/s]}}/dt}(Thermal photon angular correction factor) (cos $\alpha_i$)dα}/dt}>• d r(x,y,z)} $_{steady,state}$}/[($M_0$)[$C^2$]]}

Provided a steady state spacecraft invariant mass could be attained, the following formula provides a detailed account of the energy accrued by the space craft over a spatial interval of r(x,y,z) for which tbe spacecraft has attained steady state invariant mass and where differing invariant mass specific yield bomb models and/or bomb models having differing nuclear reaction chains are used.

$E_{kinetic,steady,state,multiple}$ = {{{∑($z_1$= 1, $z_1$ = $N_{z1}$){{∫<{d $P_{combined,incident}$/dt}>{Chargon slap-back) • d r(x,y,z)}$_{bomb,z1}$}} + {∑($z_2$= 1, $z_2$ = $N_{z2}$){{∫<{d $P_{combined,incident}$/dt}>{Chargon slap-back) • d r(x,y,z)}$_{bomb,z2}$}} + {∑($z_3$= 1, $z_3$ = $N_{z3}$){{∫<{d $P_{combined,incident}$/dt}>{Chargon slap-back) • d r(x,y,z)}$_{bomb,z3}$}} + ... + {∑($z_n$= 1, $z_n$ = $N_{zn}$){{∫<{d $P_{combined,incident}$/dt}>{Chargon slap-back) • d r(x,y,z)}$_{bomb,zn}$}}} $_{steady,state,multiple}$}

= {{{∑($z_1$ = 1, $z_1$ = $N_{z1}$){∫<{d{∫[(α = 0, α = [[f(π)] $_{max,chargon}$]]{(2){($A_{capture,chargon}$)/[(4)(π)($r^2$)]} {∑(i = 1, i = N) {{{{($k_B$)($T_{ith,species,average}$)($N_{ith,species}$)$^2$)– {[($m_{0,ith,species}$)($c^2$)]$^2$}}/($c^2$)}$^{1/2}$}}}(Chargon angular correction factor) {Chargon slap-back) (cos $\alpha_i$)dα} + {∫[(α = 0, α = [[f(π)] $_{max,neutron}$]]{{($A_{capture,neutron}$)/[(4)(π)($r^2$)]} {{{{{∑(j = 1, j = $N_{max}$){{($M_{0,neutron}$)($\gamma_j$)($c^2$)} – [($M_{0,neutron}$)($c^2$)]}} $^2$}– {[($M_{0,neutron}$)($c^2$)]$^2$}}/($c^2$)}$^{1/2}$}(Neutron,j)} (Neutron angular correction factor) (cos $\alpha_i$)dα} + {∫[(α = 0, α = [[f(π)] $_{max,photon,nuclear}$]]{{($A_{capture,photon}$)/[(4)(π)($r^2$)]} {∑(jk = 1, k = $N_{photon,nuclear}$) {($E_k$/c) = ($h\nu_k$/c) = ($h/\lambda_k$)}(Photon,nuclear,k)}} (Nuclear photon angular correction factor)(cos $\alpha_i$)dα} + {∫[(α = 0, α = [[f(π)] $_{max,photon,thermal}$]]{d{($A_{capture,photon}$) {{[5.670373 x $10^{-8}$] (J $m^{-2}$ $s^{-1}$ $K^{-4}$)}[$T_{plasma}^4$]/[3 x $10^8$ m/s]}}/dt}(Thermal photon angular correction factor) (cos $\alpha_i$)dα}/dt}>• d r(x,y,z)}$_{bomb,z1}$}}

+ {∑($z_2$ = 1, $z_2$ = $N_{z2}$){∫<{d{∫[(α = 0, α = [[f(π)] $_{max,chargon}$]]{(2){($A_{capture,chargon}$)/[(4)(π)($r^2$)]} {∑(i = 1, i = N) {{{{($k_B$)($T_{ith,species,average}$)($N_{ith,species}$)$^2$)– {[($m_{0,ith,species}$)($c^2$)]$^2$}}/($c^2$)}$^{1/2}$}}}(Chargon angular correction factor) {Chargon slap-back) (cos $\alpha_i$)dα} + {∫[(α = 0, α = [[f(π)] $_{max,neutron}$]]{{($A_{capture,neutron}$)/[(4)(π)($r^2$)]} {{{{{∑(j = 1, j = $N_{max}$){{($M_{0,neutron}$)($\gamma_j$)($c^2$)} – [($M_{0,neutron}$)($c^2$)]}} $^2$}– {[($M_{0,neutron}$)($c^2$)]$^2$}}/($c^2$)}$^{1/2}$}(Neutron,j)} (Neutron angular correction factor) (cos $\alpha_i$)dα} + {∫[(α = 0, α = [[f(π)] $_{max,photon,nuclear}$]]{{($A_{capture,photon}$)/[(4)(π)($r^2$)]} {∑(jk = 1, k = $N_{photon,nuclear}$) {($E_k$/c) = ($h\nu_k$/c) = ($h/\lambda_k$)}(Photon,nuclear,k)}} (Nuclear photon angular correction factor)(cos $\alpha_i$)dα} + {∫[(α = 0, α = [[f(π)] $_{max,photon,thermal}$]]{d{($A_{capture,photon}$) {{[5.670373 x $10^{-8}$] (J $m^{-2}$ $s^{-1}$ $K^{-4}$)}[$T_{plasma}^4$]/[3 x $10^8$ m/s]}}/dt}(Thermal photon angular correction factor) (cos $\alpha_i$)dα}/dt}>• d r(x,y,z)}$_{bomb,z2}$}}

+ {∑($z_3$ = 1, $z_3$ = $N_{z3}$){∫<{d{∫[(α = 0, α = [[f(π)] $_{max,chargon}$]]{(2){($A_{capture,chargon}$)/[(4)(π)($r^2$)]} {∑(i = 1, i = N) {{{(($k_B$)($T_{ith,species,average}$)($N_{ith,species}$)$^2$)– {[($m_{0,ith,species}$)($c^2$)]$^2$}}/($c^2$)}$^{1/2}$}}}(Chargon angular correction factor) {Chargon slap-back) (cos $\alpha_i$)dα} + {∫[(α = 0, α = [[f(π)] $_{max,neutron}$]]{{($A_{capture,neutron}$)/[(4)(π)($r^2$)]} {{{{{∑(j = 1, j = $N_{max}$){{($M_{0,neutron}$)($\gamma_j$)($c^2$)} – [($M_{0,neutron}$)($c^2$)]]}}$^2$}– {[($M_{0,neutron}$)($c^2$)]$^2$}}/($c^2$)}$^{1/2}$}(Neutron,j)} (Neutron angular correction factor) (cos $\alpha_i$)dα} + {∫[(α = 0, α = [[f(π)] $_{max,photon,nuclear}$]]{{($A_{capture,photon}$)/[(4)(π)($r^2$)]} {∑(jk = 1, k = $N_{photon,nuclear}$) {($E_k$/c) = ($h\nu_k$/c) = ($h/\lambda_k$)}(Photon,nuclear,k)}} (Nuclear photon angular correction factor)(cos $\alpha_i$)dα} + {∫[(α = 0, α = [[f(π)] $_{max,photon,thermal}$]]{d{($A_{capture,photon}$) {{[5.670373 x $10^{-8}$] (J $m^{-2}$ $s^{-1}$ $K^{-4}$)}[$T_{plasma}{}^4$]/[3 x $10^8$ m/s]}}/dt}(Thermal photon angular correction factor) (cos $\alpha_i$)dα}/dt}>• d r(x,y,z)}$_{bomb,z3}$}}

+ ... + {∑($z_n$ = 1, $z_n$ = $N_{zn}$){∫<{d{∫[(α = 0, α = [[f(π)] $_{max,chargon}$]]{(2){($A_{capture,chargon}$)/[(4)(π)($r^2$)]} {∑(i = 1, i = N) {{{(($k_B$)($T_{ith,species,average}$)($N_{ith,species}$)$^2$)– {[($m_{0,ith,species}$)($c^2$)]$^2$}}/($c^2$)}$^{1/2}$}}}(Chargon angular correction factor) {Chargon slap-back) (cos $\alpha_i$)dα} + {∫[(α = 0, α = [[f(π)] $_{max,neutron}$]]{{($A_{capture,neutron}$)/[(4)(π)($r^2$)]} {{{{{∑(j = 1, j = $N_{max}$){{($M_{0,neutron}$)($\gamma_j$)($c^2$)} – [($M_{0,neutron}$)($c^2$)]]}}$^2$}– {[($M_{0,neutron}$)($c^2$)]$^2$}}/($c^2$)}$^{1/2}$}(Neutron,j)} (Neutron angular correction factor) (cos $\alpha_i$)dα} + {∫[(α = 0, α = [[f(π)] $_{max,photon,nuclear}$]]{{($A_{capture,photon}$)/[(4)(π)($r^2$)]} {∑(jk = 1, k = $N_{photon,nuclear}$) {($E_k$/c) = ($h\nu_k$/c) = ($h/\lambda_k$)}(Photon,nuclear,k)}} (Nuclear photon angular correction factor)(cos $\alpha_i$)dα} + {∫[(α = 0, α = [[f(π)] $_{max,photon,thermal}$]]{d{($A_{capture,photon}$) {{[5.670373 x $10^{-8}$] (J $m^{-2}$ $s^{-1}$ $K^{-4}$)}[$T_{plasma}{}^4$]/[3 x $10^8$ m/s]}}/dt}(Thermal photon angular correction factor) (cos $\alpha_i$)dα}/dt}>• d r(x,y,z)}$_{bomb,zn}$}}} $_{steady,state,multiple}$}

= {{{∑($z_1$ = 1, $z_1$ = $N_{z1}$){∫<{d{∫[(α = 0, α = [[f(π)] $_{max,chargon}$]]{ (2){($A_{capture,chargon}$)/[(4)(π)($r^2$)]} {∑(i = 1, i = N) {∑(i = 1, i = N){[(($k_B$)($T_{ith,species,average}$)($N_{ith,species}$)$^2$)/($c^2$)] – [($m_{0,ith,species}{}^2$)($c^2$)] }$^{1/2}$}}} (Chargon angular correction factor) {Chargon slap-back) (cos $\alpha_i$)dα} + {∫[(α = 0, α = [[f(π)] $_{max,neutron}$]]{{($A_{capture,neutron}$)/[(4)(π)($r^2$)]} {{{{{∑(j = 1, j = $N_{max}$) {{($M_{0,neutron}$) {{1 – [($v_j{}^2$)/($c^2$)]]}$^{1/2}$} ($c^2$)} – [($M_{0,neutron}$)($c^2$)]]} $^2$}/($c^2$)} – [($M_{0,neutron}{}^2$)($c^2$)] }$^{1/2}$}(Neutron,j)} (Neutron angular correction factor) (cos $\alpha_i$)dα} +{∫[(α = 0, α = [[f(π)] $_{max,photon,nuclear}$]]{{($A_{capture,photon}$)/[(4)(π)($r^2$)]} {∑(jk = 1, k = $N_{photon,nuclear}$) {($E_k$/c) = ($h\nu_k$/c) = ($h/\lambda_k$)}(Photon,nuclear,k)}} (Nuclear photon angular correction factor) (cos $\alpha_i$)dα} + {∫[(α = 0, α = [[f(π)] $_{max,photon,thermal}$]]{d{($A_{capture,photon}$) {{[5.670373 x $10^{-8}$] (J $m^{-2}$ $s^{-1}$ $K^{-4}$)}[$T_{plasma}{}^4$]/[3 x $10^8$ m/s]}}/dt}(Thermal photon angular correction factor) (cos $\alpha_i$)dα}/dt}>• d r(x,y,z)} $_{bomb,z1}$}}

+ {∑($z_2$ = 1, $z_2$ = $N_{z2}$){∫<{d{∫[(α = 0, α = [[f(π)] $_{max,chargon}$]]{ (2){($A_{capture,chargon}$)/[(4)(π)($r^2$)]} {∑(i = 1, i = N) {∑(i = 1, i = N){[(($k_B$)($T_{ith,species,average}$)($N_{ith,species}$)$^2$)/($c^2$)] – [($m_{0,ith,species}{}^2$)($c^2$)] }$^{1/2}$}}} (Chargon angular correction factor) {Chargon slap-back) (cos $\alpha_i$)dα} + {∫[(α = 0, α = [[f(π)] $_{max,neutron}$]]{{($A_{capture,neutron}$)/[(4)(π)($r^2$)]} {{{{{∑(j = 1, j = $N_{max}$) {{($M_{0,neutron}$) {{1 – [($v_j{}^2$)/($c^2$)]]}$^{1/2}$} ($c^2$)} – [($M_{0,neutron}$)($c^2$)]]} $^2$}/($c^2$)} – [($M_{0,neutron}{}^2$)($c^2$)] }$^{1/2}$}(Neutron,j)} (Neutron angular correction factor) (cos $\alpha_i$)dα} +{∫[(α = 0, α = [[f(π)] $_{max,photon,nuclear}$]]{{($A_{capture,photon}$)/[(4)(π)($r^2$)]} {∑(jk = 1, k = $N_{photon,nuclear}$) {($E_k$/c) = ($h\nu_k$/c) = ($h/\lambda_k$)}(Photon,nuclear,k)}} (Nuclear photon angular correction factor) (cos $\alpha_i$)dα} + {∫[(α = 0, α = [[f(π)] $_{max,photon,thermal}$]]{d{($A_{capture,photon}$) {{[5.670373 x $10^{-8}$] (J $m^{-2}$ $s^{-1}$ $K^{-4}$)}[$T_{plasma}{}^4$]/[3 x $10^8$ m/s]}}/dt}(Thermal photon angular correction factor) (cos $\alpha_i$)dα}/dt}>• d r(x,y,z)} $_{bomb,z2}$}}

+ {∑($z_3$ = 1, $z_3$ = $N_{z3}$){∫<{d{∫[(α = 0, α = [[f(π)] $_{max,chargon}$]]{ (2){($A_{capture,chargon}$)/[(4)(π)($r^2$)]} {∑(i = 1, i = N) {∑(i = 1, i = N){[(($k_B$)($T_{ith,species,average}$)($N_{ith,species}$)$^2$)/($c^2$)] – [($m_{0,ith,species}{}^2$)($c^2$)] }$^{1/2}$}}} (Chargon angular correction factor) {Chargon slap-back) (cos $\alpha_i$)dα} + {∫[(α = 0, α = [[f(π)] $_{max,neutron}$]]{{($A_{capture,neutron}$)/[(4)(π)($r^2$)]} {{{{{∑(j = 1, j = $N_{max}$) {{($M_{0,neutron}$) {{1 – [($v_j{}^2$)/($c^2$)]]}$^{1/2}$} ($c^2$)} – [($M_{0,neutron}$)($c^2$)]]} $^2$}/($c^2$)} – [($M_{0,neutron}{}^2$)($c^2$)] }$^{1/2}$}(Neutron,j)} (Neutron angular correction factor) (cos $\alpha_i$)dα} +{∫[(α = 0, α = [[f(π)] $_{max,photon,nuclear}$]]{{($A_{capture,photon}$)/[(4)(π)($r^2$)]} {∑(jk = 1, k = $N_{photon,nuclear}$) {($E_k$/c) = ($h\nu_k$/c) = ($h/\lambda_k$)}(Photon,nuclear,k)}} (Nuclear photon angular correction factor) (cos $\alpha_i$)dα} + {∫[(α = 0, α = [[f(π)] $_{max,photon,thermal}$]]{d{($A_{capture,photon}$) {{[5.670373 x $10^{-8}$] (J $m^{-2}$ $s^{-1}$ $K^{-4}$)}[$T_{plasma}{}^4$]/[3 x $10^8$ m/s]}}/dt}(Thermal photon angular correction factor) (cos $\alpha_i$)dα}/dt}>• d r(x,y,z)} $_{bomb,z3}$}}

+ ... + {∑($z_n$= 1, $z_n$ = $N_{zn}$){∫<{d{∫[(α = 0, α = [[f(π)] $_{max,chargon}$]]{ (2){($A_{capture,chargon}$)/[(4)(π)($r^2$)]} {∑(i = 1, i = N) {∑(i = 1, i = N){[(($k_B$)($T_{ith,species,average}$)($N_{ith,species}$)$^2$)/($c^2$)] – [($m_{0,ith,species}{}^2$)($c^2$)] }$^{1/2}$}}} (Chargon angular correction factor) {Chargon slap-back) (cos $\alpha_i$)dα} + {∫[(α = 0, α = [[f(π)] $_{max,neutron}$]]{{($A_{capture,neutron}$)/[(4)(π)($r^2$)]} {{{{{∑(j = 1, j = $N_{max}$) {{($M_{0,neutron}$) {{1 – [($v_j{}^2$)/($c^2$)]]}$^{1/2}$} ($c^2$)} – [($M_{0,neutron}$)($c^2$)]]} $^2$}/($c^2$)} – [($M_{0,neutron}{}^2$)($c^2$)] }$^{1/2}$}(Neutron,j)} (Neutron angular correction factor) (cos $\alpha_i$)dα} +{∫[(α = 0, α = [[f(π)] $_{max,photon,nuclear}$]]{{($A_{capture,photon}$)/[(4)(π)($r^2$)]} {∑(jk = 1, k = $N_{photon,nuclear}$) {($E_k$/c) = ($h\nu_k$/c) = ($h/\lambda_k$)}(Photon,nuclear,k)}} (Nuclear photon angular correction factor) (cos $\alpha_i$)dα} + {∫[(α = 0, α = [[f(π)] $_{max,photon,thermal}$]]{d{($A_{capture,photon}$) {{[5.670373 x $10^{-8}$] (J $m^{-2}$ $s^{-1}$ $K^{-4}$)}[$T_{plasma}{}^4$]/[3 x $10^8$ m/s]}}/dt}(Thermal photon angular correction factor) (cos $\alpha_i$)dα}/dt}>• d r(x,y,z)} $_{bomb,zn}$}}} $_{steady,state,multiple}$}.

The accrued gamma factor for the spacecraft while in mass ratio burn down mode is equal to:

$\gamma_{accrued}$ = {{ 1 - {{{C Tanh {[$I_{sp,effective}$/C) ln ($M_0$/M1)}}/C}$^2$}}$^{-(1/2)}$} + {{$E_{kinetic,steady,state,multiple}$}/[($M_0$)[$C^2$]]}

= {{ 1 - {{{C Tanh {[$I_{sp,effective}$/C) ln ($M_0$/M1)}}/C}$^2$}}$^{-(1/2)}$} + {{{{∑($z_1$= 1, $z_1$ = $N_{z1}$){{∫<{d $P_{combined,incident}$/dt}>{Chargon slap-back) • d r(x,y,z)}$_{bomb,z1}$}} + {∑($z_2$= 1, $z_2$ = $N_{z2}$){{∫<{d $P_{combined,incident}$/dt}>{Chargon slap-back) • d r(x,y,z)}$_{bomb,z2}$}} + {∑($z_3$= 1, $z_3$ = $N_{z3}$){{∫<{d $P_{combined,incident}$/dt}>{Chargon slap-back) • d r(x,y,z)}$_{bomb,z3}$}} + ... + {∑($z_n$= 1, $z_n$ = $N_{zn}$){{∫<{d $P_{combined,incident}$/dt}>{Chargon slap-back) • d r(x,y,z)}$_{bomb,zn}$}}} $_{steady,state,multiple}$} /[($M_0$)[$C^2$]]}

= {{ 1 - {{{C Tanh {[$I_{sp,effective}$/C) ln ($M_0$/M1)}}/C}$^2$}}$^{-(1/2)}$}

+ {{{{∑($z_1$ = 1, $z_1$ = $N_{z1}$){∫<{d{∫[(α = 0, α = [[f(π)] $_{max,chargon}$]]{(2){($A_{capture,chargon}$)/[(4)(π)($r^2$)]}} {∑(i = 1, i = N) {{{(($k_B$)($T_{ith,species,average}$)($N_{ith,species}$)$^2$)− {[($m_{0,ith,species}$)($c^2$)]$^2$}}/($c^2$)}$^{1/2}$}}}(Chargon angular correction factor) {Chargon slap-back} (cos $\alpha_i$)dα} + {∫[(α = 0, α = [[f(π)] $_{max,neutron}$]]{{($A_{capture,neutron}$)/[(4)(π)($r^2$)]}} {{{{{∑(j = 1, j = $N_{max}$){{($M_{0,neutron}$)($\gamma_j$)($c^2$)} − [($M_{0,neutron}$)($c^2$)]}}}$^2$}− {[($M_{0,neutron}$)($c^2$)]$^2$}}/($c^2$)}$^{1/2}$}(Neutron,j)} (Neutron angular correction factor) (cos $\alpha_i$)dα} + {∫[(α = 0, α = [[f(π)] $_{max,photon,nuclear}$]]{{($A_{capture,photon}$)/[(4)(π)($r^2$)]}} {∑(jk = 1, k = $N_{photon,nuclear}$) {($E_k$/c) = ($h\nu_k$/c) = ($h/\lambda_k$)}(Photon,nuclear,k)}} (Nuclear photon angular correction factor)(cos $\alpha_i$)dα} + {∫[(α = 0, α = [[f(π)] $_{max,photon,thermal}$]]{d{($A_{capture,photon}$) {{[5.670373 x $10^{-8}$] (J $m^{-2}$ $s^{-1}$ $K^{-4}$)}[$T_{plasma}^4$]/[3 x $10^8$ m/s]}}/dt}(Thermal photon angular correction factor) (cos $\alpha_i$)dα}/dt}>• d r(x,y,z)}$_{bomb,z1}$}}

+ {∑($z_2$ = 1, $z_2$ = $N_{z2}$){∫<{d{∫[(α = 0, α = [[f(π)] $_{max,chargon}$]]{(2){($A_{capture,chargon}$)/[(4)(π)($r^2$)]}} {∑(i = 1, i = N) {{{(($k_B$)($T_{ith,species,average}$)($N_{ith,species}$)$^2$)− {[($m_{0,ith,species}$)($c^2$)]$^2$}}/($c^2$)}$^{1/2}$}}}(Chargon angular correction factor) {Chargon slap-back} (cos $\alpha_i$)dα} + {∫[(α = 0, α = [[f(π)] $_{max,neutron}$]]{{($A_{capture,neutron}$)/[(4)(π)($r^2$)]}} {{{{{∑(j = 1, j = $N_{max}$){{($M_{0,neutron}$)($\gamma_j$)($c^2$)} − [($M_{0,neutron}$)($c^2$)]}}}$^2$}− {[($M_{0,neutron}$)($c^2$)]$^2$}}/($c^2$)}$^{1/2}$}(Neutron,j)} (Neutron angular correction factor) (cos $\alpha_i$)dα} + {∫[(α = 0, α = [[f(π)] $_{max,photon,nuclear}$]]{{($A_{capture,photon}$)/[(4)(π)($r^2$)]}} {∑(jk = 1, k = $N_{photon,nuclear}$) {($E_k$/c) = ($h\nu_k$/c) = ($h/\lambda_k$)}(Photon,nuclear,k)}} (Nuclear photon angular correction factor)(cos $\alpha_i$)dα} + {∫[(α = 0, α = [[f(π)] $_{max,photon,thermal}$]]{d{($A_{capture,photon}$) {{[5.670373 x $10^{-8}$] (J $m^{-2}$ $s^{-1}$ $K^{-4}$)}[$T_{plasma}^4$]/[3 x $10^8$ m/s]}}/dt}(Thermal photon angular correction factor) (cos $\alpha_i$)dα}/dt}>• d r(x,y,z)}$_{bomb,z2}$}}

+ {∑($z_3$ = 1, $z_3$ = $N_{z3}$){∫<{d{∫[(α = 0, α = [[f(π)] $_{max,chargon}$]]{(2){($A_{capture,chargon}$)/[(4)(π)($r^2$)]}} {∑(i = 1, i = N) {{{(($k_B$)($T_{ith,species,average}$)($N_{ith,species}$)$^2$)− {[($m_{0,ith,species}$)($c^2$)]$^2$}}/($c^2$)}$^{1/2}$}}}(Chargon angular correction factor) {Chargon slap-back} (cos $\alpha_i$)dα} + {∫[(α = 0, α = [[f(π)] $_{max,neutron}$]]{{($A_{capture,neutron}$)/[(4)(π)($r^2$)]}} {{{{{∑(j = 1, j = $N_{max}$){{($M_{0,neutron}$)($\gamma_j$)($c^2$)} − [($M_{0,neutron}$)($c^2$)]}}}$^2$}− {[($M_{0,neutron}$)($c^2$)]$^2$}}/($c^2$)}$^{1/2}$}(Neutron,j)} (Neutron angular correction factor) (cos $\alpha_i$)dα} + {∫[(α = 0, α = [[f(π)] $_{max,photon,nuclear}$]]{{($A_{capture,photon}$)/[(4)(π)($r^2$)]}} {∑(jk = 1, k = $N_{photon,nuclear}$) {($E_k$/c) = ($h\nu_k$/c) = ($h/\lambda_k$)}(Photon,nuclear,k)}} (Nuclear photon angular correction factor)(cos $\alpha_i$)dα} + {∫[(α = 0, α = [[f(π)] $_{max,photon,thermal}$]]{d{($A_{capture,photon}$) {{[5.670373 x $10^{-8}$] (J $m^{-2}$ $s^{-1}$ $K^{-4}$)}[$T_{plasma}^4$]/[3 x $10^8$ m/s]}}/dt}(Thermal photon angular correction factor) (cos $\alpha_i$)dα}/dt}>• d r(x,y,z)}$_{bomb,z3}$}}

+ … + {∑($z_n$ = 1, $z_n$ = $N_{zn}$){∫<{d{∫[(α = 0, α = [[f(π)] $_{max,chargon}$]]{(2){($A_{capture,chargon}$)/[(4)(π)($r^2$)]}} {∑(i = 1, i = N) {{{(($k_B$)($T_{ith,species,average}$)($N_{ith,species}$)$^2$)− {[($m_{0,ith,species}$)($c^2$)]$^2$}}/($c^2$)}$^{1/2}$}}}(Chargon angular correction factor) {Chargon slap-back} (cos $\alpha_i$)dα} + {∫[(α = 0, α = [[f(π)] $_{max,neutron}$]]{{($A_{capture,neutron}$)/[(4)(π)($r^2$)]}} {{{{{∑(j = 1, j = $N_{max}$){{($M_{0,neutron}$)($\gamma_j$)($c^2$)} − [($M_{0,neutron}$)($c^2$)]}}}$^2$}− {[($M_{0,neutron}$)($c^2$)]$^2$}}/($c^2$)}$^{1/2}$}(Neutron,j)} (Neutron angular correction factor) (cos $\alpha_i$)dα} + {∫[(α = 0, α = [[f(π)] $_{max,photon,nuclear}$]]{{($A_{capture,photon}$)/[(4)(π)($r^2$)]}} {∑(jk = 1, k = $N_{photon,nuclear}$) {($E_k$/c) = ($h\nu_k$/c) = ($h/\lambda_k$)}(Photon,nuclear,k)}} (Nuclear photon angular correction factor)(cos $\alpha_i$)dα} + {∫[(α = 0, α = [[f(π)] $_{max,photon,thermal}$]]{d{($A_{capture,photon}$) {{[5.670373 x $10^{-8}$] (J $m^{-2}$ $s^{-1}$ $K^{-4}$)}[$T_{plasma}^4$]/[3 x $10^8$ m/s]}}/dt}(Thermal photon angular correction factor) (cos $\alpha_i$)dα}/dt}>• d r(x,y,z)}$_{bomb,zn}$}} $_{steady,state,multiple}$}/[($M_0$)[$C^2$]]}

= {{ 1 - {{{C Tanh {[$I_{sp,effective}$/C) ln ($M_0$/M1)}}/C}$^2$}}$^{-(1/2)}$}

+ {{{{∑($z_1$ = 1, $z_1$ = $N_{z1}$){∫<{d{∫[(α = 0, α = [[f(π)] $_{max,chargon}$]]{ (2){($A_{capture,chargon}$)/[(4)(π)($r^2$)]}} {∑(i = 1, i = N) {∑(i = 1, i = N){[(($k_B$)($T_{ith,species,average}$)($N_{ith,species}$)$^2$)/($c^2$)] − [($m_{0,ith,species}{}^2$)($c^2$)] }$^{1/2}$}}} (Chargon angular correction factor) {Chargon slap-back} (cos $\alpha_i$)dα} + {∫[(α = 0, α = [[f(π)] $_{max,neutron}$]]{{($A_{capture,neutron}$)/[(4)(π)($r^2$)]}} {{{{{∑(j = 1, j = $N_{max}$) {{($M_{0,neutron}$) {{1 − [($v_j^2$)/($c^2$)]}}$^{1/2}$} ($c^2$)} − [($M_{0,neutron}$)($c^2$)]}}} $^2$}/($c^2$)} − [($M_{0,neutron}{}^2$)($c^2$)] }$^{1/2}$}(Neutron,j)} (Neutron angular correction factor) (cos $\alpha_i$)dα} +{∫[(α = 0, α = [[f(π)] $_{max,photon,nuclear}$]]{{($A_{capture,photon}$)/[(4)(π)($r^2$)]}} {∑(jk = 1, k = $N_{photon,nuclear}$) {($E_k$/c) = ($h\nu_k$/c) = ($h/\lambda_k$)}(Photon,nuclear,k)}} (Nuclear photon angular correction factor) (cos $\alpha_i$)dα} + {∫[(α = 0, α = [[f(π)] $_{max,photon,thermal}$]]{d{($A_{capture,photon}$) {{[5.670373 x $10^{-8}$] (J $m^{-2}$ $s^{-1}$ $K^{-4}$)}[$T_{plasma}^4$]/[3 x $10^8$ m/s]}}/dt}(Thermal photon angular correction factor) (cos $\alpha_i$)dα}/dt}>• d r(x,y,z)} $_{bomb,z1}$}}

+ {∑($z_2$ = 1, $z_2$ = $N_{z2}$){∫<{d{∫[(α = 0, α = [[f(π)] $_{max,chargon}$]]{ (2){($A_{capture,chargon}$)/[(4)(π)($r^2$)]}} {∑(i = 1, i = N) {∑(i = 1, i = N){[(($k_B$)($T_{ith,species,average}$)($N_{ith,species}$)$^2$)/($c^2$)] − [($m_{0,ith,species}{}^2$)($c^2$)] }$^{1/2}$}}} (Chargon angular correction factor) {Chargon slap-back} (cos $\alpha_i$)dα} + {∫[(α = 0, α = [[f(π)] $_{max,neutron}$]]{{($A_{capture,neutron}$)/[(4)(π)($r^2$)]}} {{{{{∑(j = 1, j = $N_{max}$) {{($M_{0,neutron}$) {{1 − [($v_j^2$)/($c^2$)]}}$^{1/2}$} ($c^2$)} − [($M_{0,neutron}$)($c^2$)]}}} $^2$}/($c^2$)} − [($M_{0,neutron}{}^2$)($c^2$)] }$^{1/2}$}(Neutron,j)} (Neutron angular correction factor) (cos $\alpha_i$)dα} +{∫[(α = 0, α = [[f(π)] $_{max,photon,nuclear}$]]{{($A_{capture,photon}$)/[(4)(π)($r^2$)]}} {∑(jk = 1, k = $N_{photon,nuclear}$) {($E_k$/c) = ($h\nu_k$/c) = ($h/\lambda_k$)}(Photon,nuclear,k)}} (Nuclear photon angular correction factor) (cos $\alpha_i$)dα} + {∫[(α = 0, α = [[f(π)] $_{max,photon,thermal}$]]{d{($A_{capture,photon}$) {{[5.670373 x $10^{-8}$] (J $m^{-2}$ $s^{-1}$ $K^{-4}$)}[$T_{plasma}^4$]/[3 x $10^8$ m/s]}}/dt}(Thermal photon angular correction factor) (cos $\alpha_i$)dα}/dt}>• d r(x,y,z)} $_{bomb,z2}$}}

+ {∑($z_3$ = 1, $z_3$ = $N_{z3}$){∫<{d{∫[(α = 0, α = [[f(π)] $_{max,chargon}$]]{ (2){($A_{capture,chargon}$)/[(4)(π)($r^2$)]}} {∑(i = 1, i = N) {∑(i = 1, i = N){[(($k_B$)($T_{ith,species,average}$)($N_{ith,species}$)$^2$)/($c^2$)] − [($m_{0,ith,species}{}^2$)($c^2$)] }$^{1/2}$}}} (Chargon angular correction factor) {Chargon slap-back} (cos $\alpha_i$)dα} + {∫[(α = 0, α = [[f(π)] $_{max,neutron}$]]{{($A_{capture,neutron}$)/[(4)(π)($r^2$)]}} {{{{{∑(j = 1, j = $N_{max}$) {{($M_{0,neutron}$) {{1 − [($v_j^2$)/($c^2$)]}}$^{1/2}$} ($c^2$)} − [($M_{0,neutron}$)($c^2$)]}}} $^2$}/($c^2$)} − [($M_{0,neutron}{}^2$)($c^2$)] }$^{1/2}$}(Neutron,j)} (Neutron angular correction factor) (cos $\alpha_i$)dα} +{∫[(α = 0, α = [[f(π)] $_{max,photon,nuclear}$]]{{($A_{capture,photon}$)/[(4)(π)($r^2$)]}} {∑(jk = 1, k = $N_{photon,nuclear}$) {($E_k$/c) = ($h\nu_k$/c) = ($h/\lambda_k$)}(Photon,nuclear,k)}} (Nuclear photon angular correction factor) (cos $\alpha_i$)dα} + {∫[(α = 0, α = [[f(π)] $_{max,photon,thermal}$]]{d{($A_{capture,photon}$) {{[5.670373 x $10^{-8}$] (J $m^{-2}$ $s^{-1}$ $K^{-4}$)}[$T_{plasma}^4$]/[3 x $10^8$ m/s]}}/dt}(Thermal photon angular correction factor) (cos $\alpha_i$)dα}/dt}>• d r(x,y,z)} $_{bomb,z3}$}}

+ ... + {∑($z_n$= 1, $z_n$ = $N_{zn}$)}{∫<{d{∫[(α = 0, α = [[f(π)] $_{max,chargon}$]]{ (2){($A_{capture,chargon}$)/[(4)(π)($r^2$)]} {∑(i = 1, i = N) {∑(i = 1, i = N){[(($k_B$)($T_{ith,species,average}$)($N_{ith,species}$)$^2$)/($c^2$)] – [($m_{0,ith,species}{}^2$)($c^2$)] }$^{1/2}$}}} (Chargon angular correction factor) {Chargon slap-back) (cos $\alpha_i$)dα} + {∫[(α = 0, α = [[f(π)] $_{max,neutron}$]]{{($A_{capture,neutron}$)/[(4)(π)($r^2$)]} {{{{{∑(j = 1, j = $N_{max}$) {{($M_{0,neutron}$) {{1 – [($v_j{}^2$)/($c^2$)]}$^{1/2}$} ($c^2$)} – [($M_{0,neutron}$)($c^2$)]}} $^2$}/($c^2$)} – [($M_{0,neutron}{}^2$)($c^2$)] }$^{1/2}$(Neutron,j)} (Neutron angular correction factor) (cos $\alpha_i$)dα} +{∫[(α = 0, α = [[f(π)] $_{max,photon,nuclear}$]]{{($A_{capture,photon}$)/[(4)(π)($r^2$)]} {∑(jk = 1, k = $N_{photon,nuclear}$) {($E_k$/c) = ($h\nu_k$/c) = ($h/\lambda_k$)}(Photon,nuclear,k)}} (Nuclear photon angular correction factor) (cos $\alpha_i$)dα} + {∫[(α = 0, α = [[f(π)] $_{max,photon,thermal}$]]{d{($A_{capture,photon}$) {{[5.670373 x $10^{-8}$] (J $m^{-2}$ $s^{-1}$ $K^{-4}$)}[$T_{plasma}{}^4$]/[3 x $10^8$ m/s]}}/dt}(Thermal photon angular correction factor) (cos $\alpha_i$)dα}/dt}>• d r(x,y,z)} $_{bomb,zn}$}}} $_{steady,state,multiple}$}/[($M_0$)[$C^2$]]}.

Again as mentioned in the previous chapter, chargon slap-back can occur in a manner that produces reverberations and thus imply a slapping sloshing effect. We will use the term {Chargon slap-back reverberating sloshing effect) to denote this feedback mechanism. Additionally, some of the incident neutrons may be back-reflected out of the pusher plate as may be some of the impinging x-rays and gamma rays produced by the bomb blast. Both these effects can cause energy bleeding for the respective impinging neutrons and photons. We will denote these two effects by the factors, (Neutron back-reflect with possible neutron energy bleeding), (Nuclear photon back-reflect with possible neutron energy bleeding), and (Thermal photon back-reflect with possible neutron energy bleeding). The photon term is assumed to be context specific with nuclear reaction generated photons and the thermal photons.

The resulting formula for accrued gamma factor is thus the final treatment we will elaborate on in this chapter and is as follows.

$\gamma_{accrued}$ = {{ 1 - {{{C Tanh {[$I_{sp,effective}$/C) ln ($M_0$/M1)}}/C}$^2$}}$^{-(1/2)}$}

+ {{{∑($z_1$ = 1, $z_1$ = $N_{z1}$)}{∫<{d{∫[(α = 0, α = [[f(π)] $_{max,chargon}$]]{(2){($A_{capture,chargon}$)/[(4)(π)($r^2$)]} {∑(i = 1, i = N) {{{{($k_B$)($T_{ith,species,average}$)($N_{ith,species}$)$^2$)– {[($m_{0,ith,species}$)($c^2$)]$^2$}}/($c^2$)}$^{1/2}$}}}(Chargon angular correction factor) {Chargon slap-back reverberating sloshing effect) (cos $\alpha_i$)dα} + {∫[(α = 0, α = [[f(π)] $_{max,neutron}$]]{{($A_{capture,neutron}$)/[(4)(π)($r^2$)]} {{{{{∑(j = 1, j = $N_{max}$){{($M_{0,neutron}$)($\gamma_j$)($c^2$)} – [($M_{0,neutron}$)($c^2$)]}} $^2$}– {[($M_{0,neutron}$)($c^2$)]$^2$}}/($c^2$)}$^{1/2}$(Neutron,j)} (Neutron angular correction factor) (Neutron back-reflect with possible neutron energy bleeding)(cos $\alpha_i$)dα} + {∫[(α = 0, α = [[f(π)] $_{max,photon,nuclear}$]]{{($A_{capture,photon}$)/[(4)(π)($r^2$)]} {∑(jk = 1, k = $N_{photon,nuclear}$) {($E_k$/c) = ($h\nu_k$/c) = ($h/\lambda_k$)}(Photon,nuclear,k)}} (Nuclear photon angular correction factor) (Nuclear photon back-reflect with possible neutron energy bleeding) (cos $\alpha_i$)dα} + {∫[(α = 0, α = [[f(π)] $_{max,photon,thermal}$]]{d{($A_{capture,photon}$) {{[5.670373 x $10^{-8}$] (J $m^{-2}$ $s^{-1}$ $K^{-4}$)}[$T_{plasma}{}^4$]/[3 x $10^8$ m/s]}}/dt}(Thermal photon angular correction factor) (Thermal photon back-reflect with possible neutron energy bleeding) (cos $\alpha_i$)dα}/dt}>• d r(x,y,z)}$_{bomb,z1}$}}

+ {∑($z_2$ = 1, $z_2$ = $N_{z2}$)}{∫<{d{∫[(α = 0, α = [[f(π)] $_{max,chargon}$]]{(2){($A_{capture,chargon}$)/[(4)(π)($r^2$)]} {∑(i = 1, i = N) {{{{($k_B$)($T_{ith,species,average}$)($N_{ith,species}$)$^2$)– {[($m_{0,ith,species}$)($c^2$)]$^2$}}/($c^2$)}$^{1/2}$}}}(Chargon angular correction factor) {Chargon slap-back reverberating sloshing effect) (cos $\alpha_i$)dα} + {∫[(α = 0, α = [[f(π)] $_{max,neutron}$]]{{($A_{capture,neutron}$)/[(4)(π)($r^2$)]} {{{{{∑(j = 1, j = $N_{max}$){{($M_{0,neutron}$)($\gamma_j$)($c^2$)} – [($M_{0,neutron}$)($c^2$)]}} $^2$}– {[($M_{0,neutron}$)($c^2$)]$^2$}}/($c^2$)}$^{1/2}$(Neutron,j)} (Neutron angular correction factor) (Neutron back-reflect with possible neutron energy bleeding) (cos $\alpha_i$)dα} + {∫[(α = 0, α = [[f(π)] $_{max,photon,nuclear}$]]{{($A_{capture,photon}$)/[(4)(π)($r^2$)]} {∑(jk = 1, k = $N_{photon,nuclear}$) {($E_k$/c) = ($h\nu_k$/c) = ($h/\lambda_k$)}(Photon,nuclear,k)}} (Nuclear photon angular correction factor) (Nuclear photon back-reflect with possible neutron energy bleeding) (cos $\alpha_i$)dα} + {∫[(α = 0, α = [[f(π)] $_{max,photon,thermal}$]]{d{($A_{capture,photon}$) {{[5.670373 x $10^{-8}$] (J $m^{-2}$ $s^{-1}$ $K^{-4}$)}[$T_{plasma}{}^4$]/[3 x $10^8$ m/s]}}/dt}(Thermal photon angular correction factor) (Thermal photon back-reflect with possible neutron energy bleeding) (cos $\alpha_i$)dα}/dt}>• d r(x,y,z)}$_{bomb,z2}$}}

+ {∑($z_3$ = 1, $z_3$ = $N_{z3}$)}{∫<{d{∫[(α = 0, α = [[f(π)] $_{max,chargon}$]]{(2){($A_{capture,chargon}$)/[(4)(π)($r^2$)]} {∑(i = 1, i = N) {{{{($k_B$)($T_{ith,species,average}$)($N_{ith,species}$)$^2$)– {[($m_{0,ith,species}$)($c^2$)]$^2$}}/($c^2$)}$^{1/2}$}}}(Chargon angular correction factor) {Chargon slap-back reverberating sloshing effect) (cos $\alpha_i$)dα} + {∫[(α = 0, α = [[f(π)] $_{max,neutron}$]]{{($A_{capture,neutron}$)/[(4)(π)($r^2$)]} {{{{{∑(j = 1, j = $N_{max}$){{($M_{0,neutron}$)($\gamma_j$)($c^2$)} – [($M_{0,neutron}$)($c^2$)]}} $^2$}– {[($M_{0,neutron}$)($c^2$)]$^2$}}/($c^2$)}$^{1/2}$(Neutron,j)} (Neutron angular correction factor) (Neutron back-reflect with possible neutron energy bleeding) (cos $\alpha_i$)dα} + {∫[(α = 0, α = [[f(π)] $_{max,photon,nuclear}$]]{{($A_{capture,photon}$)/[(4)(π)($r^2$)]} {∑(jk = 1, k = $N_{photon,nuclear}$) {($E_k$/c) = ($h\nu_k$/c) = ($h/\lambda_k$)}(Photon,nuclear,k)}} (Nuclear photon angular correction factor) (Nuclear photon back-reflect with possible neutron energy bleeding) (cos $\alpha_i$)dα} + {∫[(α = 0, α = [[f(π)] $_{max,photon,thermal}$]]{d{($A_{capture,photon}$) {{[5.670373 x $10^{-8}$] (J $m^{-2}$ $s^{-1}$ $K^{-4}$)}[$T_{plasma}{}^4$]/[3 x $10^8$ m/s]}}/dt}(Thermal photon angular correction factor) (Thermal photon back-reflect with possible neutron energy bleeding) (cos $\alpha_i$)dα}/dt}>• d r(x,y,z)}$_{bomb,z3}$}}

+ ... + {∑($z_n$ = 1, $z_n$ = $N_{zn}$)}{∫<{d{∫[(α = 0, α = [[f(π)] $_{max,chargon}$]]{(2){($A_{capture,chargon}$)/[(4)(π)($r^2$)]} {∑(i = 1, i = N) {{{{($k_B$)($T_{ith,species,average}$)($N_{ith,species}$)$^2$)– {[($m_{0,ith,species}$)($c^2$)]$^2$}}/($c^2$)}$^{1/2}$}}}(Chargon angular correction factor) {Chargon slap-back reverberating sloshing effect) (cos $\alpha_i$)dα} + {∫[(α = 0, α = [[f(π)] $_{max,neutron}$]]{{($A_{capture,neutron}$)/[(4)(π)($r^2$)]} {{{{{∑(j = 1, j = $N_{max}$){{($M_{0,neutron}$)($\gamma_j$)($c^2$)} – [($M_{0,neutron}$)($c^2$)]}} $^2$}– {[($M_{0,neutron}$)($c^2$)]$^2$}}/($c^2$)}$^{1/2}$(Neutron,j)} (Neutron angular correction factor) (Neutron back-reflect with possible neutron energy bleeding) (cos $\alpha_i$)dα} + {∫[(α = 0, α = [[f(π)]

$_{max,photon,nuclear}$]]{{($A_{capture,photon}$)/[(4)(π)($r^2$)]} {∑(jk = 1, k = $N_{photon,nuclear}$) {($E_k$/c) = ($hv_k$/c) = (h/$λ_k$)}(Photon,nuclear,k)}} (Nuclear photon angular correction factor) (Nuclear photon back-reflect with possible neutron energy bleeding) (cos $α_i$)dα} + {∫[(α = 0, α = [[f(π)] $_{max,photon,thermal}$]]{d{($A_{capture,photon}$) {{[5.670373 x $10^{-8}$] (J $m^{-2}$ $s^{-1}$ $K^{-4}$)}[$T_{plasma}{}^4$]/[3 x $10^8$ m/s]}}/dt}(Thermal photon angular correction factor) (Thermal photon back-reflect with possible neutron energy bleeding) (cos $α_i$)dα}/dt}>• d r(x,y,z)}$_{bomb,zn}$}}}/[($M_0$)[$C^2$]]}

= {{{∑($z_1$ = 1, $z_1$ = $N_{z1}$)}{∫<{d{∫[(α = 0, α = [[f(π)] $_{max,chargon}$]]{ (2){($A_{capture,chargon}$)/[(4)(π)($r^2$)]} {∑(i = 1, i = N) {∑(i = 1, i = N){[(($k_B$)($T_{ith,species,average}$)($N_{ith,species}$)$^2$)/($c^2$)] – [($m_{0,ith,species}$ $^2$)($c^2$)] }$^{1/2}$}}} (Chargon angular correction factor) {Chargon slap-back reverberating sloshing effect) (cos $α_i$)dα} + {∫[(α = 0, α = [[f(π)] $_{max,neutron}$]]{{($A_{capture,neutron}$)/[(4)(π)($r^2$)]} {{{{{∑(j = 1, j = $N_{max}$) {{($M_{0,neutron}$) {{1 – [($v_j^2$)/($c^2$)]}}$^{1/2}$} ($c^2$)} – [($M_{0,neutron}$)($c^2$)]}} $^2$}/($c^2$)} – [($M_{0,neutron}$ $^2$)($c^2$)] }$^{1/2}$}(Neutron,j)} (Neutron angular correction factor) (Neutron back-reflect with possible neutron energy bleeding) (cos $α_i$)dα} +{∫[(α = 0, α = [[f(π)] $_{max,photon,nuclear}$]]{{($A_{capture,photon}$)/[(4)(π)($r^2$)]} {∑(jk = 1, k = $N_{photon,nuclear}$) {($E_k$/c) = ($hv_k$/c) = (h/$λ_k$)}(Photon,nuclear,k)}} (Nuclear photon angular correction factor) (Nuclear photon back-reflect with possible neutron energy bleeding) (cos $α_i$)dα} + {∫[(α = 0, α = [[f(π)] $_{max,photon,thermal}$]]{d{($A_{capture,photon}$) {{[5.670373 x $10^{-8}$] (J $m^{-2}$ $s^{-1}$ $K^{-4}$)}[$T_{plasma}{}^4$]/[3 x $10^8$ m/s]}}/dt}(Thermal photon angular correction factor) (Thermal photon back-reflect with possible neutron energy bleeding) (cos $α_i$)dα}/dt}>• d r(x,y,z)}$_{bomb,z1}$}}

+ {∑($z_2$ = 1, $z_2$ = $N_{z2}$)}{∫<{d{∫[(α = 0, α = [[f(π)] $_{max,chargon}$]]{ (2){($A_{capture,chargon}$)/[(4)(π)($r^2$)]} {∑(i = 1, i = N) {∑(i = 1, i = N){[(($k_B$)($T_{ith,species,average}$)($N_{ith,species}$)$^2$)/($c^2$)] – [($m_{0,ith,species}$ $^2$)($c^2$)] }$^{1/2}$}}} (Chargon angular correction factor) {Chargon slap-back reverberating sloshing effect) (cos $α_i$)dα} + {∫[(α = 0, α = [[f(π)] $_{max,neutron}$]]{{($A_{capture,neutron}$)/[(4)(π)($r^2$)]} {{{{{∑(j = 1, j = $N_{max}$) {{($M_{0,neutron}$) {{1 – [($v_j^2$)/($c^2$)]}}$^{1/2}$} ($c^2$)} – [($M_{0,neutron}$)($c^2$)]}} $^2$}/($c^2$)} – [($M_{0,neutron}$ $^2$)($c^2$)] }$^{1/2}$}(Neutron,j)} (Neutron angular correction factor) (Neutron back-reflect with possible neutron energy bleeding) (cos $α_i$)dα} +{∫[(α = 0, α = [[f(π)] $_{max,photon,nuclear}$]]{{($A_{capture,photon}$)/[(4)(π)($r^2$)]} {∑(jk = 1, k = $N_{photon,nuclear}$) {($E_k$/c) = ($hv_k$/c) = (h/$λ_k$)}(Photon,nuclear,k)}} (Nuclear photon angular correction factor) (Nuclear photon back-reflect with possible neutron energy bleeding) (cos $α_i$)dα} + {∫[(α = 0, α = [[f(π)] $_{max,photon,thermal}$]]{d{($A_{capture,photon}$) {{[5.670373 x $10^{-8}$] (J $m^{-2}$ $s^{-1}$ $K^{-4}$)}[$T_{plasma}{}^4$]/[3 x $10^8$ m/s]}}/dt}(Thermal photon angular correction factor) (Thermal photon back-reflect with possible neutron energy bleeding) (cos $α_i$)dα}/dt}>• d r(x,y,z)}$_{bomb,z2}$}}

+ {∑($z_3$ = 1, $z_3$ = $N_{z3}$)}{∫<{d{∫[(α = 0, α = [[f(π)] $_{max,chargon}$]]{ (2){($A_{capture,chargon}$)/[(4)(π)($r^2$)]} {∑(i = 1, i = N) {∑(i = 1, i = N){[(($k_B$)($T_{ith,species,average}$)($N_{ith,species}$)$^2$)/($c^2$)] – [($m_{0,ith,species}$ $^2$)($c^2$)] }$^{1/2}$}}} (Chargon angular correction factor) {Chargon slap-back reverberating sloshing effect) (cos $α_i$)dα} + {∫[(α = 0, α = [[f(π)] $_{max,neutron}$]]{{($A_{capture,neutron}$)/[(4)(π)($r^2$)]} {{{{{∑(j = 1, j = $N_{max}$) {{($M_{0,neutron}$) {{1 – [($v_j^2$)/($c^2$)]}}$^{1/2}$} ($c^2$)} – [($M_{0,neutron}$)($c^2$)]}} $^2$}/($c^2$)} – [($M_{0,neutron}$ $^2$)($c^2$)] }$^{1/2}$}(Neutron,j)} (Neutron angular correction factor) (Neutron back-reflect with possible neutron energy bleeding) (cos $α_i$)dα} +{∫[(α = 0, α = [[f(π)] $_{max,photon,nuclear}$]]{{($A_{capture,photon}$)/[(4)(π)($r^2$)]} {∑(jk = 1, k = $N_{photon,nuclear}$) {($E_k$/c) = ($hv_k$/c) = (h/$λ_k$)}(Photon,nuclear,k)}} (Nuclear photon angular correction factor) (Nuclear photon back-reflect with possible neutron energy bleeding) (cos $α_i$)dα} + {∫[(α = 0, α = [[f(π)] $_{max,photon,thermal}$]]{d{($A_{capture,photon}$) {{[5.670373 x $10^{-8}$] (J $m^{-2}$ $s^{-1}$ $K^{-4}$)}[$T_{plasma}{}^4$]/[3 x $10^8$ m/s]}}/dt}(Thermal photon angular correction factor) (Thermal photon back-reflect with possible neutron energy bleeding) (cos $α_i$)dα}/dt}>• d r(x,y,z)}$_{bomb,z3}$}}

+ … + {∑($z_n$= 1, $z_n$ = $N_{zn}$)}{∫<{d{∫[(α = 0, α = [[f(π)] $_{max,chargon}$]]{ (2){($A_{capture,chargon}$)/[(4)(π)($r^2$)]} {∑(i = 1, i = N) {∑(i = 1, i = N){[(($k_B$)($T_{ith,species,average}$)($N_{ith,species}$)$^2$)/($c^2$)] – [($m_{0,ith,species}$ $^2$)($c^2$)] }$^{1/2}$}}} (Chargon angular correction factor) {Chargon slap-back reverberating sloshing effect) (cos $α_i$)dα} + {∫[(α = 0, α = [[f(π)] $_{max,neutron}$]]{{($A_{capture,neutron}$)/[(4)(π)($r^2$)]} {{{{{∑(j = 1, j = $N_{max}$) {{($M_{0,neutron}$) {{1 – [($v_j^2$)/($c^2$)]}}$^{1/2}$} ($c^2$)} – [($M_{0,neutron}$)($c^2$)]}} $^2$}/($c^2$)} – [($M_{0,neutron}$ $^2$)($c^2$)] }$^{1/2}$}(Neutron,j)} (Neutron angular correction factor) (Neutron back-reflect with possible neutron energy bleeding) (cos $α_i$)dα} +{∫[(α = 0, α = [[f(π)] $_{max,photon,nuclear}$]]{{($A_{capture,photon}$)/[(4)(π)($r^2$)]} {∑(jk = 1, k = $N_{photon,nuclear}$) {($E_k$/c) = ($hv_k$/c) = (h/$λ_k$)}(Photon,nuclear,k)}} (Nuclear photon angular correction factor) (Nuclear photon back-reflect with possible neutron energy bleeding) (cos $α_i$)dα} + {∫[(α = 0, α = [[f(π)] $_{max,photon,thermal}$]]{d{($A_{capture,photon}$) {{[5.670373 x $10^{-8}$] (J $m^{-2}$ $s^{-1}$ $K^{-4}$)}[$T_{plasma}{}^4$]/[3 x $10^8$ m/s]}}/dt}(Thermal photon angular correction factor) (Thermal photon back-reflect with possible neutron energy bleeding) (cos $α_i$)dα}/dt}>• d r(x,y,z)} $_{bomb,zn}$}}}/[($M_0$)[$C^2$]]}.

# Gravitational Assists To Virtually Light Speed.

An interesting spacecraft propulsion method would involve spacecraft taking advantage of interplanetary, interstellar, intra-galactic, and inter-galactic gravitational potential rivers.

Accordingly, these rivers would involve travel paths of the greatest potentials to convert gravitational potential energy into spacecraft propulsion energy by way of gravitational assists and slingshot maneuvers.

Note that the author is not so much referring to O'berth maneuvers and relativistic analogues which would require onboard rocket fuel or heated reaction mass supplies as I am considering free energy extractable from multibody systems.

Multiple star systems, planetary and moon systems, multiple star cluster systems, multiple orbiting black holes, multiple orbiting galaxy systems, multiple orbiting galaxy clusters, and multiple orbiting galaxy super-clusters could in theory be used for the sling-shot maneuvers.

Essentially, we are considering effectively free energy here since ideally a spacecraft as such would not need to carry any fuel provided the spacecraft chose only multi-body systems for which the one or two bodies operated on were traveling faster than the overall gravitational center of mass of the system during the phase of the spacecraft approach of the faster bodies.

It is possible that a single galaxy might be able to bring a spacecraft of initial Keplerian velocities up to extremely relativistic velocities before the spacecraft would be flung out into intergalactic space.

The transit methods may take advantage of general relativistic reference frame dragging by close passage of ultra-massive black holes spinning at near the velocity of light.

A fascinating prospect would include using artificially spun-up black holes to precisely the speed of light. This likely would be very difficult to do. However, it would be a fascinating prospect if a black hole spinning at a time averaged velocity of precisely light-speed would have an occasional brief period of superluminal rotation velocity based on the Heisenberg Uncertainty Principle and/or where the black hole spin wave-function would periodically experience quantum mechanical tunneling to slightly superluminal velocity states. A spacecraft taking advantage of the associated reference frame dragging would need to travel dangerously close to the point of intake into the black hole and it would be necessary for the spacecraft to be traveling at suitably extreme Lorentz factors so that it would not fall through the event horizon of the black hole.

Another take on the gravitational assist concept would involve spacecraft coupling to occasional random densification of the empty background medium of interstellar, intra-galactic, and inter-galactic space.

Accordingly, a spacecraft might obtain a gravitational assist of ever so slightly finite extent based on random quantum-gravitational space-time compression or densification. These pockets of densification might result from random statistical mechanical thermodynamic processes for which the considered portion of background space-time would experience random energy density increases or temporary contraction.

The latter consideration may even work for cosmically remote future travel as the universe experiences a great cool-down due to expansion but where based on quantum mechanical effects, the temperature of the space-time attains a lower finite limit and thus a maximally reduced finite energy density.

Another option would include travel in hyperspaces for which hyperspatial gravitational assists would be enabled via using multiple hyperspatial orbiting island universes that may pepper one or more hyperspaces infinite extent. Analogues involving multiverses, forests, biospheres, and the like that may exist as island-like realms in greater hyperspaces might also be used for gravitational assists.

Regarding methods of enhancing gravitational assists, it is also plausible that light-sail techniques can be applied around single body luminaries but with the non-limiting option of using multi-body systems for gravitational sling-shot maneuvers.

Accordingly, a spacecraft would have a reflector sail that would be activated near the presence of a star or quasar to thus serve the role of a rocket style spacecraft with the delta velocity value of the light-sail operation being substituted in full or in part for rocket thrusting periods. For a sufficiently low area mass density light-sail, the latter mechanisms may even work with entire galaxies, or dense galaxy clusters. Thus, the galactic or super-galactic starlight and quasar light emissions may be suitable for such light-sail techniques.

Light-sails may in full or partially be replaced by ultra-efficient photo-electric propulsion mechanisms operated around stars and/or quasars, and even warm brown dwarfs for which he electrical power would then energize electro-hydrodynamic-plasma-drives, magneto-hydrodynamic-plasma-drives, electro-magneto-hydrodynamic-plasma-drives, electromagnetic-hydrodynamic-plasma-drives, magnetic field effect propulsion systems, and/or magnetic plasma bottle sails and then like. Analogues using thermo-electric conversion, or turbo-electric power generation systems are also plausible with the required heat being sources from sunlight with or without the use of sun-light concentrating optics. Sun-light concentrators may as non-limiting options include; inflatable reflectors, Fresnel lenses made of low area density membranes, and the like.

Regarding near-term use of the sun or planets in our solar system for gravitational assists and/or O'berth maneuvers, a distributed series of fuel pellets in appropriate locations with respect to these bodies may enable similar gravitational potential energy enhancements to spacecraft v velocities. Thus, the yields of the fuel pellets or pellet runways can be strongly leveraged. These pellet runway options can be employed along with the light-sail options as non-limiting examples.

Then of course, concentrated sun-light can be beamed to spacecraft to enhance the sling-shot mechanisms or O'berth maneuvers.

The radiation based enhancements considered in the previous five paragraphs may also be employed around other stars and perhaps even quasars. In cases for which interstellar and intergalactic infrastructure is eventually established, pellet stream runways can be deployed around these additional luminaries.

Regarding efficient use of cyclical motions of single galaxies or groups of galaxies, note that in theory, a spacecraft could undergo several to many revolutions around a single galaxy or galaxy cluster at ultra-relativistic velocities provided that the spacecraft would travel sufficiently close to numerous stars to obtain arcuate changes in its heading in order to make use of a single galaxy or galaxy cluster for increases in spacecraft Lorentz factors. Sufficiently safe approaches to black holes ranging from stellar mass black holes, to ultra-massive black holes would be especially helpful in maintaining a spacecraft presence within a single galaxy or galaxy cluster.

The relativistic Lorentz turning force can also be of assistance in enabling extremely relativistic spacecraft to undergo many revolutions around a galaxy or galaxy cluster. This would require the spacecraft to be suitably electrically charged or perhaps having a highly charged tow line extending out from the spacecraft. A highly charged tow-line can maximize the required electrical charge without risking electrostatic stress based failure of the charged components.

Note that the above gravitational assists and O'berth maneuvers may also be accomplished using carried aboard rocket fuels. Additionally, carried aboard fuels may also be used to energize spacecraft electrical propulsion systems. Such systems when upsweeping background magnetic energy, ionic potential energy, and the like may in principle take in more energy than the energy of onboard fuel used per differential time unit. As such, the specific impulse of the rocket fuels can grow to many multiples of light-speed without the rocket exhaust traveling faster than light.

## Some whimsical closing anecdotes.

Acceleration of a stellar white dwarf mass starship to large Lorentz factors, say between 5 and 10 billion for which segments of the ship drop off and collide with interstellar and intergalactic debris for which the collision and reaction energy is collected and converted into spacecraft kinetic energy while the spacecraft's invariant mass reduces commensurately with the drop-away segments being off-loaded.

The collision and reaction energy might conceivably be teleported into the spacecraft kinetic energy, and/or collected in a collection or reaction chamber, and/or leveraged to capture background magnetic field energy, and/or be used to collect nuclear fusion fuel to power interstellar ramjet engines.

For the scenarios where the stellar white dwarf mass starship would have a starting total invariant mass of 10 EXP 27 metric tons while the final intended payload with have mass of 10 EXP 5 tons, and the ship's initial Lorentz factor was boosted to 1010, it is conceivable that the Lorentz factor of the spacecraft could approach 10 EXP (10 + 22) = 10 EXP 32.

Consider cases for which the otherwise above considered spacecraft would leverage spacecraft kinetic energy for reactive intake of background magnetic field energy in an instantaneous power intake of nine times the kinetic energy conveyance into the spacecraft due to the collision mechanisms. The spacecraft terminal Lorentz factor would be as great as 10 EXP [10 + [(22)(10)]] = 10 EXP 230.

Consider cases for which the otherwise above considered spacecraft would leverage spacecraft kinetic energy for reactive intake of background magnetic field energy in an instantaneous power intake of 99 times the kinetic energy conveyance into the spacecraft due to the collision mechanisms. The spacecraft terminal Lorentz factor would be as great as 10 EXP [10 + [(22)(100)]] = 10 EXP 2,210.

Consider cases for which the otherwise above considered spacecraft would leverage spacecraft kinetic energy for reactive intake of background magnetic field energy in an instantaneous power intake of 999 times the kinetic energy conveyance into the spacecraft due to the collision mechanisms. The spacecraft terminal Lorentz factor would be as great as 10 EXP [10 + [(22)(1,000)]] = 10 EXP 22,010.

Consider cases for which the otherwise above considered spacecraft would leverage spacecraft kinetic energy for reactive intake of background magnetic field energy in an instantaneous power intake of 9,999 times the kinetic energy conveyance into the spacecraft due to the collision mechanisms. The spacecraft terminal Lorentz factor would be as great as 10 EXP [10 + [(22)(1,000)]] = 10 EXP 220,010.

Consider cases for which the otherwise above considered spacecraft would leverage spacecraft kinetic energy for reactive intake of background magnetic field energy in an instantaneous power intake of 99,999 times the kinetic energy conveyance into the spacecraft due to the collision mechanisms. The spacecraft terminal Lorentz factor would be as great as 10 EXP [10 + [(22)(1,000)]] = 10 EXP 2,200,010.

By now the reader can determine a pattern.

We also consider linear magnetic induction methods by which a spacecraft would produce propulsive power at ten times the Planck Power but which would bring the power down to just below the Planck Power before a black hole state could manifest. The power cut-off would need to occur before a beam of light could travel a fraction of 0.1 of half the radius of the linear induction coil.

Once the propulsive power was imprinted on the spacecraft kinetic energy wave function, the entire process could repeat any arbitrary number of times.

We may also consider linear magnetic induction methods by which a spacecraft would produce propulsive power at 100 times the Planck Power but which would bring the power down to just below the Planck Power before a black hole state could manifest. The power cut-off would need to occur before a beam of light could travel a fraction of 0.01 of half the radius of the linear induction coil.

Once the propulsive power was imprinted on the spacecraft kinetic energy wave function, the entire process could repeat any arbitrary number of times.

We may further consider linear magnetic induction methods by which a spacecraft would produce propulsive power at 1,000 times the Planck Power but which would bring the power down to just below the Planck Power before a black hole state could manifest. The power cut-off would need to occur before a beam of light could travel a fraction of 0.001 of half the radius of the linear induction coil.

Once the propulsive power was imprinted on the spacecraft kinetic energy wave function, the entire process could repeat any arbitrary number of times.

We also consider linear magnetic induction methods by which a spacecraft would produce propulsive power at ten times the Planck Power but which would bring the power down to just below the Planck Power before a black hole state could manifest. The power cut-off would need to occur before a beam of light could travel a fraction of half the radius of the linear induction coil under the condition that the generated power would be teleportively imprinted on the wave function of the spacecraft.

Once the propulsive power was imprinted on the spacecraft kinetic energy wave function, the entire process could repeat any arbitrary number of times.

We may also consider linear magnetic induction methods by which a spacecraft would produce propulsive power at 100 times the Planck Power but which would bring the power down to just below the Planck Power before a black hole state could manifest. The power cut-off would need to occur before a beam of light could travel a fraction half the radius of the linear induction coil under the condition that the generated power would be teleportively imprinted on the wave function of the spacecraft.

Once the propulsive power was imprinted on the spacecraft kinetic energy wave function, the entire process could repeat any arbitrary number of times.

We may further consider linear magnetic induction methods by which a spacecraft would produce propulsive power at 1,000 times the Planck Power but which would bring the power down to just below the Planck Power before a black hole state could manifest. The power cut-off would need to occur before a beam of light could travel a fraction of half the radius of the linear induction coil under the condition that the generated power would be teleportively imprinted on the wave function of the spacecraft.

Once the propulsive power was imprinted on the spacecraft kinetic energy wave function, the entire process could repeat any arbitrary number of times.

The process of teleportation may be classical or general relativistic of quantum mechanical via quantum entanglement processes.

Perhaps any of the spacecraft considered herein could be operated in any hyperspaces. However, this is beyond the scope of this book suffice it to state that the author has provided a suggested reading list including many of his books which delve into hyperspatial travel in a mathematical manner.

Another fascinating take on guns and mass drivers includes the concept of reducing the square-root of the product of the electrical permittivity and magnetic permeability in free space within a highly conducting or perhaps superconducting barrel vacancy. Accordingly, the speed of light within the cavity would be increased relative to that of an ordinary vacuum. Thus, in principle, pellets having ultra-high Lorentz factors within a barrel relative to the barrel wall might achieve effectively very slightly superluminal velocities. This reality has ramifications for guns spanning one or more cosmic light-cone radii. Essentially, we are considering barrels that can grow by self-assembly techniques and use these infrastructures to supply spacecraft with fuel where the spacecraft is receding from our galaxy as superluminal velocities due to space-time expansion.

Since pellets may conceivably include large masses entangled with a back at home galaxy object or person to be teleported, it is plausible that an entangled object and/or person can be entangled with a carrier pellet while a decoder classical feedback loop both travel down one or more conducting or superconducting conduits at slightly superluminal speeds. Thus, effectively superluminal travel is enabled through space-time.

Higher space-time analogues of velocity, acceleration, kinetic energy, Lorentz factors, momentum, force, power and the like can be time differentiated in the following manners.

d f/dt,1  d(d f/dt,1)/dt,2  d(d f/dt,2)/dt,1  d[d(d f/dt,1)/dt,2]/dt,3  d[d(d f/dt,1)/dt,3]/dt,2

d[d(d f/dt,2)/dt,1]/dt,3  d[d(d f/dt,2)/dt,3]/dt,1  d[d(d f/dt,3)/dt,1]/dt,2  d[d(d f/dt,3)/dt,2]/dt,1

d f/dT,1  d(d f/dT,1)/dT,2  d(d f/dT,2)/dT,1  d[d(d f/dT,1)/dT,2]/dT,3  d[d(d f/dT,1)/dT,3]/dT,2

d[d(d f/dT,2)/dT,1]/dT,3  d[d(d f/dT,2)/dT,3]/dT,1  d[d(d f/dT,3)/dT,1]/dT,2  d[d(d f/dT,3)/dT,2]/dT,1

T in each case is a time-like blend of various sorts of multiple time dimensions that appropriately and usefully enables derivation in one iteration.

We can go further to consider.

d f/dŢ,1  d(d f/dŢ,1)/dŢ,2  d(d f/dŢ,2)/dŢ,1  d[d(d f/dŢ,1)/dŢ,2]/dŢ,3  d[d(d f/dŢ,1)/dŢ,3]/dŢ,2

d[d(d f/dŢ,2)/dŢ,1]/dŢ,3  d[d(d f/dŢ,2)/dŢ,3]/dŢ,1  d[d(d f/dŢ,3)/dŢ,1]/dŢ,2  d[d(d f/dŢ,3)/dŢ,2]/dŢ,1

Ţ in each case is a time-like blend of time-like blends of various sorts of multiple time dimensions that appropriately and usefully enables derivation in one iteration.

By the same token, we can also integrate f as follows.

∫ f(dt,1)  ∫∫d f(dt,1)(dt,2)  ∫∫d f(dt,2)(dt,1)  ∭ f(dt,1)(dt,2)(dt,3)  ∭ f(dt,1)(dt,3)(dt,2)

∭ f(dt,2)(dt,1)(dt,3)  ∭ f(dt,2)(dt,3)(dt,1)  ∭ f(dt,3)(dt,1)(dt,2)  ∭ f(dt,3)(dt,2)(dt,1)

∫ f(dT,1)  ∬ f(dT,1) (dT,2)  ∬ f(dT,2) ((dT,1) )  ∭ f(dT,1) (dT,2) (dT,3)  ∭ f(dT,1) (dT,3) (dT,2)

∭ f(dT,2) (dT,1) (dT,3)  ∭ f(dT,2) (dT,3) (dT,1)  ∭ f(dT,3) (dT,1) (dT,2)  ∭ f(dT,3) (dT,2) (dT,1)

We can go further to consider.

∫ f(dŢ,1)  ∫∫ f(dŢ,1)(dŢ,2)  ∫∫ f(dŢ,2)(dŢ,1)  ∫∫∫ f(dŢ,1)(dŢ,2)(dŢ,3)  ∫∫∫ f(dŢ,1)(dŢ,3)(dŢ,2)

∫∫∫ f(dŢ,2)(dŢ,1)(dŢ,3)  ∫∫∫ f(dŢ,2)(dŢ,3)(dŢ,1)  ∫∫∫ f(dŢ,3)(dŢ,1)(dŢ,2)  ∫∫∫ f(dŢ,3)(dŢ,2)(dŢ,1)

We can continue to yet higher order derivatives and integrals suffice it to say that the general patterns continue for higher numbers of dimensions.

We can also consider higher order blends of time and thus are not limited to the blends defined by T and Ţ. We chose these two symbols for their resemblance of the letter t.

So, in closing, it is nice to consider how the pellet run-way stream enables something for just about anyone interested in deep space flight. So many aspects of relativistic spaceflight are thus enabled by the plasma gun and/or magnetic mass-driver concepts presented herein. There is something good for every theoretician of relativistic spaceflight from the various propulsion modes enabled, through extreme Lorentz factor inertial travel, through superluminal travel and hyperspatial travel.

The following list provides for additional options for the gas gun based propulsion method in clause format.

Nuclear explosives for converting background pellets, tube material, cable material, mass driver material and the like into a ionized plasma which can then be employed as an electrodynamic propulsion energy source such as in electro-hydrodynamic-plasma-drives, magneto-hydrodynamic-plasma-drives, electro-magneto-hydrodynamic-plasma-drives, electromagnetic-hydrodynamic-plasma-drives, linear induction magnetic coil plasma sails, magnetic plasma bottle sails, Mini-Magnetic-Plasma-Propulsion system and the like.

Nuclear explosives for converting background pellets, tube material, cable material, mass driver material and the like into a ionized plasma which can then be employed as an electrodynamic propulsion energy source such as in electro-hydrodynamic-plasma-drives, magneto-hydrodynamic-plasma-drives, electro-magneto-hydrodynamic-plasma-drives, electromagnetic-hydrodynamic-plasma-drives, linear induction magnetic coil plasma sails, magnetic plasma bottle sails, Mini-Magnetic-Plasma-Propulsion system and the like for travel in ordinary 4-D space-time and N-D-Space-M-D-Time where N is any integer greater than two and M is any counting number, and where the space-times can optionally positively curved, negatively curved, positively curved and torsioned at one or more scales in arbitrary patterns including but not limited to fractions, and negatively curved and torsioned at one or more scales in arbitrary patterns including but not limited to fractals.

Relativistic rocket with positively curved interior space within fuel tanks for which the effective inertia of the fuel tanks and fuel contained within is extremely reduce perhaps even to a value of zero, or less than zero but non-negative.

Spaces, space-times, or times which are negatively curved, negatively super-curved, complexly curved, N-tuple curved, R-tuple curved, Super-R-tuple curved or S-1-R-tuple curved, Super-super-R-tuple curved or S-2-R-tuple curved, Super-super-super-R-tuple curved or S-3-R-tuple curved, and so-on to S-(Aleph 0)-R-tuple curved and so-on to S-(Aleph 1)-R-tuple and so-on to S-(Aleph 2)-R-tuple curved and so-on and so-on all the way to S-(Aleph Ω)-R-tuple and onward Ho! Forevermore.

# Summary.

Herein, we considered pellet runway fuel streams for providing spacecraft propulsion energy as a primary propulsion mechanism and also as a facilitative mechanism of additional propulsion methods.

# Epilog

In some of my other writing,, attempts are made in a non-fiction presentation for how exotic effects of nuclear explosives, some of which are alleged to be very dangerous, may be applied for profoundly advanced relativistic space flight and other peaceful applications.

The effects of nuclear weapons detonations of current designs can result in peak time-averaged power levels of as great as 10 EXP 26 watts. The explosions involve extreme pressures and power generation from volumetrically small portions of space-time. Thus, the thermodynamic gradients are extreme for nuclear weapons detonations, even more extreme than that of supernova.

Note, however that the most energetic supernova have yields about 10 EXP 30 times that of the most powerful nuclear weapon design ever tested. The Tsar Bomba designed, built, and tested by the Soviet Union in the early 1960's had a yield of 59 megatons and would have caused fatal third degree skin burns on any persons standing out in the open at a distance of 100 km from the epicenter. This device had a yield about 5,000 times that of the device dropped on Hiroshima but had a weight of only about 2.5 times the weight of the Hiroshima device. A town located 600 miles from the blast was leveled by the blast which was funneled through the atmosphere acting as a wave-guide. The device was designed to have a fully fueled yield of 100 megatons with options to increase the yield to as much as 250 megatons but the Russian authorities were concerned about the huge amount of radioactive fallout that the fully fueled device would have produced and perhaps other unknown or unanticipated effects of detonation of such a powerful device.

In reality, there is no upper limit to the yield of thermonuclear weapons which can be designed by inclusion of arbitrary number of stagings

The zone of demarcation of significantly cooler space-time surrounding the peak one nano-second of power production from a nuclear weapon has a thermodynamic gradient that is many orders of magnitude sharper than that of a supernova.

Supernova explosions have plausible self-cloaking or self-screening of otherwise nuclear weapon like thermodynamic gradients based on the overlaying mass of the supernova, the strong stellar gravitational fields, and the nebulous or non-fine scale boundary of the supernova explosions.

Moreover, supernovas generally involve nuclear fusion. Nuclear fission being the opposite process might provide additional thermodynamic gradient aspects perhaps even at the level below quarks and gluons.

Any fission and/or fusion devices may enable exotic space-time, vacuum, or zero-point field effects based on phenomenon below the level of quarks and gluons that may have implications of space, space-time, time, hyperspace, and/or hyperspace-time travel. Moreover, such sub-QCD phenomenon might manifest in super-relativisitic energy yields for a given quantity of invariant fission and/or fusion fuel.

Since no such effects have ever been detected much less demonstrated, these mechanisms would likely require nuclear explosive devices with more extreme thermodynamic gradients than the devices we have deployed, tested, and designed.

There is the possibility that currently fielded nuclear weapons may affect other dimensions not currently accessible to present era human technological observation and measurement. It is plausible that any such effects may affect life-forms in these other dimensions, some of which may be of personal form. So it behooves us to be careful with the actual use and/or testing of nuclear weapons and our motives for conducting such tests.

However, hope is not lost for those hoping such a mechanism can be a power source or space-time modification mechanism for advanced space travel. It is plausible that a screening mechanism can be developed to prevent the exotic effects from harming any extraterrestrial and ultra-terrestrial personal life-forms. Such a cloaking mechanism may be required under universal law so as to not harm higher dimensional life-forms.

Regardless, responsible nuclear energy appllications promise to be a boone for deep space travel and for powering deep space infrastructure.

# References:


Ad Astra Rocket Company. ASPL Director Franklin Chang Díaz, 2005. Retrieved August 27, 2011, from http://www.adastrarocket.com/HiResImagesForPublicRelease/Franklin-ASPLHiRes.jpg

Ad Astra Rocket Company . VASIMR® Operating Principles. Retrieved March 28, 2011, http://www.adastrarocket.com/HiResImagesForPublicRelease/VASIMR_operating_principles.jpg

Ad Astra Rocket Company. Retrieved August 27, 2011, from http://www.adastrarocket.com/HiResImagesForPublicRelease/VX-200-FullPowerBothStagesHiRes.jpg

Ad Astra Rocket Company. (2009). Technology. Retrieved March 28, 2011, from http://www.adastrarocket.com/aarc/Technology

Ad Astra Rocket Company. (2009). VX-200. Retrieved March 28, 2011, from http://www.adastrarocket.com/aarc/VX200

Chuck Hawks "U.S. Navy 16" Battleship Gun Facts" retrieved August 26, 2020 from https://www.chuckhawks.com/16-50_gun_facts.html

"energysage" retrieved August 26, 2020 at https://news.energysage.com/what-are-the-most-efficient-solar-panels-on-the-market/

"Home Guide" retrieved August 26, 2020 at https://homeguide.com/costs/solar-panel-cost

Lezec, H. (2009) Left-handed metamaterials operating in the visible: negative refraction and negative radiation pressure. 2009 APS March Meeting Volume 54, Number 1

NASA.https://www.nasa.gov/centers/wstf/site_tour/remote_hypervelocity_test_laboratory/two_stage_light_gas_guns.html